BIOTECHNOLOGY AND BIOLOGICAL SCIENCES

PROCEEDINGS OF THE 3RD INTERNATIONAL CONFERENCE OF BIOTECHNOLOGY AND BIOLOGICAL SCIENCES (BIOSPECTRUM 2019), AUGUST 8-10, 2019, KOLKATA, INDIA

Biotechnology and Biological Sciences

Editors

Ramkrishna Sen

Department of Biotechnology, University of Engineering & Management, Kolkata, West Bengal, India

Susmita Mukherjee

Department of Biotechnology, University of Engineering & Management, Kolkata, West Bengal, India

Rajashree Paul

Research & Development, Internal Quality Assurance Cell (IQAC), University of Engineering & Management, Kolkata, West Bengal, India

Rajiv Narula

Department of Chemistry, State University of New York, Canton, NY, USA

CRC Press is an imprint of the
Taylor & Francis Group, an **informa** business

A BALKEMA BOOK

CRC Press/Balkema is an imprint of the Taylor & Francis Group, an informa business

Typeset by Integra Software Services Pvt. Ltd., Pondicherry, India

Published by: CRC Press/Balkema
Schipholweg 107C, 2316XC Leiden, The Netherlands

First issued in paperback 2023

ISBN: 978-1-03-257094-5 (pbk)
ISBN: 978-0-367-43161-7 (hbk)
ISBN: 978-1-003-00161-4 (ebk)

DOI: https://doi.org/10.1201/9781003001614

Publisher's Note
The publisher has gone to great lengths to ensure the quality of this reprint but points out that some imperfections in the original copies may be apparent.

Table of contents

Recent advancement on environmental biotechnology research to provide greener & safer earth

Challenges & advancement in microbial biotechnology

Biotechnology and Biological Sciences – Sen et al. (Eds)
© 2020 Taylor & Francis Group, London, ISBN 978-0-367-43161-7

Introduction

Biotechnology is an important area in the field of modern sciences to overcome shortcomings of conventional therapeutic as well as agricultural needs hence the horizon of Biotechnology is infinite. The present book comprises of diverse but interconnected ideas and techniques to strengthen modern biotechnology and is truly an application of Biotechnology to ease out human life & living.

Since the past several decades, bioactive materials have become one of the key areas of research. Surface Functionalization which is one of the most important aspects of biomaterials has been thoroughly discussed in this conference. Various nano-particle based bioactive substances like Hydroxyapatite (Hap) was incorporated to increase bioactive surface for better cellular interaction also application of Biomaterials with various mathematical and mechanical modelling that are helpful for better understanding of various tissue dynamics and integrity. Fabrication of beta-tricalcium phosphate/ crystalline silica-aluminium metal matrix, Development of 2D Biocompatible composite matrix for tissue engineering applications, identification of medical disorders in eye and biometric authentication analysis etc have also been discussed. The book provides a perspective highlights on such a multidisciplinary field of Biomaterials that will help future researchers to combine various knowledge of sciences and apply them to overcome shortcomings of modern therapeutics.

This book provides opportunity to know on industrially important enzymes which is one of the important aspect of biochemistry and microbial sciences. Apart from that, isolation of various biochemical substances from plant based source and the affects of antioxidants (i.e. green coffee beans) have been discussed in brief. It also provides cytotoxic investigation and photophysical analysis of various proteins. Such innovative work will show future pathway in the field of biochemistry and proteomics.

The book also highlights various aspects of waste water treatment and effects of heavy metals on the ecosystem. A comprehensive review for better understanding of land as well as marine environment has been demonstrated where recent developments and diverse applications of modern science and technologies have been reported. Such study is very essential in today's point of view to save our environment and furthermore how to utilise those waste products also have been pointed out. The book includes discussion on bioconverted agricultural wastes, where the fruits and vegetables, account for the largest portion of food wastes, have been reviewed. Bioconversion of waste has been emphasized to be an efficient way of utilizing perishable matters. Such research proposes biotechnological route for development and production of flavour especially from microorganism that can be a good application of agricultural biotechnology in near future. These innovative methods would be very much useful for industrial scale production of such aquatic consumables.

It also comprises chapters on various antimicrobial nanoparticles which is another interesting area of modern science and technology, where study of nano-antimicrobials using automated susceptibility testing device showed new aspects of microbial technology to control pathogenic microbes. Such innovation directly impacts on our daily life. In addition, microorganisms have been used to act as a potential source for the production of different value-added products. Such study can also be useful method for controlling bacterial diseases in a very cost effective way.

Bioinformatics, an interdisciplinary field of science has also been discussed in this book. Here bioinformatics has been used for *In silico* analyses of biological queries using mathematical and statistical techniques. Various modelling software have been used to modify and quantify various biological systems such as descriptors and docking based improvement of drugs, a molecular marker-based approach for the analysis of genetic diversity and stress, structure prediction and functional group characterization of proteins, modelling on cellulase enzymes to name a few. Such experimental results will open new probabilities in the field of modern biotechnology and medical sciences.

Plant Biotechnology is another interesting aspect of this edition where various studies have been conducted for the improvement of crops and controlling of pests. Furthermore, plant biotechnology has also developed crops that are having increased food values. such as bio-production of medicinal mushroom enriched with vitamin D. Moreover, ccomparative study of salinity tolerance mechanism in plants has also been another part of research where various comparative analysis was carried out to get a broad spectrum of their tolerance level. Such multidisciplinary techniques demonstrated in this book can be used for technology transfer to enrich our industries as well as research scientists in future that will be a piece of scope for future peer readers.

Overall, this book is an amalgamation of contemporary research works which include, application of biomaterials in medical science & tissue engineering, application of nanoparticles as antimicrobial agent, into agriculture and so on. It also include papers on isolation and characterization of industrially important enzymes, improvement of quality of food and agricultural produce, work on waste to resource, combating toxic impact of heavy metals which is a requirement of the day, research work on anti-biofilm and also a lot of in-silico studies. Holistically these diverse but connected research on Biotechnology will satisfy the need of the young and to be researchers in the varied domains, will cater to the requirement of the senior researchers working on fundamental science, researchers on medical science, interested students and academician looking for diverse fields of research at a glance.

 ISBN 978-0-367-43161-7

Preface

Hellen Keller once said "Life is either a daring adventure or nothing". The same holds true about the world of research, it is always a challenging task irrespective of whether it is undertaken by undergraduate students, PhD student, Post doctoral fellows or eminent scientists; they are all part of a journey of science which is not only contributing to gather knowledge but also helps in the betterment of human civilization. Scientific research is the capability of asking questions and persistent inquisitiveness, which is also the prerequisite of acquiring knowledge. We believe quality teaching is not possible without the unquenchable thirst of unraveling the mysteries of the infinite world of science.

It is with this spirit and enthusiasm Department of Biotechnology of University of Engineering & Management, Kolkata organizes its annual flagship event Biospectrum, the International Conference on Biotechnology and Biological Sciences. Biospectrum 2019 got immense response from all over India and abroad. We have received more than 200 research contributions. After rigorous review process we have selected seventy one research papers. This book is the culmination of the research papers discoursed in Biospectrum 2019. This book is an attempt to present the research work of the scientists, scholars, students on different aspects of application of Biotechnology and Biological Sciences with the obvious sole cause of easing out human living. This book represents research in the domain of environmental toxicology, plant biotechnology, different applications of microbiology we also received many in-silico studies which is comparatively new domain of research.

This book truly showcases an ensemble mixture of scientific works from the diverse yet interconnected research areas of both classical and applied branches of biological sciences. The editors expect that the scientific research and ideas put forward by eminent scientists and young researchers in this conference and thus incorporated in this book would help to pave a new prospect towards research and development for a better and sustainable future, which has always been one of the sole commitments of scientific research.

Finally, the editors are thankful to the publisher for their endavour to publish the research articles.

Preface

[illegible]

Biotechnology and Biological Sciences – Sen et al. (Eds)
© 2020 Taylor & Francis Group, London, ISBN 978-0-367-43161-7

Committees

Conference Chair:

- Sanjay Swarup (NUS Singapore)

Conference Co-Chair:

- Satyajit Chakrabarti (Institute of Engineering & Management, Kolkata)

Advisory Committee

Satyajit Chakrabarti (Chancellor, UEM, Kolkata)
Sajal Das Gupta (Vice Chancellor, UEM, Kolkata)
Frances Separovic (Emeritus Professor, University of Melbourne, Australia)
Anthony Scime' (Associate Professor, York University, Canada)
Soumen Basak (Professor, Saha Institute of Nuclear Physics, India)

Organising Committee:

Rajashree Paul
Susmita Mukherjee
Sonali Paul
Moupriya Nag
Biswadeep Choudhury
Pratik Talukder
Dibyajit Lahiri
Sandip Kumar Deb (Chairman, IEI, WB State Center)
Mr. Kali Pada Das (IEI, WB State Center)

Biotechnology and Biological Sciences – Sen et al. (Eds)
© 2020 Taylor & Francis Group, London, ISBN 978-0-367-43161-7

Editors biography

1. Dr. Sen's research interest focuses on Green Process & Product Development for Healthcare Energy & Environment by Microalgal Microbial Biorefinery for Biofuels &Biorenewables, Algal Biofuels with Waste valorization & Bio-CCS Bioprocess Integration Intensification, Optimization Biochemical & Bioprocess Engineering Enzymes and Biofuels Technology, Biomass & Bioenergy Environmental & Marine Biotechnology Probiotics and Nutraceuticals Biosensor development.

   Dr. Sen received Distinguished Alumnus – Jadavpur University, Kolkata (2017). He was Fulbright Visiting Faculty (2013-2014) at the Columbia University in the City of New York, Manhattan, NY 10027, USA. (2014).He was Runner up (2nd Prize) of the 3rd National Awards for Technology Innovation in Petrochemicals & Downstream Plastic Processing Industries. (2012) and also received UKIERI (British Council) Award for Exchange Visits (UK – RHUL & University of London, London and University of Ulster, Northern Ireland) (2007).

2. Dr. Susmita Mukherjee, did her Graduation and Post graduation in Zoology. Having more than ten years of teaching experience, she has participated in different national and international conferences and has authored many papers. Her research work is in environmental biotechnology i.e. environmental toxicology; bioremediation. She has done extensive work on the different ecological factors of sewage fed fishery. Presently, she is working on the phytoremediation of soil Arsenic and bioaccumulation and biomagnifications of heavy metals to understand the extent of damage on the crop plants and on food chain.

3. Ms. Rajashree Paul has obtained her B. Tech in Computer Science and Engineering from Kalyani University, West Bengal, India, and an MS in Computing Science from Simon Fraser University, Canada. She has over 10 years of experience in the software industry and is currently an Assistant Professor in the Computer Science Department at the University of Engineering and Management, Kolkata. Her present research interest is the study of electronic, optical and mechanical properties of Quantized Structures. Ms. Paul is affiliated with the conferences and workshops both in India and abroad, organizes different technical presentations and various Industry-Academia interactions.

4. Dr. Narula did his Masters in Biotechnology with Gold Medal as university topper. He did his doctoral thesis on Pathogen Reduction and Recycling of Bedding Materials on Dairy Farms. He joined State University of New York in Canton in Fall 2011, first as an instructor and finally being hired as an Assistant Professor in Environmental Science and Chemistry in Fall 2012. Teaching is his passion and he has been taking Education classes at St. Lawrence University, Canton, NY during the summer, since 2012. He has presented his pedagogy skills at several conferences including the International Conference on Education at San Diego in March 2017, The International Journal of Arts, and Sciences in Las Vegas in March 2018. Presently he is working at State University of New York, Canton.

Acknowledgement

Every year the Department of Biotechnology, University of Engineering and Management, Kolkata organizes Biospectrum - the international conference on biotechnology and biological sciences. This conference acts as a platform for discussion on recent developments in the field of biotechnological research, helping in the exchange of knowledge and expertise with opportunities of networking and collaboration across the globe. This is the third year that this conference is being organized. The success of the previous years' conferences has been reflected in this year's conference which has received an immense response from students, researchers and faculty members from different institutes of national and international repute.

The editors are thankful to all the national and international speakers who came from Southampton University; UK, National University of Singapore, K U Leuven; Belgium, Saha Institute of Nuclear Physics; Kolkata, Bose Institute; Kolkata, Calcutta University, Indian Institute of Technology, Kharagpur, India.

The editors are indebted to Chancellor Prof. Dr. Satyajit Chakrabarti, Vice-Chancellor, Pro-Vice-Chancellor and the senior faculty members of the University of Engineering & Management, Kolkata.

Editors are grateful to the researchers across India and abroad who contributed their original research work in this book, without their support this book would have been impossible.

The editors are grateful to Institution of Engineers India, Kolkata and SRL Limited for their support to the conference and also to Indian Ecological Society & Indian Microbiological Society for extending their academic collaborations.

We are immensely thankful to the national and international members of the advisory committee, all the reviewers who helped us in making the conference.

Finally, editors are thankful to the students and faculty members of the Department of Biotechnology, University of Engineering & Management, Kolkata for their tireless efforts for making the conference a success.

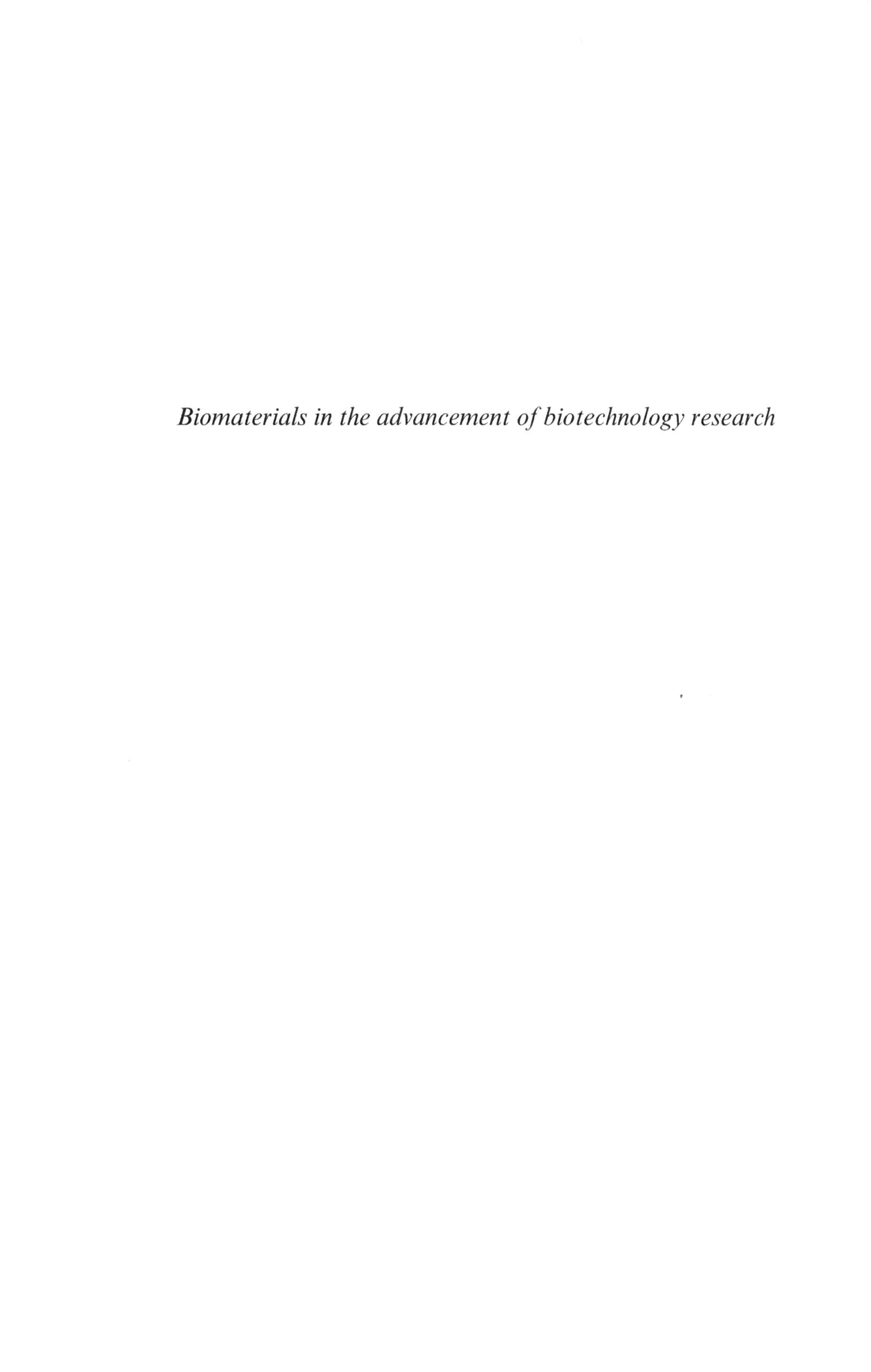

Biomaterials in the advancement of biotechnology research

Biotechnology and Biological Sciences – Sen et al. (Eds)
© 2020 Taylor & Francis Group, London, ISBN 978-0-367-43161-7

Effect of doping and surface functionalization on the conformational changes of protein upon interaction with hydroxyapatite nanoparticles

Kavita Kadu
Department of Chemical Engineering, BITS Pilani-K K Birla Goa Campus, Zuarinagar, Goa, India

Meenal Kowshik
Department of Biological Sciences, BITS Pilani-K K Birla Goa Campus, Zuarinagar, Goa, India

Sutapa Roy Ramanan*
Department of Chemical Engineering, BITS Pilani-K K Birla Goa Campus, Zuarinagar, Goa, India

ABSTRACT: Recent development in nano-biotechnology shows that protein–nanoparticle (P-NP) interactions dictate the behaviour of the NPs in the surrounding bio-systems. As NPs enter biological fluids, biomolecules (majorly proteins) adsorb to their surfaces leading to the corona formation (protein corona PC) which influences the structure and bioactivity of proteins. Hydroxyapatite (HAp; $Ca_{10}(PO_4)_6(OH)_2$, Ca/P in the molar ratio 1.67), an essential component of bones and teeth, have emerged as a new type of biomaterials due to its outstanding biological properties. Recent reports have shown that the synergistic effect of the magnetic nanomaterial in the hydroxyapatite nanostructures provide effective means for biomedical applications. The present study focusses on the synthesis of HAp NPs and Fe doped HAp NPs (DHAp-NPs). The synthesized NPs were characterized using X-Ray diffraction, Attenuated Total reflection-Fourier-transform infrared spectroscopy and Transmission Electron Microscopy. The interaction of synthesized HAP & DHAp NPs with proteins was investigated with and without prior surface functionalization by Cetyl Pyridinium Chloride (CPC) and tri-lithium citrate (TLC) having Cl^- and Li^+ as the corresponding counter-ions. Pepsin A and Hen Egg White Lysozyme (HEWL) were used as negatively and positively charged model proteins respectively. The P-NP interaction was characterized by Dynamic light scattering, Zeta-potential measurements, UV-visible absorption and fluorescence emission spectroscopy. The secondary structure of the protein was investigated using circular dichroism spectroscopy. The functionality of the protein upon interaction with HAp and DHAp was verified using enzymatic assays.

Keywords: Hydroxyapatite, surface functionalization, P-NP interaction and enzymatic assays

1 INTRODUCTION

Extensive research in nanotechnology has facilitated development of novel and multifunctional nanoparticles (NP) for bio-medical applications. NPs, upon interaction with biomolecules, initiate a sequence of nano-bio interfaces that rely on bio-physicochemical interactions leading to the formation of biomolecular coronas around the NPs.[1,2] The NP characteristics like hydrophobicity, size, radius of curvature, charge, surface coatings etc. play a role in its

*Corresponding author: E-mail: sutapa@goa.bits-pilani.ac.in

interaction with biomolecules and the subsequent cellular interactions. [3,4] Although NPs have been widely explored for various biomedical applications, it is essential to study the NP-protein (NP-P) interactions to find out the fundamental understanding of these NPs on the functionality of proteins.

Hydroxyapatite (HAp; $Ca_{10}(PO_4)_6(OH)_2$, Ca/P 1.67), an essential component of bones and teeth, have emerged as a new kind of biomaterials due to its outstanding biological properties such as non-toxicity, bioactivity, optimal bio-degradability and absence of immunological reactions.[5–7] HAp doped with magnetic phases offer biocompatible NPs for applications like magnetic resonance imaging, cancer hyperthermia therapy and targeted drug and gene delivery systems.[8,9] Of the many elements that have the capability to yield magnetic property in the HAp matrix, iron (Fe) is most commonly used for contributing different levels of magnetization.[10] While the potential of magnetic HAp NPs have been emphasized in recent years as they can remarkably improve the diagnosis and treatment, the available literature for HAp-protein interactions are scanty. These interactions, and their biological outcome needs to be deeply investigated for clear understanding of the effect on the structure of the protein.

In the present work, we synthesized rod shaped undoped and Fe doped HAp NPs and investigated the P-NP interactions using Hen egg white lysozyme (HEWL) and Pepsin A, as model proteins with positive and negative surface charges. The effect of surface functionalization with tri-lithium citrate (TLC) and cetyl pyridinium chloride (CPC), on the NP-P interactions was also examined.

2 METHOD

HAp NPs and Magnetic HAp NPs i.e. doped NPs (DHAp NPs) wherein Ca atoms were substituted by Fe were synthesized using modified sol-gel method.[11,12] Doping was carried out for 20 atomic % of calcium. Orthophosphoric acid was added dropwise to a calcium chloride solution wherein the Ca: P atomic ratio was maintained at 1.67. In case of DHAp NPs, requisite amount of dopants (Fe^{+3}: Fe^{+2} in the ratio 3:2) were dissolved in the aqueous calcium chloride solution. Subsequently, tri-ethylamine, the stabilizing agent was added and the pH was adjusted to 10 using liquid ammonia. The obtained gel was dialyzed against deionized water for 24h with frequent change of water. The dialyzed samples were oven dried at 100°C and powdered using mortar and pestle.

The structure and phase identification of HAp and DHAp NPs was carried out using an X-ray diffractometer at a scan range 2θ = 10-80° (Miniflex II Rigaku). Scherer's equation was used to calculate the crystallite size, $D = k\lambda/\beta \cos\theta$, where D is the crystallite size in Å, k is the shape parameter (0.9), λ = 1.5405 Å (monochromatic Cu K_α radiation), θ is the diffraction angle (in degrees), and β is the Full Width Half Maxima of XRD peak (in radians). Attenuated Total reflection-Fourier-transform infrared spectroscopy (ATR-FTIR) (Spectrum Two, Perkin Elmer) was performed over the wavenumber range of 4000−400 cm^{-1} to analyse the functional groups present in the NP. Transmission Electron Microscopy (TEM CM200) was used to study the morphology of the NPs. The surface of the synthesized NPs were functionalized using tri-lithium citrate (TLC) and cetyl pyridinium chloride (CPC), maintaining NP: TLC/CPC: water ratio at 1:1:1, followed by sonication for 2h. The surface functionalized NPs were then incubated with the two model proteins (HEWL and Pepsin A) for 24h. The ratio of TLC/CPC functionalized NP: HEWL/Pepsin A was maintained at 1:1. After incubation, the samples were centrifuged at 4000rpm to remove the un-adsorbed proteins. The washed NP-P conjugates were used for UV-vis, fluorescence, CD measurements, DLS & ζ-potential characterizations. The ζ-potential of functionalized NPs, both before and after protein incubation, was measured using a Malvern DLS (Zetasizer Nano-Z, UK) instrument.

The absorption and fluorescence emission spectra (excitation wavelength = 280nm and emission wavelength = 300-450nm) of HEWL/Pepsin A before and after incubation with and without surface functionalized NPs were studied using UV–visible spectrophotometer (Shimadzu, UV-1800, Japan) and Fluorescence spectrometer (JASCO (FP-8500, Japan)) respectively. The secondary conformations of proteins, were investigated using Circular dichroism spectrometer (JASCO, J-815 (Japan)) at the wavelength range of 195–260 nm. Lysozyme lysis of *E.coli* was studied to confirm the functional activity of HEWL whereas Bovine serum Albumin-Bromophenol blue substrate was developed for the enzymatic assay of Pepsin A.[13,14]

3 RESULTS AND DISCUSSION

X-Ray Diffraction was used to evaluate the phase purity of the synthesized HAp and DHAp NPs (Figure 1(i)). In both cases the XRD spectra indexed well with the JCPDS 09-0432 hexagonal crystalline phase. HAp peaks at 25.8°, 31.34° and 33.97° corresponded well to [hkl] values of [002], [211] and [300]. Minor peaks at 35° and 62.4° with indices [311] and [440] (JCPDS card no. 19-0629) were noted in addition to the above HAp peaks in DHAp NPs indicating magnetite formation.[15] The average crystallite size calculated using Scherer equation was 4.78 and 4.92 nm for HAp and DHAp, respectively. The FTIR spectra of HAp and DHAp (Figure 1(ii)) exhibited characteristic HAp bands. The band at ~560.5 cm^{-1} and ~1025 cm^{-1}corresponds to the bending mode of P-O bond and the symmetric stretching mode in PO_4^{3-} groups. The band at ~601.5 cm^{-1} and around 3220 cm^{-1} are attributed to the stretching vibration of the −OH of the hydroxyapatite.[16] In addition to these, a minor peak at 800 cm^{-1} was noted in DHAp which was due to the stretching of Fe-O. The TEM micrographs of the synthesized NPs are shown in Figure 1(iii). Both HAp and DHAp NPs possessed rod-like morphology with an average length (L) of 80 nm and diameter (D) of 20 nm, respectively. The hydrodynamic diameter and ζ-potential of unfunctionalized and TLC/CPC functionalized

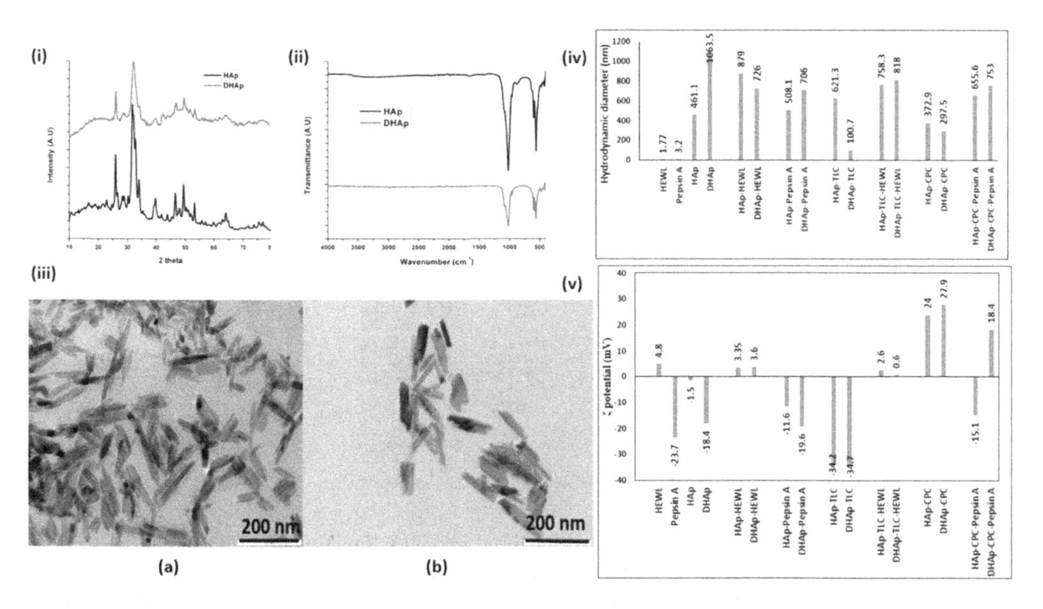

Figure 1. (i) XRD pattern of HAp and DHAp NPs showing all the characteristic peaks of HAp and additional Fe_3O_4 peaks in DHAp NPs; (ii) FTIR spectra of HAp and DHAp showing signature bands of HAp, as explained in the text; (iii) TEM micrographs of (a) HAp and (b) DHAp depicting rod like morphology; (iv) Hydrodynamic diameter of NPs in aqueous media and (v) ζ-potential of unfunctionalized and TLC/CPC functionalized NPs upon interaction with HEWL and Pepsin A.

HAp and DHAp NPs and upon subsequent incubation with HEWL/Pepsin A are shown in Figure 1(iv) and 1(v) respectively. The hydrodynamic diameter was seen to increase after incubation with proteins. HAp NPs had a net negative charge (Figure 1v) due to the PO_4^- and OH^- groups on the outer surface of the unit cell.[17] The ζ-potential of the NPs changed to net negative and positive values upon surface functionalization with TLC and CPC (negatively and positively charged molecules), respectively. Upon protein interaction, the ζ-potential appropriately changed confirming the electrostatic binding of the proteins on the surface of NPs thus leading to formation of the P-NP conjugates.[18]

UV-vis and Fluorescence spectroscopy are the most commonly used analytical techniques for understanding the binding of proteins to the NPs. Figure 2(i and ii) shows the UV absorption spectra of HEWL and Pepsin A upon interaction with and without surface functionalized HAp and DHAp NPs respectively. Native proteins exhibited the characteristic maximum absorbance at 280nm due to tryptophan (W) amino acid in the protein with certain contributions from tyrosine (Y) and phenylalanine (F) groups. The absorption intensity increased upon interaction of both HAp and DHAp NPs with proteins. The band at 260nm (Figure 2ii) corresponded to the CPC molecule.[19] The absorption maxima was observed to be almost same for all the samples. The increased absorption intensity confirms the interaction of respective proteins with the functionalized and unfunctionalized NPs. The fluorescence emission spectra for both proteins showed maximum fluorescent intensity for native proteins Figure 2(iii and iv) (λ_{ex}= 280 nm). In case of unfunctionalized NPs (both HAp and DHAp), insignificant change was observed in the fluorescence intensity as compared to native proteins. However, TLC

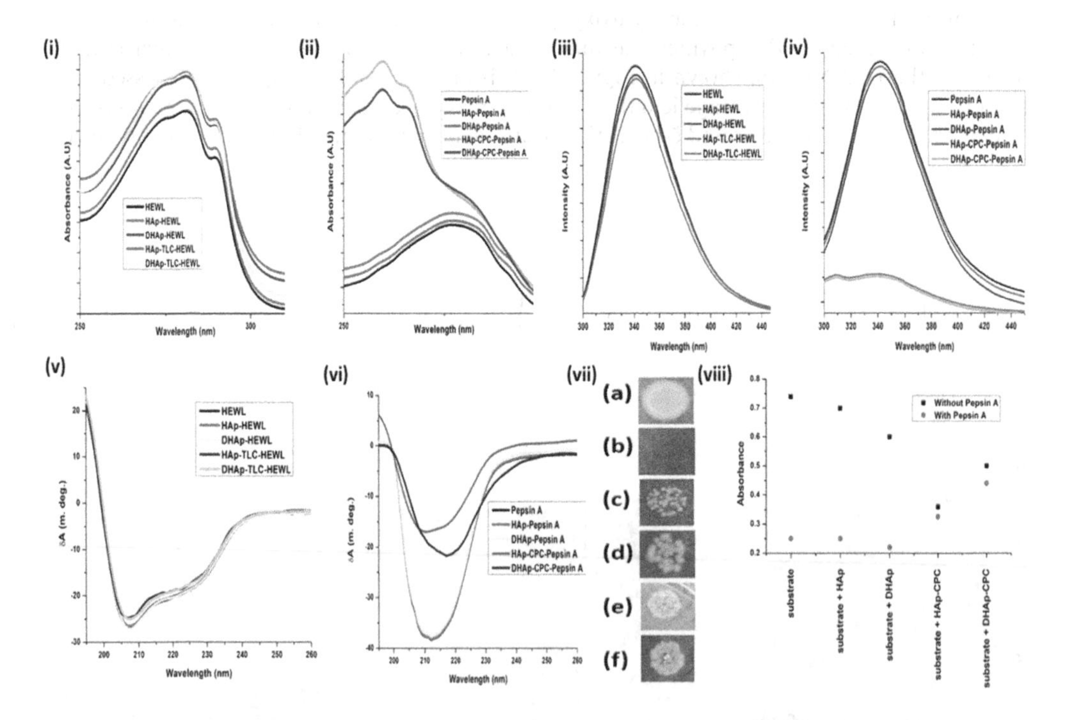

Figure 2. UV absorption spectra of (i) unfunctionalized and TLC functionalized NPs after interaction with HEWL and (ii) unfunctionalized and CPC functionalized NPs after interaction with Pepsin A; Fluorescence emission spectra of (iii) HEWL and (iv) Pepsin A on interaction with and without surface functionalized HAp and DHAp NPs; CD spectra of (v) HEWL and (vi) Pepsin A with and without surface functionalized HAp and DHAp NPs, respectively; (vii) Plate spot assay depicting Lysozyme lysis of DH5α strain *E. coli* (a) control *E.coli*, (b) HEWL, (c) HAp- HEWL, (d) DHAp- HEWL, (e) HAp-TLC-HEWL and (f) DHAp-TLC- HEWL, respectively and (viii) The Absorbance of BSA-BrB substrate at 605nm after action of Pepsin A.

functionalized NPs showed a slight decrease in intensity. This suggests that there was no major change in the secondary structure of HEWL. In contrast, the emission band at 340 nm was completely quenched after interaction of Pepsin A with CPC functionalized HAp and DHAp NPs, indicating significant conformational changes of Pepsin A (Figure 2iv). This verifies the role of the ions present in the surface functionalizing group in the unfolding of the protein. To confirm this, circular dichroism spectroscopy, an extensively used analysis for monitoring secondary structural changes in protein, was used to study the NP induced conformational changes in the adsorbed protein molecules. Figure 2(v and vi) shows the CD spectra of the proteins upon interaction with unfunctionalized and TLC/CPC functionalized HAp and DHAp NPs. The interaction of HEWL with unfunctionalized NPs did not show any major change in the CD spectra indicating no significant unfolding of HEWL whereas TLC functionalized HAp and DHAp NPs showed a minor effect on the secondary structure of HEWL (Figure 2(v)) The % change in α helix of HEWL was observed to be 3.17% and 2.65% for HAp and DHAp NPs functionalized with TLC respectively. Similarly unfunctionalized HAp and DHAp NPs upon interaction with Pepsin A did not show any secondary conformations (Figure 2 (vi)). However, drastic change in the CD spectra was seen in the CPC functionalized HAp and DHAp NPs after interaction with Pepsin A as compared to that of the native Pepsin A (Figure 2(vi)), corroborating the corresponding fluorescence results substantiating the role of counter ions in the surface functionalization groups in the protein unfolding. For the NP-TLC-HEWL interactions, Li^+, the counter ion, did not show any major unfolding of HEWL. This could be due to the fact that the Li^+ ions were immobilized on the negatively charged NP surface (both HAp and DHAp) affecting their availability for diffusion into the protein. In case of CPC functionalized HAp and DHAp NPs, the Cl^- counter ion of CPC possessed a higher tendency to diffuse away from the negatively charged NP surface thereby contributing towards the significant unfolding of Pepsin A. The change in the secondary structure of Pepsin A (i.e % change in α helix) was observed to be 63.43% for CPC functionalized HAp NP and 43.59% for CPC functionalized DHAp NPs. This may be due to the fact that Fe ions in the DHAp NPs have a higher affinity toward the Cl- compared to that of Ca ions thereby reducing its availability for diffusion into the protein.

Enzyme activity, was carried out to confirm the changes in protein conformation. Figure 2 vii shows the lysis activity of native HEWL and NP interacted HEWL. Native HEWL is known to lyse the cell walls of *E. coli* DH5α as a result of which when native HEWL was added to *E. coli* culture, no bacterial growth was seen (Figure 2(vii a)).[13] HEWL interacted NPs showed growth inhibition of *E.coli* in the order of HAp>DHAp>HAp-TLC>DHAp-TLC. This indicated that the activity of HEWL was slightly higher in the unfunctionalized NPs wherein the protein did not unfold as compared to TLC functionalized NPs which showed a slight unfolding of the protein. Albumin -bromophenol blue complex (BSA-BrB substrate), was developed as a substrate for the assay of Pepsin A. The yellow colour of Bromophenol changes to greenish blue after combining with BSA. Pepsin A, an endopeptidase, hydrolyses the BSA from the complex thus regenerating the free bromophenol blue dye.[14] 65.44% decrease in substrate absorbance (from 0.74 to 0.25) was noted upon the action of native Pepsin A thus confirming its activity (Figure 2viii). Both unfunctionalized NPs upon interaction with Pepsin A did not unfold the protein and hence contributed to the enzymatic activity (Figure 2viii). CPC functionalized NPs (both HAp and DHAp) upon interaction with Pepsin A significantly changed its secondary structure (Figure 2vi) hence its enzyme activity should be minimum. Greater decrease in enzymatic activity noted for HAp-CPC NPs as compared to DHAp-CPC NPs, confirmed the CD results mentioned above.

4 CONCLUSION

This study showed that the counter-ions associated with the surface functional groups on the NPs has a vital role in the unfolding of protein structure. The unfunctionalized HAp and

DHAp NPs upon interaction with the respective model proteins did not show any changes in the protein secondary structure. For NP-TLC-HEWL conjugates a small change was noted whereas for the NP-CPC-Pepsin A conjugates a significant unfolding of the protein structure was observed. This was attributed to the availability of the respective counter ions present in the functional groups to diffuse into the protein. Presence of Fe in the DHAp NPs reduced the protein unfolding as compared to that of the HAp NPS due to reduced availability of the Cl- ions most probably due to its higher affinity towards Fe as compared to that of Ca.

ACKNOWLEDGEMENT

We would like to thank UGC-DAE Consortium for Scientific Research for providing a research fellowship; SAIF, IIT Mumbai for the TEM micrographs and Prof. Shyamalava Mazumdar, Department of Chemical Sciences, TIFR, Mumbai for providing Circular Dichroism facility.

REFERENCES

1. Gagner, J.E., Shrivastava, S., Qian, X., Dordick, J.S. & Siegel, R.W. Engineering nanomaterials for biomedical applications requires understanding the nano-bio interface: A perspective. *J. Phys. Chem. Lett.* **3**, 3149–3158 (2012).
2. Jasmin, Š. A review on Nanoparticle and Protein interaction in biomedical applications. **4**, (2016).
3. Lundqvist, M. *et al.* Nanoparticle size and surface properties determine the protein corona with possible implications for biological impacts. *Proc. Natl. Acad. Sci. U. S.A.* **105**, 14265–70 (2008).
4. Mahmoudi, M., Bertrand, N., Zope, H. & Farokhzad, O.C. Emerging understanding of the protein corona at the nano-bio interfaces. *Nano Today* (2016). doi:10.1016/j.nantod.2016.10.005
5. Vallet-Regí, M. Ceramics for medical applications. *J. Chem. Soc. Dalt. Trans.* **1**, 97–108 (2001).
6. Okada, M. & Matsumoto, T. Synthesis and modification of apatite nanoparticles for use in dental and medical applications. *Jpn. Dent. Sci. Rev.* **51**, 85–95 (2015).
7. Wu, H.C., Wang, T. W., Sun, J.S., Wang, W.H. & Lin, F.H. A novel biomagnetic nanoparticle based on hydroxyapatite. *Nanotechnology* **18**, (2007).
8. Hervault, A. & Thanh, N.T.K. Magnetic nanoparticle-based therapeutic agents for thermo-chemotherapy treatment of cancer. *Nanoscale* **6**, 11553–11573 (2014).
9. Gudovan, D. *et al.* Functionalized magnetic nanoparticles for biomedical applications. *Curr. Pharm. Des.* **21**, 6038–6054 (2015).
10. Tampieri, A. *et al.* Magnetic bioinspired hybrid nanostructured collagen-hydroxyapatite scaffolds supporting cell proliferation and tuning regenerative process. *ACS Appl. Mater. Interfaces* **6**, 15697–15707 (2014).
11. Deshmukh, K., Shaik, M.M., Ramanan, S.R. & Kowshik, M. Self-Activated Fluorescent Hydroxyapatite Nanoparticles: A Promising Agent for Bioimaging and Biolabeling. *ACS Biomater. Sci. Eng.* **2**, 1257–1264 (2016).
12. Kadu, K., Ghosh, G., Panicker, L., Kowshik, M. & Roy Ramanan, S. Role of surface charges on interaction of rod-shaped magnetic hydroxyapatite nanoparticles with protein. *Colloids Surfaces B Biointerfaces* **177**, (2019).
13. Voss, J.G. Lysozyme Lysis of Gram-Negative Bacteria Without Production of Spheroplasts. *J. Gen. Microbiol.* **35**, 313–317 (1964).
14. Gray, S.P. & Billings, J.A. Kinetic assay of human pepsin with albumin-bromphenol blue as substrate. *Clin. Chem.* **29**, 447–451 (1983).
15. Veerla, S.C., Kim, D.R., Kim, J., Sohn, H. & Yang, S. Y. Controlled nanoparticle synthesis of Ag/Fe co-doped hydroxyapatite system for cancer cell treatment. *Mater. Sci. Eng. C* doi:S0928493117344715
16. Chen, C. *et al.* Pressure effecting on morphology of hydroxyapatite crystals in homogeneous system. *CrystEngComm* **13**, 1632–1637 (2011).
17. Corno, M., Busco, C., Civalleri, B. & Ugliengo, P. Periodic ab initio study of structural and vibrational features of hexagonal hydroxyapatite Ca10(PO4)6(OH)2. *Phys. Chem. Chem. Phys.* **8**, 2464 (2006).
18. Casals, E., Pfaller, T., Duschl, A., Oostingh, G.J. & Puntes, V. Time evolution of the nanoparticle protein corona. *ACS Nano* **4**, 3623–3632 (2010).
19. Wang, J., Lu, J., Zhang, L. & Hu, Y. Determination of cetylpyridinium chloride and tetracaine hydrochloride in buccal tablets by RP-HPLC. *J. Pharm. Biomed. Anal.* **32**, 381–386 (2003).

Biotechnology and Biological Sciences – Sen et al. (Eds)
© 2020 Taylor & Francis Group, London, ISBN 978-0-367-43161-7

Mathematical modeling of cardiovascular muscle dynamics using lumped parameter approach

Soumyendu Bhattacharjee
Department of ECE, KIEM, Mankar, W.B, India

Biswarup Neogi
Department of ECE, JISCE, Kalyani, W.B, India

ABSTRACT: In this work, a mathematical model of human cardiovascular muscle dynamics has been developed using lumped parameter approach. Basically linear control system based approach is used here to develop this model. The model is comprised of three parts. Fluid viscosity, mass and young's modulus of elasticity. The proposed model has been simulated in MATLAB software after finding overall transfer function when two cells are connected in series has taken into consideration. The beauty of this work is describing the system based on the lumped parameters and then simulating mathematical model based equations with active electrical elements. Terminology of human physical system and required physical data like cardiac muscle mass based on age, constant of elasticity etc., which are required to calculate in terms of electrical circuit parameters like inductance, capacitance and resistance are taken from reputed medical books. The proposed model will be very much helpful to understand the dynamics of human cardiovascular muscle and related syndromes. Another objective of this work is the designing of controller. Here a PID controller is designed and explained in detail using Z-N tuning method to improve the system response of the above described model which will be beneficial for future research work.

1 INTRODUCTION

A distributed network can easily be simplified using lumped parameter model. Here some assumptions are considered. The parameters of the proposed circuit like resistance, capacitance, inductance or gain are concentrated into a particular point and resistors, capacitors, and inductors are connected by a network with perfectly conducting wires, the lumped element model of electronic circuits can be formed [3]. This paper represents a mathematical modeling of human cardiovascular muscle dynamics by lumped parameter and then simulation has been done by MATLAB software. The mechanical model of cardiovascular muscle dynamics can completely be remodeled using lumped parameter analogous to the electrical circuits. In fact, for every closed system, there is an electrical circuit whose behavior can be observed by finding transfer function which has actually been done in this work. A lumped parameter based model had been used by Rideout et al. [13][7] where the various parts of human cardiovascular systems in their work has been depicted. Snyder et al. [3][5] had used an equal volume type modeling feature in their proposed research work. According to the above stated model [14][15], the human arterial system has been divided into some compartments in which length of artery and cross sectional area was in reversely proportional [10][11]. Olfusen et al. had proposed a lumped parameter model for arteries of human cardiovascular system by the various fluid dynamic equations [6], [13]. Liang and Liu, Formaggia, et al. [9]; had also incorporated an OD models to observe the flow in the larger arteries and cardiac circulation. Then Avolio had modeled multi-branch of the human arterial system based on the branching structure of arterial tree [1]. So from the previous research work, it has been found that, there

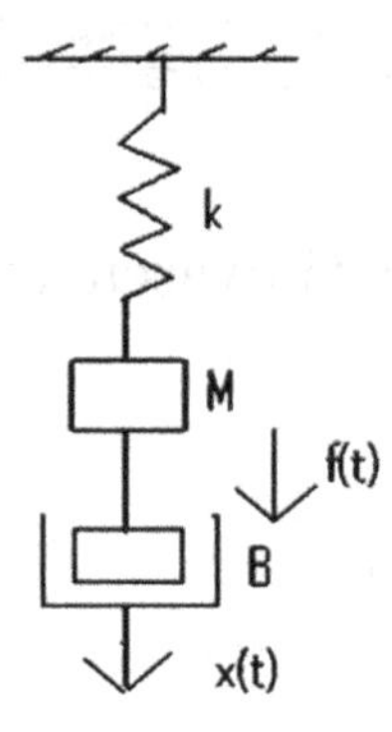

Figure 1. Basic mechanical model of cardiac muscle.

are so many lumped parameter based models had been proposed by the researcher to accomplish there jobs.

2 MODELING OF CARDIAC MUSCLE

The control system based modeling is used to analyze the various parameters of cardiac muscle. To analyze the different parameters, a basic mechanical model of a dynamic system is taken. Figure 1 represents the basic mechanical model of cardiac muscle. From the above figure it is possible to derive the dynamical equation [1][4] for the movement of cardiac muscle with the help of control modeling as given below.

$$M\frac{d^2x}{dt^2} + B\frac{dx(t)}{dt} + k.x(t) = f(t) \tag{1}$$

Let, M is the mass of cardiac muscle, B is the constant of viscous drag of myocardial cell. x(t) is the movement of cardiac wall which is generated due to exerted force f(t) by electrical and electrochemical activity effects on the cardiovascular system. The viscous damping is proportional to the muscle wall movement of the specimen, so that the contribution to this viscous damping may be represented by the expression $B\frac{dx(t)}{dt}$. Tensional drag is proportion to the displacement of the specimen, so that its contribution is given by the expression k.x(t),k being the constant of proportionality. Taking the Laplace transform of equation (1) the following equation can be written

$$Ms^2X(s)+BsX(s)+kX(s)= F(s) \tag{2}$$

Here we have taken the Laplace Transform of the equation, where X(s) = L[x(t)] and F(s) = L[f(t)]. So from equation (2) it is possible to write

$$\frac{X(s)}{F(s)} = \frac{1}{Ms^2 + Bs + k} \tag{3}$$

$$\frac{X(s)}{F(s)} = \frac{K_1}{s^2 + K_2 s + K_3} = T(s) \tag{4}$$

2.1 *Model of one cells of cardiac muscle*

The quantities which are analogous of the mechanical system and electrical system are given in following table shows these analogous quantities. Here Mass is equivalent to inductance.

But in case of DC voltage source (which is equivalent to force) inductance behaves like a short circuit. Hence in case of modeling inductance can be neglected.

Now the clerical equivalent circuit of above mechanical model can be represented by the following Figure 2 given below which is nothing but a representation of single cell of cardiac muscle.

As transfer function represents the behavior of any system, so we are going to calculate the transfer function of above single cell model of cardiac muscle. The mathematical equations of model are given as follows. From the input loop current i has been calculated and given in the equation (5),

$$i = \frac{V_i}{R + 1/s.C} \tag{5}$$

Again Vo is calculated from output loop and given in equation (6),

$$\mathrm{V_O} = \mathrm{i.(1/s.C)} \tag{6}$$

Substituting the value of i, equation (6) can be rearranged in terms of Vo as follows.

$$\mathrm{V_i} = \mathrm{V_O(1 + s.RC)} \tag{7}$$

The transfer function has been formed from the above equation and given below.

$$\frac{\mathrm{V_o}}{\mathrm{V_o}} = (1 + \mathrm{s.RC}) \tag{8}$$

Table 1. Represents the analogous quantities of the mechanical system and electrical system.

Quantity of translational mechanical system	Electrical system
Frictional Coefficient (B)	Resistance (R)
Spring Constant (K)	Reciprocal of Capacitance (C)

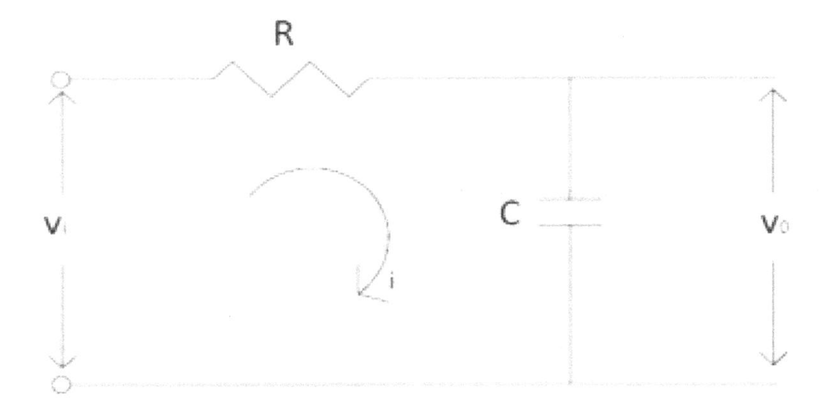

Figure 2. Electrical model of a single cell of cardiac muscle.

2.2 *Model of two cells connected in series of cardiac muscle*

When two cells are connected in series of cardiac muscle, loading effect occurs, due to which transfer function will be critical to calculate. Simple multiplication of two stage is not possible due to loading effect. In this work, the transfer function has also been calculated considering two cells are connected.

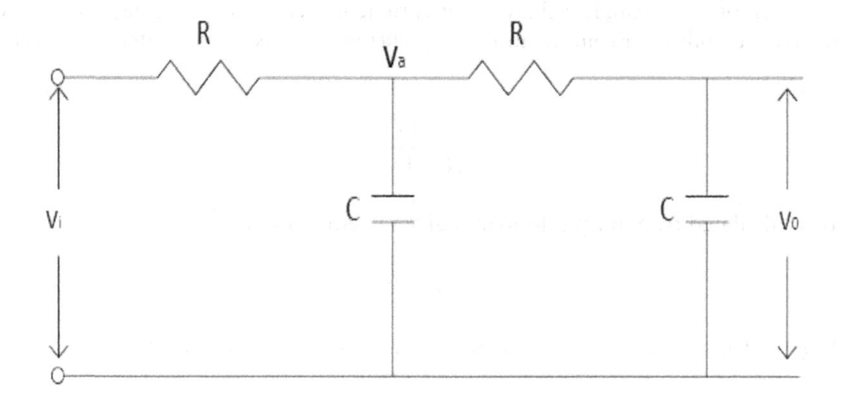

Figure 3. Electrical model of two cells of cardiac muscle connected in series.

Let the voltage at node A is Va. Applying KCL at node A, the following equations can be obtained.

$$\frac{V_a - V_i}{R} + \frac{V_a - V_o}{R} + \frac{V_a}{1/sC} = O \tag{9}$$

$$V_a\left[\frac{1}{R} + \frac{1}{R} + \frac{1}{sC}\right] - \frac{V_i}{R} - \frac{V_o}{R} = O \tag{10}$$

$$V_a\left[\frac{2}{R} + \frac{1}{sC}\right] - \frac{V_O}{R} = \frac{V_i}{R} \tag{11}$$

Again Va can be obtained by applying the concept of one cell calculated above and given below.

$$V_a = V_O(1 + s.RC) \tag{12}$$

Putting the value of Va in equation (11), equation (11) can be rearranged as follows.

$$\frac{V_o(1 + s.RC)(2 + s.RC)}{R} - \frac{V_O}{R} = \frac{V_i}{R} \tag{13}$$

Transfer function of two stage or overall transfer function can be written as

$$\frac{V_o}{V_i} = \frac{1}{(1 + s.RC)(2 + s.RC) - 1} \tag{14}$$

A generalized expression can be written from the above calculation is given below.

$$\frac{V_o}{V_i} = \frac{1}{(1 + s.RC)(2 + s.RC)\ldots\ldots\ldots\ldots(1 + N.RC) - 1} \tag{15}$$

Considering N = 3 and RC = T, equation (15) can be reformed as follows.

$$\frac{V_o}{V_i} = \frac{1}{(1 + s.RC)(2 + s.RC)(3 + N.RC) - 1}.$$

$$\frac{V_o}{V_i} = \frac{1}{(s^2\tau^2 + 3s\tau + 1)(3 + s\tau) - 1}.$$

$$\frac{V_o}{V_i} = \frac{1}{(s^3\tau^3 + 6s^2\tau^2 + 10s\tau + 2)}.$$

$$\frac{V_O}{V_i} = \frac{1/\tau^3}{s^3 + \frac{2}{\tau^3} + s.\frac{10}{\tau^2} + s^2\frac{6}{\tau}} \tag{16}$$

For typical value R and C, the numerical value of T is equal to 1. So the equation is changed into simple form and given below.

$$\frac{V_o}{V_i} = \frac{1}{(s^3 + 6s^2 + 10s + 2)} \tag{17}$$

3 MODELING OF CARDIAC MUSCLE

Control System and Technology has nowadays been one of the most important parts of development in the field of technology as well as research [6][4]. The step response method is the best tuning method for PID controllers which are basically based on an open loop step response method and requires the system to be stable. The process parameters have been shown in Figure 4.

The different parameter values of P, PI and PID controllers obtained from the Ziegler-Nichols open loop step response method has been tabulated in Table 2.

For sustain oscillation or closed loop method shown in Figure 5

Different parameter values of P, PI, PID controller observed from Z-N tuning closed loop method shown below in tabular form in Table 3

4 PID CONTROLLER FOR OUR PROPOSED DESIGN

Initially it is considered that Ti = ∞, Td = 0. So the transfer function of PID Controller is Gc(s) = Kp. The closed loop Transfer Function of system and different parameters of the controller has been calculated as follows.

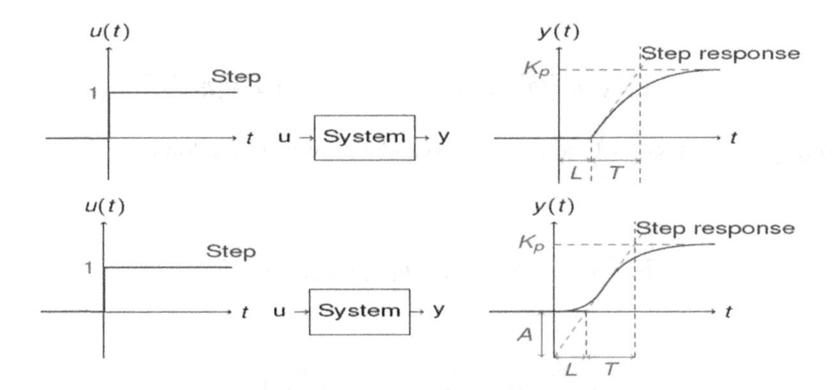

Figure 4. Response curve for Ziegler-Nichols tuning rule.

Table 2. Parameter values from Z-N step response method.

Controller Type	Kp	Ti	Td
P	T/L		
PI	0.9 T/L	L/0.3	
PID	. 1.2 T/L	2L	0.5L

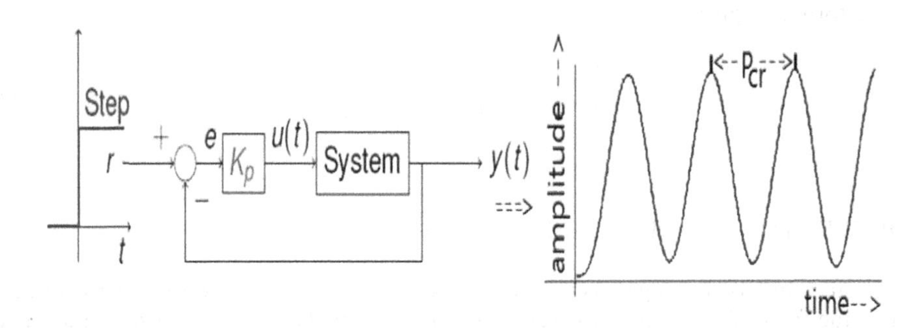

Figure 5. Closed loop method for Ziegler-Nichols tuning.

Table 3. Parameter values from Z-N sustain oscillation method.

Controller type	Kp	Ti	Td
P	0.5 Kcr		
PI	.45Kcr	1/1.2 Pcr	
PID	.6Kcr	0.5Pcr	.125Pcr

$$\frac{C(s)}{R(s)} = \frac{Kp}{s^3 + 6s^2 + 10s^1 + (2 + Kp)} \quad (18)$$

Characteristic Polynomial can be obtained from equation (18),

$$q(s) = s3 + 6s2 + 10s^1 + (2 + Kp) \quad (19)$$

The value of Kp is responsible for making the proposed system marginally stable. So the sustain oscillation occurrence can be obtained by Routh criterion.

$$\begin{array}{cccc} s^3 & 1 & 10 & 0 \\ s^2 & 6 & 2+Kp & 0 \\ s^1 & \frac{60-(2+Kp)}{6} & 0 & 0 \\ s^0 & 2+Kp & . & . \end{array}$$

To find the critical gain (Kcr) of the system the s1 row is made zero. Thus,

$$\frac{60-(2+Kp)}{2+Kp}=0 \tag{20}$$

The above equation gives Kp = 58 which means Kp = Kcr = critical gain = 58.
The auxiliary equation can be written from the R-H array as.

$$6s^2+\ 2+Kp=\ 0 \tag{21}$$

Putting the value of Kp = 58 in equation (21) we get,

$$6s^2+60\ =\ 0 \tag{22}$$

So, s^2 = -10. As it is known s = jω, the equation can be replaced with the value of s. So the equation becomes

$$(j\omega)2=-10$$

$$\omega 2=10$$

or,

$$\omega=3.16 \tag{23}$$

This ω is called ω_{cr}.

$$\text{Now calculating Pcr}=\ \frac{2\pi}{\omega_{cr}}\ =\ \frac{2\pi}{3.16}\ =\ 2 \tag{24}$$

According to Z-N tuning rule we can calculate the value of Kp, Ti, Td using the values of Kcr & Pcr.

$$Kp=\ 0.6\,Kcr=\ 58^*\ 0.6\ =35 \tag{25}$$

$$Ti\ =\ 0.5\ Pcr\ =\ 0.5^*\ 2\ =\ 1 \tag{26}$$

$$T_d=\ .125\ Pcr\ =\ .125\ ^*2\ =2.5 \tag{27}$$

As the value of Kp, Ti, Td has been calculated, the transfer function of the PID controller can also be calculated using as

$$Gc(s)=\ K_p\left(1+\ T_d s+\frac{1}{T_i s}\right)=58\left(1+s+\frac{1}{2.5s}\right) \tag{28}$$

The above equation gives the transfer function of PID controller when two cells are considered in series.

5 SIMULATION RESULT

The simulation result is depicted in this section considering the step response of cardiac muscle without controller and integrator. Here the system has also been simulated in MATLAB software and is given in the figure below.

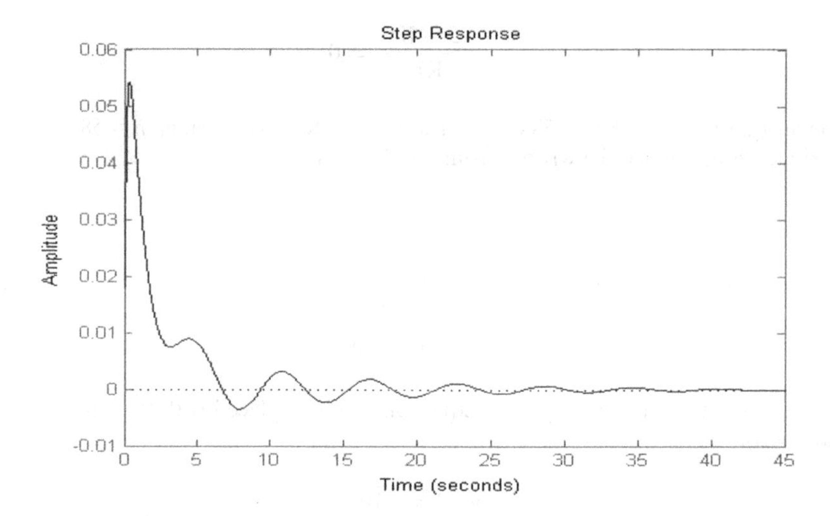

Figure 6. Step Response of the system without controller.

The step response of the plant with Z-N tuned PID controller and integrator has also been simulated and is given in the figure below.

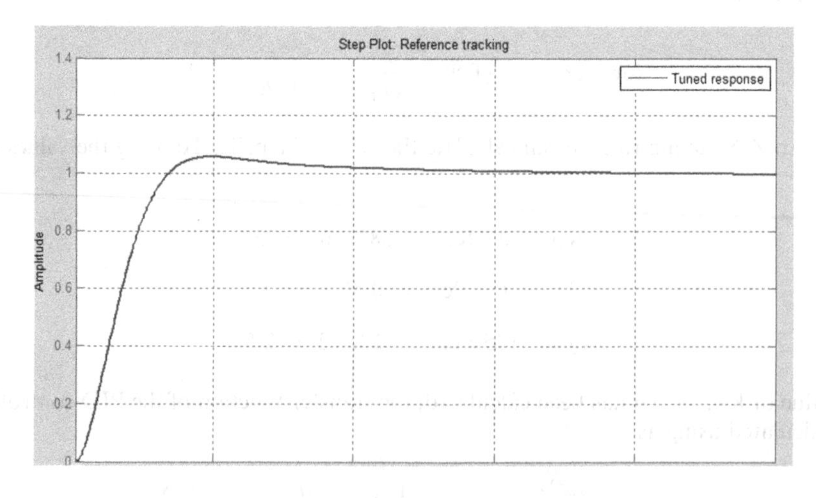

Figure 7. Step Response of overall transfer function with PID controller.

6 CONCLUSION

This paper represents the mathematical modeling of cardiovascular muscle dynamics by lumped parameter and then simulated in MATLAB software. The mechanical model of cardiovascular muscle dynamics has completely been transformed into the electrical model using lumped parameter. From the lumped parameter model design of single cardiac cell is developed, but due to loading effect transfer function of two cells connected in series has also been calculated separately. Finally a generalized transfer function has been formulated for future study. Another objective of this work is designing of controller. Here a PID controller is designed and explained in detail using Z-N tuning method to improve the system response.

REFERENCES

[1] A. Avolio, (1980), Multi-branched model of the human arterial system, Medical & Biological engineering & Computing, pp. 709-718, November 1980.

[2] Ang, K.H., Chong, G.C.Y., and Li, Y. (2005). PID control system analysis, design, and technology, IEEE Trans Control Systems Tech, 13(4),pp.559-576.

[3] Beneken J.E.W.; Rideout V.C. 1968 The Use of Multiple Models in Cardiovascular System Studies: Transportation and Perturbation Methods, IEEE Transactions on Biomedical Engineering, October, Vol. BME-15.

[4] Bouallegue S, Haggege J, Ayadi M, Benrejeb M. 2012 "ID-type fuzzy logic controller tuning based on particle swarm optimization";Vol.25 (3), pp. 484–93,.

[5] F. Y. Liang, S. Takagi, R. Himeno, and H. Liu., 2009 Biomechanical characterization of ventricular-arterial coupling during aging: A multiscale model study. Journal of Biomechanics, 42:692-704.

[6] K. Hassani, M. Navidbakhsh, M. Abdolrazaghi, 2010. Mathematical Modelling and Electrical Analog Equivalent of the Human Cardiovascular System, Cardiovascular Engineering, Springer, pp .45-451.

[7] K. Hassani, M. Navidbakhsh, M. Rostami, 2014 Simulation of the cardiovascular system using equivalent electronic system, J biomedical papers of medical faculty of the university Palacky, Olomouc. 150(1):pp. 105-112.

[8] K. Astrom and T. Hagglund, "The future of PID control,2012" Control Engineering Practice, vol. 9, no. 11, pp. 1163–1175.

[9] M. S. Olufsen, A. Nadim, 2004 On deriving lumped models for blood flow and pressure in the systemic arteries. J Math Biosci Eng. ;1(1):61.

[10] Mohammad Reza Mirzaee, Omid Ghasemalizadeh, and Bahar Firoozabadi 2008 Simulating of Human Cardiovascular System and Blood Vessel Obstruction Using Lumped Method World Academy of Science Engineering and Technology 41, Page(s) No.267-274.

[11] M. Sahib, 2015 "A novel optimal PID plus second order derivative controller for AVR system", Eng. Sci. Technol. Int. J. Vol.18, pp. 194.

[12] M.J. Mahmoodabadi, H. Jahanshahi, 2016 "Multi-objective optimized fuzzy-PID controllers for fourth order non linear systems", Engineering Science and Technology, an International Journal, Vol. 19, pp.1084–1098.

[13] Rideout V.C.; and Katra J.A. 1969 Computer Study of the Pulmonary Circulation, Simulation, Vol. 12, No. 5, Page(s) 239-245.

[14] Rideout V.C.; Dick D.E. 1967 Difference-Differential Equations for Fluid Flow in Distensible Tubes, IEEE Trans Biomed Eng, Vol. BME-14, No. 3, Page(s) 171-177.

[15] Rideout, V.C, 1972 Cardiovascular System Simulation in Biomedical Engineering Education, Biomedical Engineering, IEEE Transactions on, March, Vol. BME-19, No. 2, Page(s) 101-107 73.

Biotechnology and Biological Sciences – Sen et al. (Eds)
© 2020 Taylor & Francis Group, London, ISBN 978-0-367-43161-7

Influence of physical activity on electrodermal response

M.N. Valli*, S. Sudha & R. Kalpana
Department of Biomedical Engineering, Rajalakshmi Engineering College, Thandalam, Chennai

ABSTRACT: Physical exercise is usually the primary steps in the management of chronic disease. It increases oxygen flow to all cells and also helps in more perspiration. Therefore influence of physical activity (PA) on sweat gland is studied in this work. Because of the presence of elements like Na^+, Cl^- and Ca^+ etc in sweat, any change in its secretion leads to change in electrical conductance of the skin which can be picked up by electrodermal activity (EDA). Such a response of skin activity to electrical stimulus is studied through EDA for group of volunteers who are with and without PA. A total of 52 healthy subjects 29 male and 23 female in the age group between 35 and 65 are recruited for this study. In these 27 subjects are termed as GROUP-I who are continuously involved in PA and the remaining 25 are termed as GROUP-II who are not into any PA. Since these signals are low in magnitude and frequency, for better feature extraction they are convolved with morlet wavelet. The statistical analysis performed on these features shows higher skin conductivity response and distinct rate of change of conductivity during electrical stimuli. This confirms the influence of PA on sweat gland and hence better activity.

1 INTRODUCTION

Habitual physical exercise helps to maintain good physical and mental health. There are studies establishing the reduction of cardiovascular disease, diabetes and neurological disorder [1,3] after regular practice of PA. Craig [14] said that PA improves number of metabolic factors like cholesterol levels, resting blood pressure and insulin resistance. He has also established the association between cardio respiratory fitness and PA. Thus regular PA results in sweating, which is a bodily function in which sweat glands release salt based fluids, in a way reducing the load on kidney.

Sweating normally happens due to the co- ordination of brain, autonomous nervous system and hence excitation of nerve fibers in the sweat gland. This process in addition to maintaining thermoregulation, helps to maintain balanced temperament [2].

Physical exercise is not only related to internal bodily function, but also relates the physiological function with the environment. Hence, some amount of oxygen and nutrients are delivered to skin via blood flow thus reducing wrinkles due to ageing process [3]. This helps to keep fit and healthy. Thus the effect of PA could be observed in skin through two parameters. First is the skin conductance- due to sweat gland activity [4,5] second skin appearance - due to environmental interaction [6].

This study is conceived to analyse the effect of physical exercise on thermoregulation. Hence dermal response are collected from group of volunteers (few are into regular physical exercise - who are physically active for more than 150 minutes per week according to American Diabetes Association (ADA) and American Heart Association (AHA) and other few are not into any physical exercise - white collared jobs and who are not physically active 150 minutes per week). By analysing the response to EDA results confirms better thermoregulation and physical fitness among the volunteers who are into regular PA.

*Corresponding author: M.N. Valli

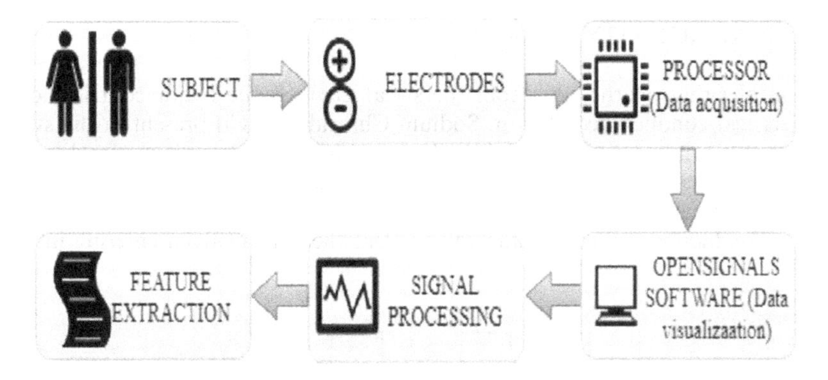

Figure 1. Block diagram of the proposed work.

2 PROTOCOL

A total of 52 healthy subjects 29 male and 23 female in the age group between 35 and 65 with mean of 43.772 and standard deviation of 7.818 are recruited for this study. In these volunteers 27 subjects are termed as GROUP-I who are continuously involved in PA and the remaining 25 are termed as GROUP-II who are not into any PA. For this study the subjects are selected under certain criteria like who have normal cardiovascular, renal and neurological function with controlled blood parameters.

3 DATA COLLECTION

Before examination participants were asked to relax for 15 minutes in a sound proof air conditioned room. The electrodermal activity is acquired through the AgCl electrode placed in the palm with the reference in forearm. This acquired signal is amplified and processed using Bit-alino [16] revolutionary kit and interfaced to the PC. The signals are visualized using opensignals software. The signals are recorded for first 30 seconds without stimulation and then 4V DC is applied for 20 seconds, then again for 30 seconds without stimulation is withdrawn. Again electrical stimulation is applied for next 20 seconds followed by 30 seconds of final recording after the removal of stimuli. The experimental setup is shown in Figure 2.

Figure 2. Experimental setup.

4 ELECTRICAL ACTIVITY OF SKIN

The excretion of sweat on the skin from the sweat gland is the major contribution for the property, electrical conductance of skin. Sodium Chloride (NaCl) present in the sweat when increases then the conductivity of the skin also increases [7]. This when stimulated through external agency, either inhibits or aid the nerve fibers that results in change in electrical conductivity. When DC voltage is applied on skin the current must pass extracellularly, because of the poorly conducting cell membrane will prevent the current from entering into the cell.

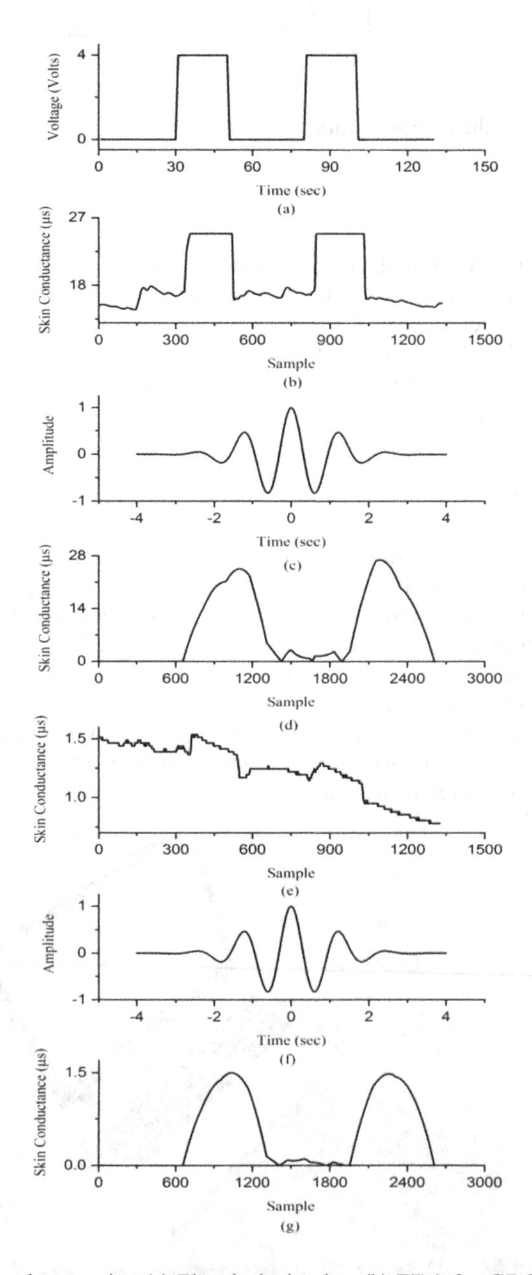

Figure 3. EDA signal and processing (a) Electrical stimulus, (b) EDA for GROUP-I, (c) Morlet wavelet, (d) Convolved signal for GROUP-I, (e) EDA for GROUP-II, (f) Morlet wavelet, (g) Convolved signal for GROUP-II.

As the frequency increases the current will pass through the cell due to the capacitive shunting of cell membrane [8].

In the present study a constant source of 4V is applied as a stimulus to find the changes happening in the conductivity. Figure 2 illustrates the method adapted to acquire EDA data from the volunteers. Since the interest of the study is in analysing the sweat gland activity, measurement of conductance is done by connecting AgCl electrodes on palm and forearm. EDA signals are recorded for a period of 130 seconds. In that 30-50 seconds and 80-100 seconds are under electrical stimulation of 4V DC. The remaining period are without stimulation (Figure 3a). The sample of EDA is shown in Figure 3b.The sample of EDA signals are generally of less than 20 microsiemens in magnitude and frequency is less than 1 Hz [4,5] differentiating among different individuals under different circumstance there comes the difficulty. However, high frequency stimulus is also not recommended as the current may pass the internal structure of the cells. Hence a morlet wavelet (Figure 3c) is generated and convolved with the EDA signal [9]. Figure 3d and 3g shows the convolved signals for GROUP-I and GROUP-II respectively.

5 INFLUENCE OF PHYSICAL ACTIVITY ON SKIN

PA increases oxygen flow to all cells and helps to maintain all glands in active state. This increases physical and mental energy [1,2]. Specifically skin is one most influenced organ, where it sweats thus releasing waste via skin pores. In general due to muscular contractions happening during dynamic physical exercise elevates internal temperature that is followed by increase in sweat rate [15]. Thus, regular PA would stimulate sweat gland and helps it to be active, thus increasing volume of sweat secreted by sweat glands. Even the sports athletes with spinal cord injuries shows normal sweating in the region above the injury [10].

Using sweat lactate concentrations, Green et al. [11] showed that sweat gland metabolism is different between high and low fitness participants, reflecting sweat gland alteration to training. Thus Change in sweating on the skin or accumulation of sweat in the ducts will cause change in the skin conductivity [4,5].

6 MODELING EDA

Modeling is one important aspect for predicting or for early diagnosis. Based on the observation made on the type of pattern of EDA obtained for two groups- GROUP-I and GROUP-II. The model proposed by chong et.al., [13] is found to be suitable for the type of stimulation and related EDA response.

$$f(t) = \left\{ g \frac{\exp -\left(\frac{t-t_{onset}}{t_d}\right)}{\left[1 + \left(\frac{t-t_{onset}}{t_r}\right)^{-2}\right]^2} \right\} \tag{1}$$

Where,

g is gain
t_{onset} is onset time
t_r is rise time
t_d is decay time

The plot of this model for the data obtained from GROUP-I & GROUP-II are presented in Figure 4. Due to lack of PA, sweat gland activity also reduces and hence takes longer recovery time on removal of stimuli as in Figure 4. This is what exactly obtained through the model.

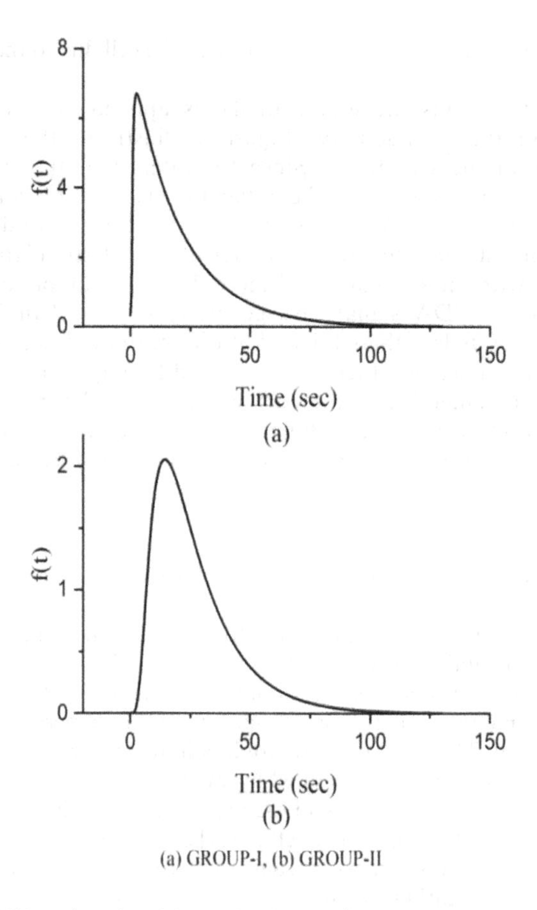

Figure 4. Variation of f(t) against time (a) For GROUP-I, (b) For GROUP-II.

7 FEATURE EXTRACTION

To distinguish the obtained signals as two groups (with and without PA). Discriminative information called unique features are extracted from each of the EDA signal [4,5,12]. This transforms the original data to a dataset with reduced number of variables. By carefully examining the convoluted EDA signal, 6 features are selected and extracted. The features are obtained from the stimulation duration of the signal that is (30-50 sec,80-100 sec). The features are,

i. *Peak I & II*: This is the maximum skin conductance from the base line when the stimulus is applied.
ii. *Rate of rise I & II:* This is the speed at which maximum rise of skin conductance from the base line when an stimuli is applied.
iii. *Rate of recovery I & II:* This indicates the speed at which the skin conductance drop from the maximum value to the base line.
iv. *Onset time:* It is defined as a period between stimulus onset and the first significant deviation in the signal
v. *Rise time*: The time required for a pulse to rise from base value to a maximum value.
vi. *Recovery time:* The time required for a pulse to rise from maximum value to base value

The obtained values of are listed in the Table 1

Table 1. Normalised values of extracted features.

	GROUP-I	GROUP-II
Peak I	0.098 ± 0.02	0.203 ± 0.12
Peak II	0.97 ± 0.03	0.1931 ± 0.13
Rate of rise I	0.98 ± 0.02	0.1259 ± 0.065
Rate of rise II	0.978 ± 0.022	0.1573 ± 0.135
Rate of recovery I	0.99 ± 0.01	0.1562 ± 0.083
Rate of recovery II	0.972 ± 0.0028	0.1233 ± 0.0068
Rise time (t_r)	9.9 ± 2.64	3.5 ± 3.29
Recovery time (t_d)	10.83 ± 3.16	28.2 ± 6.44

8 CONCLUSION

Table 1 gives the mean and standard deviation of the features extracted from EDA in GROUP-I and GROUP- II subjects. It can be seen that the skin conductivity increases and decreases at higher rate in GROUP-I when compared with GROUP-II, the skin conductivity of GROUP-I is higher than GROUP-II on the application of electrical voltage, stimulation of sweat gland happens through Autonomic Nervous System (ANS) leading to change in skin conductance and the response time. It is interesting to note that GROUP-I subjects, who are physically active shows elevated skin conductance and higher rate of rise and recovery during external stimulation. Therefore, it could be said that the rate of sweat gland activity is higher in GROUP-I than in GROUP-II, since the difference between these two groups is only related to PA, the difference in response of sweat glands happening in the two groups is due to regular PA. Therefore it can be said that PA would improve sweat production and hence would also reduce the burden on kidney.

REFERENCES

[1] Janssen, I., & Leblanc, A. G. (2010). Systematic review of the health benefits of physical activity and fitness in school-aged children and youth. Theinternational journal of behavioral nutrition and physical activity, 7, 40.

[2] Shibasaki, M., & Crandall, C. G. (2010). Mechanisms and controllers of eccrine sweating in humans. Frontiers in bioscience (Scholar edition), 2, 685–696.

[3] McPhee, J. S., French, D. P., Jackson, D., Nazroo, J., Pendleton, N., & Degens, H. (2016). Physical activity in older age: perspectives for healthy ageing and frailty. Biogerontology, 17(3), 567–580.

[4] Benedek, M., & Kaernbach, C. (2010). A continuous measure of phasic electrodermal activity. Journal of neuroscience methods, 190(1), 80–91.

[5] Benedek, M., & Kaernbach, C. (2010). Decomposition of skin conductance data by means of nonnegative deconvolution. Psychophysiology, 47(4), 647–658.

[6] Udhayarasu, M., Ramakrishnan, K., & Periasamy, S. (2017). Assessment of chronic kidney disease using skin texture as a key parameter: for South Indian population. Healthcare technology letters, 4(6), 223–227.

[7] Sonner, Z., Wilder, E., Heikenfeld, J., Kasting, G., Beyette, F., Swaile, D., Naik, R. (2015). The microfluidics of the eccrine sweat gland, including biomarker partitioning, transport, and biosensing implications. Biomicrofluidics, 9(3).

[8] Pavlin, M., Kanduser, M., Rebersek, M., Pucihar, G., Hart, F. X., Magjarevic, R., & Miklavcic, D. (2005). Effect of cell electroporation on the conductivity of a cell suspension. Biophysical journal, 88(6), 4378–4390.

[9] Feng, Huanghao & M. Golshan, Hosein & Mahoor, Mohammad. (2018). A Wavelet-based Approach to Emotion Classification using EDA Signals. Expert Systems with Applications.

[10] Pritchett, R. C., Al-Nawaiseh, A. M., Pritchett, K. K., Nethery, V., Bishop, P. A., & Green, J. M. (2015). Sweat gland density and response during high-intensity exercise in athletes with spinal cord injuries. Biology of sport, 32(3), 249–254.

[11] Green, James & Pritchett, Robert & C Tucker, D. & R Crews, T. & R McLester, J. (2004). Sweat lactate response during cycling at 30°C and 18°C WBGT. Journal of sports sciences. 22. 321-7.
[12] Posada-Quintero, H. F., Reljin, N., Mills, C., Mills, I., Florian, J. P., VanHeest, J. L., & Chon, K. H. (2018). Time-varying analysis of electrodermal activity during exercise. PloS one, 13(6).
[13] Lim, Chong Lee & Rennie, Chris & Barry, Robert & Bahramali, Homayoun & Lazzaro, Ilario & Manor, Barry & Gordon, Evian. (1997). Decomposing skin conductance into tonic and phasic components. International journal of psychophysiology: official journal of the International Organization of Psychophysiology. 25. 97-109.
[14] Stump C. S. (2011). Physical Activity in the Prevention of Chronic Kidney Disease. Cardiorenal medicine, 1(3), 164–173.
[15] Shibasaki, Manabu & Crandall, Craig. (2010). Mechanisms and controllers of eccrine sweating in humans. Frontiers in bioscience (Scholar edition). 2. 685-96.
[16] D. Batista, H. Silva and A. Fred, "Experimental characterization and analysis of the BITalino platforms against a reference device," 2017 39th Annual International Conference of the IEEE Engineering in Medicine and Biology Society (EMBC), Seogwipo, 2017, pp. 2418-2421.

Biotechnology and Biological Sciences – Sen et al. (Eds)

Microstrip patch antenna with fractal structure for on-body wearable medical devices

Shreema Manna
University of Engineering and Management, Kolkata, India

Tanushree Bose & Rabindranath Bera
Sikkim Manipal Institute of Technology, Sikkim, India

ABSTRACT: A unique planner rectangular microstrip patch antenna operating in Industrial, Scientific and Medical (ISM) band of frequency 2.45 GHz is presented in this article. Cross type fractal and groove structure is being incorporated in the structure during different iterations in order to improve its performance in terms of gain, directivity, bandwidth, return loss etc and also in terms of effective size reduction of the patch. Different iterations of the designed antenna are simulated using IE3D. The resultant antenna structure with patch and groove operates exactly in ISM band with centre frequency 2.42 GHz, return loss of -36.37 dB and a gain of around 4 dB. A comparison table of return loss, bandwidth, gain, directivity, VSWR, radiation pattern and overall area of the patch is also presented in this paper. From the table it is clearly observed that the antenna performance is much improved after the introduction of groove in the structure

Keywords: Fractal design, Groove, ISM band, Patch antenna, Return loss

1 INTRODUCTION

Recently the interest grows rapidly for the wearable technology. This technology is useful for monitoring the surroundings in military, medical and security services. Wearable and implantable medical devices are useful for remote monitoring of patients by sending biological signals to distant places, such as glucose concentration, blood pressure etc. This helps doctors diagnose the patient from remote place and give proper treatment. The antenna is essential component of all wearable devices. Industrial scientific and medical band (ISM band) of frequency 2.4-2.5 GHz with centre frequency 2.45 GHz is recommended by FCC (Federal Commission Communication) for wearable medical devices. The structure of different antenna mainly depends on different structures of the radiating patch. In this paper a material having permittivity 10.2 is used as dielectric and copper is used as metal for radiating patch and ground plane. Several studies are going on around the globe on the design of different off body communication system. In [1] the research work was on the design of wearable antenna operating both as on-body and off-body communication system. In on-body the antenna operates in 10 GHz and in off-body the antenna operates in 2.45 GHz ISM band. In [2] authors presented an investigation on on-body performance of a range of wearable antennas. Mainly the research work was on to study the performance of compact higher mode microstrip patch antenna. In [3] authors presented in their paper a miniature antenna resonating in medical implant communication service (MICS 402-405 MHz) and ISM bands (433.1-434.8, 868.0-868.6 and 902.8-928.0 MHz). In [4] authors presented Π shaped microstrip antenna operating in 2.3 GHz WiMAX, 2.4GHz WiFi and 2.45GHz ISM band for medical applications. The designed antenna provides SWR of 1.045 and gain of 15.92 dBi at 2.45 GHz. Researchers of [5] presented wearable patch antenna design using biodegradable plastic material. Since these substrates are eco-friendly and reduce environmental contaminations, these are more suitable as substrate in antenna designs. In [6] authors have presented metamaterial based inductor loaded wearable patch antenna in which radiation pattern can be reconfigured. The antenna resonates at two different modes (+1 and 0) and hence can provide omnidirectional and broadside radiation pattern. In both operating modes the antenna resonates at around 2.45 GHz. In other subsequent papers [7-10] authors have used different substrate like textile, 3D printing material etc for designing the antenna in order to make the antenna more flexible, bendable, reduced size, increased gain etc for designing antennas. Authors are more focused in wearable antennas since with implanted antennas more riak is involved during implantation.

2 ANTENNA DESIGN

The proposed antenna structures are designed and simulated in IE3D platform. The final antenna structure is achieved in three iterations and are simulated using IE3D. S parameter display illustrates that the final antenna covers ISM band from 2.37 – 2.47 GHz with centre frequency of 2.42GHz. The antennas are simulated using a substrate material having permittivity 10.2 and thickness of 1.9mm. The structure is assembled as patch on one side and ground plane on the other side of the substrate. Microstrip line feed has been used in all the iterations. The overall dimension of patch used is 24×24 mm^2. The performance of the antenna in different iterations shows that there is much improvement in the return loss, which means the antenna input impedance is getting perfectly matched with the generator input impedance. Also there is much improvement in the VSWR and overall size of the antenna. A VSWR of 0.52 has been achieved at the final level, where groove structure has been incorporated. The overall dimension of the patch is much reduced in different iterations. As there is cutting in the ground plane also, there is much reduction in the overall structure of the whole antenna.

3 RESULTS & DISCUSSION

Initially in the zero iteration a 24×24 mm^2 square copper plate is used as radiating patch and 40×40 mm^2 square copper plate is used as ground plane. The result achieved for zero iteration is slightly shifted from that of the ISM band, which ranges between 2.4 – 2.5 GHz. Hence in order to improve the performance, fractal and triangular groove is incorporated in the structure as shown in Figure 3 and 4. The table of comparison gives the performance comparison of different antenna parameters achieved through simulation at different iterations.

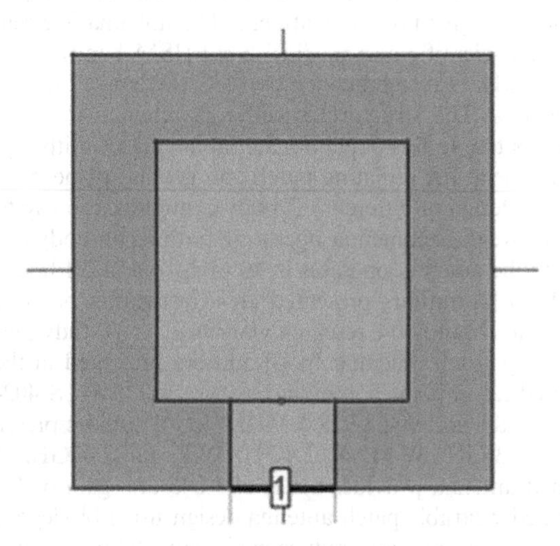

Figure 1. Antenna of zeroth iteration.

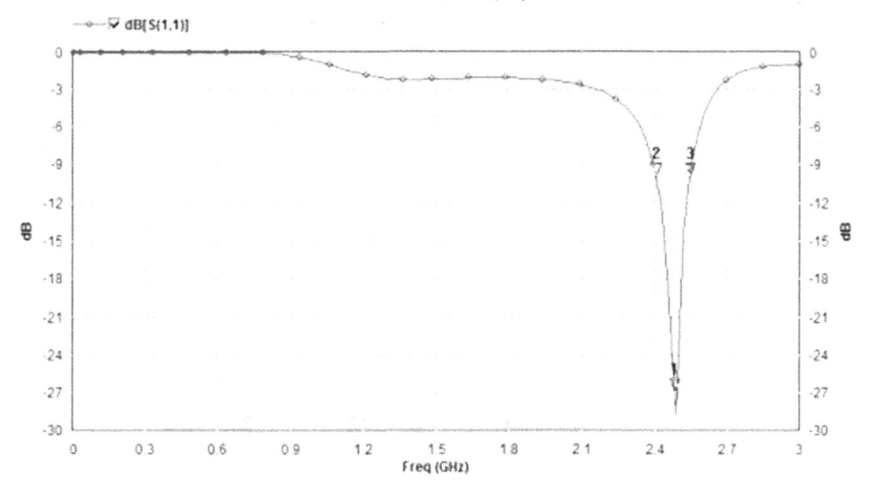

Figure 2. S11 of antenna of zeroth iteration.

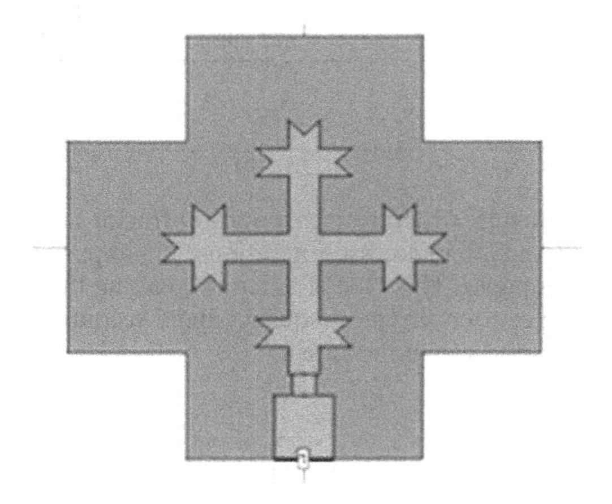

Figure 3. Final antenna with groove structure.

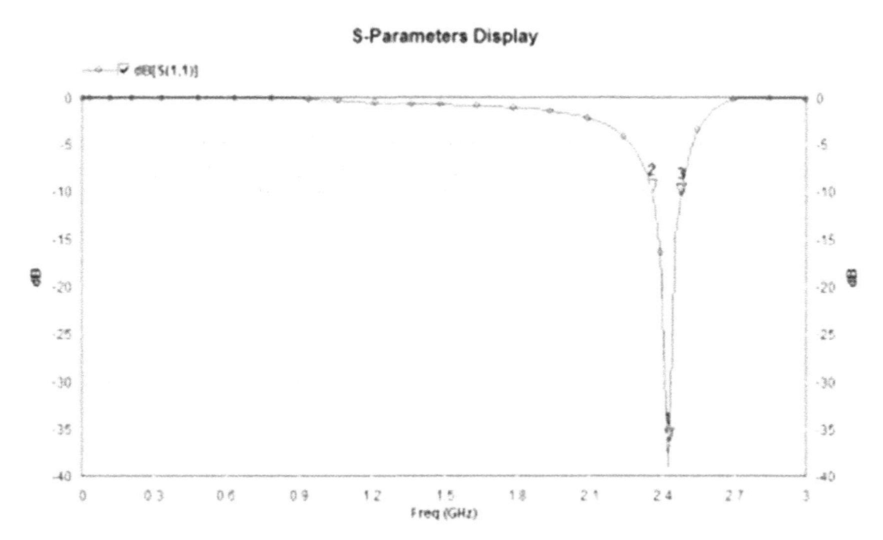

Figure 4. S11 of Final antenna with groove structure.

TABLE OF COMPARISON

Parameters	Iteration 0	Iteration 1	Iteration 2	Groove Cut
Frequency Band (GHz)	2.40-2.54	2.43-2.58	2.32-2.43	2.35-2.47
Band Width (GHz)	0.14	0.15	0.11	0.12
Centre frequency (GHz)	2.47	2.51	2.38	2.42
Return loss (dB)	26.9	12.6	21.9	36.37
Gain	3.02	2.53	3.46	4.3
Directivity	3.56	2.9	3.94	4.9
2D pattern				
3D pattern				
VSWR	1.3	4.13	4.13	0.52
Overall area (mm^2)	576	323.208	179.56	159.8

4 CONCLUSION

This paper presented an antenna structure incorporating fractal and groove structure. The final antenna design is achieved after successive iterations. A comparison table of all the structures is also presented in this paper. This type of antenna covering ISM band can be useful for wearable medical devices since it covers the frequency band recommended by FCC. Further work can be done to improve the size of the antenna.

REFERENCES

[1] Ito, K., & Lin, C. H. (2013, March). Dual-mode wearable antenna for medical applications. In *Antenna Technology (iWAT), 2013 International Workshop on* (pp. 372–375). IEEE.

[2] Conway, G. A., & Scanlon, W. G. (2009). Antennas for over-body-surface communication at 2.45 GHz. *IEEE Transactions on Antennas and Propagation, 57*(4), 844-855.

[3] Kiourti, A., & Nikita, K. S. (2012). Miniature scalp-implantable antennas for telemetry in the MICS and ISM bands: design, safety considerations and link budget analysis. *IEEE Transactions on Antennas and Propagation, 60*(8), 3568-3575.

[4] Rahman, M. M., Hossain, M. M., & Karmakar, K. K. (2013, February). Π-shape microstrip antenna design for WiMAX, Wi-Fi and biomedical application at 2.45 GHz. In *Advance Computing Conference (IACC), 2013 IEEE 3rd International* (pp. 546-549). IEEE.

Biotechnology and Biological Sciences – Sen et al. (Eds)
© 2020 Taylor & Francis Group, London, ISBN 978-0-367-43161-7

Identification of medical disorders in eye and biometric authentication analysis with iris retina scan using machine learning

Sukanya Roy, Pritaam Dutta, Arnab Bhowmik, Bipasha Roy, Kumar Sourav & Lovely Kumari
University of Engineering & Management, Kolkata, India

ABSTRACT: This paper presents the importance of iris recognition for disease identification and security analysis. Clustering methods are used to segment the image of the eye for feature extraction: iris, pupil, collarette, scelera. We have used iris cluster for disease identification and the combination of pupil, collarette, scelera for security analysis. The project is to recognize the uniqueness of the iris. The execution might get affected due to some minor deflections in the eye due to reflection of light and squeaking. So image processing is done in order to evade different types of noise and other light sources which might affect the iris recognition process.

Keywords: Clustering, feature extraction, image processing, iris recognition

1 INTRODUCTION

Authentication toward biometric field and as a faultless method recognition of iris is visualized.Unlike other human body parts, the eyes get affected by various diseases this lead to distort the accuracy. This paper resolves the problem of distorted accuracy by taking a database of images of diseases affecting the eyes and three methods: Mirlin, Osiris, VeriEye based on iris recognition and demonstrating types of vision loss and diseases based on cholesterol levels [1].

Iris plays a vital role in automatic recognition of an individual i.e. biometric system. The recognition is based on some unique feature of an individual. This paper is to study the real life application of security system based on iris recognition and the steps towards real life implementation. This methodology helps in intensifying security and limiting access to system. This also helps the sectors where access to specific resources is highly controlled. The technique is used in prevention from forgery in various examinations, bank systems and for boarding purpose in airport area. Also helps in access in phone, building, confidential sectors or departments. This technique for biometric verification is proven to be much more effective compared to fingerprint technique.

2 MEDICAL DISORDERS AFFECTING THE IRIS

1. Cataract: when protein builds up in the lens of your eyeand makes it cloudy. This keeps light from passing through clearly. It can cause you to lose some of your eyesigh
2. Acute glaucoma: In this disease the optic nerve of the eye get effected and this leads ti loss of vision. The open angle glaucoma is the most frequent disease rather than closed and normal tension glaucoma.
3. Retinal detachment: In this disease the retina ditaches from the layering under it. It causes small spots or pinpricks of light in the field of our vision. For about 10% of such disease both the eyes are affected. Proper care and treatment is needed to cure this disease or else it lead us to vision loss.

3 RELATED WORKS

In this paper three well know iris detection methods are used one of them are open source solution and other two are commercially available [2]. There is no need for examine the sub processes of these three methods so there are being utilized as "black boxes" only compared outputs are considered.

One of the method used is MIRLIN i.e Monro Iris Recognization Library. It is derived from the features of iris by the point at which a function crosses the horiontal axis [3] as its value passes through zero and changes sign of the difference between finite sequence of data points in terms of a sum of cosine functionws oscillating at different frequencies calculated in a division of region of a rectangular image of iris. [1].

The second method used is VeriEye which is offered by Neurotechnology, with the help of active shape modelling the manufacturer can demand the active shaping by the use of segmentation of iris with right off-axis which is very different from particular approximation of circular boundaries of the iris. NIST IREX project is the application of this methodology [4].

The last method used is the OSIRIS, gabor filter is used in this software using the ideology of Daugman increasing particular features of iris and rest of the features losses their value. After the filtering is done the composite vectors are quantized to two bit. Used calculating the distance between feature vector [5].

4 PROPOSED METHODOLOGY

4.1 *Disease*

4.1.1 *Algorithm*

STEP 1: Image is imported.

STEP 1.1: Image is resized to 125x125.

STEP 2: The image is converted from RGB scale to a grayscale image.

STEP 3: Image clustering is done using a modified C-mean clustering [3] technique using Hamming distance instead of Euclidean distance.

STEP 3.1: Each image clusters are given a particular colour.

STEP 4: The best clusters are selected based on clarity. If there is any repetition of clusters, it is omitted.

STEP 4.1: The selected clusters are fused to get the final image.

STEP 5: In this step now verify the output is available or not.

STEP 6: According to the value of database we plot the graph to categories between clear, geometry, tissue or obstruction.

STEP 7: Now according to modified c clustering method we differentiate to determine disease with 98% accuracy.

4.1.2 *Input Images*

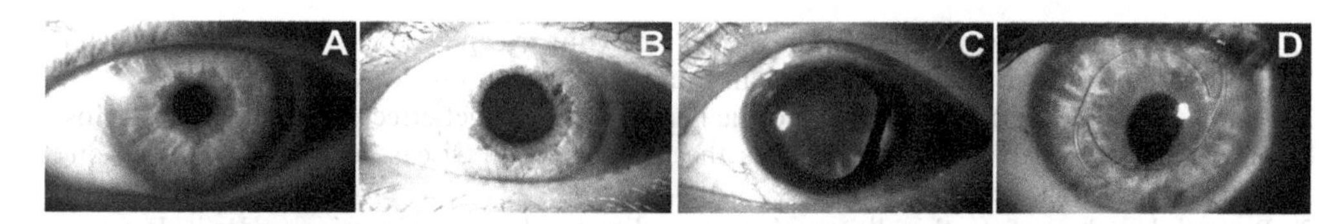

Figure 1. Images representing some examples of ocularness conditions that occurs due to actions of the eye they are namely *clear*, *geometry*, tissue or *obstructions*.

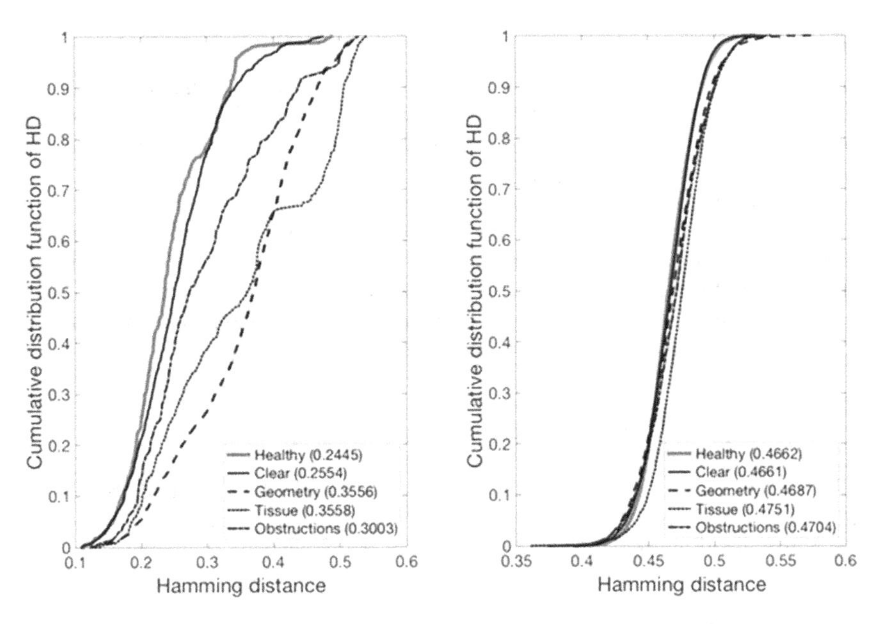

Figure 2. The 1st image shows the cumulative distributions of genuine, and the 2nd image shows the impostor scores. It is shown for independent data gathered in five subsets: *Healthy, Clear, Geometry, Tissue* and *Obstructions*. Comparison scores are calculated by MIRLIN matcher, *i.e.*, lower score denotes a better match.

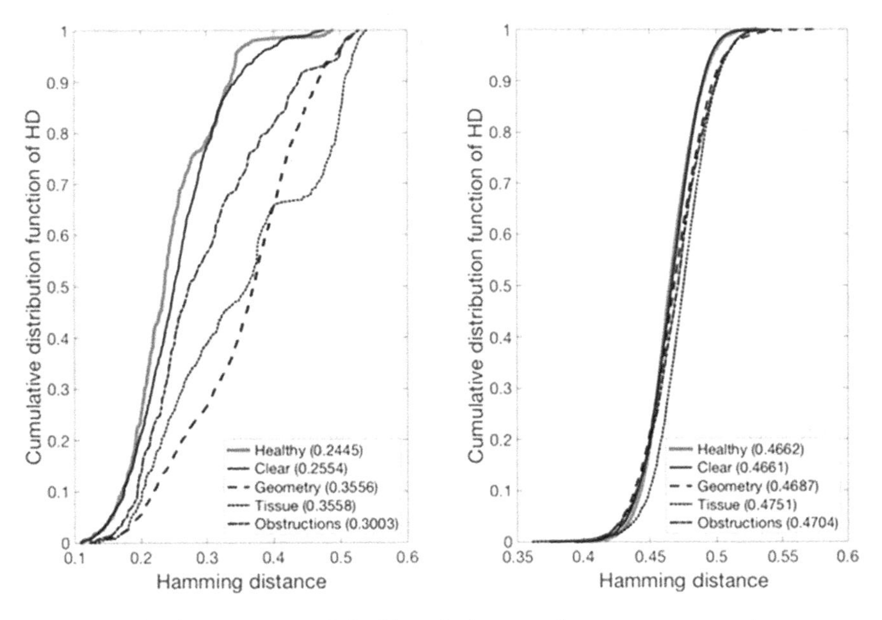

Figure 3. Same as in Figure 2., except OSIRIS method was used to generate comparison scores.

4.1.3 *Result*

([0.984653 'Cataract' 'geometry'])

This result shows the accuracy, type of disease and the subset the image of an eye belongs to.

4.2 *Security*

Iris contains significantly distinctive features and via a distance it can be identified. This makes iris a challenging field of security due to lower false rate. Since it's the most authentic ans sucured thus such a technology is choosen by various institutes,companies, hospitals, and many other highly secured places. [2][6].

The variability of iris is about 245 degrees of freedom and a 3.3 bits/mm^2 due to higher randomness in the iris pattern. Thus a mixed complexity makes the iris unique. This lead to easier recognition. If the biometric system is enrolled with the charecterization of an individual ten due t higher degree of freedom the authentication by iris biometric becomes easier. An instant job is done by just standing firmly infront of camera. For the mass the process is easy and leads to comfortness which doesn't require further skills and techniques.

4.2.1 *Algorithm*

STEP 1: Image capturing/data collection using opencv and pixel analysis.

STEP 2: Quality testing by slayhem method

STEP 2.1: Pixels are prioritizing on the basis of iris recognition method (SVM, adagore, softmachine) which will help to determine the depth of cornea in STEP 5.

STEP 3: Normalization [4]

$$N_i(x)^n = \sum_{k=0}^{n} \frac{(x_i - \mu)}{x_0 + x_n}$$

Where, i = 1,2,3,4,5. . . . . .n

N_i = Normalization Function, x_i = pixel values, μ = mean value.

STEP 4: Feature Extraction by digitalization method

STEP 4.1: Each pixel is converted into 8bit binary value which contains 0 or 1.

STEP 5: Depth analysis:

STEP 5.1: Identification. The output from the slayhem method are stored and used for training purpose.

STEP 5.2: matching: Extracted features from user given input image are matched for testing purpose.

4.2.2 *Training Data*

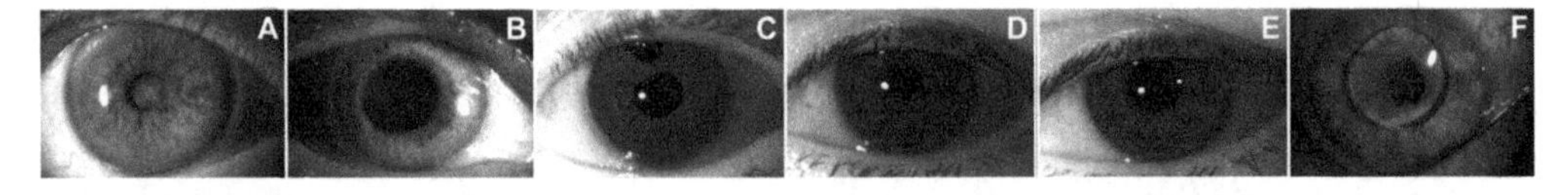

Figure 4. These samples represents ocular pathologies or conditions such as: cataract (A), acute glaucoma (B), iridectomy (C), synechiae before and after lens replacement (D, E), silica oil in the anterior chamber of the eye (F).

5 CONCLUSION

The iris recognition system is used in areas like airport, crime detection, business application, various research firm and industries thus experts foresee the growth of iris recognition system. Its uses are expanding world-wide as the people have been oriented about the benefits of this system. This method is even used in banks where it is incorporated into the Automated Teller Machines(ATMs) . Various countries in this world usess iris biometric for authentication purpose for several critical situations in this world. Since this technology is way to simple and easy so it is difficult to copy fraudulently. This advanced technology is used in various projects and accessways like health departments, companies for it unique features. Iris authentication is highly scalable due to project volume. Projects like immigration and healthcare services iris biometric is widely used due to its scalability feature. Due to distant scanning facility iris provides the factors representing hygiene. Thus it is hygiene and a stable process. Helps in liveness detection and highly secured and also tractable. The paper also serves us in various important and interesting way by giving us various important results. It is not very usual to watch an unhealthy eye which get affected by a single disease only.

REFERENCES

[1] Govinda, S. M. (2011) Immunomodulatory Potential of Methanol Extract of Aegle marmelos in Animals. Indian Journal of Pharmaceutical Sciences, 235.

[2] Hall, C. W., Mah, T. F. (2017). Molecular mechanisms of biofilm-based antibiotic resistance and tolerance in pathogenic bacteria. FEMS Microbiology Reviews, 41(3),276–301.

[3] Hahn-Hagerdal B., Galbe M., Gorwa-Grauslund M.F., Liden G., Zacchi G. (2006) Bioethanol—the fuel of tomorrow from the residues of today. Trends Biotechnol, 24:549–556.

[4] Hajhashemi, V., Ghannadi, A., Sharif, B. (2003), Anti-inflammatory and analgesic properties of the leaf extracts and essential oil of Lavandulaangustifolia Mill. J. Ethnopharmacol. 89, 67–71.

[5] Hall C. W., Mah T. F. (2017). Molecular mechanisms of biofilm-based antibiotic resistance and tolerance in pathogenic bacteria. FEMS Microbiology Reviews, 41(3),276–301.

[6] Handral H.K., Pandith A., Shruthi S.D. (2012), A review on Murrayakoenigii: Multipotential medicinal plant. Asian J Pharm Clin Res 5:5.

Biotechnology and Biological Sciences – Sen et al. (Eds)
© 2020 Taylor & Francis Group, London, ISBN 978-0-367-43161-7

Development of 2D biocompatible composite matrix for tissue engineering applications with antifungal properties

Sayak Roy Chowdhury, Debasmita Deb, Soumyajit Sen & Biswadeep Chaudhuri*
Department of Biotechnology, University of Engineering & Management, Kolkata (Newtown), West Bengal, India

ABSTRACT: In parallel to conventional medical therapeutics, Tissue Engineering (TE) has immerged as a therapy to cope up with certain drawbacks of modern medical facilities. In this connection, fabrication of biocompatible matrix for TE applications draws certain interests to the researchers working in these areas. In this work a 2D biocompatible matrix is prepared in addition of Silver Nanoparticles (AgNPs) and Hydroxyapatite (HAp) for TE applications. The developed matrix can be successfully used for *in-vivo* applications because of its strong antifungal and cell-supportive properties.

Keywords: Tissue Engineering, scaffolds, 2D biocompatible matrix, medical therapeutics

1 INTRODUCTION

Silver Nano-particles (AgNPs) and Hydroxyapatite (HAp) are among the major bio-implantable materials that are in use today. HAp having a chemical formula $Ca_{10}(PO_4)_6(OH)_2$, has properties like bioactivity and biocompatibility which allows the growth of cells like fibroblast and osteoblasts on it [1-3]. Moreover, it being osteoconductive is extensively used as a bone graft and scaffold material [4]. On the other hand, use of AgNPs in bio-composite either helps in the process of wound healing, or it acts as a carrier of the biologically active compound used for wound healing process [5]. AgNPs is also widely used in Tissue Engineering applications for its excellent anti-microbial properties [6]. But both of the biomaterials have certain limitations when administered directly into the *in-vivo* system. HAp has an inherent brittle nature [7] and silver coming in contact with blood plasma gets inactivated and hence partial improvements are observed. Therefore, the material is combined with good biodegradable polyester and polymers to produce ceramic composite or scaffold which are found to be more durable and effective. In this experiment, HAp samples and AgNPs were synthesized and then different composite solutions were produced by incorporating the samples into polymeric solutions of polycaprolactone (PCL) and chloroform. Finally, the solutions were casted to produce 2D scaffolds and characterization of the samples and the produced scaffolds were made to evaluate their properties for supporting Tissue regeneration and *in-vivo* applications. Further assays were also made to observe the biodegradation aspects of the scaffolds and the antifungal properties of AgNPs [8-12].

2 MATERIALS AND METHOD

Preparation of HAp: Egg shells being a calcium rich, economical and biodegradable source was used in the experiment for producing HAp. HAp was synthesized in the laboratory by

*Corresponding author: chaudhuri.biswadeep@googlemail.com

four different methods. In the first three, the compound was formed from raw egg shell and in the last method it was formed chemically.

A. ***Wet Precipitation Method*** [8]: Initially, washed raw egg shells were treated with Hydrochloric Acid (35%; 0.5 N) and the inner membrane of eggshells were separated. Then the hard-outer covering was dried and grinded. Dilute Nitric Acid (70%; 1:10) was then slowly added to the powder resulting in froth formation, which was ultimately dissolved by warming the solution followed with continuous stirring. Then, 0.6 M Diammonium Hydrogen Phosphate solution was added. Finally, dilute sodium hydroxide (0.1g/L) solution was added dropwise and after addition of a particular amount of the solution, milky white color precipitate of HAp was produced. pH of the solution was checked and it was found to be in between 9 to 11. The mixture was stirred in a shaker for about 6 to 8 hours at a speed of about 100 to 120 rpm and then was kept and aged for about 24 hours at room temperature. Finally, the aged precipitate was filtered and purified by continuous washing with distilled water and repeated centrifugation. Purified precipitate was then taken in a sterilized petri dish and dried in a hot air oven at about 100 °C for 24 hours producing pure and dried HAp. This sample was denoted as HAp sample 1 [12-14].
B. ***Acidic Precipitation Method*** [9]: In this method firstly cleaned and uncrushed egg shells were taken in bulk in a porcelain crucible and calcined in a furnace at a temperature of about 900°C for 1 hour. Beyond 850°C the egg shells evolved carbon dioxide and calcium oxide was produced [10].Then, the calcined powder was then taken in a measured amount in a beaker and thereafter dispersed in distilled water. In this reaction CaO is converted into $Ca(OH)_2$ [11]. Thereafter 0.6M orthophosphoric acid was added and the formation of the precipitate was observed. The pH of the solution turned out to be in between 8 to 9 and the solution was kept in normal room temperature for about 24 hours which helped in proper settling and hardening of the precipitate. Finally, the solution was again stirred for about 30 minutes and kept for aging for another 24 hours which resulted in the complete precipitation. Purification of the precipitate was done in the similar way and the produced HAp sample was denoted as sample 2.
C. ***Non-Acidic Precipitation Method*** [12]: Like before, here also eggshells were initially calcined in a furnace to produce calcium oxide.Then, proper amounts of potassium hydrogen phosphate(K_2HPO_4) was dissolved in distilled water and to this solution calcium oxide powder was added slowly along with continuous stirring for about 10 minutes. The pH was adjusted by adding sodium hydroxide and was kept around 11 to 12. Finally precipitates of HAp appear and start setting at the bottom of the beaker. Similar aging and purification techniques were followed and the pure HAp obtained was denoted as sample 3.
D. ***Chemical Synthesis/biomimetic precipitation Method*** [13]: Among all the four processes this is the only process in which synthetic HAp is produced from chemical sources. Firstly, a supersaturated calcium solution was prepared by dissolving 0.55 gram of calcium chloride, 0.150 gram of sodium dihydrogen phosphate and 0.073 gram of Sodium Bicarbonate in 500 ml of distilled water. Thereafter the solution was continuously stirred in a shaker for 24 hours. In the end a white color precipitate was observed. Finally, purification of the precipitate was done and lastly the purified precipitate was heated at a temperature of about 110°C for 2 hours and as a result a very thin, rough and flaky layer of pure HAp was produced. Similarly, it was denoted as HAp sample 4.

Preparation of AgNPs [14]: Silver (Ag) nanoparticles were synthesized by chemical reduction method. 8.45 mg of 0.001M $AgNO_3$ was mixed with 50ml of water and boiled. A magnetic stirrer was introduced to continuously stir the solution. To the boiling solution 5ml of 1% w/v of Tri-sodium citrate was introduced dropwise. After a few minutes change in color was observed in the solution. Firstly, it turned pale yellow then it turned greenish and finally it changed to greyish color. After stirring for about an hour the solution was kept at room temperature. Then, a part of the solution was centrifuged and it was observed that after 30 minutes of centrifugation the silver nanoparticles settled down. The particles were then collected carefully using a micro-pipette.

Preparation of Polymeric Solution: For this experiment polymeric solution of Chloroform and Polycaprolactone (PCL) Beads was used. For each 25 ml of chloroform 1 gram of PCL was used. After mixing the chemicals the solution was stirred for about 30 to 45 minutes. Thereafter it was sonicated for about 10 to 15 minutes and in the end a translucent white solution was prepared.

Preparation of Composite Mixtures with HAp and AgNPs: After the preparation of the polymeric solution, composite mixtures were prepared. Firstly, along with 25 ml of of the polymeric solution, 0.0125 gram of the finely grinded or powdered HAp sample was mixed. The mixture was sonicated twice for 30

minutes and finally a composite was produced. The same procedure was repeated for each HAp sample and hence a set of composite mixtures were prepared.

Using the similar polymeric solution and similar technique, composites with AgNPs was prepared.

Preparation of 2D Scaffolds from HAp composite mixtures: For the preparation of the 2D scaffolds, the composites with different HAp samples were casted onto small metallic prototypes of actual bioimplants. Firstly, small metallic sheets of stainless steel were taken and the surface was roughened by rubbing with a sandpaper. Thereafter the composite mixtures were added dropwise on the surface of the sheets forming a coating. Then, the sheets were kept for drying at room temperature for about 10 to 15 minutes. The same process of was repeated for about 2 to 3 times and a uniform coating with proper thickness was formed on the metallic surfaces.

Preparation of 2D Scaffolds from AgNPs composite mixture: Scaffold from the AgNPs composite was also formed by the similar process, but in this case the composites were casted onto pre sterilized petri dishes which were sterilized with 70% ethanol and by overnight exposure to UV radiation.

In the similar way composite mixtures were produced by mixing both HAp powder and silver nanoparticle solution into the polymeric solution of PCL and chloroform. There after even this composite mixture was used to produce 2D scaffold.

Biodegradation Assay: In order to check the biodegradability properties of the scaffolds, a biodegradation test was done. For the test Simulated Body Fluid (SBF) was used. After every 7 days the scaffolds were taken out of the solution, dried and weighed. This process was repeated for about a month and the results were plotted.

Antifungal Assay of AgNPs: For identifying the antifungal properties of Silver Nanoparticles, proper experiment was conducted. Fungal inoculum from fruit sources were taken and grown in two different Sabouraud Dextrose Agar (SDA) plates, one of which contained sufficient amounts of AgNPs. Fungi was allowed to grow for 7 days and then colonies from both the plates were observed under the light microscope following proper identification techniques.

Characterization: Phase composition of the HAp Powder and the Silver Nanoparticles was studied by X-ray diffraction analysis. The surface morphology of the HAp layers and AgNPs and the 2D scaffolds were primarily revealed from the Light Microscope images taken under magnifications of 20X.finally characterization of the 2D scaffolds was done using Scanning Electron Microscope under a magnification of 1000x. graphs were plotted on the basis of the biodegradation assay data and antifungal properties of AgNPs were checked through proper microscopic investigations.

3 RESULTS AND OBSERVATION

3.1 *X- ray diffraction pattern*

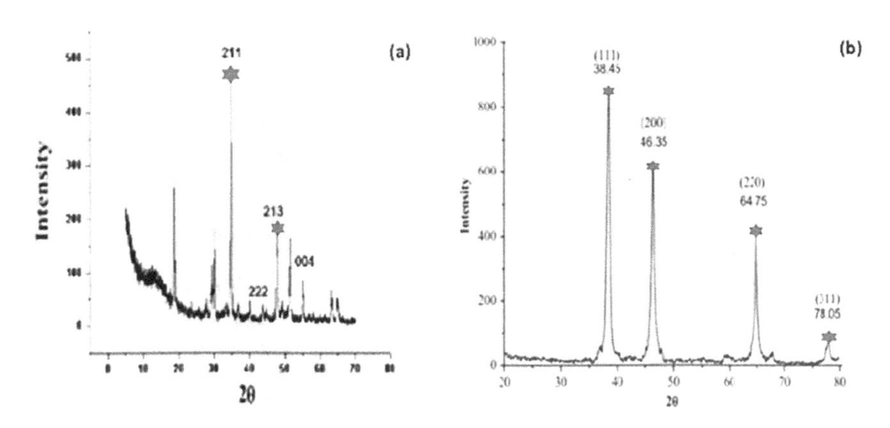

Figure 1. (a): X-ray diffraction pattern of synthesized HAp sample 2, (b) AgNPs sample.

3.2 *Surface of HAp and AgNPs samples observed under light microscope at 20X magnification*

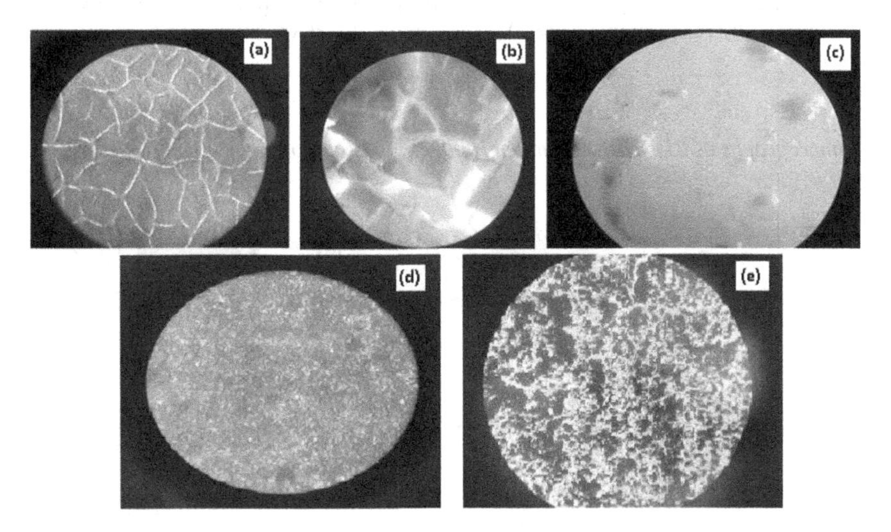

Figure 2. (a): Microscopic Images of Surface of HAp sample1, (b) HAp sample 2, (c) HAp sample 3, (d) HAp sample 4, (e) AgNPs sample.

3.3 *2D scaffolds observed under light microscope at 20X magnification*

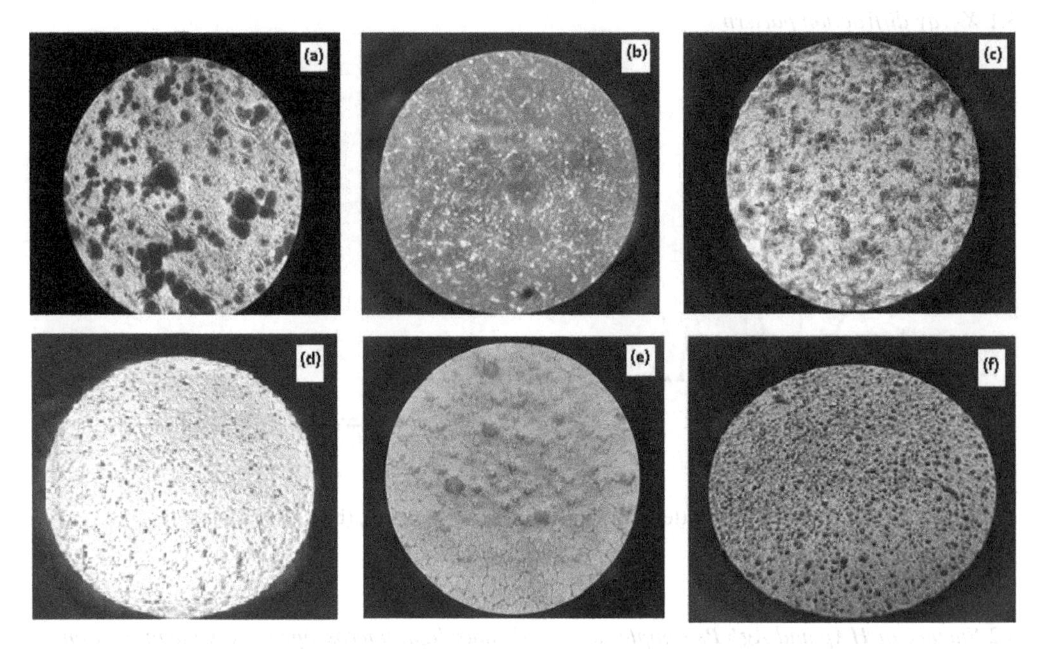

Figure 3. (a): Microscopic Image of 2D scaffold formed with HAp sample 1, (b)HAp sample 2, (c) HAp sample 3, (d) HAp sample 4, (e) AgNPs sample, (f) HAp sample 2 and AgNPs sample.

3.4 *SEM micrographs of 2D scaffolds under 1000X magnification*

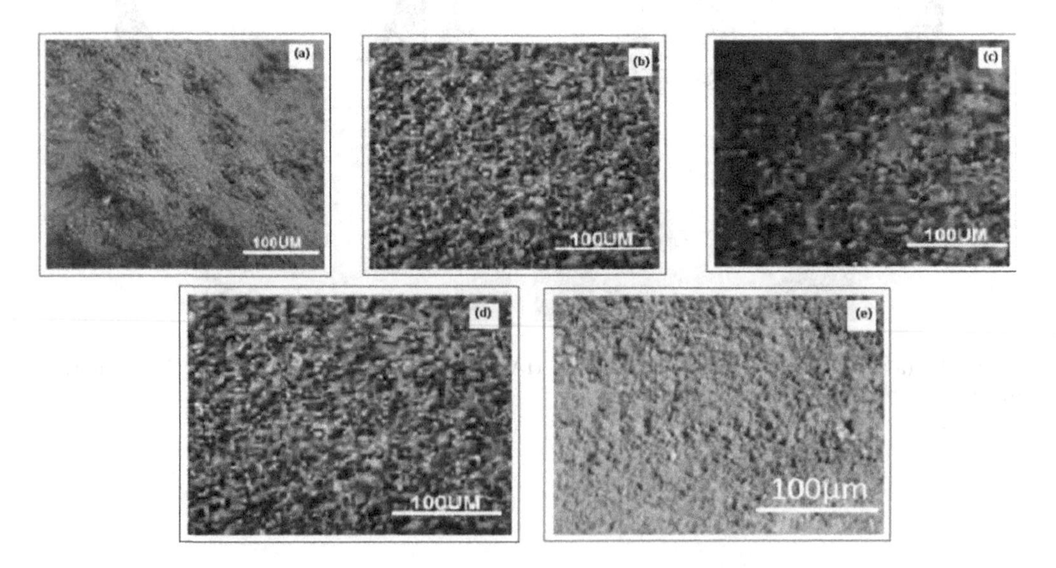

Figure 4. (a): SEM Micrograph of 2D scaffold with HAp sample 1, (b): HAp sample 2, (c): HAp sample 3, (d): HAp sample 4, (e): AgNPs sample.

3.5 *Biodegradation assay data*

No of Days	Weight of Control material	Weight of Scaffold material
0	1.24	1.24
7	1.24	1.09
14	1.24	0.992
21	1.24	0.981
28	1.24	0.97

(a)

No of Days	Weight of Control material	Weight of Scaffold material
0	1.223	1.223
7	1.223	0.996
14	1.223	0.908
21	1.223	0.88
28	1.223	0.801

(b)

Table 1. Table (a): Biodegradation Assay data for Scaffold with HAp sample 2 and control, Fig 5(a) Graph plotted from the data of Table (a), Table (b): Biodegradation Assay data for Scaffold with AgNPs sample and control, Fig5(b) Graph plotted from the data of Table (b).

3.6 *Observations from the antifungal assay*

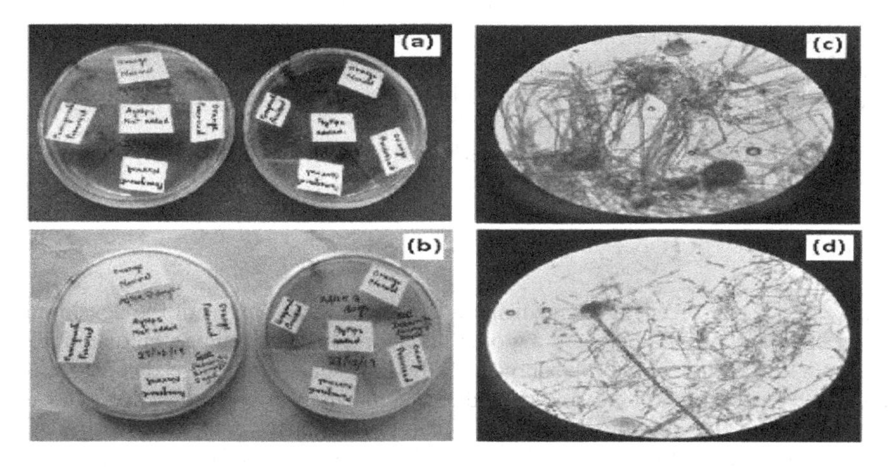

Figure 6. (a) Agar plates with and without AgNPs on the 1st day after inoculation, (b) (a) Agar plates with and withoutAgNPs on the 8 th day after inoculation (c): Microscopic images of fungal colonies isolated from Agar plate withoutAgNPs, (d) With AgNPs.

4 DISCUSSION

X-Ray Diffraction Analysis: The X-ray diffraction pattern of HAp sample 2 is represented in Figure 1(a). This method of characterization was mainly used to confirm that the particles produced are of HAp and to know the structural information. The diffraction pattern shows intense reflection peaks in between 30°- 40° of 2θ values. Moreover, the peaks are observed at

(211), (213), (222), and (004) plane. This indicates the presence of the apatite phase.Similarly, in the Figure 1(b), we see the diffraction pattern of the AgNPs sample.

Analysis of The Microscopic Images of The Surface of The HAp and AgNPs Samples: Figures 2(a), 2(b), 2(c), 2(d) and 2(e) are the microscopic images of the surface. The micro-cracks present in the sample resemble interconnectivity between the particles. It indicates that the particles within the samples are well distributed and it also proves the fact that these materials when used to produce 3D systems will provide proper surface for the cellular adherence.

Analysis of The Microscopic Images of 2D scaffolds: Figures 3(a), 3(b), 3(c), 3(d), 3(e) and 3 (f) are the microscopic images of the 2D scaffolds formed with HAp It has been observed that HAp has nicely dispersed and uniformly spread in the scaffolds. Finally, we can infer that the dispersion is enhanced to a great extent in the scaffold which consists of the HAp sample along with AgNPs. Thus, we can say that due to the presence of AgNPs the rates of dispersion of the HAp particles increases.

Inferences from the SEM Micrographs: Scanning electron microscope images shown in the Figure 4(a), 4(b), 4(c), 4(d) and 4(e) show the nanocrystalline nature of the HAp and silver particles present on the scaffold. Agglomerates with irregular sHApes are observed and the material is found to be bulky and porous. This porosity is advantages as it will allow tissue growth and cellular regeneration on the surface of the implants when the material would be used as a coating.

Biodegradation Assay Observations: On observing thedata present in Tables 1 and 2, and graphs plotted in Figures 5(a) and 5(b), for the biodegradation assay. Finally, it is observed that the scaffold which consists of both HAp and AgNPs also got the degraded when kept in the SBF solution. Thus, we can say that presence of silver nanoparticles did not influence the biodegradation properties.

Antifungal Assay: From Figure 6(a) and 6(b) it is clearly visible that the Fungal growth in the agar plate without AgNPs was maximal, whereas in the plate containing AgNPs the growth was comparatively less. Moreover, on observing isolated Fungal colonies under the Light Microscope as represented by Figure 6(c) and 6(d) it is seen that in the absence of the Nanoparticles, the growth and morphology of the fungus is normal. Whereas in its presence the fungi became fragmented and it could not grow properly. These results prove that Silver Nanoparticles (AgNPs) possess remarkable antifungal properties and hence can be effectively used for curing *in-vivo* infections and skin related problems.

Fungal colonies from different sources were also observed under the microscope and the results were found to follow similar trends.

5 CONCLUSION

Importance of TE and regenerative medicine is gradually increasing with the development of modern science and technologies as a potential parallel therapeutic technique. Previously what was a matter of concern, TE has now already done successfully with the help of various interdisciplinary expertises of advanced medicine and engineering knowledge. The present work demonstrates the development of PCL based 2D matrix embedded with HAP and AgNPs. As both of these materials are biocompatible, fabrication of interconnectivity among the prepared samples would further enhance the cellular interaction, hence, accelerating the cellular interaction and new osseous tissue regeneration. Presence of HAp is also advantageous for osseous tissue regeneration as it provides essential microenvironment required for the osteogenic cells. Moreover, the presence of AgNPs showed excellent antifungal property that is beneficial to support cellular interaction (in-vivo & in-vitro) by maintaining aseptic condition. All of these results predict excellent bioactivity and cell supportive nature of the developed matrix. There is also high scope for future research and development especially in the field of Biomedical Engineering.

REFERENCES

[1] Saiz, L. Gremillard, G. Menendez, P. Miranda, K. Gryn, A.P. Tomsia, Preparation of porous hydroxyapatite scaffolds, Materials Science and Engineering C 27 (2007) 546–550.

[2] K. de Groot, Bio ceramics consisting of calcium phosphate salts, Biomaterials 1 (1980) 47–53.

[3] Hench L.L., Wilson J. An introduction to bio ceramics. London: World Scientific, 1993. p. 8, 165, 170, 331, 335.

[4] Sun, L., Berndt, C.C., Gross, K.A. and Kucuk, A. (2001) Material Fundamentals and Clinical Performance of Plasma- Sprayed Hydroxyapatite Coatings: A Review. *Journal of Biomedical Materials Research*, 58, 570–592.

[5] Kalashnikova, I., Das, S., Seal, S. 2015. Nanomaterials for wound healing: scope and advancement. Nanomedicine. Doi:10.2217/NNM.15.82.

[6] www.wikilectures.eu/w/Scaffolds_in_tissue_engineering

[7] Zhang R., Ma P.X. Porous poly (l-lactic acid)/apatite composites created by biomimetic process. J Biomed Mater Res 1999; 45:285–293.

[8] R. Bardhan, S. Mahata and B. Mondal Processing of natural resourced hydroxyapatite from eggshell waste by wet precipitation method.

[9] Himanshu Khandelwal, Satya Prakash Synthesis and Characterization of Hydroxyapatite Powder by Eggshell.

[10] Ahmed, S. and Ahsan, M. (2009) Synthesis of Ca-Hydroxyapatite Bio ceramic from Eggshell and Its Characterization.

[11] Dasgupta, P., Singh, A., Adak, S. and Purohit, K.M. (2004) Synthesis and Characterization of Hydroxyapatite Pro- duced from Eggshell. Proceedings of the International Symposium of Research Students on Materials Science and Engineering, 1–6.

[12] Chaudhuri, B., Mondal, B., Modak, D.K., Pramanik, K. and Chaudhuri, B.K. (2013) Preparation and Characterization of Nanocrystalline Hydroxyapatite from Egg Shell and K2HPO4 Solution. Materials Letters, 97, 148–150 https://doi.org/10.1016/j.matlet.2013.01.082.

[13] Adrian Paz; Dainelys Guadarrama; Mónica López; Jesús E. González; Nayrim Brizuela; Javier Aragón A comparative study of hydroxyapatite nanoparticles synthesized by different routes http://www.scielo.br/scielo.php?script=sci_arttext&pid=S0100-40422012000900004

[14] Essam N. Sholkamy, Maged S. Ahamd, Manal M. Yasser, Noor Eslama. Anti-microbiological activities of bio-synthesized silver Nano-stars by Saccharopolyspora hirsute. Saudi Journal of Biological Sciences. 26, 1, January 2019, 195–200.

Biotechnology and Biological Sciences – Sen et al. (Eds)
© 2020 Taylor & Francis Group, London, ISBN 978-0-367-43161-7

Fabrication of alginate/poly (γ-glutamic acid) 3D-bioprinted scaffolds and investigating their mechanical physicochemical and biological properties

S. Datta*, R. Barua & A. Das
Centre for Healthcare Science and Technology, Indian Institute of Engineering Science and Technology, Shibpur, Howrah, India

A. RoyChowdhury
Centre for Healthcare Science and Technology, Indian Institute of Engineering Science and Technology, Shibpur, Howrah, India
Department of Aerospace Engineering and Applied Mechanics, Indian Institute of Engineering Science and Technology, Shibpur, Howrah, India

P. Datta
Centre for Healthcare Science and Technology, Indian Institute of Engineering Science and Technology, Shibpur, Howrah, India

ABSTRACT: Three dimensional bio-printing is mainly energetic and prompt procedure that is needed for constructing the in-vivo and in-vitro biological purposeful tissues. It is very important for an active scaffold printing, the printed construct should have suitable strength and rigidity for maintaining fundamental accuracy. For bio-printing process mainly alginate are used as the bio-inks. Bio-ink which is used should not contain high shear force because high shear force results more extrusion pressure through the printer nozzle which causes cell death and fracture. In the paper we studied the cellular bio-ink properties alginate scaffolds by 3D bio printing technique. We have taken one concentration of poly(γ-glutamic acid) mixing fixed alginate percentage. Poly amino acid has great importance in the area of drug delivery, tissue engineering and biomaterials due to its exceptional bio-degradability and bio-compatibility. We characterization the constructs by using SEM, Surfacetensiomete MTT and UTM. Osteoblast-like cells (MG63) cells were cultured on the printed scaffolds showed this is a very good method for construction of 3D printed scaffolds for bone tissue engineering applications.

Keywords: Bio-inks, Scaffolds, Alginate, UTM, SEM

1 INTRODUCTION

The technique of 3D bioprinting is one of the additive manufacturing process in which the constructs are created layer by layer i.e polymer, plastic, ceramic and powder and living cells are used [1], [2]. First a CAD file is generated and uploaded in the 3D printer then the printer prints the constructs accordingly like the CAD structure file. The printer prints first the first layer by moving in X and Y axis after the first layer is complete the printer head moves in the Z axis and then prints the second layer above the first layer. The process repeats until the desired final construct is needed until the desired final structure is generated [3]. By using the CAD file we can convert 2D images like x-rays, computerized tomography and magnetic resonance imagining into

* Corresponding author: dattadip440v@gmail.com;

3D objects [4], [5]. To print an efficient and useful biological constructs the selection of desired bio-ink is very much important, alginate is one such biopolymer which is mainly used for printing various scaffolds because of its various advantages. [6] The advantages of alginate bioink compared to other bioinks are as follows a) naturally available, b) non-toxic, c) biodegradable, d) non-immunogenic etc. Alginate is composed of guluronic and manuronic acids [7] Alginate is very biocompatible the cost is low and is found in the brown algae cell wall [8] it forms hydrogels by crosslinking it with CaCl2 in normal condition. Because of all these advantages of alginate, it is mainly used as bio-ink for bio-printing technique. There are few cons of alginate bio-ink like fast gelation, high hydrophilic nature of alginate causes less cell material interaction and inadequate degradation. For overcoming few limitations here we are using Poly(γ-glutamic acid) (PGA) low concentration mixture with alginate solution. Poly amino acid has great importance in the area of, tissue engineering, drug delivery and biomaterials due to its exceptional bio-degradability and biocompatibility. PGA can be considered as a perfect biodegradable polymer in the field of medical science, as glutamic acid is being formed by degradation which is a vital component for human body. The carboxyl group of PGA inclines to associate with the drugs. PGA has some shortcomings also in its uses, like the uncontrolled degradation rate. Biomaterials depends upon PGA are generally modified through copolymerization with other monomers like L-aspartic acid [9].

2 MATERIALS & METHODS

2.1 *Alginate "bio-ink" preparation*

Alginate (Alginate, viscosity 2000 cP, 2%) purchased from (Sigma-Aldrich, USA). PGA Molecular Weight 15,000–50,000) was purchased from Sigma-Aldrich, USA. Here we prepared two solutions alginate 5% (w/v) in typeII water and alginate 5%-PGA 2.5% (w/v) in typeII water. After mixing the alginate and PGA in the typeII water the solution is mechanically stirred for 3 hours until it is homogenously mixed.

2.2 *Alginate-PGA bioinks physico-chemical characterization*

The conductivity as well as the PH of the solutions was measured by Thermo Oakton meter at a temperature of 26°C. The viscosity was measured by Rhamy-rheology France viscometer in temperature of 24°C at rpm 4 for 1 minute. On the glass slide the Surface tension was calculated using SURFTENS 4.5 instrument Figure 2. Represents the pH, conductivity, viscosity and surface tension of the two solutions.

2.3 *Extrusion 3D Bio-printer*

A 4 axis 3D bio-printer which is designed by us was mainly used for output scaffolds printing. The three axis X axis, Y axis, Z axis and the controlled pump syringe extruder completes the printer which helps to make different scaffolds structures as per the program which is uploaded in the printer. X, Y axis are controlled by the stepper motors while the vertical Z axis is controlled by screw control on linear bearing. The processor which is used in our printer is a SAM controller 32-bit, which regulates the 3D bio-printer for precise scaffold printing along with a compatible computer CAD software. The computer CAD software that is used for slicing the solid scaffold and for conversion of .stl file to .GCODE file is CURA open source version 15.04.4 after that the upload of the designed scaffold GCODE file was done into the printer for printing the desired output construct.

2.4 *Scaffold design*

SketchUp 2017 was used to create the 3D solid object i.e. square in this case having dimension 35×35 mm X, Y. The file was in .stl then the .stl file was uploaded to CURA 15.04.4 software for slicing the solid scaffold by decreasing the fill density starting 100 % to 22 % which results in a mesh like structure. Finally, the designed scaffold was converted in .GCODE format and was

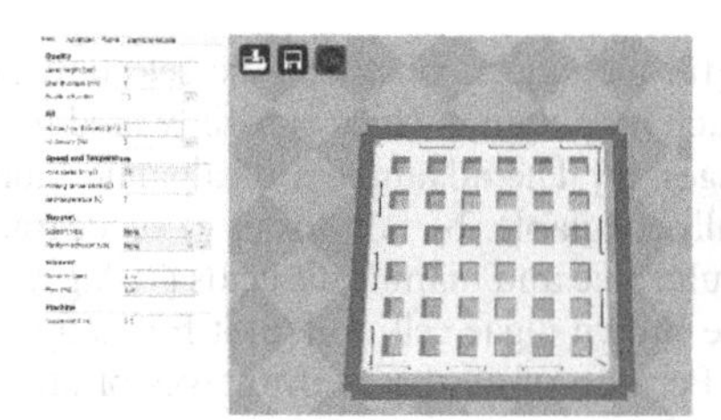

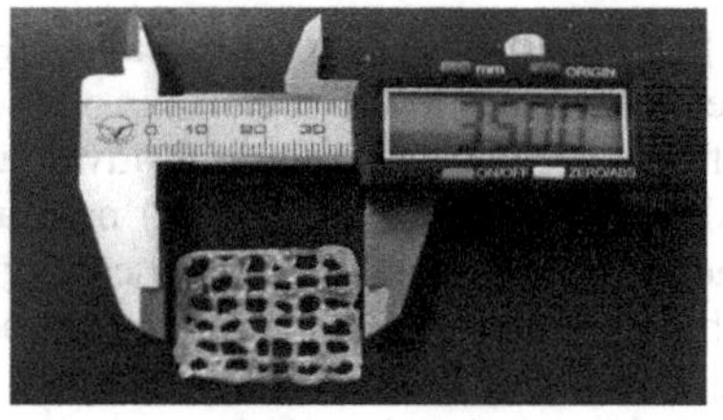

Figure 1. (a) The desired structure of the scaffold generated by CURA version 15.04.4. (b) Output printed scaffold of alginate-PGA.

loaded in the 3D bio-printer for scaffold printing. Figure 1a. Denotes the computer designed scaffold CAD input structure by using both SketchUp 2017 as well as slicing software CURA software [12].

2.5 *Alginate-PGA bio-ink extrusion printing*

The two solutions Alginate and the Alginate-PGA was loaded in the 3D printer 5ml syringe in which a microtip was attached to the syringe from which the extruding the bio-inks occurred. The printing speed was kept constant 50mm/s at room temperature 30°C temperature. The mesh like scaffolds was printed on glass Petri dish. During the extrusion of the bio-ink from the syringe CaCl2 1 molar was sprayed on the printed scaffolds for crosslinking the extruded post printed bio-inks, which results in a stable solid scaffold like the input structure which was uploaded in the Bio-printer. Finally, after the solid scaffold was formed the scaffold was cleaned by 70% ethanol and was dried by vacuum drier for 1 hour in 35°C. Figure 1b. Shows the output printed scaffold of size 35 × 35 mm.

2.6 *Printability analysis of the printed scaffolds*

The study of the post printed scaffolds were done to study the structural scaffold stability. Oyang et al [10] [11] proposed two structural stability calculation procedures by mathematical formulas. S Datta et.al [12] studied the printability analysis of the alginate-honey scaffolds by bio 3D printing using the formulations.

$$C = \frac{4\pi A}{P^2} \tag{1}$$

$$P = \frac{\pi}{4C} \tag{2}$$

Table 1. Printability analysis of the two bio-ink solutions.

Samples	Strut Diameter (mm)	Pore Area (mm^2)	Perimeter (mm)	Circularity	Printability
A5	1.3 ± 0.6	17.0 ± 4.6	15.6 ± 1.8	0.77	1.01
A5PGA2.5	2.3 ± 0.3	8.02 ± 0.6	12.6 ± 1.0	0.63	1.24

Intended pore area 15mm^2

Here C represents circularity when the pores of the scaffolds are similar to circle then its value is closer to 1 i.e. $C \leq 1$, and P represents whereas P represents printability when scaffold pores are like square its value is closer to 1 i.e. $P \leq 1$.

2.7 *SEM images, cell culture and alizarin red staining assay of the scaffolds*

Cell adhesion and proliferation for the samples was analysed under scanning electron microscope (HitachiS-3400N). At the end of 5 days culture the samples were harvested from the culture plates and rinsed with PBS solution. The samples were then transferred to a 2.5% glutaraldehyde solution for 1h. After cell fixation, the samples were washed with PBS solution and dehydrated by immersion in ethanol solutions of growing concentrations from 30 to 99.9%, with a residence time of 5 min in each ethanol solution. SEM observation was performed at a magnification of 500X and 15kV energy. Alizarin Study on done by seeding 1×10^5 osteosarcoma MG63 cells on it for 3 days in DMEM glucose free medium. Then staining in 40mM alizarin red staining for 20 min then dissolving the two scaffolds in 10% acetic acid for nearly 1 hour and the absorbance reading was taken in 405nm. Scaffolds 10mm × 10mm ×100μm was used MTT cell viability study. The MG63 cells culture was done in DMEM with supplemented with 1% penicillin-streptomycin antibiotic solution and 10% fetal bovine serum. Cells concentration was 10^4 cells per mL and the UV sterilization was done on the two samples and kept temp 37°C at with CO_2 5 percent. 1 Day cell viability was measured using [4,5-Dimethylthiazol-2-yl]-2,5-diphenyltetrazolium bromide MTT assay in a 96 well plate. After 1 day in DMEM medium serum free reagent of MTT was added and kept for 4 hours. In solubilisation buffer the formazan crystals were solubilized and the absorbance was done at 570 nm and 670 nm.

3 RESULTS & DISCUSSION

3.1 *Alginate:PGA physico-chemical properties*

One of the biggest device in tissue engineering is the extrusion 3D bio-printing. There are many challenges of 3D bio-printing among them one big problem is to create a proper bio-ink for 3D bio-printing such as the cells are not harmed during and after printing as well as the pH of the solution such be optimized for better cell viability. Shear thinning property should also be present in the bio-Ink. Alginate which is mainly used in almost all extrusion 3D bio-printing is mostly used in regenerative medicine and tissue engineering bio-printing purpose. Here we measured the pH, conductivity, surface tension and the viscosity of the two solutions.

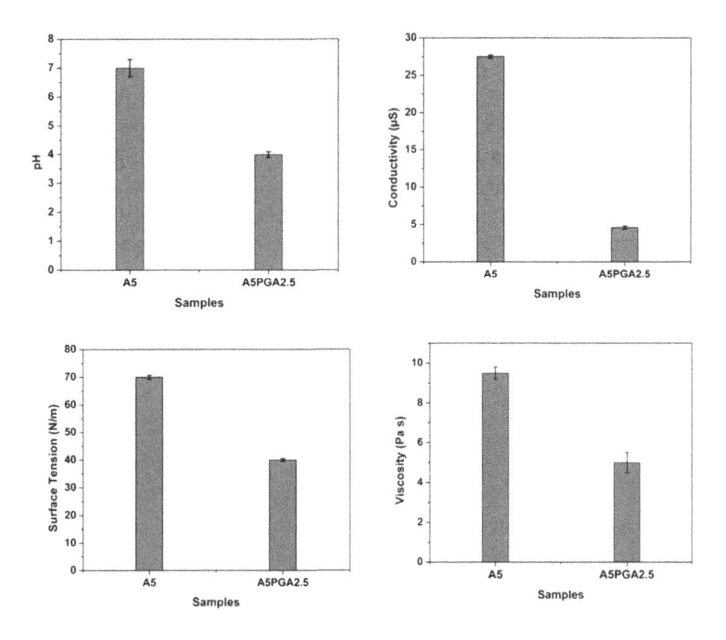

Figure 2. (a) pH; (b) Conductivity; (c) Surface tensions; (d) Viscosity of the two samples.

From the Figure 2 we can see as we can see as we added PGA in the solution of alginate the viscosity, surface tension, pH and conductivity falls.

3.2 *Printed scaffold characterization*

The mechanical tensile testing was done by universal testing machine (UTM) Titinus-Olsen dimensions of the scaffolds were taken 15 × 15 mm X, Y length for the experiment. Figure 3a. Shows the ultimate tensile stress (UTS) of the two scaffolds it can be seen from the Figure 3a ultimate tensile stress of pure alginate is more compared to the blend solution of alginate-PGA. The young's modulus of the SURFTENS 4.5 Germany was used for measuring the contact angles of the two scaffolds. The contact angle of alginate 5% was 14.9/14.9° while for Alginate 5%-PGA2.5% the contact angle was 82.1/82.1°. The contact angle decreases as we added PGA in the alginate solution.

3.3 *Cells-Scaffold interactions*

Biocompatibility study of the scaffolds were done by culturing osteoblast-like cells (MG63) cells on the scaffolds and culturing them for 1 day. Viability of cells of sample alginate-PGA were was found out comparatively more than normal alginate sample. It is seen that from the Figure 3b that the viability of the cells increased on Alg5PGA2.5 on day 1 compared to the other samples. The Alg5PGA2.5 was found significantly different from Alg5 i.e. ($p<0.05$). PGA scaffold is more firm compared to the alginate scaffold. This result indicates that the modification of alginate with PGA dramatically increases the cell cellular affinity and compatibility for the alginate scaffold. The calcium deposition was found out more in Alg5PGA2.5 by using alizarin staining absorbance measurement at 405nm. Figure 4c shows the osteogenic differentiation of cells is more in Alg5PGA2.5.

3.4 *Cell-Scaffolds morphologies*

SEM observation in Figure 4 demonstrates cell growth and attachment on the two scaffolds. In Figure 4a, 4b cell morphologies on the alginate scaffolds are shown. The cell as shown in the SEM images at magnification x500 looks like spheres on both scaffolds types after 5 days culture. Morphology of the cells are almost like shuttle-like. The cells on the alginate-PGA scaffolds are more stretched out and elongated. The cells on the alginate-PGA scaffold is more firm compared to the alginate scaffold. This result indicates that the modification of alginate with PGA dramatically increases the cell cellular affinity and compatibility for the alginate scaffold. Figure 4c shows the deposition of calcium on the scaffolds here A5PGA2.5 calcium deposition is more.

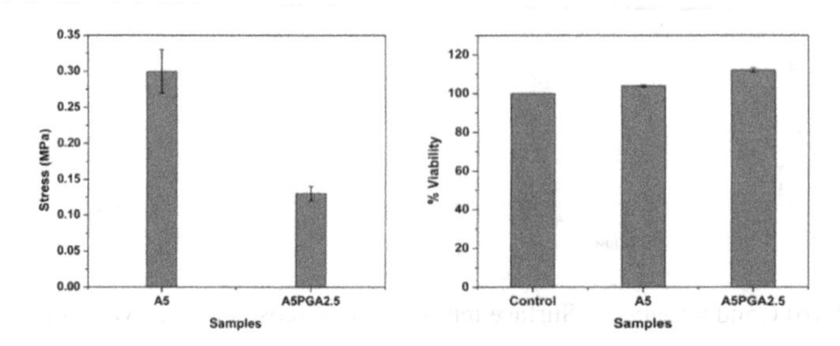

Figure 3. (a) Ultimate tensile stress vs sample, (b) Cell viability of alginate-PGA acid scaffolds on day 1 of osteoblast-like cells (MG63) cell culture.

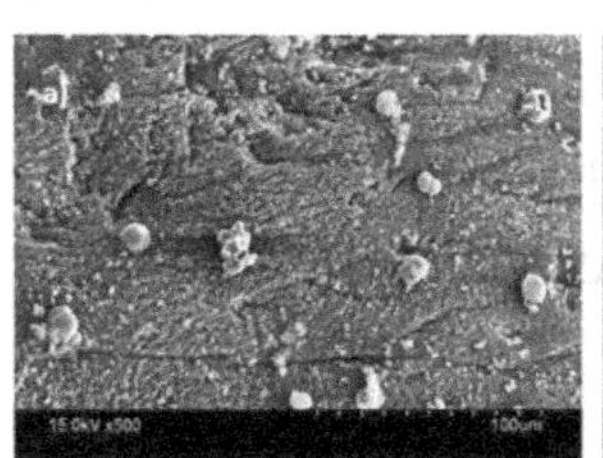

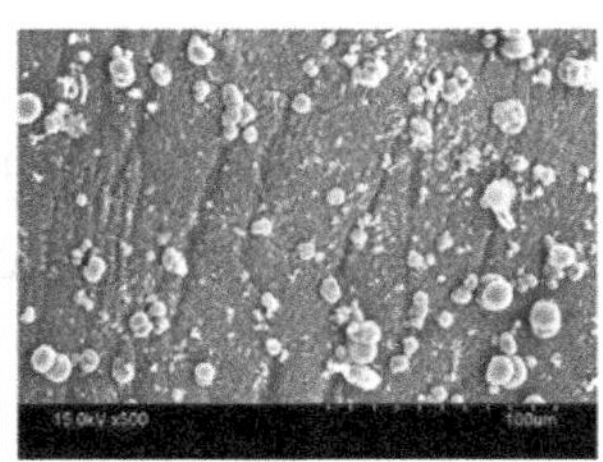

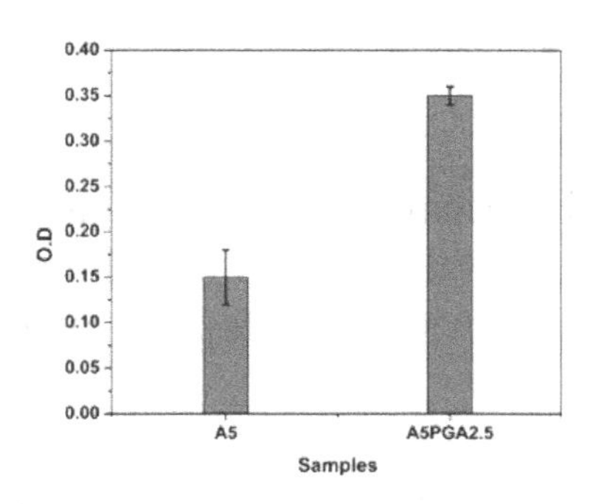

Figure 4. SEM images of osteoblast-like cells (MG63) cultured on Alginate 5% scaffolds at 5 days (a) Without 2.5% PGA; (b) With 2.5% PGA (c) Alizarin red staining of the samples.

4 CONCLUSION

The objective of the current study was to fabricate novel 3D bio-printed alginate/PGA scaffolds and study their morphological, mechanical and cell viability properties for tissue engineering applications. For soft tissue engineering applications, a reduction of mechanical strength is obvious. pH reduction with PGA addition is also obvious for bone tissue engineering application. The morphology of the scaffold changed as we added PGA 2.5% with alginate. The results propose that modification of alginate with PGA efficiently increases the alginate scaffold biocompatibility. Our work suggests a simple and efficient procedure to modify the alginate scaffold which creates and develops the potential application of alginate in the bone tissue engineering field.

REFERENCES

[1] Klein, G.T, Lu, Y & Wang, M.Y. 2013. 3D printing and neurosurgery–ready for prime time?, *World Neurosurg.*, 80 (3): 233–235.

[2] Barua, R. Datta, S. Datta, P & RoyChowdhury, A. 2019. Scaffolds and Tissue Engineering Applications by 3D Bio-Printing Process, *Design, Development, and Optimization of Bio-Mechatronic Engineering Products*, IGI Global Publishers: 78–99.

[3] Barua, R. Datta, S. RoyChowdhury, A & Datta, P. 2019. Importance of 3D Printing Technology in Medical Fields, *Additive Manufacturing Technologies from an Optimization Perspective*, IGI Global Publishers: 21–40.

[4] Banks, J. 2013. Adding value in additive manufacturing: Researchers in the United Kingdom and Europe look to 3D printing for customization, IEEE Pulse, 4 (6) 22–26.

[5] Cui, X. Boland, T. D'Lima, D.D. & Lotz, M.K. 2012. Thermal inkjet printing in tissue engineering and regenerative medicine, *Recent Patents Drug Deliv. Formul.*, 6 (2) 149–155.

[6] Stanton, M.M. Samitier, J. & Sánchez, S. 2015. Bioprinting of 3D hydrogels, *Lab Chip*, 15 (15) 3111–3115.

[7] Pawar, S.N. & Edgar, K.J. 2012. Alginate derivatization: A review of chemistry, properties and applications, *Biomaterials*, 33 (11) 3279–3305.

[8] Tønnesen, H.H. & Karlsen, J. 2002. Alginate in drug delivery systems. *Drug Dev. Ind. Pharm.*, 28 (6) 621–630.

[9] Wang, L. 2008. SYNTHESIS AND CHARACTERIZATION OF POLY (γ-GLUTAMIC ACID-co-L-ASPARTIC ACID), 26(4) 381–391.

[10] Ouyang, L. Yao, R. Zhao, Y. & Sun, W. 2016. Effect of bioink properties on printability and cell viability for 3D bioplotting of embryonic stem cells, *Biofabrication*, 8(3) 1–12.

[11] Datta, S. Sarkar, R. Vyas, V. Barui, A. Bhutoria, S. RoyChowdhury, A. & Datta, P. 2018.Alginate-honey bioinks with improved cell responses for applications as bioprinted tissue engineered constructs, *J. Mater. Res.*, 1–11.

[12] Datta, S. Das, A. Sasmal, P. Bhutoria, S, RoyChowdhury, A. Datta, P. 2018 Alginate-poly(amino acid) extrusion printed scaffolds for tissue engineering applications, International Journal of Polymeric Materials and Polymeric Biomaterials, 0(0) 1-9.

Biotechnology and Biological Sciences – Sen et al. (Eds)
© 2020 Taylor & Francis Group, London, ISBN 978-0-367-43161-7

Fabrication of beta-tricalcium phosphate/crystalline silica-aluminium metal matrix composites: physical, mechanical, and in vitro biological evaluations

Sourav Debnath & Akshay Kumar Pramanick
Metallurgical & Material Engineering Department, Jadavpur University, Kolkata

ABSTRACT: In this study beta- tricalcium phosphate (β TCP) has prepared for incorporating it into crystalline silica- aluminium powder to fabricate aluminium metal matrix composites. In this work, sintering of the compact powder is carried out in different environments, keeping the same sintering temperature, to investigate suitable optimum property for the fabricated composites. Optimized property of the fabricated composites is decided by observing its surface morphology and its various physical as well as mechanical properties. Suitable composite has considered for in vitro biological evaluation as per ISO and ASTM standards. Changes of surface morphology has observed for the sample during SBF analysis. Elemental composition for the bioactive surfaces of the soaked sample is also examined by EDX. Blood compatibility of the composite has also examined and reported.

Keywords: Beta-tricalcium phosphate (β TCP), aluminium metal matrix composite/Al based hybrid composite, sintered structure, in vitro bioactivity, blood compatibility

1 INTRODUCTION

Modern technology has not yet been able to provide a bone substitute which have equal or better properties than natural bone. However, natural bone cannot always be obtained in adequate quantities to repair severe bone trauma to fill large bone gap or large defects caused by trauma or various bone diseases. The biocompatible osteoinductive or conductive materials offer adequate promise for both implantable biomedical devices and to improve quality of life of the affected patients. In recent years, several calcium phosphate based synthetic ceramic particles, powders, blocks have been used as fillers in oral and orthopedic surgery. The materials include ceramics and glasses such as hydroxyapatite, tri-calcium phosphate, alumina calcium phosphates oxide (ALCAP) or bioglass [1]. But these implants are not suitable for load bearing applications due to insufficient mechanical stability and load bearing capabilities.

In previous literature, scientists are tried to develop doped tri calcium phosphate/MgO, ZnO, CaO, bio glass, TiO_2 composites [2-5] through various process but the fabrication of tri calcium phosphate- crystalline silica- aluminium metal matrix composite is rare.

In this study, first we have synthesized β TCP via wet chemical precipitation method [2] and then we have to fabricate beta tri calcium phosphate- crystalline silica- aluminium metal matrix composite through powder metallurgy technique (P/M technique) at different sintering environment. Finally optimized the suitable sintering environment by observing its various physical, mechanical properties and surface morphology. The optimized sample is undertaken for various in vitro biological evaluation followed by ISO/FDIS 23317 and ASTM F756 standards.

2 EXPERIMENTAL PROCEDURE

2.1 *Fabrication of Al based hybrid composite*

In the present study, 10 Wt. % β TCP, 20 Wt. % Crystalline SiO_2 and rest Al metal matrix composite, also referred as Al based hybrid composite, is fabricated followed by mixing and grinding of the powder (mesh size -250 μm), pressing and finally sintering of the compact powder [8-9]. In this experiment, sintering is carried out into chemically inert box furnace and controlled hot press (Nascor Technogies Private Limited, West Bengal, India), maintaining temperature 600°C for 1 hour.

2.2 *Testing & characterization*

2.2.1 *XRD analysis*

X-ray diffraction pattern were analyzed and reported for the β TCP powder and Al based hybrid composite using Rigaku Ultima III analytical diffractometer (with Cu-Kα radiation, λ = 1.54059 Å) at 30 kV and 15 mA. Various phases are identified by comparing standard JCPDS file.

2.2.2 *SEM/EDX analysis*

SEM/EDX analysis are carried out using JEOL MAKE SEM model JSM 6360, operated by PCSEM software.

2.2.3 *Measurement of physical properties*

Density of the fabricated composites were calculated by measuring the weight and dimension of each sample. Apparent Porosity is measured by Archimedes' principle.

2.2.4 *Measurement of hardness*

Micro- hardness has taken for each metallographic specimen of bed- on- plate sample at different positions by employing 100 gf loads with 10 sec dwell time. Hardness survey is carried out randomly in three different positions and finally average the hardness numbers employing Leco Micro Hardness tester, model LM248SAT.

2.3 *In vitro biocompatibility study*

First of all swelling was measured for all samples under study, procedures are described elsewhere [10] and consequently optimized sample was found out. The optimized sample was considered for 28 days SBF analysis followed by ISO/FDIS 23317, standards for implants [6]. The optimized sample has also undergo for blood compatibility analysis as per ASTM F756 standards, described in details somewhere [10].

3. RESULTS & DISCUSSION

3.1 *XRD Analysis*

Intensity vs. 2θ value of synthesized β TCP and its metal matrix composite are plotted in Figure 1. From the XRD pattern, synthesized β TCP shows highest peaks at 2θ angle of ~ 17.080 (d = 5.1869Å), 25.820 (d = 3.4345 Å), 27.920 (d = 3.1928 Å), 290 (d = 3.0763 Å), 29.480 (d=3.0273 Å), 31.840 (d = 2.8081 Å), 32.240 (d = 2.7742 Å), 32.960 (d = 2.7152 Å), 34.840 (d = 2.5729 Å), 370 (d = 2.4275 Å), 38.30 (d = 2.3282 Å), 39.880 (d = 2.2586 Å), 46.760 (d = 1.9410 Å), 49.520 (d = 1.8391 Å), 52.160 (d = 1.7521 Å), 53.120 (d = 1.7226 Å) and 62.160 (d = 1.4921 Å) as reported in Figure 1(a). From Figure 1(b)., it is seen that Al based hybrid composite shows same peak for β TCP with very low abundance in the addition of crystalline SiO_2 peaks at 2θ angle of ~ 20.8020 (d = 4.2668 Å), 26.6220 (d = 3.3456 Å), 36.5510 (d = 2.4564 Å), 40.2910 (d = 2.2366 Å), 50.1520 (d = 1.8175 Å), 59.9180 (d = 1.5425 Å),

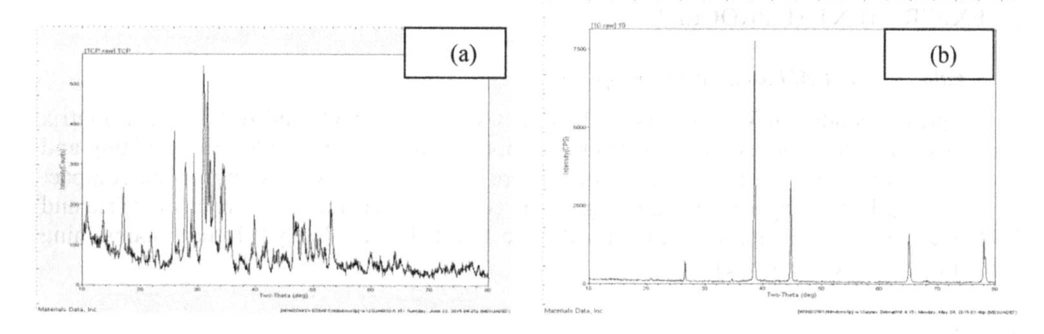

Figure 1. XRD pattern of (a) synthesized β TCP, (b) Al based hybrid composite.

68.0910 (d = 1.3759 Å) and Al at 2θ angle of ~ 38.4930 (d = 2.3369 Å), 44.7390 (d = 2.0240 Å), 65.0120 (d = 1.4334 Å), 78.0190 (d = 1.2238 Å). Presence of the individual peaks for β TCP, crystalline SiO_2 and Al after sintering implies that no reaction takes place between the matrix and the reinforcement in this sintering environment.

3.2 *Microstructure*

From the surface morphology after sintering shows that the sample sintered into hot press have better surface structure than the other one. Crystalline silica and β TCP are well distributed into Al matrix for the sample sintered in hot press as shown in Figure 2. On the other side, the sample sintered into inert box furnace shows dull, porous and unstable structure as shown in Figure 3.

3.3 *Measurement of porosity*

Significant value of apparent porosity is observed for the sample sintered in muffle furnace as shown in Figure 4. From the figure, it is found out that the sample developed in hot press gives tolerable porosity for further study.

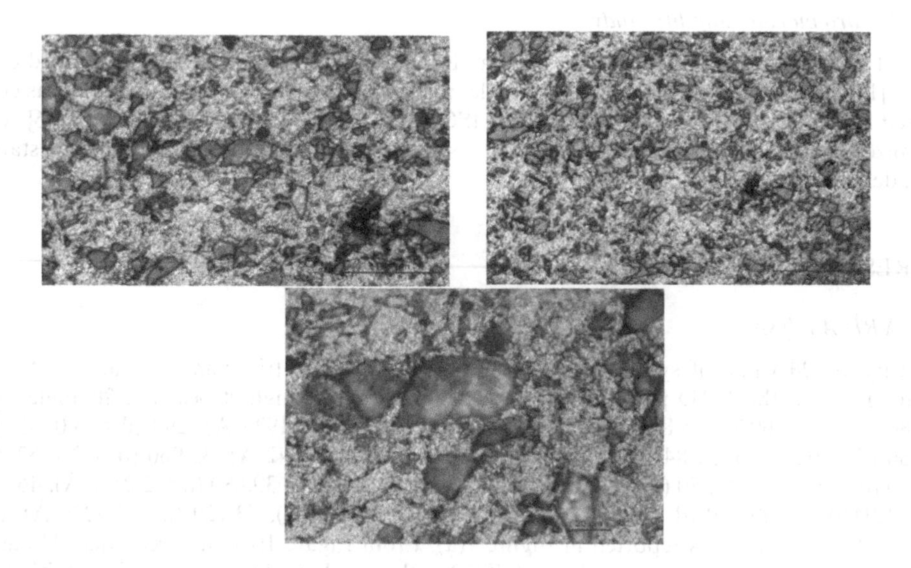

Figure 2. Microstructure of Al based hybrid composite (fabricated into hot press) at different magnifications.

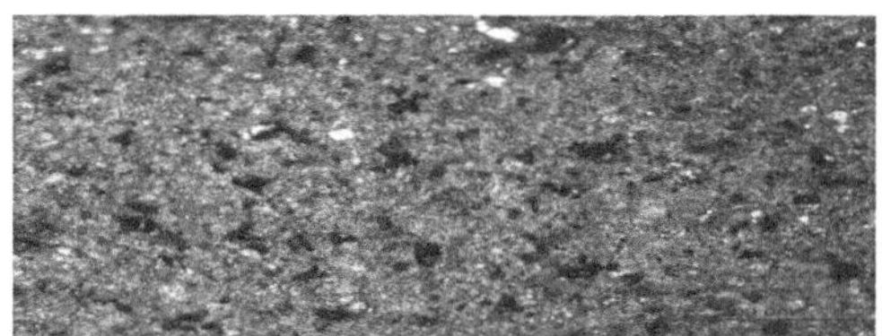

Figure 3. Microstructure at different magnifications of Al based hybrid composite (sintered in inert box furnace).

Figure 4. Apparent Porosity of sintered composite samples.

3.4 *Measurement of density*

Measurement of both green and sintered densities of pure sintered Al, 20 Wt. % crystalline silica- aluminium composite and Al based hybrid composite have measured and reported in Table 1. From the measured data, it has seen that the value of density has increased after sintering which implies formation of bonding after sintering. It is also noticed that sintered density improves slightly with incorporating β TCP into crystalline silica- Al composite.

3.5 *Micro-hardness survey*

From the data obtained, it is observed that hardness value is improved for incorporating β TCP into crystalline silica- aluminium composite, shown in Table 2. Due to insufficient mechanical property, hardness survey cannot be performed for the sample sintered into inert muffle furnace.

3.6 *In vitro biocompatibility study*

3.6.1 *Swelling analysis*

The composite fabricated into hot press which has shown only acceptable swelling ratio, typical value 0.1851 % and considered further in vitro biocompatibility analysis.

Table 1. Green and sintered density.

Sample	Green density (in gm/cc)	Sintered density (in gm/cc)
Sintered Al	1.869	2.705 [8]
20 Wt. % Crystalline Silica- Al composite	1.79	2.615 [8]
Al based hybrid composite	1.83	2.645

Table 2. Vickers micro-hardness.

Sample	Hardness (in MPa)
Sintered Al	386.4 [8]
20 Wt. % Crystalline Silica- Al composite	642 [8]
Al based hybrid composite	756.7

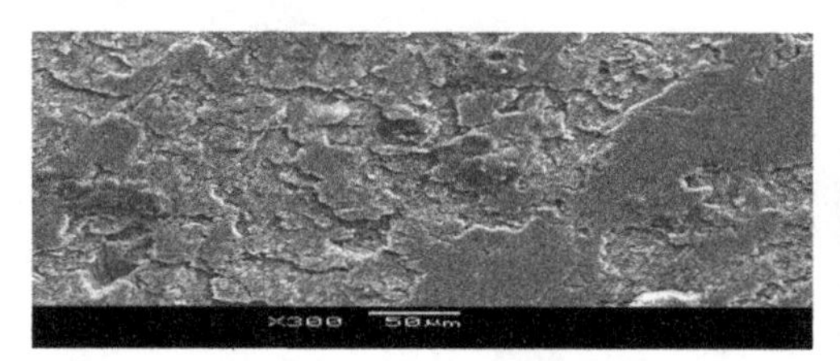

Figure 5. SEM image of Al based hybrid composite during SBF study at different magnifications.

Table 3. Biocompatibility parameters.

Sample	% Haemolysis	Blood clot weight (mg)
Al based hybrid composite	3.617	2

3.6.2 *SBF study*

The Al based hybrid composite is suspended into Simulated Body Fluid (SBF solution), prepared via Kokubo's method as described somewhere [6-7], for assuming the nature of interaction after implantation, in vitro. Formation of bioactive layer on the surface of the composite denote the influence of bone growth into it. SEM image shows the random formation of bioactive layer around all the surfaces of Al based hybrid composite during SBF study as shown in Figure 5.

3.6.3 *Blood compatibility study*

The composite fabricated into hot press which was considered for blood compatibility study and the results obtained are placed in Table 3. The sample has considered as highly haemocompatible (as per ASTM F756) and the value of clot formed is being tolerable limit.

4 CONCLUSIONS

The significant conclusions of the studies are as follows:

- Al based hybrid composite is fabricated successfully by using hot press, followed by P/M technique.
- The experiment shows that an extruded mixture of crystalline silica, β TCP, and commercially available pure aluminium powder heated at 600°C for one hour using hot pressing promote no reaction between matrix and reinforcing phases.
- From the surface morphology after sintering shows the uniform distribution of second phases in to aluminium matrix for the sample fabricated into hot press.
- Incorporation of β TCP into crystalline silica- aluminium composite effects a little bit on its physical and mechanical properties, concluded with consulting previous studies done by these authors [8].
- The fabricated composite shows bioactivity properties during SBF study, under normal physiological conditions, in vitro. Therefore, it is hoped that this composite will promote bone growth after implantation, in vivo.
- The response of composite is also satisfactory for both haemolysis study and clot test.

REFERENCES

[1] Hench, L.L. 1998. Bioactive materials: The potential for tissue regeneration. Journal of Biomedical Materials Research 41(4): 511-518.

[2] Samanta, S.K. and Chanda, A. 2016. Study On Different Characteristics Of Doped Tri Calcium Phosphate At Different Sintering Temperatures. AIP Conf. Proc. 1724: 020042-1– 020042-6. http://dx.doi.org/10.1063/1.4945162

[3] Kalita, S.J., Bose, S., Hosick, H.L., Martinez, S.A. and Bandyopadhyay, A. 2002. Calcium Carbonate Reinforced Natural polymer Composite for Bone Grafts. Mat. Res. Soc. Symp. Proc. 724 (N8.18): 185-190.

[4] Kalita, S.J., Rokusek, D., Bose, S., Hosick, H.L. and Bandyopadhyay, A. 2004. Effects of MgO-CaO-P2O5-Na2O-based additives on mechanical and biological properties of hydroxyapatite. Journal of Biomedical Materials Research 71A(I): 35-44.

[5] Kalita, S.J. and Ferguson, M. 2006. Fabrication of 3-D Porous Mg/Zn doped Tricalcium Phosphate Bone-Scaffolds via the Fused Deposition Modeling. American Journal of Biochemistry and Biotechnology 2(2): 57-60.

[6] Mizuno, M. 2014. Implants for surgery: in vitro evaluation for apatite- forming ability of implant materials, ISO/DIS 23317. International Organization for Standardization (ISO), Geneva, Switzerland.

[7] Kokubo, T., Kushitani, H., Sakka, S., Kitsugi, T. and Yamamoto, T. 1990. Solutions able to reproduce in vivo surface- structure changes in bioactive glass- ceramic A-W. J Biomed Mater Res. 24: 721-734.

[8] Debnath, S. and Pramanick, A.K. 2016. Development and Evaluation of Various Properties of Crystalline Silica-Aluminium Metal Based Composites. IJERGS 4(2): 236-245.

[9] Debnath, S. and Pramanick, A.K. 2016. DEVELOPMENT AND STUDY ON DIFFERENT PROPERTIES OF ALUMINIUM- CRYSTALLINE SILICA CERAMIC MATRIX COMPOSITES AT DIFFERENT SINTERING TEMPERATURES. IJERGS 4(3): 415-423.

[10] Bajpai, A. and Saini, R. 2005. Preparation and characterization of spongy cryogels of poly(vinyl alcohol)– casein system: water sorption and blood compatibility study. Polymer International 54: 796–806. DOI: 10.1002/pi.1773

Multidisciplinary research in the field of biochemistry

for human need

Biotechnology and Biological Sciences – Sen et al. (Eds)
© 2020 Taylor & Francis Group, London, ISBN 978-0-367-43161-7

Identification and characterization of an industrially important enzyme laccase from *Fusarium* sp. FW2PhC1

Rinku Debnath* & Tanima Saha
Department of Molecular Biology and Biotechnology, University of Kalyani, Kalyani, India

ABSTRACT: A laccase has been isolated from *Fusarium* sp. FW2PhC1 and characterized. The highest extracellular laccase production from the fungus was observed as 43.41 U/ml using sodium acetate buffer (100 mM, pH-5.0) on day 14 in static condition. The pH and temperature optima of the enzyme were 5.0 and 30°C, respectively. The enzyme was stable in the pH range from 4.0 to 10.0 and at temperatures below 40°C. Protein concentration of the enzyme was estimated to be 0.31 mg/ml by using Lowry method. The laccase showed the highest substrate specificity for DMP (2,6-dimethoxyphenol) as a substrate. The K_m and V_{max} of the laccase were 3.27 mM and 196.07 U/mg, respectively for the substrate DMP. About 20 U/ml of the crude laccase, efficiently decolorized three industrial dyes neutral red (67%), indigo carmine (63%) and methyl green (58%) at a concentration of 50 μM after 24 h of incubation at 30°C.

1 INTRODUCTION

Laccases are multi copper oxidases which catalyze oxidation of various phenolic compounds with concomitant reduction of oxygen into water (Rivera-Hoyos et al., 2013). Laccases are widely distributed among bacteria, fungi and plants. But the majority of laccases so far characterized are derived from fungi. Due to a wide range of substrate specificity, it has considerable applications in pulp and paper industry, food and beverages, denim washing, dye decolorization, biosensor formulation, environmental bioremediation (Pezzella et al., 2015) etc. Our study aims to isolate a novel laccase producing fungi, produce laccase from the fungi in submerged culture condition and characterize the laccase. The potential dye decolorization activity of the crude laccase towards various textile dyes was also evaluated.

2 MATERIALS & METHODS

2.1 *Chemicals*

All media ingredients, 2,6-dimethoxyphenol (DMP), guaiacol, tannic acid and dyes were purchased from HiMedia. All other chemicals were of analytical grade and were obtained locally.

2.2 *Screening for laccase production*

Leaf decaying soil samples were collected from different locations of University of Kalyani and screened for laccase producing fungi. 1 ml of soil solutions were transferred to potato dextrose agar (PDA) plates containing 0.01% guaiacol or 0.5% tannic acid as laccase indicator (Kiiskinen et al., 2004). Plates were incubated at 28 ± 2°C until reddish-brown and dark-brown zones in

*Corresponding author: debnath.rinku87@gmail.com

the presence of guaiacol and tannic acid, respectively were appeared on the plates. The isolated fungus was maintained at 4°C for long term storage. The isolated laccase producing fungus was identified through 18S rDNA sequence analysis method (Forootanfar et al., 2011).

2.3 *Culture conditions*

For laccase production study, 0.5 ml of fungal mycelial culture was used to inoculate 50 ml of culture medium in a 250 ml Erlenmeyer flask. The *Fusarium* sp. was cultured in modified potato dextrose broth which contained 0.5% yeast extract, 0.5% peptone and 1% tannic acid. The basal medium pH was adjusted to 5.0. To stimulate laccase production, 200 μM final concentration of $CuSO_4$ was added to the medium as an inducer. Inoculated cultures were incubated at 28 ± 2°C for 20 days in static condition. 1 ml of aliquot was taken at 24 h of interval from broth culture and centrifuged at 13000 rpm for 15 min followed by extracellular laccase activity measurement. The experiment was done in triplicate.

2.4 *Laccase assay and protein estimation*

The enzyme activity was measured by oxidation of DMP as a substrate. The reaction mixture contained 1 mM DMP, 100 mM sodium acetate buffer (pH 5.0) and enzyme solution. The enzymatic activity was calculated using the molar extinction coefficient of DMP (ε_{525} = 65,000 M^{-1} cm^{-1}) (Sengupta and Mukherjee, 1997). One unit (U) of activity was defined as the amount of enzyme that can oxidize 1 μmol of substrate per minute under standard condition (Younes and Sayadi, 2011). Protein concentration was estimated by using Lowry method.

2.5 *Effect of pH and temperature*

The pH and temperature optima of the laccase was measured by performing enzymatic assays at different pH (3.0-10.0) and temperatures (10-55°C). 0.1 M citrate buffer pH 3, 0.1 M acetate buffer pH (4-5), 0.1 M phosphate buffer pH (6-8) and 0.1 M glycine buffer pH (9-10) were used to adjust the pH levels. Effect of pH on laccase stability was investigated by determining the enzyme activity after incubating the crude enzyme in different pH ranges for 8 h. Stability of the laccase at various temperatures was determined by pre-incubating the crude enzyme at different temperatures between 4-50°C for 8 h followed by determination of enzyme activity (Park and Park, 2014).

2.6 *Substrate specificity and kinetic constants*

Substrate specificity of the laccase was investigated by using a range of substrates including DMP, ABTS, syringaldazine, guaiacol, catechol and vanillin at a final concentration of 1 mM. Oxidation of the substrates were measured at their respective maximum absorbance wavelength followed by calculating the enzyme activity by using molar extinction coefficient of each substrate. Kinetics constants of the enzyme were determined for DMP in sodium acetate buffer pH 5.0 and calculated by using Lineweaver–Burk plot (Park and Park, 2014).

2.7 *Dye decolorization study of crude laccase*

Dye decolorization ability of the crude laccase was tested for methyl green, neutral red and indigo carmine (Lu et al., 2013). Approximately 20 U/ml of enzyme was incubated separately with 50 μM final concentration of the textile dyes at 30°C for 24 h. Decolorization of dyes was monitored at λ_{max} of each dye using a UV-visible spectrophotometer. The following equation was used for calculation of dye decolorization percentage (Manavalan et al., 2013).

$$\text{Decolorization } (\%) = \frac{\text{Initial absorbance } - \text{ Final absorbance}}{\text{Initial absorbance}} \times 100$$

3 RESULT AND DISCUSSION

3.1 *Extracellular laccase production in submerged culture*

The isolated fungus was identified as a novel laccase producing strain *Fusarium* sp. FW2PhC1. The *Fusarium* sp. produces laccase extracellularly in the culture medium. Laccase production was gradually increased up to day 14 and then decreased during the 22 days of incubation. The highest laccase production was recorded as 43.41 U/ml on 14th day (Figure 1). During laccase production, the laccase activity is affected and varies considerably by various parameters including media composition, fermentation condition and inducers used. $CuSO_4$ induction in the fermentation medium increased the laccase activity of *Aspergillus flavus* up to 51.84 U/mL and the result is comparable with our study (Gomaa et al., 2015).

3.2 *Effect of pH and temperature*

The laccase showed highest activity at pH 5.0 and 30°C. The enzyme remained active in the pH range from 3.0 to 10.0 but it showed maximum activity at pH 5.0. (Figure 2a and 2b). Several fungal laccases show highest activity at a pH ranging from 4.0 to 6.0, thus confirming our results (Min et al., 2001; Robles et al., 2002; Litthauer et al., 2007). Stability studies showed that laccase was unstable at pH below 4.0, and 30.4% of activity remained after 8 h. However, excellent stability was observed at pH 5.0–8.0 and

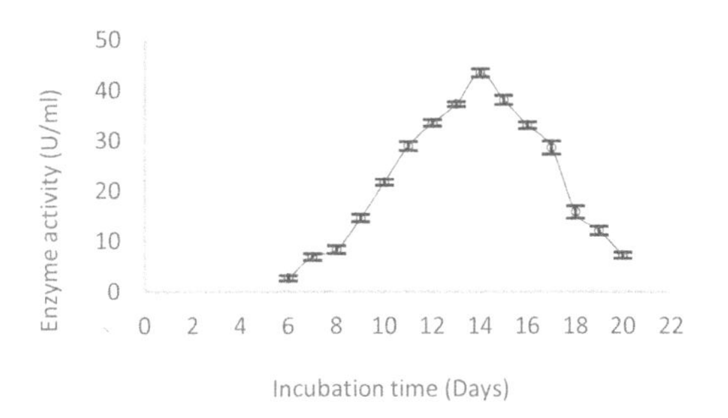

Figure 1. Extracellular laccase production by *Fusarium* sp. FW2PhC1.

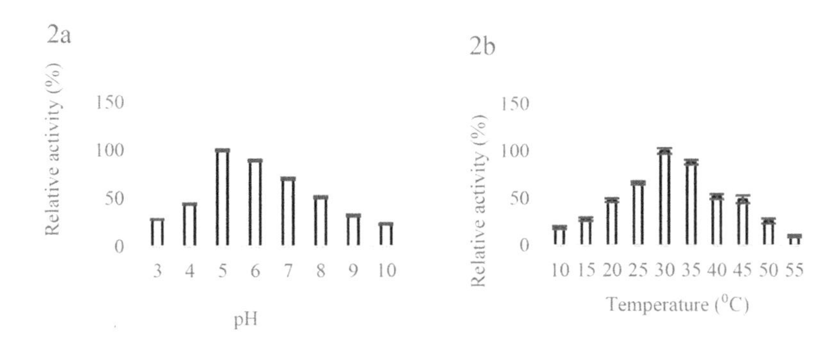

Figure 2. Effect of pH (2a) and temperature (2b) on laccase activity.

95.8–76.7% of laccase activity remained after 8 h. From the pH stability experiment it was observed that the enzyme was more stable at alkaline pH than acidic pH (Figure 3a). Laccase from *Fusarium proliferatum* also showed high stability at pH ranges from 5.0-8.0 confirmed our study (Fernaud et al., 2006). The enzyme was most stable in the temperature range 4-25°C and retained its 34% of activity even after 8 h of incubation at 30°C. But at high temperature, the laccase was inactivated rapidly. Only 10% activity was detected after 6 h when incubation temperature was raised to 40°C and at 50°C, almost no enzyme activity was detected after 2 h (Figure 3b). Although the laccases from *F. prolifertum* and *F. incarnatum* were investigated as thermostable enzyme but temperature optima for the laccase activity greatly varies from the temperature ranges 10-75°C (Fernaud et al., 2006; Chhaya and Gupte, 2013; Litthauer et al., 2007; Perez et al., 1996; Dedeyan et al., 2000; Robles et al., 2000). *Chalara* (syn. *Thielaviopsis*) *paradoxa* CH32 laccase showed the temperature optima at 30°C (Robles et al., 2002) comparable with our result.

3.3 *Substrate specificity and kinetic constants*

Substrates specificity of The *Fusarium* laccase was investigated for the phenolic compounds including (ABTS) 2,2′-azino-bis (3-ethylbenzothiazoline-6-sulphonic acid), (SGZ)

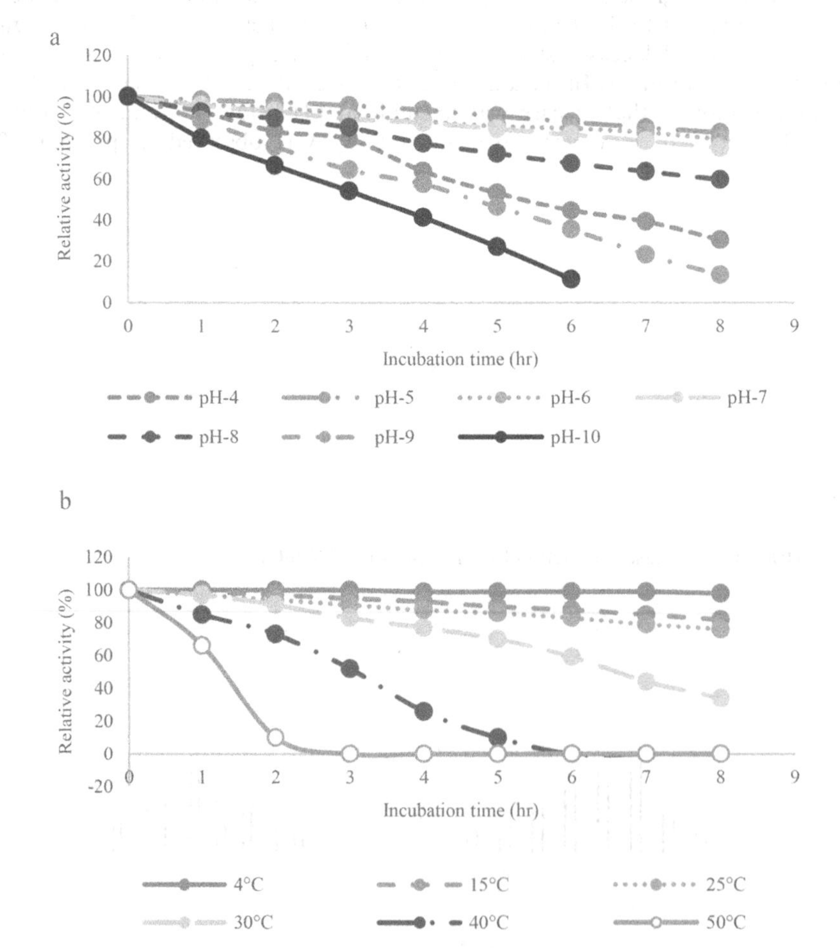

Figure 3. Effect of pH (3a) and temperature (3b) on stability of laccase.

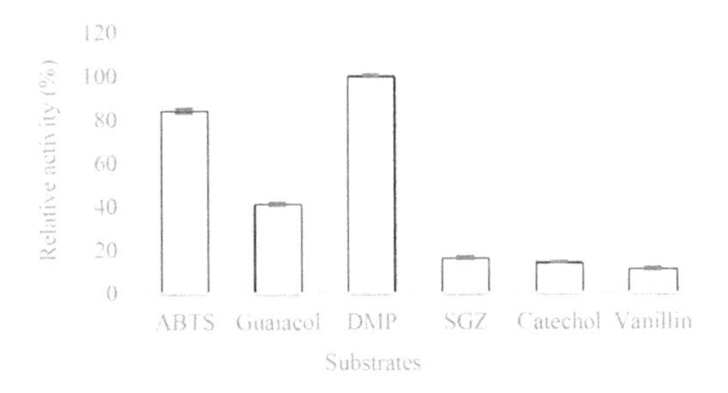

Figure 4. Substrate specificity of *Fusarium* sp. laccase.

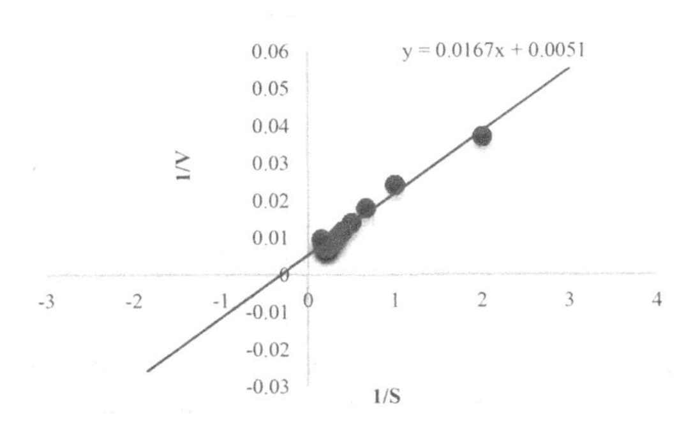

Figure 5. Lineweaver-Burk plot for laccase from *Fusarium* sp.

syringaldazine, DMP, guaiacol, catechol and vanillin (Figure 4). The enzyme oxidized the phenolic compounds with highest activity towards DMP, with about 84% and 41% activity towards ABTS and guaiacol, respectively. Usually fungal laccases showed highest affinity for ABTS although laccase from *Perenniporia tephropora* also has higher affinity for DMP (Younes et al., 2007). The affinity of the laccase for SGZ was very low only 16%. In this study DMP is the most preferred substrate for the laccase rather than ABTS and SGZ. The laccase showed only 14% and 11% activity towards catechol and vanillin, respectively. Km and V_{max} for the laccase were determined as 3.27 mM and 196.07 U/mg, respectively (Figure 5). The Km was quite low which shows that the laccase had good catalytic properties. The K_m value varies with organism to organism but K_m 2.9 mM with DMP has been reported for the laccase from *Ceriporiopsis subvermispora* (Fukushima and Kirk, 1995). This data is in agreement with the apparently same K_m value with DMP in the present study.

3.4 *Dye decolorization study with crude laccase*

The physical or chemical methods used for decolorization of synthetic dyes and their effluents released from textile industries are expensive. Therefore, enzyme based dye decolorization method is efficient process and is of great interest. In this study, the *Fusarium* sp. was investigated for dye decolorization of three industrial dyes. Crude laccase of 20 U/ml was evaluated for enzymatic decolorization of industrial dyes by using neutral red, indigo carmine and methyl green. After 24 h of incubation, the maximum decolorization of 67% was achieved for neutral red. Whereas, in case of indigo carmine and methyl green, the maximum decolorization was 63% and 58%, respectively (Figure 6). From our study it was revealed that the crude enzyme is a potential decolorizing agent and efficiently decolorizes various textile dye without any mediator. Based on this result, the selected dyes would be used for further studies on *Fusarium* sp. laccase.

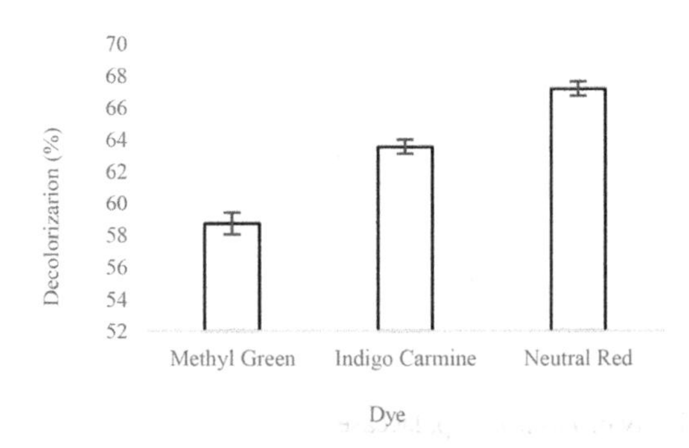

Figure 6. Decolorization of industrial dyes with crude laccase.

4 CONCLUSION

The Ascomycetes fungus, *Fusarium* sp. FW2PhC1 produced extracellular laccase in static condition with high activity. The enzyme was also highly active at basic pH and was stable upto 35°C. The laccase showed decolorization efficiency against textile dyes, therefore possess important properties for industrial applications. Further research will be focused on the purification of laccase, characterization of the purified laccase and compare it with the crude laccase characterization as well as investigation of other biotechnological applications of laccase.

ACKNOWLEDGMENT

This study was supported by the Personal Research Grant and DST-PURSE programme provided by University of Kalyani for the financial assistance necessary for the paper.

REFERENCES

[1] Chhaya U., Gupte A. (2013). Possible role of laccase from *Fusarium incarnatum* UC-14 in bioremediation of Bisphenol A using reverse micelles system. *J. Hazard. Mater.* 254–255:149–156.

[2] Dedeyan B., Klonowska A., Tagger S., Tron T., Iacazio G., Gil L., Petit J. (2000). Biochemical and Molecular Characterization of a Laccase from *Marasmius quercophilus. Appl. Environ. Microbiol.* 6:925–929.

[3] Fernaud J.R.H., González A.M.K., Vázquez J., Falcón M.A. (2006). Production, partial characterization and mass spectrometricstudies of the extracellular laccase activity from *Fusarium proliferatum. Appl Microbiol Biotechnol.* 70: 212–221.

[4] Forootanfar H., Faramarzi M.A., Shahverdi A.R., Yazdi M.T. (2011). Purification and biochemical characterization of extracellular laccase from the ascomycete *Paraconiothyrium variabile. Bioresour. Technol.* 102:1808–1814.

[5] Fukushima Y., Kirk T.K. (1995). Laccase Component of the *Ceriporiopsis subvermispora* Lignin-Degrading System. *Appl. Environ. Microbiol.* 61872–61876.

[6] Kiiskinen L.L., Ratto M., Kruus K. (2004). Screening for novel laccase-producing microbes. J. Appl. Microbiol. 97:640–646.

[7] Litthauer D., Vuuren M.J.V., Tonder A.V., Wolfaardt F.W. (2007). Purification and kinetics of a thermostable laccase from *Pycnoporus sanguineus* (SCC 108). *Enzyme Microb. Technol.* 40:563–568.

[8] Lu L., Wang T.N., Xu T.F., Wang J.Y., Wang C. L., Zhao M. (2013). Cloning and expression of thermo-alkali-stable laccase of *Bacillus licheniformis* in *Pichia pastoris* and its characterization. *Bioresour. Technol.* 134:81–86.

[9] Manavalan T., Manavalan A., Thangavelu K.P., Heese K. (2013). Characterization of optimized production, purification and application of laccase from *Ganoderma lucidum. Biochem Eng. J.* 70:106–114.

[10] Min K-L, Kim Y-H, Kim Y.W., Jung H.S. and Yung Chil Hah. (2001). Characterization of a Novel Laccase Produced by the Wood-Rotting Fungus *Phellinus ribis*. *Arch. Biochem. Biophys.* 392 279–286.

[11] Park N., Park S-S. (2014). Purification and characterization of a novel laccase from Fomitopsis pinicola mycelia. *Int J Biol Macromol.* 70:583–589.

[12] Perez J., Martinez J., Rubia D.L.T. (1996). Purification and Partial Characterization of a Laccase from the White Rot Fungus *Phanerochaete flavido-alba*. *Appl. Environ. Microbiol.* 62:4263–4267.

[13] Pezzella C. Guarino L. and Piscitelli A. (2015). How to enjoy laccases. *Cell. Mol. Life Sci.* 72:923–940.

[14] Robles A., Lucas R., Alvarez de Cienfuegos G., Gálvez A.. (2000). Phenol-oxidase (Laccase) Activity in Strains of the Hyphomycete *Chalara Paradoxa* Isolated from Olive Mill Wastewater Disposal Ponds. *Enzyme Microb. Technol.* 26:484–490.

[15] Robles A., Lucas, R., Martínez-Cañamero, M., Omar, N.B. Pérez R., Gálvez A. (2002). Characterization of Laccase Activity Produced by the Hyphomycete *Chalara* (syn. Thielaviopsis) *paradoxa* CH32. *Enzyme Microb. Technol.* 31:516–522.

[16] Rivera-Hoyos C.M., Morales-Álvarez E.D., Poutou-Piñales R.A., Pedroza-Rodríguez A.M., Rodríguez-Vázquez R., Delgado-Boada J. M. (2013). Fungal laccases. *Fungal Biol. Rev.* 27:67-82.

[17] Younes S.B., Mechichi T., Sayadi S. (2007). Purification and characterization of the laccase secreted by the white rot fungus *Perenniporia tephropora* and its role in the decolourization of synthetic dyes. *J. Appl. Microbiol.* 102:1033–1042.

[18] Younes S.B., Sayadi S. (2011). Purification and characterization of a novel trimeric and thermotolerant laccase produced from the ascomycete *Scytalidium thermophilum* strain. *J Mol Catal B- Enzym.* 7335–42.

Biotechnology and Biological Sciences – Sen et al. (Eds)
© 2020 Taylor & Francis Group, London, ISBN 978-0-367-43161-7

Unprecedented redox scavenging signature along with antioxidant action of silver nanoparticle coupled with *Andrographis paniculata* (AP-Ag NP) against carbon tetrachloride (CCl_4) induced toxicity in mice

Soumendra Darbar & Atiskumar Chattopadhay
Faculty of Science, Jadavpur University, Kolkata

Kausikishankar Pramanik
Department of Chemistry, Jadavpur University, Kolkata

Srimoyee Saha
Department of Physics, Jadavpur University, Kolkata

ABSTRACT: Nano technology possesses several branches including nanomedicine, which is the most promising field in the future medicine and is a probable therapeutic agent in prevention and medication of life threatening diseases through ROS inhibition. Therapeutic potential and antioxidant activity of Silver Nanoparticle coupled with *Andrographis paniculata* (AP-Ag NP) was assessed against CCl_4 induced oxidative stress at tissue level. The main aim and objective of the study is to find out the comparative efficacy of AP-Ag NP against carbon tetrachloride (CCl_4) induced oxidative stress model. Carbon tetrachloride (CCl_4) was administered upon Swiss albino mice (male) for 28 days concurrently with AP-Ag NP (50 mg/kg body weight) orally to evaluate the therapeutic effects on hepatic oxidative injury, antioxidant potential and heme synthesis pathway. Serum ROS level was significantly elevated and blood and liver superoxide dismutase (SOD), catalase (CAT) activity and GSH level also significantly decreased after exposure of carbon tetrachloride (CCl_4). Treatment with AP-Ag NP, as nano-antioxidant significantly increased SOD, CAT activity and GSH levels which indicate the recovery of oxidative injury and indicates restoring inhibited aminolevulinate dehydratase (ALAD) activity. In conclusion our results suggest that Silver Nanoparticle synthesized using *Andrographis paniculata* (AP-Ag NP) have the potential antioxidant effect in experimental animals.

1 INTRODUCTION

Application of nanotechnology in medicine is needed to threat various disease. Synthesis of nanomaterials through green route can be useful for both in vivo and in vitro in biomedical research. Chronic diseases such as diabetes, cardiovascular disease and liver fibrosis increase oxidants and decrease antioxidants in patients 2,3. Clinical and experimental studies have shown that disturbing the balance of the oxidant–antioxidant system can contribute to the pathogenesis of liver and kidney fibrosis 4-6.

Green route synthesis promotes the use of biologically active microorganisms and plants for the development of novel nanoparticles with therapeutic benefits 7. Andrographis paniculata leaf extract was used in this study as a reducing agent and the concentration of the plant extract upon silver ions for the synthesis of the nanoparticle were also assayed 8.

Radical scavengers kidnap hydrogen atoms from the activated biomaterials under variable physiological parameters and play a vital defending role by ROS (Reactive oxygen species) degradation along with prohibition of oxidative damage 9, 10. The need of the hour is to formulate a novel drug that can combat against ROS and augment redox sanation.

The present study was aimed at investigating the efficacy of Silver Nanoparticle Coupled with Andrographis paniculata (AP-Ag NP) against CCl 4 induced oxidative injury. Liver & Kidney was selected as the major organ for investigation as these are the primary target organs for free radical scavenging activity progressing to hyper lipid peroxoidation activity.

2 MATERIALS AND METHODS

The study initiated by the formation of silver nanoparticles using AP as a reducing agent. Chemical route was taken for the formation of 50 mM (AgNO3) solution. The plant extract was added in a drop wise manner followed by continuous stirring. The change of color from yellow to brown along with a brownish precipitate of silver indicated formation of nanoparticles. The setup was left untouched for half an hour for complete degradation of silver nitrate. Nanoparticles were obtained finally when the precipitate was centrifuged at 15000 rpm for 10min at a temperature of 4°C. Characterization of synthesized AP-Ag NP was carried out by UV-VIS spectra, XRD patterns and FTIR. In the in vivo experiment we developed liver and kidney fibrosis in animal model using CCl_4.

2.1 *Experimental design*

Table 1. Study design.

Groups	Treatment
Control	Normal Saline for 28 days
CCl_4 treated	Received 1:1 (v/v) CCl4 in olive oil for 28 days
CCl_4 + AP-AgNP	Received 1:1 (v/v) CCl4 in olive oil along with 0.50 g/kg/day for 28 days
AP-AgNP	0.50 g/kg/day for 28 days

After the experimental period liver and kidney function enzymes like AST, ALT, ALP, GGT BUN, creatinine and uric acid were measured using biochemical kits. Liver MDA content and different antioxidant enzymes like SOD, CAT, GPx and GSH were measured. Histology of the liver was done as per standard method. H&;E Masson's trichrome (MT) and Sirius red (SR) staining procedures were undertaken to estimate the extent of fibrotic degeneration in liver. Microscopic examinations revealed (Olympus BX51) histopathological modulations.

Characterization of Silver nanoparticles (AgNPs)

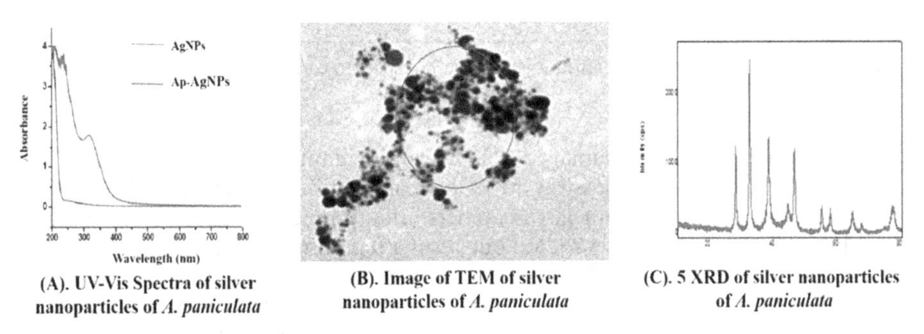

Figure 1. Characterization of Ag nanoparticle.

3 RESULTS & DISCUSSION

Leaf extract of Andrographis paniculata was used as a novel bio-reductant to synthesize silver nanoparticle (AgNPs). The whole process is eco-friendly and cheap, thus can be used for the large scale synthesis of AgNPs in nanotechnology processing industries. AgNPs synthesized from *Andrographis paniculata* leaf extract possesses unprecedented antioxidant activity which can be utilised for a novel drug synthesis in biomedical industries.

Table 2. Effect of AP-AgNP on body weight changes on CCl_4 Exposure in mice.

Groups	BW (initial) (g)	BW (final) (g)	BW gain or loss (g)
Control	25.14 ± 0.13	29.37 ± 0.11	4.23 ± 0.042
CCl_4	24.98 ± 0.14	26.01 ± 0.14	1.03 ± 0.051
CCl_4+AP-AgNP	25.22 ± 0.11	28.99 ± 0.12	3.77 ± 0.039
AP-AgNP	25.31 ± 0.12	29.48 ± 0.11	4.17 ± 0.044

Values are means ± SEM (n = 6)

It was observed that AP-AgNP significantly attenuate CCl_4 induced alteration in the activities of liver function enzymes more effectively. A significant recovery of liver function enzymes was observed after co-administration of AP-Ag NP. Administration of CCl_4 significantly decrease the essential antioxidant effect in the liver. A significant recovery in GSH, CAT and SOD was observed after co-administration of AP-Ag NP. Histology strongly proved that AP-Ag NP attenuate CCl_4 induced redox healing in the in-vivo model.

During the Cellular oxidative stress elevation of cellular ROS level occurred followed by reduced glutathione level along with increased lipid peroxidation and impaired antioxidant defense status. Our study reported elevated ROS levels in CCl_4-treated group, suggesting free radical generation leading to oxidative stress conditions. Increased lipid peroxidation like elevated TBARS and reduced GSH, further signifies oxidative stress condition. Concomitant administration of Ag nanoparticle coupled with *Andrographis paniculata* led to pronounced recovery, suggesting it become a more effective scavenger of free radicals.

Table 3. Effect of AP-AgNP on blood biochemical and antioxidant variables on CCl_4 Exposure in mice.

Blood	Control	CCl_4	CCl_4 + AP-AgNP	AP-AgNP
ROS (FIU)	412.59 ± 14.3	736.13 ± 17.8$^{\#}$	492.81 ± 11.24**	462.14 ± 12.19**
SOD (U/mg)	106.14 ± 8.1	72.05 ± 6.9$^{\#}$	98.34 ± 7.1**	100.02 ± 6.2**
CAT (U/mg)	232.62 ± 21.3	161.1 ± 19.4$^{\#}$	212.4 ± 16.2**	224.9 ± 13.1**
GSH (mg/g)	38.39 ± 0.49	21.92 ± 0.51$^{\#}$	33.12 ± 0.81**	33.66 ± 0.41**
MDA (nmol/g)	34.16 ± 2.14	76.39 ± 3.02$^{\#}$	40.12 ± 3.44**	36.13 ± 2.91**

Values are means SEM (n=6), P<0.05 significant change with respect to control group, P<0.001 significant change with respect to CCl_4 group.

ROS generation proceeds to damaged cellular antioxidant immune system. GSH levels decreased after CCl_4 treatment, possibly to its increased utilization in neutralizing free radicals. As reported by several prominent researchers Glutathione is considered to be the major form of cellular glutathione which validate our results regarding the safety profile of the synthesized nanoparticle. Damage of biological molecules such as lipids are implicated by elevated ROS generation, which are altered by peroxidation. Under the oxidative stress elevation in TBARS is an indicator of lipid peroxidation. We observed a significant elevation in TBARS level following CCl_4 exposure and back to normal level after NP administration.

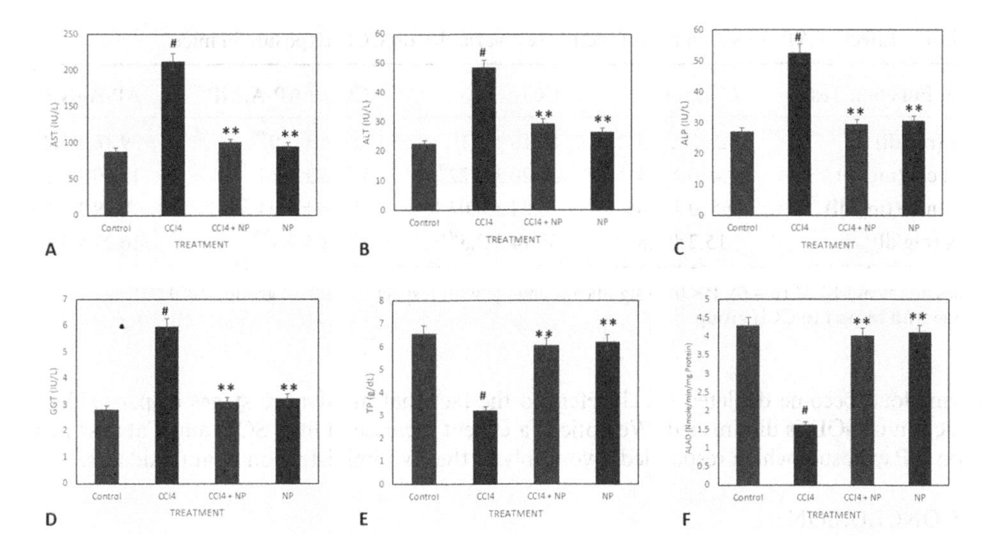

Figure 2. Effect of AP-AgNP on Liver biochemistry on CCl_4 exposed mice. Values are means SEM (n = 6), P<0.05 significant change with respect to control group, P<0.001 significant change with respect to CCl_4 group.

Interestingly the level of plasma AST, ALT, ALP and GGT were restored to normal in the animals co-exposed to AP-Ag NP.

Antioxidant profile was determined to detect the antioxidant status. The detrimental signatures of the superoxide ion was prohibited by SOD which transformed them into less toxic hydrogen peroxides which consequently breaks down into nontoxic water and oxygen molecule by catalase action.Catalase is another major antioxidant enzyme whose activity decreases during oxidative stress, leading to H_2O_2 accumulation and finally peroxidation of lipids. We observed decreased hepatic SOD activity in our study. Intrinsic antioxidant defense systems are regulated by various components like SOD is culpable for dissemination of Superoxide radicals. Concurrently in times of oxidative stress the body uses its defense mechanism to nullify the activity of lipid peroxidation by using the antioxidant enzymes such as SOD, therefore, the activity of this enzyme become higher in initiation of damage, but if the insult prolongs, the

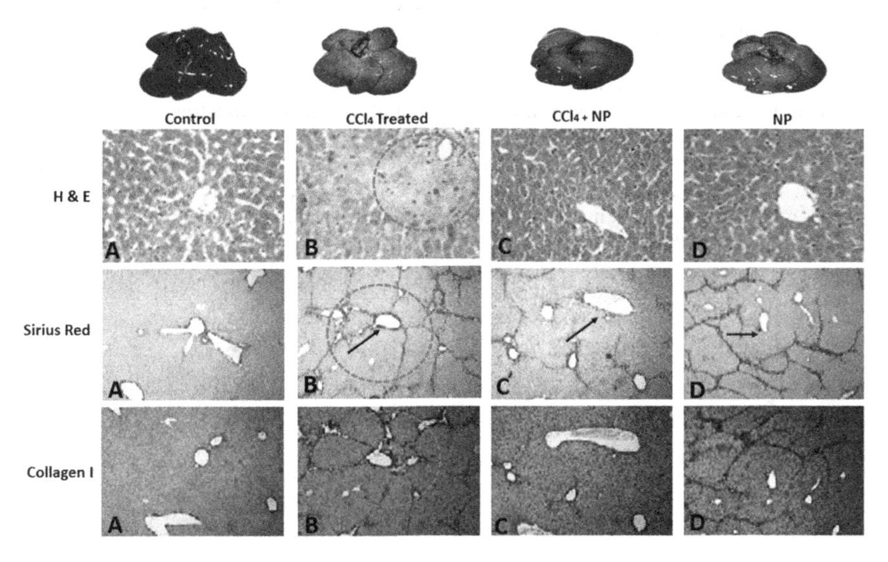

Figure 3. Effect of AP-AgNP Histopathology of Liver on CCl_4 exposed mice.

Table 4. Effect of AP-AgNP on renal function test variables on CCl_4 Exposure in mice.

Renal Function Test	Control	CCl_4	CCl_4+ AP-AgNP	AP-AgNP
Urea (mg/dl)	32.52 ± 2.5	64.16 ± 3.1$^{\#}$	38.06 ± 2.9**	39.16 ± 3.4**
Uric acid (mg/dl)	1.410 ± .24	2.920 ± 0.22$^{\#}$	1.850 ± 0.14**	1.510 ± 0.31**
Creatinine (mg/dl)	0.510 ± 0.05	0.84 ± 0.04$^{\#}$	0.56 ± 0.04**	0.59 ± 0.06**
BUN (mg/dl)	15.2 ± 1.8	35.14 ± 2.3$^{\#}$	17.03 ± 2.1**	16.22 ± 1.8**

Values are means SEM (n = 6), $P < 0.05$ significant change with respect to control group, $P < 0.001$ significant change with respect to CCl_4 group.

enzyme load become depleted which refers to the fact that in advance stages of per-oxidation the action of SOD is diminished. We noticed a cogent increase in liver SOD and Catalase activity on NP exposure which responded favourably to the co-administration of antioxidants.

4 CONCLUSION

The AgNPs were synthesized successfully using the leaf extract of *Andrographis paniculata* following inexpensive and better bioavailable green route technique.It can be summarized from this study, that the synthesized green nanohybrid (AP-Ag NPs), a prospective nanomedicine, ameliorates oxidative stress in mice most effectively. The synthesis process is not only facile but cost and time effective as well.

REFERENCES

AHMADVAND, H., MABUCHI, H., NOHARA, A., KOBAYAHI, J. & KAWASHIRI, M.-A. 2013. Effects of coenzyme Q10 on LDL oxidation in vitro. *Acta Medica Iranica*, 12–18.

AHMADVAND, H., TAVAFI, M. & KHOSROWBEYGI, A. 2012. Amelioration of altered antioxidant enzymes activity and glomerulosclerosis by coenzyme Q10 in alloxan-induced diabetic rats. *Journal of Diabetes and its Complications*, 26, 476–482.

CSÁNYI, G. 2014. Oxidative stress in cardiovascular disease. Multidisciplinary Digital Publishing Institute.

GANNIMANI, R., PERUMAL, A., KRISHNA, S., SERSHEN, M., MISHRA, A. & GOVENDER, P. 2014. Synthesis and antibacterial activity of silver and gold nanoparticles produced using aqueous seed extract of Protorhus longifolia as a reducing agent.

HOET, P.H., BRÜSKE-HOHLFELD, I. & SALATA, O.V. 2004. Nanoparticles–known and unknown health risks. *Journal of nanobiotechnology*, 2, 12.

PATRA, J.K. & BAEK, K.-H. 2014. Green nanobiotechnology: factors affecting synthesis and characterization techniques. *Journal of Nanomaterials*, 2014, 219.

RÍOS-SILVA, M., TRUJILLO, X., TRUJILLO-HERNÁNDEZ, B., SÁNCHEZ-PASTOR, E., URZÚA, Z., MANCILLA, E. & HUERTA, M. 2014. Effect of chronic administration of forskolin on glycemia and oxidative stress in rats with and without experimental diabetes. *International journal of medical sciences*, 11, 448.

Biotechnology and Biological Sciences – Sen et al. (Eds)
© 2020 Taylor & Francis Group, London, ISBN 978-0-367-43161-7

A study on different biochemical components of papaya (*Carica papaya*) leaves consequent upon feeding of citrus red mite (*Panonychus citri*)

Soma Karmakar
Ex-research Scholar, Entomology Laboratory, Department of Zoology, University of Kalyani, Nadia

ABSTRACT: With the objective to find out the most harmful and injurious mite pests occurring on fruit trees, surveys were made throughout the different districts of South Bengal and *Panonychus citri* (McGregor) commonly known as citrus red mite, appeared to be the major pest of Papaya and showed different degrees of damage symptoms. Hence, a biochemical component study was undertaken to examine the extent of damage done by this mite. Different minerals, inorganic and organic compounds were taken into consideration for experiments and they were examined according to the standard methodologies. The result revealed huge mechanical damage and depletion in amount of chlorophyll a and chlorophyll b (76.47% and 77.93% respectively). The nitrates and nitrites were also damaged by considerable amount (61.53% and 70.93% respectively). Among the minerals, Calcium and Magnesium were affected with the percentage of depletion of 36.12% and 35.83% respectively. Depletion in case of Total Protein and Carbohydrate were 26.01% and 48.73% respectively.

Keywords: citrus red mite, biochemical component, major pest

1 INTRODUCTION

Papaya (*Carica papaya)* belongs to the family Caricaceae and is native to Southern Mexico and Central America. Now a days it is cultivated many tropical countries like Brazil, Peru, Nigeria, Philippines, Thailand and China. Since ancient times, it is being used as a nutritive and delicious fruit. Besides it has medicinal values also. It has high medicinal value and India stands first in the world in the production of papaya as it solely produce 3 million tones of papaya per year which is the half of the total world production of papaya i.e. 6 million tones. Different districts of South Bengal which includes Nadia, North 24 Parganas, South 24 Parganas, Kolkata, Howrah, Hooghly, Burdwan, East Midnapur and West Midnapur were surveyed during the years 2009-2013, with the objective to find out the harmful and potentially injurious mite pest of the fruit plants. A total number of 22 fruit plant species were studied during the study. During the study, *Panonychus citri* came out to be a major pest of papaya plant. It was collected from nearly all the papaya plants from different areas of different districts and it showed different degrees of damage symptoms as well. Thatswhy it was thought to undertake a biochemical component assay, to measure the extent of damage done by this mite.

2 MATERIAL AND METHODS

2.1 *Experimental parameters*

The selected parameters for experimental procedure belonged mainly to 3 groups-
Minerals: Magnesium (Mg), Iron(Fe), Calcium (Ca), Copper (Cu) and Zinc (Zn) were selected.
Inorganic compounds: nitrates and nitrites

Organic compounds: the selected parameters were chlorophyll, total protein, sugar, and carbohydrate.

2.2 *Methodologies*

2.2.1 *Minerals*

The minerals as mentioned earlier were estimated by the help of Atomic Absorption Spectrophotometer (AAS). The infected and uninfected leaves were subjected to laid down procedure as required for AAS estimation. The leaves (both infected and uninfected) were digested with HNO_3 before estimation.

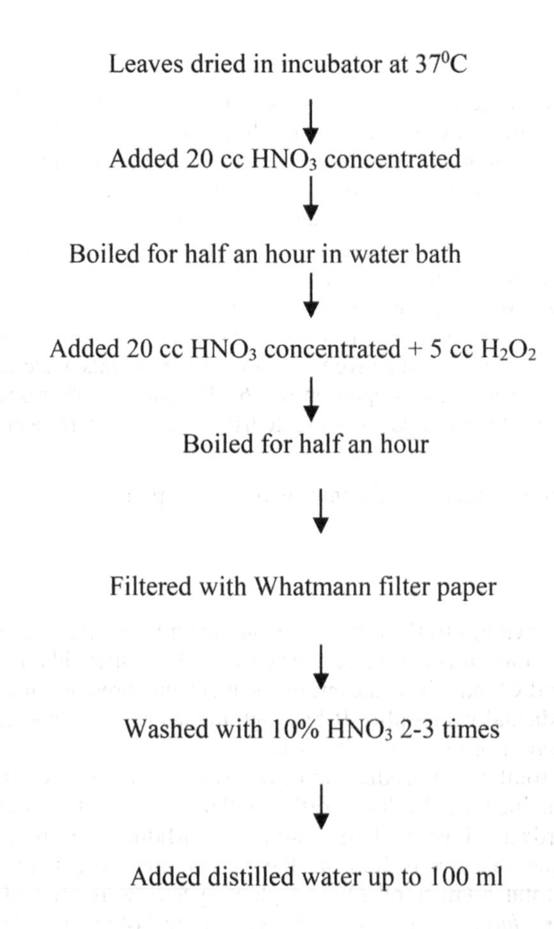

Leaves dried in incubator at 37⁰C
Added 20 cc HNO3 concentrated
Boiled for half an hour in water bath
Added 20 cc HNO3 concentrated + 5 cc H2O2
Boiled for half an hour
Filtered with Whatmann filter paper
Washed with 10% HNO3 2-3 times
Added distilled water up to 100 ml

2.2.2 *Chlorophyll*

The methodology described by Arnon (1949) was followed to estimate the chlorophyll content of the leaves. Fresh leaves were homogenized with excess of 80% acetone. A standard curve was prepared against a reagent blank read at 645 nm (nanometer) and 663 nm in the UV visible spectrophotometer.

2.2.3 *Total Carbohydrate*

The method of McCready (1959) was followed to estimate the total carbohydrate. In this case also, to measure the carbohydrate content, a standard curve was prepared against a reagent blank of D-glucose and it was read at 650 nm in UV visible spectrophotometer.

2.2.4 *Nitrate and Nitrite*

Nitrates and nitrites were estimated according to the procedure of Bray and Thrope (1954). The leftover of papaya leaves extract after carbohydrate estimation was taken for this experiment. In a test tube, the extract was mixed with 1ml of Folin - Ciocalteu reagent and 2 ml of 20% Sodium Carbonate ($NaCO_3$) solution was added to it. This solution was diluted with water to make it upto 25 ml. Then a standard curve was prepared against a reagent blank to estimate the levels of nitrates and nitrites.

2.2.5 *Total Protein*

The method of Lowry and Foline (1951) was followed to estimate the total protein. UV visible spectrophotometer read at 750 nm was used for this. The soluble proteins were taken into separate pellets where it was suspended in 5% Tri Carboxylic Acid (TCA) solution in a suitable volume in an ice bath (0-5°C) for a time period of 10 minutes. Then Alliquotes of 1 ml each were taken in centrifuge tube where it was added with 1 ml of 10% TCA solution. The tubes were centrifuged at 5000 rpm for 45 mins. After discarding the supernatants, the pellets were again re-extracted with alkaline Cu reagent (alkaline $NaCO_3$: CuSO4 –Na-K tartrate solution 50: 1 in v/v). Then a standard curve was prepared against a reagent blank of BVA (Bovine Serum Albumin) solution added with Folin - Ciocalteu reagent and it was read at 750nm in the UV visible spectrophotometer

3 RESULT

The result of biochemical component study is summarized as follows:

Name of the experimental parameter	Control	Infested	Percentage of Depletion
Calcium (ca)	501.00 (mg/kg)	320.60(mg/kg)	-36.12
Magnesium (mg)	854.00 (mg/kg)	548.00(mg/kg)	-35.83
Iron (fe)	58.50 (mg/kg)	52.35 (mg/kg)	-10.51
Zinc (ZN)	5.32 (mg/kg)	4.93 (mg/kg)	-7.33
Copper (cu)	4.08 (mg/kg)	3.56 (mg/kg)	-12.74
Nitrates	2.34 (mg/kg)	0.90 (mg/kg)	-61.53
Nitrites	2.89 (mg/kg)	0.84 (mg/kg)	-70.93
Chlorophyll a	3.06 (mg/kg)	0.72 (mg/kg)	-76.47
Chlorophyll b	1.45 (mg/kg)	0.32 (mg/kg)	-77.93
Total protein	20.89 (mg/lit)	10.71 (mg/lit)	-48.73
Carbohydrate	3.96 mg/100 mg sample	2.93 mg/100 mg samplc	-26.01

Table: Different experimental parameters and their percentage of depletion.

4 DISCUSSION

4.1 Very high degree of depletion in minerals like Ca and Mg

4.2 Depletion in Mg content resulted in heavy damage in case of Chlorophyll a and chlorophyll b as these were highly depleted

4.3 Reasonable degree of depletion in case of total protein, nitrates and nitrites and total carbohydrate

4.4 Depletion in minerals like Cu, Fe and Zn was not that much high

From these experiments, it is amply clear that feeding of *P. citri* on papaya caused various degrees of depletion in the biochemical components comprising minerals, inorganic and organic compounds. The degree of depletion in case of Chlorophyll a and b was 76.47% and 77.93% respectively and the reason behind this depletion is due to heavy mechanical damage which the mite does at the cellular level especially to the damage of chloroplast. According to Tomezyk and Kropzynska (1985), the probable reason also may be water stress induced by mite feeding causing stomatal opening and that resulting reduction of chlorophyll metabolism. Another probable reason may be the damage and reduction of chloroplast level.

As far as nitrates and nitrites are concerned the percentage of depletion was 61.53% and 70.93% respectively and those are with close proximity with the observations made by Chatterjee and Gupta (1997) who reported 73.13% and 70.93% of depletion incase of nitrates and nitrites respectively.

So far as, total protein was concerned, the extent of depletion was very high, which was 48.73% which was very close to the observations made by Nangia *et al.* (1999) where the extent of loss was 57.5% in case of Mulberry (Mysore local variety) and 38.8% (RFS-175 variety) as a result of feeding of *Eetetranychus sexmaculatus*. The reason which was attributed was the same i.e. due to breakdown of protein by prteolytic enzyme secreted by mites. The present observation also supports the reports of Agarwal (1992) and Goyal and Sadana (1983) as they also reported loss of protein due to mite feeding.

In the case of total carbohydrate, the decline was 26.01%. Nangia *et al.* (1999) also reported decline in total carbohydrate level. A decline in case of reducing sugar, non reducing sugar and total sugar, was reported by Usha *et al.* (1999) on plants due to feeding of the mites.

As regards depletion in mineral levels, a reduction in Ca, K and Mg contents in leaves was reported by Sadana and Goyal (1984). In Pineapple, due to feeding of *Dolichotetranychus floridanus,* the Fe and Zn contents were reduced by 42.9% and 31.11% respectively as reported by Das (1987). The present observations are very close to those reported by Das (1987).

5 CONCLUSION

5.1 *Panonychus citri* is a major pest of papaya and is able to cause substantial economic loss to the growers.

5.2 A control measure is needed at the earliest to prevent the economic loss caused by this mite.

REFERENCES

Agarwal, A.A. 1992, Leaf damage and associated cues induced aggressive ant recruitment in a neotropical ant – plant. *Ecology*.80: 1713–1723.

Arnon, D.I. (1949) Copper enzymes in isolated chloroplasts. Polyphenoloxidase in *Beta vulgaris. Plant Physiol.*, 24: 1–15.

Bray, H.G. and Thrope, W.V. (1954) Methodologies for total phenol estimation of plant tissues. *Meth. Biochem. Anal.*, 1: 7–52.

Chatterjee, K. and Gupta, S.K. (1997). An overview of mites occurring on vegetables, fruit trees and ornamental plants in West Bengal, India with their importance as pests and predators. *J. Beng. Nat. Hist. Soc.* NS 15(2): 18–27.

Das. T.K. (1987). *Studies on mites found in association with pineapple plantation. Ph.D. Thesis.* Bidhan Chandra Krishi Viswavidyalaya, Kalyani, (West Bengal), pp. 1–238.

Goyal, M. And Sadana, G.L. (1983). Quantitative changes in some biochemical components of *Coleus* sp. in response to infestation of *Brevipalpus obovatus* (Tenuipalpidae: Acarina) and factors affecting it's suitability as host. *Indian. J. Acarol.*, 8: 22–30.

Lowry, W. And Folin, J. (1951). Estimation of total protein. *Ann. Biochem.*, 14: 15–32.

McCready, R.M, J. Plant Pectin Analysis, Determination of Pectic Substances by Paper Chromatography. *Agric. Food Chem.*, 1959, 8 (6), pp 510–510513.

Nangia, N., Jagadish, P.S. and Nageshchandra, B.K. (1999). Biochemical changes in different varieties of mulberry infested by *Eotetranychus sujinamensis. J. Acarol.*, 15: 29–31.
National Horticultural Board, India (nhb.gov.in).
Sadana, G.N. and Goyal, M. (1984). Influence of morphological and biochemical characteristics of host plants in the life-cycle of *Brevipalpus obovatus* Donnadieu (Acari: Tenuipalpidae). *Indian J. Acarol.*, 8: 49–56.
Tomezynsk, A. and Kropzynska, M. (1985). Physiological and biochemical changes in three cultivars of *Chrynsanthemum* after feeding by *Tetranychus urticae. Proc. Int. Symp. Insect. Plant Relationship*, Washington, pp. 391–392.
Usha, R.V., Mallik, B. and Harishkumar, M. (1999). Biochemical changes in French bean plant grown under different water stress level and their effect on population of (*Tetranychus urticae*) (Acari: Tetranychidae) *Biol. 3 Lief*, 3: 1–112.
www.google.co.in

Biotechnology and Biological Sciences – Sen et al. (Eds)
© 2020 Taylor & Francis Group, London, ISBN 978-0-367-43161-7

Effect of roasting on antioxidants and related compounds in green coffee beans

S. Tripathi, R. Srivastava, S. Agarwal & S. Shrimali
Centre of Food Technology, University of Allahabad, Allahabad, India

N. Mishra*
Centre of Food Technology, University of Allahabad, Allahabad, India
**Department of Home Science, University of Allahabad, Allahabad, India*

ABSTRACT: Coffee owns a refreshing and stimulating taste due to numerous chemical compounds generated during the roasting process. Green coffee beans (GCB), as such, do not possess any characteristic taste or aroma which appeals palate, but these are gaining attention due to its various health benefits. The present study aimed to explore the anti-oxidant activity of GCB in different extracts at a different roasting temperature (170°C, 190°C, 210°C) for 5 minutes. The result showed that Total Polyphenolic Content (TPC) and Total Flavonoid Content (TFC) content was highest in ethanolic extract and increased with increasing temperature. Ferric reducing anti-oxidant potential was maximum in the ethanolic extract at 170°C. Anti-radical activity and reducing capacity remained unaffected by roasting temperature and solvent.

Keywords: Green Coffee Beans, Anti-oxidant activity, Polypenolic Content, Flavonoid Content, Anti-radical activity

In the past few years, the quest of knowledge regarding green coffee beans is spreading due to the growing appreciation of its various health benefits. Traditionally, the worthwhile outcome of coffee on human health were mostly attributed to the presence of an ingredient, known as caffeine; however, the other appreciated qualities like antioxidant activity is also influenced by additional constituents (Pandey and Rizvi., 2009). A variety of chemical compounds is present in green coffee beans with complex composition (Franca *et al.*, 2005). Antioxidant activity of coffee seeds depends on the characteristics of phenolic compounds (Skowron et al., 2016) and studies suggest that Green coffee beans have higher level of 5-O-caffeoylquinic acid (5-CQA), which is the major polyphenolic compound in green, even twofold more quantity than in roasted coffee depending on the time of roasting (Perrone et al., 2008). The chemical structure and biological activity of coffee changes during the roasting process and the loss of natural phenolic compounds takes place, whereas due to Maillard reaction some other antioxidant compounds are formed (Gouvêa et al., 2005) and maintain the antioxidant activity. However, as the amount of roasting increases, it leads to much more loss of phenolic compounds. The present study aimed to explore the effect of roasting temperature and solvent on the total polyphenols, total flavonoid content and antioxidant properties of green coffee beans.

1 MATERIAL AND METHODS

All the chemicals and reagents used were purchased from Sigma-Aldrich and SRL. The analysis was carried out in Research laboratory of Foods and Nutrition, Centre of Food Technology, University of Allahabad, Allahabad. The extract was prepared by the method adopted from Tamilmani and Pandey, 2015 with slight modification. For the extract preparation, unroasted seed and lightly roasted seeds (170°C, 190°C and 210°C for 5 min) were used in three solvent, aqueous, ethanol and methanol. TPC was determined using Prifits et al., 2015 method. Flavonoid content was calculated by method taken from Marinova et al., 2005. Ferric Reducing antioxidant power, Anti-radical scavenging activity and reducing capacity was estimated by methods explained by Sutharut and Sudarat, 2012, Sanja *et al.,* 2005 and Oktay *et al.,* 2003 respectively. The experimental data were examined by one-way analysis of variance (ANOVA) at 95% confidence level and the means were equated using Tukey's test of SPSS software version 16.0.

2 RESULTS AND DISCUSSION

The popular radicals identified in coffee were stated to be moulded through roasting route (Goodman et al.,1998). The sum of radicals rises on light roasting which shows involvement of Maillard reaction, supposing formation of pyrazinum radical cations in the initial phase of Maillard reaction (Hofmann et al., 1999) and especially melanoidins are responsible for radical content of coffee beans. About 40% of total anti-oxidant activity decreases on increasing roasting temperature (C¨ammerer and Kroh., 2006). It is well illustrated from Table 1 and Table 2 that the maximum extraction of Total Polyphenols and total Flavonoid content was at 210 °C in methanol, which is 37.60 ±3.43mgGAE/g and 186 ±2.27mg QE/g respectively. Most frequently used solvent for the extraction of bioactive component is methanol and in coffee seeds methanol came out to give greater yield of antioxidants than hexane, chloroform or acetone (Ramalakshmi et al., 2008).

Antiradical activity is the capability of constituents to react with free radicals (Table 3). Our finding reveals that coffee seeds retain higher percentage of anti-radical activity and roasting does not affect the activity, whereas solvent does, which could be due to the greater extraction of polyphenols and flavonoids. Flavonoids can act as antioxidants by scavenging free radicals, in which polyphenol interrupt the free radical chain reaction (Croft, 1998). Ferric reducing anti-oxidant power and reducing capacity of the coffee sample is shown in Table 4 and Table 5.

Table 1. Total Polyphenolic Content (mgGAE/g).

Green Coffee Beans	Unroasted	170°C	190°C	210°C
Extracts				
Aqueous	14.82 ± 1.27^{aP}	$23.72 \pm .94^{aQ}$	12.93 ± 2.15^{aP}	14.32 ± 1.50^{aP}
Ethanol	31.95 ± 1.98^{bP}	$35.06 \pm .78^{bP}$	37.44 ± 1.77^{cP}	37.60 ± 3.43^{cP}
Methanol	$33.72 \pm .94^{bQ}$	24.45 ± 2.66^{aP}	23.53 ± 2.52^{bQ}	$22.11 \pm .20^{bQ}$

Average values with similar letter in the same row (P,Q) or in the same column (a,b,c) for a specific temperature and extract type do not change significantly by the Tukey test at $p < 0.05$.

Table 2. Total Flavonoid Content (mgQE/g).

Green Coffee Beans	Unroasted	170°C	190°C	210°C
Extracts				
Aqueous	13.50 ± 1.09^{aPQ}	$11.43 \pm .63^{aP}$	$12.93 \pm .08^{aPQ}$	15.02 ± 1.27^{aQ}
Ethanol	68.70 ± 1.68^{cP}	$143.09 \pm .611^{cQ}$	$144.00 \pm .33^{cQ}$	186 ± 2.27^{cQ}
Methanol	55.12 ± 1.50^{bQ}	$45.09 \pm .62^{bP}$	$59.12 \pm .69^{bQ}$	65.42 ± 1.50^{bR}

Average values with the similar letter in the same row (P,Q, R) or in the same column (a,b,c) for a specific temperature and extract type do not change significantly by the Tukey test at $p < 0.05$.

Table 3. % Antiradical Activity.

Green Coffee Beans	Unroasted	170°C	190°C	210°C
Extracts				
Aqueous	78.76 ± .36^{aQ}	70.88 ± .63^{aP}	83.20 ± .04^{aR}	86.21 ± .25^{aS}
Ethanol	96.38 ± .66^{cP}	97.07 ± .09^{cP}	96.11 ± .75^{cP}	96.32 ± .38^{cP}
Methanol	89.41 ± .38^{bP}	90.28 ± .47^{bP}	90.57 ± 1.28^{bP}	89.50 ± .31^{bP}

Average values with the similar letter in the same row (P,Q, R, S) or in the same column (a,b,c) for a specific temperature and extract type do not change significantly by the Tukey test at $p < 0.05$.

Table 4. Ferric Reducing Anti-oxidant Power (mmol Fe II eq/g).

Green Coffee Beans	Unroasted	170°C	190°C	210°C
Extracts				
Aqueous	.023 ± .000^{bS}	.019 ± .000^{aR}	.015 ± .000^{aQ}	.011 ± .000^{aP}
Ethanol	.088 ± .002^{cPQ}	.113 ± .000^{cQ}	.081 ± .027^{bP}	.059 ± .003^{aP}
Methanol	.011 ± .000^{aP}	.034 ± .001^{bQ}	.038 ± .004^{aR}	.040 ± .002^{aS}

Average values with the similar letter in the same row (P,Q, R, S) or in the same column (a,b,c) for a specific temperature and extract type do not change significantly by the Tukey test at $p < 0.05$.

Table 5. Reducing Capacity (mmol AA eq/g).

Green Coffee Beans	Unroasted	170°C	190°C	210°C
Extracts				
Aqueous	1.25 ± .001^{cP}	1.25 ± .000^{bP}	1.25 ± .007^{aP}	1.26 ± .003^{bP}
Ethanol	1.19 ± .028^{bP}	1.24 ± .032^{abPQ}	1.24 ± .025^{aPQ}	1.29 ± .006^{cQ}
Methanol	1.13 ± .002^{aP}	1.20 ± .001^{aQ}	1.23 ± .012^{aQ}	1.24 ± .003^{aQ}

Average values with the similar letter in the same row (P,Q) or in the same column (a,b,c) for a specific temperature and extract type do not change significantly by the Tukey test at $p < 0.05$.

which was maximum at 170 °C. It has also been found that medium and dark roasted coffee seeds indicated lower FRAP that could be due to the formation of compound evolved in the course of initial roasting such as melanoidins and Maillard products have inadequate iron reducing activity and thereby do not contribute to FRAP (Moreira et al., 2005).

Literature reveals that total antioxidant activity rises on light roasting (Daglia et al., 2000) which could be owing to liberation of extremely active low molecular weight phenols from the coffee constituents (Singleton et al., 1999; Montavon et al., 2003). Antioxidant activity of coffee beans is reported to be influenced by presence of phenolic complexes, exclusively chlorogenic acids (Shahidi and Chandrasekara, 2010).

Caffeolquinic acid (CQA) is a natural phenolic complex produced by esterification of quinic acid with caffeic acid, as a group called Chlorogenic acid (Gouvêa et al., 2005). In the course of roasting process, concentration of chlorogenic acids decreases especially of 5-Caffeolquinic Acid (5-CQA), but the antioxidant activity during processing did not change as much as chlorogenic acid concentration changes (Pilipczuk et al., 2015). After roasting, significant alteration in the CQA composition ensues in coffee seeds, concentration of 5-CQA declines whereas 4-CQA and 3-CQA increases on light roasting and decreases in medium and dark roasting (Gouvêa et al., 2005).

Coffee can turn into functional food by means of antioxidant properties and these are not only because of chlorogenic acid but are also associated to bioactive compounds like caffeine, trigonelline, cafestol, kahweol and tocopherols (Skowron et al., 2016). Caffeine is an important psychoactive alkaloid and boosts psychomotor performance, recovers memory and cognitive function (Corley et al., 2010; Desbrow et al., 2012). Recently coffee beverage is included in the list of "Foods that Fight Cancer" by American Institute for Cancer Research due to its anti-carcinogenic activity (Pilipczuk et al., 2015). The health benefits of chlorogenic acid, caffeine and other bioactive compounds are very well known but whether melanoidins exhibits such properties are not known, which are formed due to roasting. Our finding reveals that the Polyphenolic compounds and anti-oxidant activity of coffee increases on light roasting. Our results are supported by several other studies (Gouvêa et al., 2005; Moreira et al., 2005).

3 CONCLUSION

Coffee, being a rich source of anti-oxidants could be a source for prevention and treatment of diseases for which allopathy has no hope. Since coffee seeds are roasted at very high temperature and are used only for preparing beverages, however it retains anti-oxidant potential if lightly roasted. Therefore by making changes in roasting process, coffee seeds could be used by food industries to prepare food products other than beverages and thereby consumption of coffee seeds could be increased in diet which can provide protection against several diseases. Light roasted coffee seeds may also be used by pharmaceutical industries in the form of nutraceuticals or dietary supplements against metabolic syndrome and cancer.

ACKNOWLEDGMENT

The research study is funded by UGC, combined research entrance test (CRET) fellowship of University of Allahabad.

REFERENCES

Cämmerer, B. Kroh, L.W. 2006. Antioxidant activity of coffee brews. *Eur Food Res Technol* 223: 469–474.

Corley, J. Jia, X.L. Kyle, J.A.M. Gow, A.J. Brett, C.E. Starr, J.M. McNeill, G. Deary, I.J. 2010. Caffeine consumption and cognitive function at age 70: the Lothian Birth Cohort 1936 study. *Psychosom Med* 72(2):206–214.

Croft, K.D. 1998. The chemistry and biological effects of flavonoids and phenolic acids. *Ann N Y Acad Sci* 435-442.

Daglia, M. Papetti, A. Gregotti, C. Bertè, F. Gazzani, G. 2000. *J Agric Food Chem* 48: 1449–1454.

Desbrow, B. Biddulph, C. Devlin, B. Grant, G.D. Kumar, A. Dukie, S. Leveritt, M.D. 2012. The effects of different doses of caffeine on endurance cycling time trial performance. *J Sports Sci* 30(2):115-120.

Franca, A.S. Mendonça, J.C.F. Oliveira, S.D. 2005. Composition of green and roasted coffees of different cup qualities. *LWT - Food Sci Techn* 38: 709-715.

Goodman, B.A. Glidewell, S.M. Deighton, N. Morrice, A.E. 1994. Free radical reactions involving coffee. *Food Chem* 51:399–403.

Gouvêa, C.M.C.P. Duarte, S.M.S. Abreu, C.M.P. Menezes, H.C. Santos, M.H.D. 2005. Effect of processing and roasting on the antioxidant activity of coffee brews. *Ciênc Tecnol Aliment* 25(2): 387-393.

Hofmann, T. Bors, W. Stettmaier, K. 1999. Studies on radical intermediate in the early stage of non enzymetic browning reaction of carbohydrate and amino acids. *J Agric Food Chem* 47:391–396.

Marinova, D. Ribarova, F. Atanassova, M. 2005. Total phenolics and total flavonoids in bulgarian fruits and vegetables. *J of Uni of Chem Tech Metall* 40(3):2255-2260.

Montavon, P. Mauron, A.F. Duruz, E. 2003 Changes in Green Coffee protein profiles during roasting. *J Agric Food Chem* 51:2335–2343.

Moreira, D.P. Monteiro, M.C. Alves, M.R. Donangelo, C.M. Trugo, L.C. 2005. Contribution of chlorogenic acids to the iron-reducing activity of coffee beverages. *J Agric Food Chem* 53:1399-1402.

Oktay, M. Gülçin, I. Küfrevioglu, Ö.˘I. 2003. Determination of in vitro antioxidant activity of fennel (*Foeniculum vulgare*) seed extracts. *Lebenson Wiss Technol* 36:263–271.

Pandey, M.C. Tamilmani, P. 2015. Optimization and Evaluation of phenolic compounds and their antioxidant activity from coffee beans. *Int J of Adv Res* 3(4): 296-306.

Priftis, A. Stagos, D. Konstantinopoulos, K. Tsitsimpikou, C. Spandidos, D.A. Tsatsakis, A.M. Tzatzarakis, M.N. Kouretas, D. 2015. Comparison of antioxidant activity between green and roasted coffee beans using molecular methods. *Mol Med Rep* 12,7293-7302.

Pandey, K.B. Rizvi, S.I. 2009. Plant polyphenols as dietary antioxidants in human health and disease. *Oxid Med Cell Longev* 2: 270-278.

Perrone, D. Farah, A. Donangelo, C.M. Paulis, T. Martin, P.R. 2008. Comprehensive analysis of major and minor chlorogenic acids and lactones in economically relevant Brazilian coffee cultivars. *Food Chem* 106: 859–867.

Pilipczuk, T. Kusznierewicz, B. Zielińska, D. Bartoszek, A. 2015. The influence of roasting and additional processing on the content of bioactive components in special purpose coffees. *J Food Sci Technol* 52(9):5736–5744.

Ramalakshmi, K. Kubra, I.R. Rao, L.J.M. 2008. Antioxidant potential of low-grade coffee beans. *Food Res Int* 41(1):96–103.

Skowron, M.J. Sentkowska, A. Pyrzyn´ska, K. Paz De Peña, M. 2016. Chlorogenic acids, caffeine content and antioxidant properties of green coffee extracts: influence of green coffee bean preparation. *Eur Food Res Technol* 242:1403–1409.

Sutharut, J. Sudarat, J. 2012. Total anthocyanin content and antioxidant activity of germinated colored rice. *Int Food Res J* 19(1): 215-221.

Sanja, S.D. Sheth, N. Patel, N. K. Patel, D. Patel, B. 2009. Characterization and evaluation of antioxidant activity of Portulucaoleracea. *Int J Pharma Pharm Sci* 1: 74-84.

Singleton, V.L. Orthofer, R. Lamuela-Raventos, R.M. 1999. Analysis of total phenols and other oxidation substrates and antioxidants by means of Folin-Ciocalteu reagent. *Methods Enzymol* 299:152– 178.

Shahidi, F. Chandrasekara, A. 2010. Hydroxycinnamates and their in vitro and in vivo antioxidant activities. *Phytochem Rev* 9:147–170.

Biotechnology and Biological Sciences – Sen et al. (Eds)
© 2020 Taylor & Francis Group, London, ISBN 978-0-367-43161-7

Cytotoxic investigation and photophysical analysis of serum protein binding with beta-carboline alkaloids

T. Ghosh & K. Bhadra
Department of Zoology, University of Kalyani, Kalyani, Nadia, West Bengal, India

ABSTRACT: The work focused on interaction of beta-carboline alkaloids, harmalol and harmine, with HSA by biophysical and biochemical assays. Serum protein in the form of FBS in the culture media negatively alters the cytotoxicity of the alkaloids. MTT assay indicates concentration dependent growth inhibitory effect of the alkaloids on A375, MDA-MB-231, HeLa and ACHN cell, having maximum cytotoxicity with minimum GI_{50} value of 6.5 μM on ACHN by harmine in presence of 1% of FBS. Detail cytotoxic investigation on ACHN cell highlight the apoptotic induction ability of harmine.

The binding constant was found to be 2.5×10^3 M^{-1} and 4.3×10^4 M^{-1}, respectively, for harmalol and harmine by UV spectroscopy, the trend was also supported by fluorescence spectroscopy with HSA. Site markers demonstrated warfarin binding sites of the alkaloids to the protein.

The results serve as data for the future development of serum protein based targeted drugs.

Keywords: beta-carboline alkaloids, human serum protein albumin HSA, cytotoxicity, spectroscopic analysis

1 INTRODUCTION

The importance of plasma protein binding in modulating the effective drug concentration at pharmacological target sites has been the topic of significant discussion over the past few decades. Albumin, α-acid glycoprotein, lipoproteins and erythrocytes are the major drug binding constituents of plasma (Bohnert & Gan, 2013). Among them, albumins are the most abundant (52-60 %), non-glycosylated, multifunctional plasma protein which is negatively charged. The overall pharmacokinetic property of a drug *i.e.* its distribution, half-life inside the body, and elimination outside the body is depend upon the binding with the plasma proteins, especially with the albumin. Albumin functions as a cargo protein for several exo- & endo-genous compounds (Carter & Ho, 1994). High binding level with serum protein (> 85%) have the significant inhibitory impact on the penetration of antibiotics to extra vascular tissue sites (Bergan, Engeset & Olszewski, 1987). The fraction of drug which is unbound and is free from serum proteins are the only biologically available molecules to diffuse from circulatory system and accumulate in tissues, thus enabling interaction with the therapeutic targets, which is the soul objective of this work.

HSA, a 585-residue protein, is monomeric but contains three structurally similar α-helical domains (I, II & III), each divided into two subdomains- A (Contain six α- helices) and B (Contain four α-helices) (Petitpas et al., 2001; Carter and Ho, 1994; He and Carter, 1992). The targeted delivery of drugs by the albumins is effected through principal binding in hydrophobic cavities located in Sudlow's sites I (located in subdomain IIA), and II (located in subdomain IIIA) in the three structurally similar α-helical domains of the protein (Sudlow, Birkett & Wade, 1975; Petitpas et al., 2001).

The beta-carboline alkaloids are a large group of natural and synthetic indole alkaloids that possess a common tricyclic pyrido [3, 4-b] indole ring structure, which is well known for its several pharmacological functions (Mahmoudian, Jalilpour & Salehian, 2002; Cao et al., 2007; Ghosh et al., 2019).

Much study has been done on the interaction of beta-carboline alkaloids with the nucleic acids, comparatively less study has been done on the interaction of beta-carboline alkaloids with the serum protein (Galecki et al., 2012; Hemmateenejad et al., 2012; Nafisi, Panahyab & Sadeghi, 2012; Ghosh et al., 2019).

This study helps to elucidate the cytotoxic property of the beta-carboline alkaloids on different human cancer cell lines and how their cytotoxic property is being affected after binding with the serum protein molecules.

2 MATERIALS & METHODS

HSA, 99% pure, EFA and globulin free, (Mw. 68.563 kDa), harmalol (Mw. 272.73 g/mol), harmine (Mw. 212.25 g/mol), warfarin (C0.97 mass fraction purity) and ibuprofen (C0.98 mass fraction purity) were purchased from Sigma-Aldrich Corporation (St. Louis, MO, USA). All samples were prepared in citrate–phosphate (CP) buffer, pH 6.8. Concentration of the samples was determined by UV spectroscopic measurements using molar extinction coefficient values of 36600 $M^{-1}cm^{-1}$ for HSA at 280 nm, 19000 $M^{-1}cm^{-1}$ for harmalol at 371 nm, 14600 $M^{-1}cm^{-1}$ for harmine at 318 nm, 346 $M^{-1}cm^{-1}$ for ibuprofen at 272 nm and 13900 $M^{-1}cm^{-1}$ for warfarin at 308 nm.

2.1 *MTT & other in vitro apoptotic assay*

All cell lines were purchased from NCCS, Pune. Standard protocol was used for the cell culture and cell growth (Ghosh et al., 2019).

MTT assay was performed for all the cell lines at the concentration of 1 mg/ml to evaluate the percent growth inhibition (GI) of the two beta-carboline alkaloids.

To examine the apoptotic induction ability of the alkaloid through FITC-Annexin V/PI, flow cytometry was performed (Mallick et al., 2010).

2.1.1 *Statistical analysis*

50% growth inhibition values are statistically analyzed by ANOVA and significant values are calculated against both untreated as well as solvent control (*$P < 0.05$ vs. solvent control).

2.2 *Spectroscopic and calorimetric study of the interaction*

UV and fluorescence spectral analysis for the binding of beta-carboline alkaloids with the HSA molecule were performed as described by our previous study (Ghosh et al., 2019).

The energetics of the binding was studied by isothermal titration calorimetry (ITC) using a GE Microcal ITC 200, (Northampton, USA) microcalorimeter.

3 RESULTS & DISCUSSION

Efficacy of the two beta-carboline alkaloids *viz.* harmalol, and harmine (Figure 1) on four human cancer cell line *viz.* A375, MDA-MB-231, HeLa, and ACHN, was performed by cell viability or MTT assay in presence of 1 % and 10 % FBS in the culture media (Figure 2).

The MTT assay showed that the presence of serum protein in the form of FBS supplemented in the growth media negatively affects the cytotoxicity of the two alkaloids and among the above four cell lines, harmine showed maximum cytotoxicity with minimum GI_{50} values of 6.5 μM on ACHN (renal adenocarcinoma) cells at 1% FBS. The effect of various

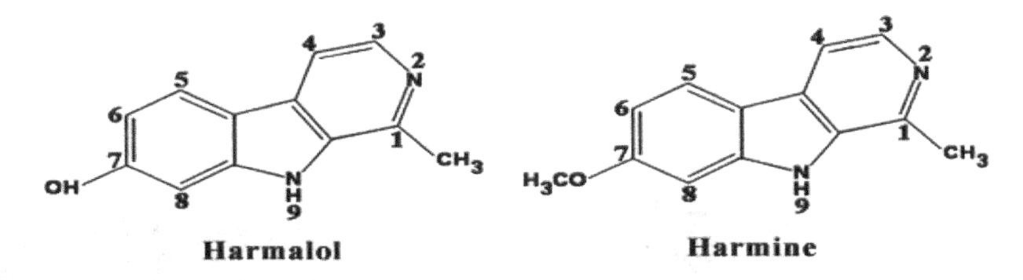

Figure 1. Chemical structure of harmalol, and harmine.

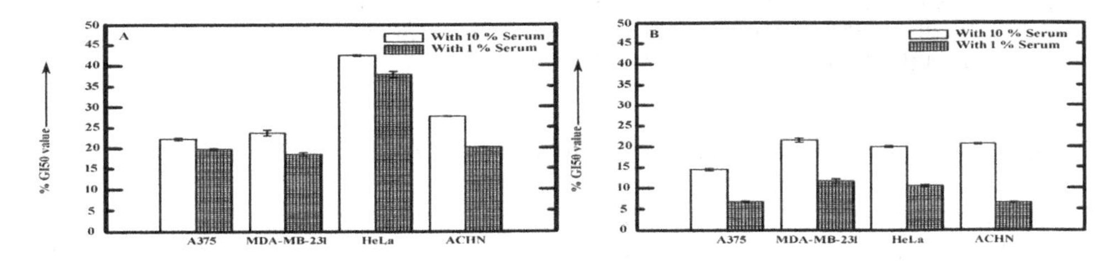

Figure 2. % GI_{50} values of four different human cancer cell lines in presence of 1% FBS (stripe) and 10 % FBS (blank) treated with (A) harmalol, and (B) harmine for 48 h through MTT assay.

percentages of FBS in the culture media viz. 3.0%, 7.5% and 15%, on the GI_{50} values after treatment with harmine on ACHN cell line have found to be 10.2 ± 0.2, 18.3 ± 0.2 and 28.9 ± 0.2, respectively.

However, in case of the normal mutated liver cell line, WRL-68, no cytotoxicity were reported TILL 60 μM of concentration.

Henceforth, *in vitro* cytotoxicity of harmine, the most potent drug, was studied in detail on ACHN.

Treatment with harmine, induced phosphatidylserine externalization in ACHN cells, which was characterized by bivariate FITC-Annexin V/PI flow cytometric analysis at GI_{50} concentrations of 6.5 μM (Figure 3A-B). The $FITC^{+}/PI^{-}$ apoptotic cell population (lower right quadrant) increased gradually from 0.39±0.1 % to 20.67±1.0 % after treatment with harmine. On the other hand FITC+/PI+ (upper right quadrant) necrotic cell population shows negligible change after 48 h of treatment with harmine.

The morphology of the cells as observed by phase contrast and SEM has been shown in Figure 4A-D. The morphological changes includes cell rounding (as observed under phase contrast microscope), nucleolus fragmentation, cell shrinkage (as observed by SEM) as well as irregularities seen in cell contour and size in harmine treated cell line, whereas, in control, the cells remain spread out in normal shape.

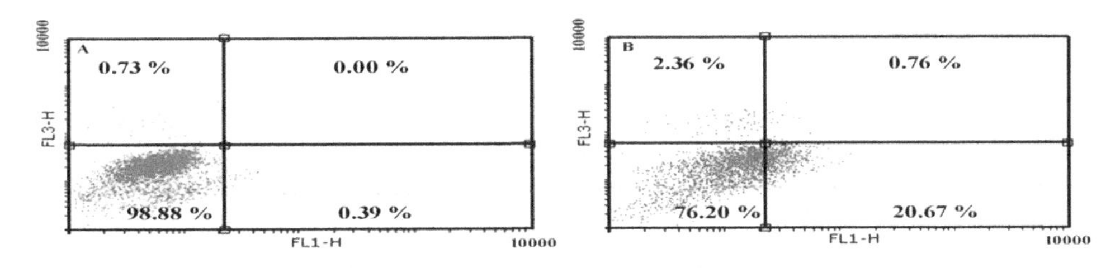

Figure 3. Contour diagram of FITC-Annexin V/PI flow cytometry of ACHN cells after 48 h of incubation at (A) untreated, and (B) GI_{50} concentrations of 6.5 μM. Data are representative of three independent experiments.

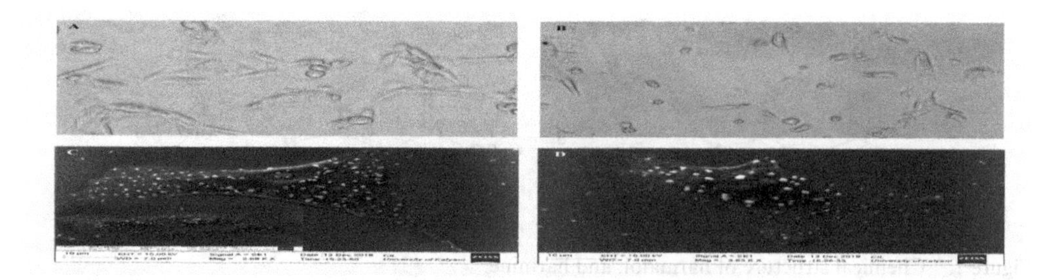

Figure 4. Phase contrast microscopic images (A-B), Olympus CKX53 (10×), of control and harmine treated ACHN cells at GI_{50} concentration of 6.5 μM. Changes in cell shape and contour after 6.5 μM of harmine treatment (C-D) as observed under ZEISS EVO LS 10 (SEM).

3.1 *Binding constant and stoichiometry analysis of alkaloid-HSA complex*

3.1.1 *UV absorption spectral study of the HSA–beta–carboline interaction*

The absorption spectrum of the two beta-carboline alkaloid *viz.* harmalol and harmine (Figure 5A-B.) showed hypochromic effect in presence of increasing concentration of the HSA protein in the 300–500 nm region, that eventually leading to saturation at a P/D value of 4.4 and 3.5 for harmalol and harmine, respectively. Benesi-Hildebrand plot of the absorption data gave a 1:1 stoichiometry for the HSA-harmalol and HSA-harmine complex, with binding constants of 2.5×10^3 M^{-1}, and 4.3×10^4 M^{-1}, respectively.

3.1.2 *Fluorescence spectral study of the HSA-beta-carboline alkaloid interaction*

Upon excitation at 295 nm, only the single tryptophan residue at position 214 (Trp 214) of the HSA fluoresces. Addition of the two beta-carboline alkaloids *viz.* harmalol and harmine led to the remarkable quenching in the fluorescence intensity. The emission fluorescence intensity was seen maximum around 342 nm. The steady state fluorescence spectra of HSA molecule in presence of increasing concentration of two alkaloids are presented in Figure 6A-B. Beyond wavelength 400 nm, both the two beta-carboline alkaloids has strong emission maxima and hence these regions are not presented. Saturation was achieved at D/P 12.5, and 8.2 for harmalol, and harmine, respectively. The quenching mechanism was analyzed by the traditional Stern–Volmer equations (Lakowicz and Weber, 1973) to yield K_{sv} values. A plot of F_o/F versus $[Q]$ was found to be linear, indicating the occurrence of only one type of quenching. The K_{sv} values elucidated from the plots at 298 K were 2.88×10^3 M^{-1}, and 3.68×10^4 M^{-1} with harmalol, and harmine, respectively, for HSA.

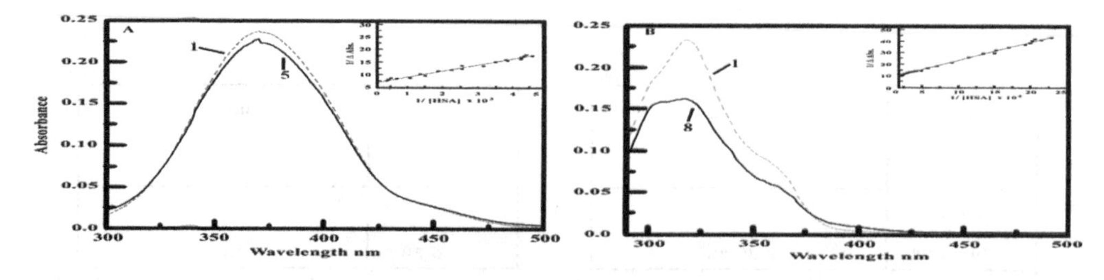

Figure 5. UV absorption spectra of the beta-carboline alkaloids (curve 1 in each case) treated with HSA in 15 mM CP buffer, pH 6.8 at 25 ±0.5 °C. (A) harmalol (10.52 μM) treated with 47 μM (curve 5) of HSA, (B) harmine (13.69 μM) treated with 48.5 μM (curve 8) of HSA. Inset: Benesi-Hildebrand plots for binding.

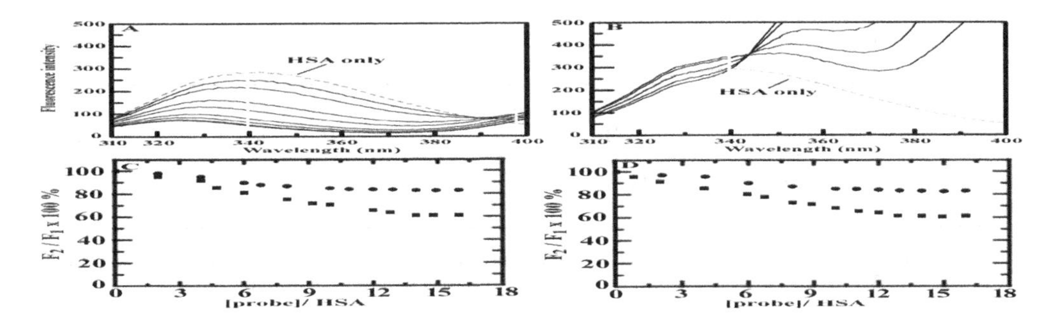

Figure 6. Fluorescence spectra of HSA (5 μM) treated with increasing concentrations of (A) Harmalol, and (B) Harmine, respectively. In C-D the graphical presentations that shows the effect of two site specific probes, warfarin (■) and ibuprofen (•) on the fluorescence of HSA molecule binding with (C) harmalol, (D) harmine. Where, [alkaloid]: [HSA]= 4:1 ratio.

3.1.3 *Site specific binding of beta-carboline alkaloids*

Figure 6C-D. shows the graphical presentation of the effect of two site specific probes, warfarin and ibuprofen, added on the fluorescence of HSA binding with harmalol, and harmine where alkaloid and HSA concentrations were kept at 4:1 ratio. These site specific studies suggested both the beta-carboline alkaloid binds to warfarin site, *i.e.*, Sudlow's site I in the subdomain IIA of the HSA molecule.

3.1.4 *Calorimetric evaluation of beta-carboline alkaloids-HSA interaction*

Figure 7A-B. represents calorimetric profiles of the titration of harmalol, and harmine with the HSA molecule. In both the cases, bindings were exothermic processes and have single binding event. The equilibrium constant (K), binding stoichiometry (N), enthalpy change ($\Delta H°$), entropy contribution ($T\Delta S°$), and Gibbs energy change ($\Delta G°$) obtained from the calorimetric data are summarized in Table 1. The binding constants of harmalol and harmine to HSA at 298 K were evaluated to be $2.55 \pm 0.06 \times 10^3$ M^{-1}, and $55.32 \pm 0.18 \times 10^3$ M^{-1}, respectively, which eventually supports the spectroscopic results.

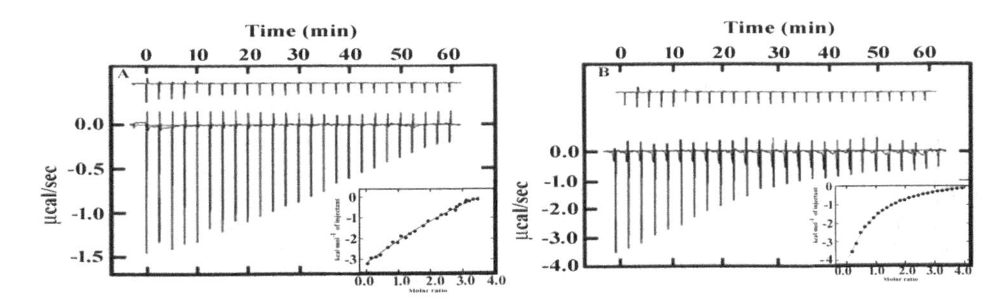

Figure 7. ITC profile for binding of 1 mM of HSA to (A) harmalol (22 μM), (B) harmine (20 μM) at 298.15 K, pH 6.8. Each heat burst curve (in the bottom part of upper panel) is due to the 1.5 μL sequential injection of the HSA into the alkaloid (curves at the bottom). The top part of upper panel show the heat burst for the injection of the HSA into the same buffer as control in each experiment (curves offset for clarity). Inset of lower panel represent the corresponding normalized heat data against the molar ratio (P/D).The data points (•-•) reflect the experimental injection heats while the solid line represents the calculated best fit of the data.

Table 1. ITC derived profiles at 25±0.5 °C for the binding interaction of HSA to the beta-carboline alkaloids [a].

Alkaloids	N	K_b ($\times 10^3 M^{-1}$)	$\Delta G°$ (kcal/mol)	$\Delta H°$ (kcal/mol)	$T\Delta S°$ (kcal/mol)
Harmalol	0.95±0.04	2.55±0.06	-4.67±0.13	-3.72±0.04	0.95
Harmine	0.98±0.02	55.32±0.18	-6.51±0.11	-3.89±0.03	2.62

a Average of three independent experiments in CP Buffer, pH 6.8. Values of $\Delta G°$ were determined using the equation $\Delta G° = -RT \ln K_b$ and $T\Delta S° = \Delta H° - \Delta G°$. Where, N denotes the binding site size.

4 CONCLUSION

In summary, the study showed that the presence of fetal bovine serum in the cell culture media negatively alters the growth of the four human cancer cell lines in presence of harmalol, and harmine. Among these two, the affect was more prominent with harmine indicating maximum cytotoxicity on ACHN cell line, in presence of 1% FBS. Furthermore both spectroscopic and calorimetric results suggested that harmine has the maximum binding affinity towards the HSA molecule, which is followed by harmalol. Competitive binding using two site markers *viz.* warfarin and ibuprofen demonstrated that both harmalol and harmine binds to site I, subdomain IIA, of HSA protein.

ACKNOWLEDGMENT

KB is indebted to DST-RFBR 2017-19 (DST/INT/RUS/RFBR/P-254) for partial funding. TG, Senior Research Fellow, supported by NET-UGC.

REFERENCES

[1] Bohnert, T., Gan, L.S. (2013). Journal of Pharmaceutical Sciences, 102(9), 2953–2994.
[2] Carter, D.C., Ho, J.X. (1994). Advances in Protein Chemistry, 45, 153–203.
[3] Bergan, T., Engeset, A., Olszewski, W. (1987). Reviews of Infectious Diseases, 9(4), 713–718.
[4] Petitpas, I., Bhattacharya, A.A., Twine, S., East, M., Curry, S. (2001). Journal of Biological Chemistry, 276, 22804–22809.
[5] He, X.M., Carter, D.C. (1992). Nature, 358(6383), 209–215.
[6] Sudlow, G., Birkett, D.J., Wade, D. N. (1975). Molecular Pharmacology, 11(6), 824–832.
[7] Mahmoudian, M., Jalilpour, H., Salehian, P. (2002). Iranian Journal of Pharmacology & Therapeutics, 1(1), 1–4.
[8] Cao, R., Peng, W., Wang, Z., Xu, A. (2007). Current Medicinal Chemistry, 14(4), 479–500.
[9] Ghosh, T., Sarkar, S., Bhattacharjee, P., Jana, G.C., Hossain, M., Pandya, P., Bhadra, K. (2019). Journal of Biomolecular Structure and Dynamics, DOI: https://doi.org/10.1080/07391102.2019.1595727.
[10] Galecki, K., Despotovic, B., Galloway, C., Ioannidis, A.G., Janani, T., Nakamura, Y., Oluyinka, G. (2012). Biotechnology and Food Science, 76(1), 3–12.
[11] Hemmateenejad, B., Shamsipur, M., Samari, F., Khayamian, T., Ebrahimi, M., Rezaei, Z. (2012). Journal of Pharmaceutical and Biomedical Analysis, 67-68, 201–208.
[12] Nafisi, S., Panahyab, A., Sadeghi, G.B. (2012). Journal of Luminescence, 132(9), 2361–2366.
[13] Mallick, S., Ghosh, P., Samanta, S.K., Kinra, S., Pal, B.C., Gomes, A., Vedasiromoni, J.R. (2010). Cancer Chemotherapy and Pharmacology, 66(4), 709–719.
[14] Lakowicz, J.R., Weber, G. (1973). Biochemistry 12(21), 4171–4179.

Biotechnology and Biological Sciences – Sen et al. (Eds)
© 2020 Taylor & Francis Group, London, ISBN 978-0-367-43161-7

A survey on control theoretic research paradigms of insulin signaling pathways study

Darshna M. Joshi
Instrumentation and Control Department, Government Polytechnic Ahmedabad, Ahmedabad, India

Jignesh Patel
Instrumentation and Control Department, Nirma University, Ahmedabad, India

ABSTRACT: Signaling pathways play a pivotal role to maintain the homeostasis of the biological system, alterations in which, lead to the onset of severe diseases such as hypertension, diabetes and cancer. Dysregulation of the glucose metabolism causes malfunction of the insulin signaling pathways. Insufficient or no insulin production leads to type 1, while, insulin resistance leads to type 2 diabetes mellitus. Recent advances in the field of systems biology encourages use of control theoretic approach to investigate the interactions within the insulin signaling pathways. The present paper provides a detailed survey and critics on various control theoretical approaches like parameter optimization, sensitivity and robustness analysis adopted by system biologists. Additionally, unsatisfactory drug treatments may lead to need of an alternative therapy like body exercise and acupuncture. Last section of the paper discusses on alternative therapies in type 2 diabetes as an open area for the researchers working from control theoretic perspective.

Keywords: Control Theory, Systems Biology, Insulin Signaling Pathways, Parameter Estimation, Sensitivity Analysis, Modeling, Optimization, Robustness

1 INTRODUCTION

Signaling is one of the most important tasks performed by the biological system to maintain the homeostasis. Interruptions in the signaling pathways results in the development of various diseases like diabetes, hypertension and cancer. Out of the number of biochemical pathways, insulin signaling pathway is responsible for the regulation of the key functions in the human body, especially glucose metabolism. Disruptions in the insulin signaling pathway introduces insulin resistance that leads to the onset of diabetes. People with type 1 diabetes lack the production of sufficient insulin, while, that with type 2 diabetes fail to respond to the insulin produced and thus disturbs the glucose homeostasis in the body. Analysis of these insulin signaling pathways is gaining attention in the field of systems biology. Each entity in the pathways performs its work in an autonomous manner, but they interact with each other in one or the other way. These interactions are of enormous importance to perform the analysis of insulin signaling pathway [1], [2], [3]. Here comes the role of control system specialists to apply their system theory knowledge to the biological pathways. As the application of control theory to such pathways is not straightforward, the insulin signaling pathway is fragmented and then analyzed by various possible approaches.

1.1 *Insulin signaling pathways*

To unravel the problem of diabetes, significant knowledge of underlying mechanisms of insulin signaling pathways is required. Traditionally, the signaling pathways are simple linear cascade structure with a combination of a receptor and an effector. Whereas contemporary view shows it to be a complex structure of multiple inputs, outputs and interactions. To resolve the issues behind such complex insulin signaling pathways, researchers have revealed the basic mechanisms behind the pathways. Figure 1 shows the basic mechanisms involved in the insulin signaling pathways known so far. As an initiating process, insulin first binds insulin receptor.

This binding causes the auto-phosphorylation and activation of the insulin receptor. Activated insulin receptor then causes the phosphorylation of insulin receptor susbtrate-1(IRS-1), subsequently forming a complex with phosphatidylinositol-3- kinase (PI3K) [3]. The IRS1-PI3K complex catalysis the production of phosphatidylinositol triphosphate (PIP3) which then interacts allosterically with phosphoinositide dependent kinase 1 (PDK1). The PIP3-PDK1 complex phosphorylates protein kinases AKT and protein kinase C (PKC ζ). Activated AKT and PKC ζ through an unknown mechanism then trigger glucose transporter (GLUT4) translocation from an internal compartment to the cell membrane [4]. With GLUT4 at the cell membrane, a cell can uptake glucose from its environment. This pathway is regulated by the action of a few other proteins. In addition, SHIP2 and PTEN (lipid phosphatases) deactivate PIP3 into PI (3, 4) P2 and PI (4, 5) P2, respectively [5]. Analyzing these pathways from engineering point of view is gaining vital attention of the researchers. A critical survey has been carried out to obtain the possible approaches in the area of insulin signaling pathways. The following sections thus provides exploration of the same. The subsequent sections then provide the detailed survey on the control theoretical approaches adopted by the researchers for understanding the mechanisms behind the insulin signaling pathways [2], [6]. Additionally, a brief review on alternative traditional therapies like acupuncture and body exercise for treating diabetes is also discussed.

2 CLASSIFICATION OF CONTROL THEORETIC STUDIES

The necessity for more specific understanding of insulin signaling pathways for type 2 diabetes mellitus plays a tremendous role to understand the overall phenomenon as well for new drug

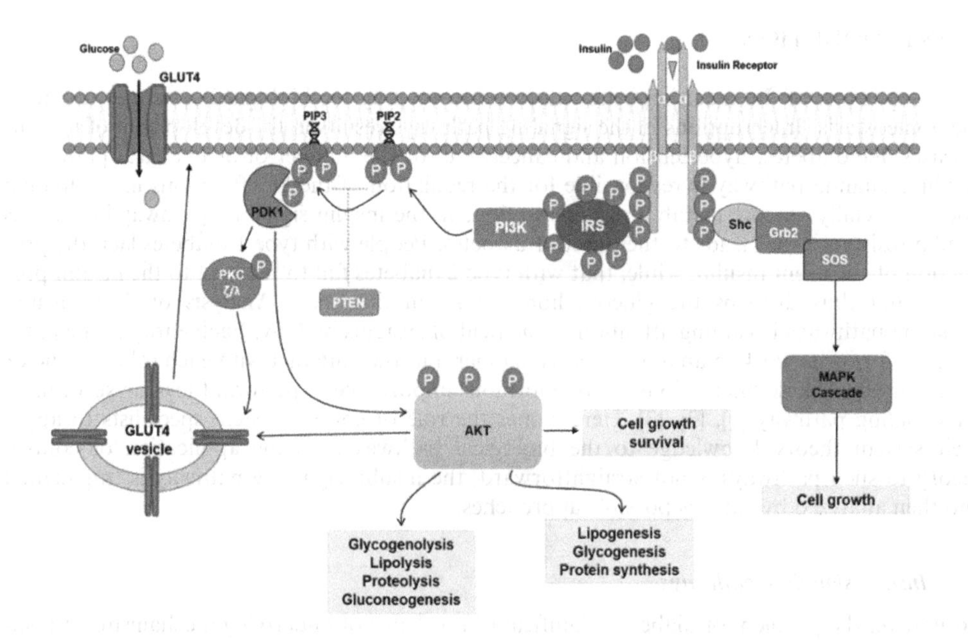

Figure 1. Insulin signaling pathway.

discovery. Advancement in control theory for biological systems provides the results which are of inestimable importance. These results with sufficient accuracy are helpful in the better understanding of the disease mechanisms. Number of control theoretical approaches is adopted by various researchers to analyze the insulin signaling pathway for diabetes [2], [7], [8]. These approaches help in clarifying the insulin signaling pathways in a detailed satisfactory manner. Various software tools/packages are available to analyze the signaling pathways. This paper is an attempt to help by offering a classification of various available control theoretical approaches adopted by researchers for analyzing the insulin signaling pathways for diabetes. Figure 2 shows the classification of control theoretical approaches for insulin signaling pathways. The classification is completely based on referring work of researchers in recent studies. Henceforth, each section describes the detailed review and the outcome achieved by adopting various control theoretical approaches for insulin signaling pathways.

2.1 *Parameter optimization*

Optimization refers to the method of obtaining the best values of some objective function that maximizes or minimizes it in the most feasible manner. Parameter optimization is one of the most commonly used control theoretical approach in scientific research in the field of systems biology. It is a fact that with evolution, all the biological structures are optimized [9]. Although diverse parameter optimization techniques are available in the literature, there is always a scope of development for new algorithms as no single one performs best for all problems. The evolvement of numerical optimization from local, global, stochastic and deterministic has yet not reached up to that extent where we can simply rely on a single method of parameter optimization for analyzing the complex insulin signaling pathways [10], [11]. Thus, it is necessary to apply a series of optimization techniques to obtain the best possible solution. A systems approach is needed to examine and understand insulin resistant state of the human adipocytes in type 2 diabetes. The insulin receptor (IR)-insulin receptor substrate (IRS) pathway model is implemented with a feedback from mammalian target of rapamycin complex (mTORC1) and then analyzed under normal and insulin resistant states. The feedback from mTORC1 is not modelled as positive or negative explicitly. Instead, both variants in the optimization were allowed. When the attenuation of a positive feedback is made while explaining the type 2 diabetic state, best agreement between model simulation and the quantitative data is obtained. Hence, the most important change affecting the whole signaling network is the attenuation of mTORC1 activation and feedback from mTORC1 to IRS1 [9], [10], [12]. Global method of simulation annealing approach is chosen for carrying out the optimization with local downhill simplex approach using systems biology toolbox of MATLAB [13]. Thus, overall observation shows that the parameter optimization is done by following the least square regression method for optimizing the cost function of the ordinary differential equations (ODE's) obtained from the model in general by the researchers.

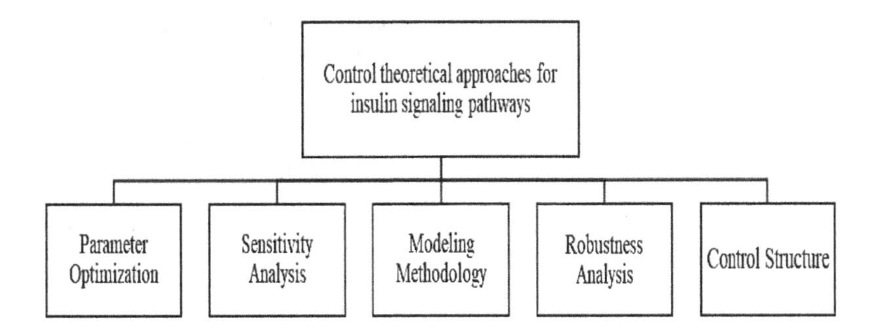

Figure 2. Classification of control theoretical approaches for insulin signaling pathways.

2.2 *Sensitivity analysis*

With the advancement in the field of systems biology, sensitivity analysis is chosen as one of the promising approaches for better understanding the signaling pathways and biochemical networks. The development of a dynamic model is difficult to build with the uncertainties of the model parameters. Feedback loops and PI3K branch present in the signaling pathways is proven to enhance the robustness of the pathway by being one of the most sensitive parameters of the pathway. Thus, sensitivity analysis plays an important role in providing valuable insights to test robustness of the model as well as to provide the most contributing factors affecting pathways [14], [15]. Local sensitivity analysis covers the impact of small perturbations on the model outputs. On the other hand, global sensitivity analysis covers the large perturbations on the model outputs [16]. Many important signaling control mechanisms of type-2 diabetes and their interactions are explained in the Sedaghat, Sherman and Quon model with certain limitations. Based on the time integral of GLUT4 expression at the plasma membrane, a local Parametric Sensitivity Analysis (PSA) is used to investigate the Sedaghat model. Over a range of insulin concentrations, input profiles and parameter perturbations, sensitivity profiles for all rate constants and initial conditions are presented. The PSA revealed that there was an obvious saturation phenomenon at high insulin levels that affected much of the network [16], [17]. It is suggested that if the arbitrary components in the Sedaghat model are reduced to almost nil number, then sensitivity analysis might give more satisfactory analyzed results for the insulin signaling pathways. Further, Multi-parametric sensitivity analysis (MPSA) is employed for identifying the network components and rate constants that are most critical to GLUT4 translocation in the model developed considering AS160 and PKC ζ as the convergence points. The sensitivity is evaluated by a direct measure of the separation of the two cumulative frequency distributions of Kolmogorov-Smirnov (K-S) statistic. Thus, the sensitivity analysis shows the parameters having the strongest effect on the overall behavior of the insulin signaling pathways and hence give the control points in the system. [1], [18], [19]. Thus, it is possible to neglect the effect of the parameters which are less affecting the system and hence it reduces the complexity of the model. Also, the most sensitive parameter or part of the system can be focused for drug discovery [18].

2.3 *Modeling methodology*

The ongoing advancement in the field of systems biology has enabled us to construct the mathematical model of the biological network. Modeling is thus an ongoing process of analysis and experimental validation. [3], [17]. The basic step in constructing a model is the realization of the parameters involved and the interactions among them. These parameters are related to each other via certain mathematical expression and that leads to the development of mathematical model [20]. It can be designed using the simple Ordinary Differential Equation (ODE), Partial Differential Equation (PDE) or Stochastic Differential Equation (SDE), Discrete or Continuous, Delay Differential Equation (DDE), Fred holm Integral Equations (FIES), Integro-Differential Equations (IDEs) and so on [21]. Insulin signaling pathways for type 2 diabetes is modelled mathematically by various researchers. The minimal model comprising of ordinary differential equation is modelled in majority of the analysis as it is well suited. Also, the signal transfer in insulin mediated PI3K/AKT pathway and GLUT4 translocation is modelled using ordinary differential equations. Effect of perturbations on the insulin signaling pathway with positive or negative feedback is analyzed using ODEs [3], [5], [12]. Thus, after studying the different perspective of control theoretic approaches from the available scientific literature, it is observed that most of the insulin signaling pathways uses ordinary differential equations to model the system dynamics.

2.4 *Robustness analysis*

Robustness is an important phenomenon of signaling pathways ensuring signal propagation with high fidelity in the event of perturbations. Scientific literature has revealed that under parameter uncertainties and fluctuations, negative feedback ensures the robustness. Also, when robustness of the system is increased against certain perturbations, it may increase the instability in the system. Increased instability demands more resources which degrades the system's performance. Thus, a trade-off is to be maintained to achieve the best possible robustness and performance of the system against uncertain perturbations. Various tools are available for robustness analysis like μ-analysis, optimization-based robustness analysis, sum of squares method, Monte Carlo simulation and so on [8], [22]. Further, time dependent analysis is done to indicate the role of PI3K branch in the coupled pathway to enhance the robustness of the MAPK pathway. It is obvious that the biological networks are noisy. And thus, high sensitivity leads to the acceptance of even small fluctuations. To avoid this behavior, the robustness of the MAPK pathway is measured using the integrated sensitivities analysis. This helps in making the MAPK pathway less sensitive to noise by measuring the integrated sensitivities of the cascade effectors in MAPK pathway. Thus, it makes the pathway robust in nature by avoiding the noise interventions [23], [24].

2.5 *Control structure*

The role of systems and control theory in the field of biology is growing in an inevitable manner. To infer the knowledge about the systems dynamics, control theoretical approaches are adopted. Signal transduction pathways are consisting of various cascades and feedback loops that is regulating the pathways. Number of feedback loops are observed in the insulin signaling pathways and many of them are still unknown [7], [8]. It is a complex network with many interacting molecules, multiple regulatory sites, different feedback loops and cross talk to other signaling systems. Feedforward loops may serve as filters, sign-sensitive delays, and pulse generators in transcription networks [16], [24]. Insulin signaling pathways consists of positive feedback loops or double negative feedback loops embedded with nonlinearity. Thus, it exhibits switch-like bistable responses. It ensures that GLUT4 settles between two discrete, but mutually exclusive stable steady states. Thus, a steady state analysis indicates that negative feedback regulation of phosphatase PTP1B by AKT elicits bi-stability in insulin-stimulated GLUT4 translocation [6], [25], [26] In system and control theory, Kalman filter provides the estimation of behavior of unmeasured components when partial measurements are available. This is analogous to a classic problem in automatic control, namely, the state regulator problem (SRP), which penalizes deviations of the state variables as well as large control action. The SRP estimator calculates an optimal set of state trajectories via a constrained convex programming problem in which the available measurements are the constraints [2], [15], [22]. Further, based on the evidences it is suggested the AKT and PKC ζ may participate in positive and negative feedbacks in the insulin signaling pathways. Linear effect of AKT is considered while implementing the positive feedback, while explicitly a time lag is incorporated in the negative feedback loop. To test the pathway for the biphasic response, a comparison is made for the pathways with and without feedback. It is concluded that to generate biphasic activation of PKC ζ in response to insulin, it is essential to implement the feedback [3], [12]. Using the same approach and just adding the effect of regulation of exercise in the pathways for type 2 diabetes could provide new insights in the area.

3 ALTERNATIVE THERAPIES FOR DIABETES AFFECTING INSULIN SIGNALING PATHWAYS

Tremendous efforts have been spent in finding alternate therapies for treating diabetes. Miscellaneous approaches like acupuncture, exercise, medicinal herbs and hydrotherapy is used to treat insulin resistance prevailing in type 2 diabetes. Acupuncture is known to act on the pancreas to enhance the synthesis of insulin as well as increase the insulin receptors on the target cells. This accelerates the utilization of glucose and thus maintains the glucose metabolism [27], [28], [29]. Traditionally, many drugs for treating diabetes have been derived from medicinal herbs. Furthermore, exercise is also gaining attention to increase the number of insulin receptors and thus treating diabetes by elimination of insulin resistance by prevention of the accumulation of fat via AMPK signaling. Clinical evidences show that such alternative therapies are helpful in changing expression of vital components of insulin signaling pathways such as PI3K, AKT and GLUT4. Thus, Integration of scientific advancement with other control research techniques would further develop treatment for diabetes by adopting various available approaches [29].

4 CONCLUSION

Systems biology provides insight into the root causes of complex diseases, contributing to the drug discovery by exploring the underlying mechanisms. This paper provides an insight on the available control theoretical approaches used by various researchers in the analysis of insulin signaling pathways. Insulin signaling pathway is a complex structure and hence difficult to analyze the whole at once. To overcome this difficulty, the pathways are divided, and various control theoretical approaches are applied. The classification of the approaches is done by performing critical survey and the use of typical approach by various researchers is shown. Selection of typical approach according to need could be obtained via comparison of researchers work according to justification and research gap provided. Sensitivity analysis results in the development of predictive models. Parameter optimization provides the best suitable parameters for the model analysis. Also, various modeling methodologies are used, out of which, analysis based on ordinary differential equations (ODEs) is most common. Interestingly, role of feedback and feedforward loops in the insulin signaling pathways is revealed. Additionally, alternative therapies such as exercise, acupuncture and medicinal herbs, available for the treatment of diabetes, induces a change in expression of certain components of insulin signaling pathways. Modeling a combination of the effect of various therapies on the insulin resistance is still an open area of research. Also, time-dependent analysis of the effect of exercise, acupuncture and medicinal herbs on the insulin signaling pathways is an important area to touch upon. Integrating the scientific advancement with the control theoretical techniques is a wide area for the researcher to work on. Limited work is found on analysis of alternative therapies and their interactions for insulin signaling pathways in control theoretic framework. The present survey reveals different outcomes obtained by the system biologists for insulin signaling pathways from control standpoint. The study also provides comprehensive critics on the research of analyzing the insulin signaling pathways from the viewpoint of control theorists.

REFERENCES

[1] C. Colmekci, "The insulin signalling pathway in skeletal muscle: in silico and in vitro," Technische Universiteit Eindhoven, Eindhoven, 2015.

[2] E. Sontag, "Some new directions in control theory inspired by systems biology," *IEE Proceedings-Systems Biology*, vol. 1, no. 1, pp. 9-18, 2004.

[3] C. Azzurra, "Modelling and simulation of insulin signalling," Padova, 2013.

[4] N. Bergqvist, E. Nyman, G. Cedersund and K. Stenkula, "A systems biology analysis connects insulin receptor signaling with glucose transporter translocation in rat adipocytes," *Journal of Biological Chemistry*, vol. 292, no. 27, pp. 11206-11217, 2017.

[5] M. Wong, "Unravelling the insulin signalling pathway using mechanistic modelling," April 2016. [Online].

[6] L. Giri, M. V and K. Venkatesh, "A steady state analysis indicates that negative feedback regulation of PTP1B by Akt elicits bistability in insulin-stimulated GLUT4 translocation," *Theoretical Biology and Medical Modelling*, vol. 1, p. 2, 2004.

[7] P. Wellstead, E. Bullinger, D. Kalamatianos, M. Verwoerd and O. Mason, "The role of control and system theory in systems biology," *Elsevier: Annual Reviews in Control*, vol. 32, no. 1, pp. 33-47, 2008.
[8] C. Cosentino and D. Bates, "An introduction to feedback control in systems biology," CRC Press, London, 2011.
[9] Y. Wang, X. Zhang and L. Chen, "Optimization meets systems biology," *BMC Systems Biology*, vol. 4, p. Suppl S2: S1, 2010.
[10] H. Greenberg, W. Hart and G. Lancia, "Opportunities for combinatorial optimization in computational biology," *Informs Journal on Computing*, vol. 16, no. 3, pp. 211-231, 2004.
[11] P. Mendes and D. Kell, "Non-linear optimization of biochemical pathways: Applications to metabolic engineering and parameter estimation," *BIoinformatics*, vol. 14, no. 10, pp. 869-883, 1998.
[12] C. Brannmark, "Insulin signalling in human adipocytes: A systems biology approach," Liu-tryck Linkoping, Sweden, 2012.
[13] H. Schmidt and M. Jirstrand, "Systems Biology Toolbox for MATLAB. A computational platform for research in systems biology," *Bioinformatics*, vol. 22, no. 4, pp. 514-515, 2006.
[14] Z. Zi, "Sensitivity analysis approaches applied to systems biology models," *IET Systems Biology*, vol. 5, no. 6, pp. 336-346, 2011.
[15] B. Ingalls, "Sensitivity Analysis: From model parameters to system behaviour," *Essays in Biochemistry*, vol. 45, pp. 177-193, 2008.
[16] C. Gray, "Sensitivity analysis of the insulin signalling pathways for glucose transport," South Wales, 2013.
[17] A.S. edaghat, A. Sherman and M. Quon, "A mathematical model of insulin signalling pathways," *American Journal of Physiology-Endocrinology and Metabolism*, vol. 283, no. 5, pp. 1084-1101, 2002.
[18] T. Sumner, "Sensitivity analysis in system biology modeling and its application to a multiscale model of glucose homeostasis," Centre for Mathematics and Physics in the Life Sciences and Experimental Biology, London, 2010.
[19] S. Marino, I. Hogue, C. Ray and D. Kirschner, "A methodology for performing global uncertainity and sensitivity analysis in systems biology," *Journal of theoretical biology*, vol. 254, no. 1, pp. 178-196, 2008.
[20] S. Dhar, "Modelling methodologies for systems biology," in *Systems and Synthetic Biology*, Dordrecht, Springer, 2014, pp. 43-62.
[21] P. Rangamani and R. Iyengar, "Modelling cellular signalling systems," *Essays in Biochemistry*, vol. 45, pp. 83-94, 2008.
[22] H. Kitano, "Towards a theory of biological robustness," *Molecular Systems Biology*, vol. 3, no. 1, p. 167, 2007.
[23] D. Hu and J. Yuan, "Time-dependent sensitivity analysis of biological networks: Coupled MAPK and PI3K signal transduction pathways," *Journal of Physical Chemistry*, vol. 110, no. 16, pp. 5361-5370, 2006.
[24] "Dynamic sensitivity and control analyses of metabolic insulin signalling pathways," *IET Systems Biology*, 2008.
[25] E. Ferrell, "Tripping the switch fantastic: how a protein kinase cascade can convert graded inputs into switch-like outputs," *Trends in Biochemical Sciences*, vol. 21, no. 12, pp. 460-466, 1996.
[26] S. Kulkarni, S. Sharda and M. Watve, "Bi-stability in type 2 diabetes mellitus multiorgan signalling network," *PLOS One*, vol. 12, no. 8, 2017.
[27] F. Liang and D. Koya, "Acupuncture: is it effective for treatment of insulin resistance?," *Diabetes, Obesity and Metabolism*, vol. 12, pp. 555-569, 2010.
[28] A. S. Kumar, A. Maiya, B. Shashtry, K. Vaishali, N. Ravishankar, A. Hazari, S. Gundmi and R. Jadhav, "Exercise and insulin resistance in type 2 diabetes mellitus: A systematic review and meta-analysis," *Annals of Physical and Rehabilitation Medicine*, vol. 62, no. 2, pp. 98-103, 2019.
[29] L. Dey and A.S. Attele, "Alternative therapies for type 2 diabetes," in *Textbook of Complementary and Alternative Medicine*, New York, The Parthenon Publishing Group, 2003, pp. 267-278.

Biotechnology and Biological Sciences – Sen et al. (Eds)
© 2020 Taylor & Francis Group, London, ISBN 978-0-367-43161-7

Preparation and characterization of sweetened pineapple cheese ball (rasgulla)

Tanmay Sarkar, Pritha Nayak, Molla Salauddin, Sudipta Kumar Hazra & Runu Chakraborty*
Department of Food Technology and Biochemical Engineering, Jadavpur University, Kolkata, India

ABSTRACT: Sweetened Cheeseball or Rasgulla (as known in India) is a delicious sweet originated from West Bengal, India. It has nutritive value that provides several health benefits. Pineapple, the queen of fruits as termed for its excellent flavor and aroma, is one of the most important fruit crops in India. In this experiment, we have combine the values that pineapple provides as a fruit with normal rasgulla that is prepared from channa or cottage cheese. Antioxidant profiling through HPLC and sensorial attributes has been analyzed for pineapple rasgulla prepared with 10%, 20%, 30% and 50% pineapple powder (with respect to channa or cottage cheese weight) respectively and compared it with normal rasgulla available in shops selling these sweets. It has been found that the rasgulla All the pineapple rasgulla undergone same type of preparation process, the pineapple rasgulla with 20% pineapple powder got higher hedonic rating therefore higher acceptability from the panellists and also shown higher polyphenol content in HPLC. which have been lacking in standard rasgulla.

Keywords: Pineapple rasgulla, Antioxidants, Sensory evaluation, HPLC

1 INTRODUCTION

Pineapple (Ananas comosus L.), also known as the queen of fruits, is one of the most important commercial fruit crops with several health benefits [1]. It belongs to Bromeliaceae (a large family of American tropics) family having its root from South America. The fresh pineapple fruit has approximately 60% edible portion and contains 80-85% moisture, 12-15% sugars, 0.6% acid, 0.4% protein, 0.5% ash (mainly K), 0.1% fat, fibre, vitamin A, C and β-carotene and antioxidants mainly flavonoids, citric and ascorbic acid [2,3,4]. The mature fruit contains a proteolytic digestive enzyme named Bromelin, which if taken with meals aids in digesting protein by breaking proteins to amino acids. The fruit is usually consumed fresh or as squash, syrup, jelly and vinegar [5].

India is the largest single milk producing country in the world (828 million tonnes in the year 2017) [6] and almost 54% of milk produced in India is converted into traditional and western products [7]. Rasgulla or the sweet syrupy cheese ball, is one of the most popular and delicious sweets of India. It is also sold in the Indian sweet shops outside India as canned rasgulla. Rasgulla is prepared from chhana (casein mass, analogous to cottage cheese) and considered as a nutrient-dense food providing nutrients for growth, development, body maintenance and protection against major chronic diseases. As per literature review normal Rasgulla is shelf stable for 20–25 h at 25–30°C in sweet shops. However, it is likely to suffer from both hydrolytic and oxidative rancidity and undergoes several physicochemical changes during storage like any water-oil emulsion. However, antioxidants are the compounds which can delay, retard or prevent the development of rancidity in food or other flavour deterioration due to oxidation

*Corresponding author: crunu@hotmail.com

processes and also scavenge the free radicals in tissues and thus protect us from ageing, cancer, heart disease and several other diseases. Increasing consumer demand for natural products has encouraged us to select edible, nutritious fruit pineapple as safer and more effective natural antioxidant source in rasgulla [8].

In this study, our aim is to improve the nutritional quality of traditionally prepared rasgulla by incorporating freeze dried pineapple powder at different concentration levels.

2 MATERIALS & METHODS

Pineapples has been procured from Agri-Horticultural Society of India, Kolkata, India. Fresh premium quality cow milk (Fat 3.8, Solid not fat 8.7) has been collected from local dairy plant certified with HACCP and ISO 9622:2013. Sugar, wheat flour and all other ingredients were purchased from local market of Jadavpur, West Bengal, India.

2.1 *Preparation of pineapple powder*

Pulp from pineapples has been collected after peeling and coring of the fruit. It is then frozen to solid at -30°C for 12 hours in deep freezer (New Brunswick Scientific, England; Model no: C340-86) followed by freeze drying for 7-8 h in a laboratory freeze dryer (FDU 1200, EYELA, Japan) at -20°C under 0.1mbar pressure.

2.2 *Preparation of chhana and chhana balls*

The milk has been heated to boil for a few seconds and then allowed to cool to 70–75°C followed by addition of freshly prepared 1% (w/v) hot lactic acid solution (70°C). The milk has been added slowly with gentle stirring until the entire mass of milk coagulated and the contents are left undisturbed for 5–10 min. The coagulated mass (channa) is collected in a stainless steel strainer and washed in running tap water for 1–2 min. The chhana is separated from water by pressing and kneading in a vertical dough mixer (Khare & Associates, Delhi, India) resulting in a smooth dough. Several chhana additives such as wheat flour (25 g/kg of chhana) and arrowroot starch (15 g/kg of chhana), pineapple powder (in case of pineapple rasgulla 10, 20, 30, 40, and 50% w/w on the basis of raw chhana) has been added during kneading. Chhana dough as well as chhana–pineapple dough is then cut into small pieces for preparing rasgulla (sweetened cheese ball).

2.3 *Preparation of rasgulla (sweetened cheese ball)*

For the preparation of rasgulla, cooking syrup (55–60°B) and soaking syrup (35–40°B) have been prepared separately (by dissolving sugar in water). About 2-3 L of freshly clarified 55–60°B sugar syrup has been brought to boil and chhana balls as well as pineapple powder incorporated chhana balls is dropped into the boiling syrup separately. After every 2 min, 20–30 mL of hot water was added to the boiling syrup for compensating the amount of evaporated water. The balls have been kept in the boiling solution for 10 to 15 min and at the end of cooking, rasgulla balls have been transferred to a covered container containing warm soaking sugar syrup (35–40°B, temperature 60–70°C). Finally, the rasgullas have been stored in ambient condition (25±2°C) and used for further analysis.

2.4 *Sample extraction procedure*

1g of pineapple rasgulla is extracted in 30 mL mixture (3:1 ratio) of isopropyl alcohol and deionized water by ultra-sonication (Trans-O-Sonic, Mumbai) for 30 minutes.

2.5 *HPLC analysis*

Phenolic compounds have been analyzed via reversed-phase high-performance liquid chromatography (Alliance 2695 HPLC system, Waters Corporation, Massachusetts, USA). Separations were achieved using Symmetry® C-18 reversed-phase column (250 mm × 4.6 mm length, 5 μm particle size) at 30°C. Detection and quantification has been carried out with a binary pump, a dual λ absorbance UV detector 2487, inline degasser and Empower 2 software from Waters Corporation. The eluates has been detected at 278 nm using dual λ absorbance UV detector 2487. Two solvents were used for the separation in a gradient system that ran for 40 mins. Solvent A consisted of 0.5% phosphoric acid in water while Solvent B was 90% methanol. For analysis, the samples were dissolved in methanol and 10 μL was injected into the column. The elution gradient applied at a flow rate of 1 mL/min was: 100% A/0%B for 0-8 min, 70%A/30%B in 8-15 min, 50%A/50%B in 15–20 min, 40%A/60%B in 20–25 min, 30%A/70%B in 25–35 min and 100%B for 5 min until the end of the run [9].

Seventeen phenolic compounds has been considered as standard. These compounds are gallic acid, protocatechuic acid, (+)-catechin, p-coumaric acid, chlorogenic acid, caffeic acid, epicatechin, syringic acid, vanillic acid, ferulic acid, sinapinic acid, rutin, rosmarinic acid, trans-cinnamic acid, quercetin, kaempferol and apigenin. Identification and quantitative analysis has been done by comparison with standards. The amount of each phenolic compound has been expressed as mg/g of the extract.

2.6 *Sensory analysis*

The semi-trained panel for sensory analysis consisted of 30 members including 19 male and 11 female members of the institute and the evaluation has been carried out in the laboratory itself. The different rasgulla samples have been stored at 7°C and is taken out 3 h before serving. Color, mouthfeel, texture, flavour, chewiness and overall acceptability of rasgulla samples have been evaluated following nine point hedonic scale (9 = like extremely, 8 = like very much, 7 = like moderately, 6 = like slightly, 5 = neither like nor dislike, 4 = dislike slightly, 3 = dislike moderately, 2 = dislike very much, 1 = dislike extremely) [11]. All samples are presented before the panelists at room temperature under normal lighting conditions in cups coded with random, three-digit numbers. Crackers and spoons are provided to the panelists. Drinking water has been provided for oral rinsing. In each session, the panelists evaluated 3–5 samples. The average values of the sensory scores (color, taste, odor, texture and overall acceptability) have been used in the analysis.

2.7 *Statistical analysis*

Principal Component Analysis (PCA) has been performed and a dendrogram of variations with RStudio software version 3.4.4 (2018-03-15) software has been obtained.

3 RESULTS AND DISCUSSION

3.1 *HPLC analysis*

The composition of phenolic compounds present in the extract prepared from the pineapple rasgulla as well as normal rasgulla has been interpreted by the HPLC analysis. As the compounds present in the sample extract is more on the polar side, reversed phase HPLC has been chosen. Out of the 17 standard phenolic compounds around 6-7 compounds have been found depending on the composition of rasgulla, the predominant of which is dihydroxy benzoic acid. The other phenolic compounds that were detected in the chromatogram are gallic acid, catechin, vanillic acid, cholorogenic acid, dihydrocaffeic acid and rutin. Dihydroxy benzoic acid content has been found highest (0.488mg/g Fresh Weight) in the pineapple rasgulla where 20% of pineapple powder (Figure 1) is incorporated whereas the normal rasgulla contains

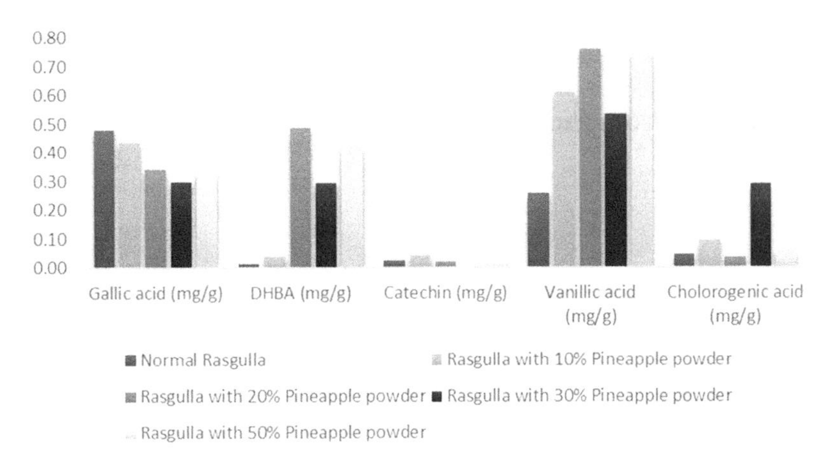

Figure 1. Graphical representation of polyphenolic compounds present in different types of pineapple rasgulla.

0.012mg/g FW. Apart from dihydroxy benzoic acid, there is another polyphenol which has been found in significant amount in pineapple rasgulla containing 30% of pineapple powder namely vanillic acid (0.532 mg/g Fresh Weight). The polyphenol content in all the samples of pineapple rasgulla to be found higher than the normal rasgulla.

3.2 *Sensory evaluation*

Average scores of sensory attributes in respect of colour, mouthfeel, texture, flavour, chewiness and overall acceptability of all the rasgulla have been reported in Table 1. Though the overall acceptability has been found best (hedonic rating 9) in case of pineapple rasgulla constituting 30% pineapple powder, rasgulla which contains 20% pineapple powder has performed better if we consider all the sensory attributes.

The PCA divulges a precise and robust correlation between sensory properties and proportion of pineapple in sweetened cheese-ball or rasgulla (Figure 2). These variations has been described by the dimension 1 of the PCA (at 77.01%), be influenced by on percentage of pineapple in sweetened cheese-ball, which affect the sensory attributes. It has been observed that mutually axis 1 and 2 showed a very high correlation (77.01% and 14.07%) among colour, mouth feel and overall acceptability. The flavour profile of sweetened pineapple cheese ball has been accommodated at the positive quadrant of dimension 1, while it is the only sensory attribute which has been occupied the negative quadrant of dimension 2 (Figure 2). It can be inferred that, the flavour attribute has contributed differently in sensory profile, from the other sensory components.

Subsequent dendrogram of variation of sensory properties and composition of sweetened cheese-ball (Figure 3), which describe about the correlations and has been, elucidated the relations between sensory properties and composition of sweetened cheese-ball more specifically

Table 1. Sensory evaluation of normal rasgulla and pineapple rasgulla samples.

Sample name	Colour	Mouthfeel	Texture	Flavour	Chewiness	Overall acceptability
Normal Rasgulla	8	7	9	5	9	8
10% Pineapple	6	6	9	6	8	7
20% Pineapple	6	7	8	7	7	8
30% Pineapple	8	8	6	8	6	9
50% Pineapple	9	9	5	9	5	8

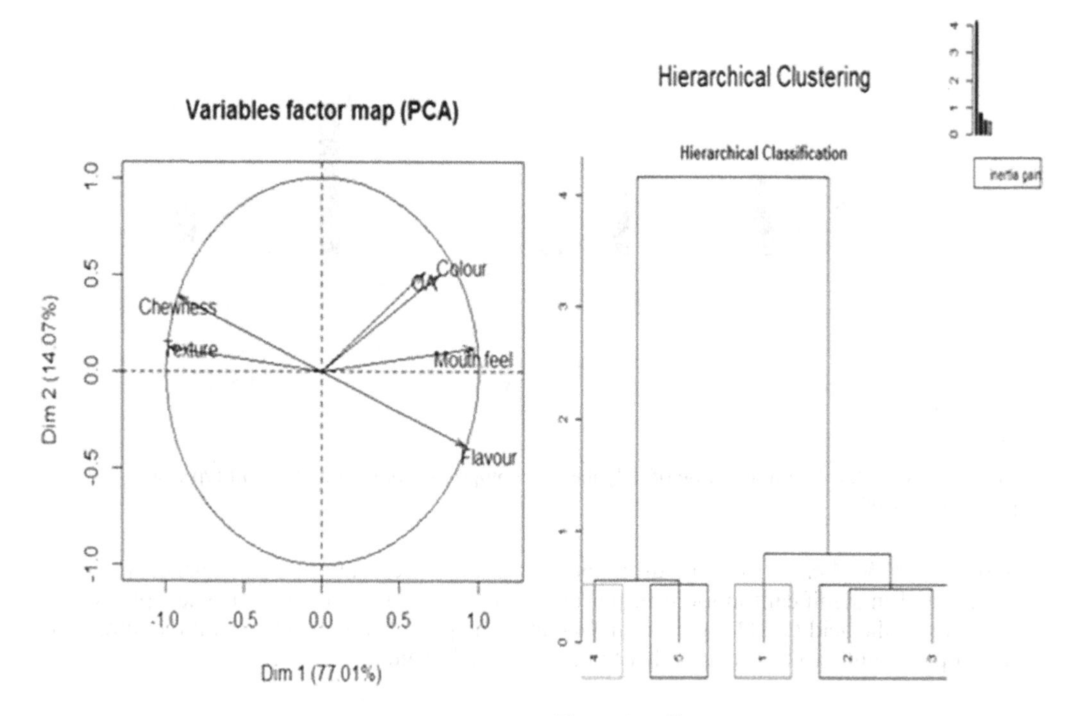

Figure 2. Factor map (PCA).

Figure 3. Cluster map.

(1-flavour, 2-mouthfeel, 3-color, 4-chewiness, 5-texture). The sweetened pineapple cheese ball with 10%, 20%, 30% and 50% freeze dried pineapple powder have been analysed for consumer sensory potentials. It has been perceived that variations of mouth-feel and texture are mostly correlated to colour, while flavour explicated separately of these correlations along with chewiness [12]. The hierarchical clustering analysis for the sensory attributes (colour, mouth feel, texture, flavour, chewiness, overall acceptability) have been studied with agglomerative or bottom-up approach, where each of the observations for sensory attributes twitches in its individual cluster and pairs of clusters are amalgamated as one travels up the hierarchy. Euclidean distance has been considered in order to achieve the appropriate metric and complete linkage clustering has been considered for linkage criteria [13, 14 and 15].

4 CONCLUSION

The pineapple rasgulla has been found as scrumptious as normal rasgulla if added to a certain concentration. Pineapple contains several nutrients that can confer to betterment of health. Pineapple rasgulla brings taste to the expectation of consumers in terms of sensorial attributes and also provides several antioxidants that are higher (invitro) in quantity that of a normal rasgulla. In our experiment, it is seen that 20% pineapple powder containing rasgulla has polyphenols such as dihydroxy benzoic acid that are of higher quantity than that of normal rasgulla and also accepted by the panelists. There are polyphenols like rutin, quercetin, and kaempferol also present in considerable amount in pineapple rasgulla.

ACKNOWLEDGEMENT

None.

REFERENCES

[1] Hossain, M.F., Shaheen, A., Anwar, M. (2015). Nutritional Value and Medicinal Benefits of Pineapple. International Journal of Nutrition and Food Sciences, 4(1), 84-88.

[2] Silva, D.I.S., Nogueira, G.D.R., Duzzioni, A.G., Barrozo, M.A.S. (2013). Changes of antioxidant constituents in pineapple (Ananus comosus) residue during drying process. Industrial Crops and Products, 50, 557-562.

[3] Samson, J.A. (1980). Tropical Fruits. Longman Group Limited, London.

[4] Offia Olua, B.I., Edide, R.O. (2013). Chemical, Microbial and Sensory Properties of Candied-Pineapple and Cherry Cakes. Nigerian Food Journal, 31(1), 33-39.

[5] Sarkar, T., Nayak, P., Chakraborty, R. (2018). Pineapple [Ananas comosus (L.)] Product processing techniques and packaging: A review. The IIOAB Journal, 9(4), 6-12.

[6] FAO (2017). www.fao.org

[7] Vaswani, L.K. (2002). Market survey and analysis. In Technology of Indian Milk Products, 23. Aneja RP, Mathur BN, Chandan RC and Banerjee AK, eds. New Delhi: India Dairy India publication.

[8] Bandyopadhyay, M., Chakraborty, R., Raychaudhuri, U. (2008). Effect of carrot on quality improvement of sweet syrupy cheese ball (Rasgulla). International Journal of Dairy Technology, 61 (3), 290-299.

[9] Saha, S. K., Dey, S., Chakraborty, R. (2019). Effect of microwave power on drying kinetics, structure, color, and antioxidant activities of corncob. Journal of Food Process Engineering, e13021.

[10] Basu, S., Shivhare, U.S., Singh, T.V., Beniwal, V.S. (2011). Rheological, textural and spectral characteristics of sorbitol substituted mango jam. Journal of Food Engineering, 105, 503-512.

[11] Ferreira, E.A., Siqueira, H.E., Boas, E.V. V., Hermes, V.S., Rios, A.D.O. (2016). Bioactive compounds and antioxidant activity of pineapple fruit of different cultivars. Rev. Bras. Frutic, 38(3).

[12] Mahieddine, B., Amina, B., Faouzi, S.M., Sana, B. (2018). Effects of microwave heating on the antioxidant activities of tomato (Solanum lycopersicum). Annals of Agricultural Sciences, 63(2), 135-139.

[13] Ghosh, D., & Chattopadhyay, P. (2012). Application of principal component analysis (PCA) as a sensory assessment tool for fermented food products, Journal of food science and technology, 49 (3), 328-334.

[14] Šnirc,M., Kral, M., Ošťádalová, M., Golian, J., & Tremlová, B. (2017). Application of principal component analysis method for characterization chemical, technological, and textural parameters of farmed and pastured red deer, International Journal of Food Properties, 20(4), 754-761.

[15] Zarantonello, L. & Schmitt, B. J Brand Manag (2010). Using the brand experience scale to profile consumers and predict consumer behaviour. Journal of Brand Management, 17(7), 532-540.

Biotechnology and Biological Sciences – Sen et al. (Eds)
 ISBN 978-0-367-43161-7

A comparative analysis of how biochemically varied commonly consumed allergic foods are

S. Sengupta[#], A. Roy[#], N. Chakraborty & M. Bhattacharya*
Department of Biotechnology, Techno India University, India

ABSTRACT: Food allergies are manifested by a wide variety physiological responses including fatality. They have also been found to have correlation with various types of Non-communicable diseases (NCDs). Thus, delineation of the molecular basis of food allergies is important for getting a better understanding of their mechanism of action as well as their role in aggravating NCDs. This study looked into variations in various biochemical parameters of a few food items which are consumed regularly and are also known to cause allergic reactions. We observed significant variations in these parameters among equivalent quantities of samples of food materials.

Keywords: Food allergy, Biochemical characteristics, Differential allergic reactions, Non-communicable disease

1 INTRODUCTION

Food allergy is turning out to be one of the vital causes of death in all countries irrespective of the socio-economic status of the people. The extent of mortality from food allergens has increased several folds in the last decade. This has brought research on food allergy and its causative properties to the front. Human beings have the natural power to fight with an allergen, ranging from viruses to food. But in the case of immune-compromise or some other defects in the immune system, individuals can react against some food allergic component(s). Various studies have indicated that cereal or grain particles are more allergic materials than fruits or vegetables.

2 MATERIALS & METHODS

In this study, a few food materials (cucumber, peanut, papaya and yoghurt) known to elicit allergic reactions in some individuals were selected and comparative analysis was performed in an attempt to understand the basis of their differential responses.

2.1 *Determination of presence of sugar in samples*

Benedict's Test was used to test for simple carbohydrates. The Benedict's test identifies reducing sugars (monosaccharide's and some disaccharides), which have free ketone or aldehyde functional groups. For performing this test, approximately 1 ml of sample was placed into a clean test tube. 2 ml (10 drops) of Benedict's reagent (CuSO4) was added to it. The solution was then heated in a boiling water bath for 5 minutes and color change was observed in the solution of test tubes. The observed data were collected and tabulated.

[#]Contributed equally
*Corresponding author

2.2 *Determination of presence of vitamin C in samples*

In this test, a blue substance called 2, 6-dichlorophenolindophenol (or DCPIP for short) acts as an indicator. Its colour changes from blue to red with acids but is lost in the presence of certain chemicals, one of which is ascorbic acid (vitamin C). Thus, DCPIP solution can be used to test for the presence of vitamin C in foods. For performing this test, a small amount of the sample was put into a test tube to a depth of about 2cm. An equivalent similar amount of distilled water was added to it and the mixture was stirred with a glass rod. Next, it was allowed to stand for a few minutes. Subsequently, a small amount of the clear liquid was transferred into to a test tube and DCPIP solution was added to it dropwise. If the extract is acid, the colour will change from blue to red. More DCPIP solution was added to check whether the colour disappeared altogether. Decolourisation of DCPIP showed that a vitamin C is probably present. Other chemicals can do this in food and drink, but vitamin C is the main one.

2.3 *Determination of presence of starch in samples*

Iodine solution is used to test for the presence of starch. If starch is present in a food item, it turns an intense blue-black colour upon addition of aqueous solutions of tri iodide anion, due to the formation of an intermolecular charge-transfer complex. In the absence of starch the brown colour of the aqueous solution remain same. This interaction between starch and tri iodide is also the basis for iodometry. For performing this test, the test tubes containing samples were placed in a water filled beaker and the beaker was placed over a hot plate. The mixture was heated for 5 minutes while continuously stirring the water in beaker with a glass rod. The solutions were filtered from one test tube to another using filter papers. 4 clean test tubes were taken and some of the filtrate was poured into it. A few drops of iodine solution was taken using a dropper and added to the filtrates and color change was observed.

2.4 *Determination of presence of vitamin C in samples*

2 ml of ethanol was added to the food sample and the mixture was shaken well. It was allowed to settle in a test tube rack for 2 minutes for the food to dissolve in ethanol. Any clear liquid was emptied into a test tube containing 2 ml of distilled H2O. Appearance of a milky-white emulsion indicated a positive result (presence of lipid). If the mixture remained clear, it indicated that no fats are present in the sample.

3 RESULTS & DISCUSSION

The studies indicated difference in various biochemical parameters between equivalent quantities of the food samples.

3.1 *Determination of presence of sugar, vitamin C, starch and lipids*

All the samples were tested for presence of sugar, vitamin C, starch and lipids in their contents. All food samples were found to be rich source of starch and sugar, whereas vitamin C was not found in peanut. Lipids were not detected in papaya. Figure 1 summarizes the findings. Figure 2 shows photographs of the tests.

Result Of Qualitative Analysis Of The Samples

Sample	Sugar Test	Vit C	Starch	Lipid Test
Cucumber (CU)	✓	✓	✓	✓
Peanut (P)	✓	✗	✓	✓
Papaya (PA)	✓	✓	✓	✗
Yoghurt (Y)	✓	✓	✓	✓

Figure 1. Summary of qualitative analysis (presence/absence) of the samples for detection of sugars, vitamin C, starch and lipids.

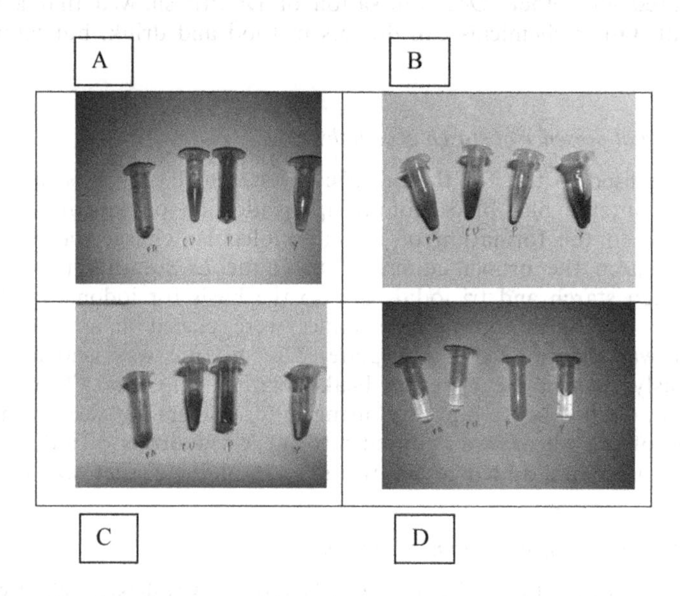

Figure 2. Photographs of qualitative analysis of the samples for detection of (A) sugars, (B) vitamin C, (C) starch and (D) lipids.

4 CONCLUSION

Our study has revealed differential levels of nutrient contents between equivalent quantities of the samples of allergic food materials. These findings can be used for further research on the underlying mechanisms of their action.

ACKNOWLEDGMENT

The authors are thankful to Chancellor, Techno India University, West Bengal for providing the necessary infrastructural facilities.

REFERENCES

Bousquet, J., Lockey, R. & Malling, H. J. (1998) J Allergy Clin Immunol, 102(4 Pt 1), 558.

D' Auria, E., Mameli, C., Piras, C., Cococcioni, L., Urbani, A., Zuccotti, G. V. & Roncada, P. (2018) J Proteomics, pii: S1874-3919(18)30049-6. doi: 10.1016/j.jprot.2018.01.018. [Epub ahead of print] PubMed PMID: 29408543.

Devdas, J. M., Mckie, C., Fox, A. T. & Ratageri, V. H. (2017) Indian J Pediatr, doi: 10.1007/s12098-017-2535.
Dye, C. (2014) Philos Trans R Soc Lond B Biol Sci, 369(1645):20130426.
Gilles, S. & Traidl-Hoffmann, C. (2014) J Dtsch Dermatol Ges, 12(5), 395.
Li, D. (2014) J Sci Food Agric, 94(2), 169.
Miranda, J. J., Kinra, S., Casas, J. P., Davey Smith, G. & Ebrahim, S. (2008) Trop Med Int Health, 13 (10), 1225.
Sicherer, S. H. & Sampson, H. A. (2014) J Allergy Clin Immunol, 133(2), 291.
Upadhyay, R. P. (2012) Iran J Public Health, 41(3), 1.
van Ree, R., Poulsen, L.K., Wong, G.W., Ballmer-Weber, B. K., Gao, Z. & Jia, X. (2015) Zhonghua Yu Fang Yi Xue Za Zhi, [Chinese] 49(1):87.

Biotechnology and Biological Sciences – Sen et al. (Eds)
© 2020 Taylor & Francis Group, London, ISBN 978-0-367-43161-7

Phenotypes of vitamin D binding protein in healthy sybjects of Kuwait

Abu S. Mustafa
Department of Microbiology, Faculty of Medicine, Kuwait University, Kuwait

Suhail Al Shammari & Arpita Bhattacharya
Department of Medicine, Faculty of Medicine, Kuwait University, Kuwait

ABSTRACT: This study was undertaken to determine the major phenotypes of Vitamin D Binding Protein (VDBP) in Kuwaiti population. The genomic regions containing the nucleotide bases responsible for two point mutations (T to G and C to A) in the VDBP gene were amplified in vitro and sequenced. The sequence data were used to deduce the VDBP phenotypes from the genotypes. The results showed the presence of four VDBP phenotypes (Gc1S, Gc1F, Gc2 and GcA) in Kuwaiti nationals. Three of the phenotypes (Gc1S, Gc1F and Gc2) have been described in the literature in other populations, e.g. Africans, Asians and Caucasians etc., but GcA is unique and it has not been described in those populations.

1 INTRODUCTION

Vitamin D binding protein (VDBP, 474 aa in length)) is the primary carrier of vitamin D and its metabolites in the human blood. The VDBP exists into three major phenotypes due to two single nucleotide polymorphisms in exon 11 (1). The resulting VDBP phenotypes are known as Gc1F, Gc1S and Gc2 (2). Significant differences in frequencies of VDBP phenotypes are reported in various populations (3). The people living in northern climates possess more Gc1S and less Gc1F phenotype, and the frequency of Gc2 is higher in Caucasians but rare in Africans (2, 3). However, the frequency of VDBP phenotypes is not known in Kuwaiti population. The aim of this study was to determine the frequency of VDBP phenotypes in healthy Kuwaiti nationals.

2 MATERIALS & METHODS

Genomic DNA were isolated from the blood of 127 randomly selected adult healthy Kuwaiti nationals using DNA isolation kits according to standard procedures (4). The quality and quantity of isolated genomic DNA were determined using a low volume (2 μl sample) Epoch Microplate Spectrophotometer (BioTek Instruments, Inc., Winooski, VT, USA) and measuring the absorbance at 260 nm and 280 nm (5).

A 596 bp DNA region of the VDBP gene (6), covering the two specific mutation sites in exon 11, giving rise to three known Gc phenotypes (Gc1F, Gc1S and Gc2), were amplified by polymerase chain reaction (PCR) using forward (F) (5'-gatctcgaagaggcatgtttc-3') and reverse (R) (5'-gttgcctgtgttcacagactc-3') primers. These primers annealed at the sites in the VDBP gene, which encompassed the two specific mutation sites, i.e. T and C at the locations 35706 and 35717 in the VDBP gene, respectively (Figure 1).

*Corresponding author: Abu S. Mustafa

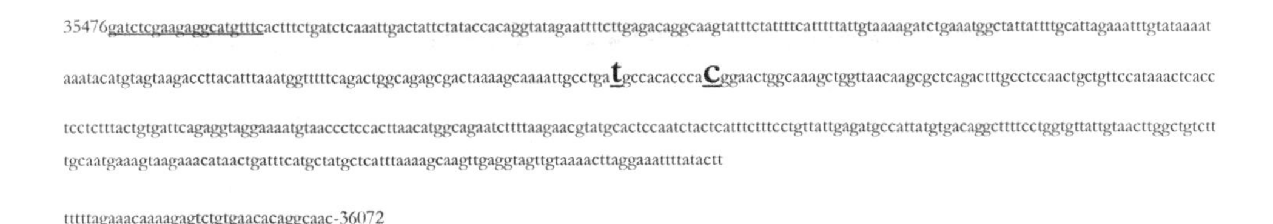

Figure 1. The forward and reverse primer annealing sites in the VDBP gene are underlined at the beginning and end of the 596 bp sequence. The mutation sites (t and c) in the sequence giving rise to Gc phenotypes are marked in bold and underlined.

Table 1. The possible alleles at the VDBP gene locations 35706 and 35717 and the resulting genotypes and phenotypes.

Alleles at Location 35706	Alleles at Location 35717	Genotypes	Phenotypes
T/T	C/C	T-C/T-C	Gc1F/Gc1F
T/T	A/A	T-A/TA	Gc2/Gc2
T/T	C/A	T-C/T-A	Gc1F/Gc2
T/G	C/C	T-C/G-C	Gc1F/Gc1S
T/G	A/A	T-A/G-A	Gc2/GcA
T/G	C/A	T-C/T-A/G-C/GA	Gc1F/Gc2/Gc1S/GcA
G/G	C/C	G-C/G-C	Gc1S/Gc1S
G/G	A/A	G-A/G-A	GcA/GcA
G/G	C/A	G-C/G-A	Gc1S/Gc1A

PCR reactions were performed in 0.2 ml microtubes with 50-µl of reaction mixture containing 100 ng of genomic DNA, 250 µM of each dNTP, 10 mM tris–HCl (pH 8·3), 50 Mm KCl, 2 mM MgCl2, 2·5 units of *AmpliTaq* Gold® DNA polymerase (Perkin-Elmer Applied Biosystem) and 25 pmol of each F and R primer. PCR cycles were performed with an initial denaturation step of 10 min at 95°C for the activation of Ampli-*Taq*Gold, followed by 30 cycles of 94°C for 30 seconds, 60°C for 30 seconds, and 72°C for 30 seconds with a final extension step of 72°C for 5 minutes. The PCR amplified DNA were analysed for the presence of 596 bp specific product by agarose gel electrophoresis using standard procedures (7).

The PCR amplified products were sequenced by Sanger method using an automated DNA Sequencing System (ABI 3130xl Genetic Analyzer) (8, 9). The obtained DNA sequences were searched for the specific mutations (T to G and C to A at VDBP gene locations 35706 and 35717, respectively) using the in-built software within the Genetic Analyzer. The VDBP phenotypes were deduced on the basis of the presence of the genotypes (Table 1). The frequencies of VDBP phenotypes were determined from the mutation data.

3 RESULTS & DISCUSSION

The results of PCR, followed by agarose gel electrophoresis, showed that the product of specific size (596 bp DNA) was amplified from the genomic DNA of all individuals (Figure 2, results shown for 14 samples).

The sequencing of PCR amplified DNA was successful in all cases to identify the nucleotide bases present in the VDBP gene at the specific mutation sites, i.e. locations 35706 (T to G) and 35717 (C to A) in the VDBP gene. A representative result is shown in Figure 3. The results in this figure show that both alleles at position 35706 are homozygous and have G/G (position 321 in the given sequence), whereas the alleles at position 37717 are heterozygous and have C/A (position 332 in the given sequence), which give rise to genotypes G-C/G-A and phenotypes Gc1S/GcA, as given in Table 1.

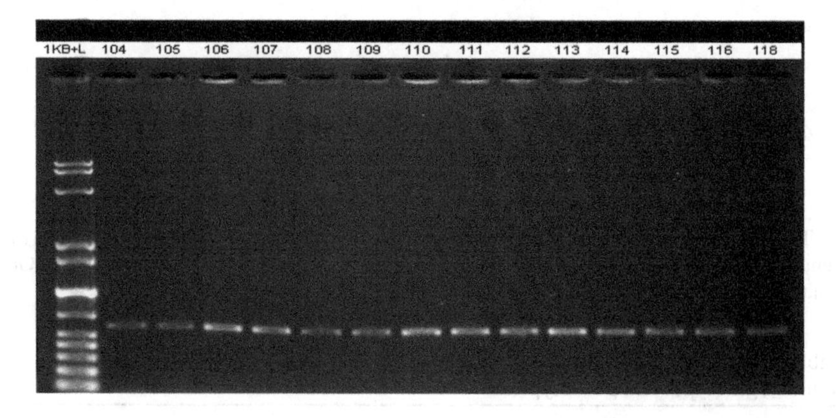

Figure 2. Agarose gel electrophoresis results of 14 samples from an equal number of healthy Kuwaiti nationals. The first lane on the left side represents 100 bp DNA ladder. The remaining 14 wells are showing the PCR amplification product (596 bp DNA) from genomic DNA samples from 14 different individuals.

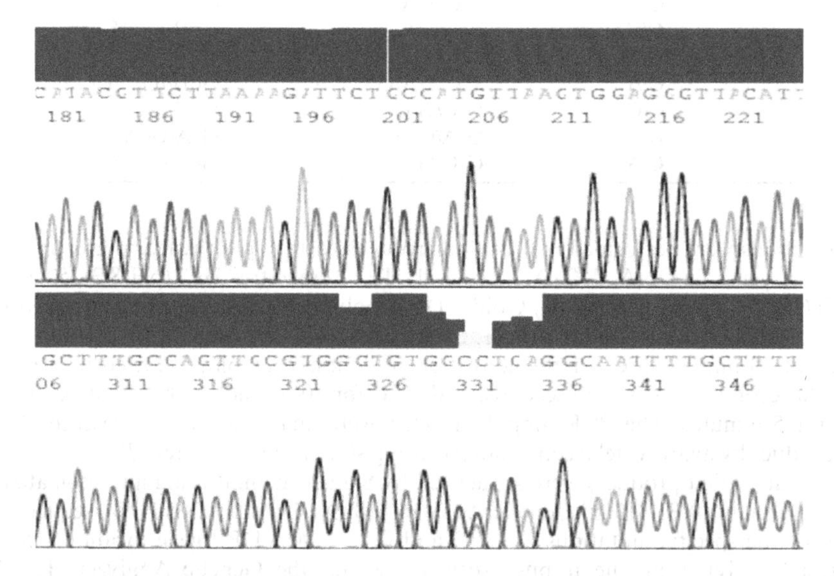

Figure 3. DNA sequencing results of a representative sample with genotype G-C/GA and phenotype Gc1S/Gc3. The sites for mutation in the above sequence are at position 321 (G/G) and 332 (C/A).

The sequencing of PCR products from all samples included in this study lead to the identification of four different VDBP phenotypes, i.e. Gc1F, Gc1S, Gc2 and GcA in the Kuwaiti population based on the detection of mutations at the specific sites in exon 11 of the VDBP gene (Table 2).

The overall results showed that Gc1S (n=106, 83.5%) was the most common phenotype among the population studied. The other phenotypes were Gc1F (n=80, 63.0%), Gc2 (n=47, 37.0%) and Gc3 (n=31, 24.4%) (Table 2). Considering the possibility of four different alleles in every individual based on two mutations in each chromosome, Gc1F, Gc1S and Gc2 occurred in both homozygous and heterozygous combinations, whereas GcA was found only in heterozygous combinations (data not shown). The identification of Gc1F, Gc1S and Gc2 phenotypes has been reported previously in various African, Asian and Caucasian populations (2, 3, 10 - 14) using kits containing specific probes to detect PCR amplified DNA or restriction

Table 2. Gc phenotypes and their FREQUENCIES IN Kuwaiti population.

Gc phenotype	Frequency
Gc1S	106/127 (83.5%)
Gc1F	80/127 (63.0%)
Gc2	47/127 (37.0%)
GcA	31/127 (24.4%)

enzyme digestion of the PCR products. However, to our knowledge, this is the first study that used DNA sequencing of PCR products and identified GcA as a unique phenotype in Kuwaiti population.

4 CONCLUSION

Kuwait nationals express four phenotypes of VDBP gene, i.e. Gc1F, Gc1S and Gc2 and GcA. The GcA phenotype is unique and has not been reported in African, Asian and Caucasian populations.

ACKNOWLEDGMENT

The present research work was supported by Kuwait University, Grant No. MM02/17.

REFERENCES

[1] Malik, S., Fu, L., Juras, D.J., Karmali, M., Wong, B.Y., Gozdzik, A., Cole, D.E. (2013). Common variants of the vitamin D binding protein gene and adverse health outcomes. Crit Rev Clin Lab Sci. 5, 1-22.

[2] Bouillon, R. (2017). Genetic and racial differences in the vitamin D endocrine system. Endocrinol Metab Clin N Am. 46, 1119–1135.

[3] Braithwaite, V.S., Jones, K.S., Schoenmakers, I., Silver, M., Prenti, A., Hennig, B.J. (2015). Vitamin D binding protein genotype is associated with plasma 25OHD concentration in West African children. Bone. 74, 166-70.

[4] Nasser, K., Mustafa, A.S., Khan, M.W., Purohit, P., Al-Obaid, I., Dhar, R., Al-Fouzan, W. (2018). Draft genome sequences of six multidrug-resistant clinical strains of *Acinetobacter baumannii*, isolated at two major hospitals in Kuwait. Genome Announc. 6, e00264-18.

[5] Khare P, Raj V, Chandra S, Agarwal S. (2014). Quantitative and qualitative assessment of DNA extracted from saliva for its use in forensic identification. J Forensic Dent Sci. 6, 81-5.

[6] Witke WF., Gibbs PE., Zielinski R, Yang F, Bowman BH, Dugaiczyk A. (1993). Complete structure of the human Gc gene: differences and similarities between members of the albumin gene family. Genomics. 16, 751-4.

[7] Mustafa AS, Abal AT, Chugh TD, 1999, Detection of *Mycobacterium tuberculosis* complex and non-tuberculous mycobacteria by multiplex polymerase chain reactions. East Mediterr Health J. 5, 61-70.

[8] Mustafa, AS, Habibi, N., Osman A., Shaheed, F., Khan, M.W. (2017). Species identification and molecular typing of human Brucella isolates from Kuwait. PLoS One. 12, e0182111.

[9] Shaban K, Amoudy HA, Mustafa AS (2013). Cellular immune responses to recombinant *Mycobacterium bovis* BCG constructs expressing major antigens of region of difference 1 of *Mycobacterium tuberculosis*. Clin Vaccine Immunol. 20, 1230-37.

[10] Cleve H., Constans J. (1988). The mutants of the Vitamin-D-binding protein: more than 120 variants of the GC/DBP system. Vox Sang. 54, 215-25.

[11] Lafi Z.M., Irshaid Y.M., El-Khateeb M., Ajlouni K.M., Hyassat D. (2015). Association of rs7041 and rs4588 Polymorphisms of the Vitamin D Binding Protein and the rs10741657 Polymorphism of CYP2R1 with Vitamin D Status Among Jordanian Patients. Genet Test Mol Biomarkers. 19, 629-36.

[12] Park Y., Kim Y.S., Kang Y.A., Shin J.H., Oh Y.M., Seo J.B., Jung J.Y., Lee S.D. (2016). Relationship between vitamin D-binding protein polymorphisms and blood vitamin D level in Korean patients with COPD. Int J Chron Obstruct Pulmon Dis. 11, 731-8.
[13] Al-Azzawi M.A., Ghoneim A.H., Elmadbouh I. (2017). Evaluation of Vitamin D, Vitamin D Binding Protein Gene Polymorphism with Oxidant - Antioxidant Profiles in Chronic Obstructive Pulmonary Disease. J Med Biochem. 36, 331-40.
[14] Zhou J.C., Zhu Y., Gong C., Liang X., Zhou X., Xu Y., Lyu D., Mo J., Xu J., Song J., Che X., Sun S., Huang C., Liu X.L. (2019). The *GC2* haplotype of the vitamin D binding protein is a risk factor for a low plasma 25-hydroxyvitamin D concentration in a Han Chinese population. Nutr Metab (Lond). 14, 16:5.

Biotechnology and Biological Sciences – Sen et al. (Eds)
© 2020 Taylor & Francis Group, London, ISBN 978-0-367-43161-7

Structure, salt-bridge's energetics and microenvironments of nucleoside diphosphate kinase from halophilic, thermophilic and mesophilic microbes

D. Mitra & A.K. Bandyopadhyay*
Department of Biotechnology, University of Burdwan, Burdwan, India

R.N.U. Islam
Department of Zoology, University of Burdwan, Burdwan, India

S. Banerjee
Department of Biological Sciences, Indian Statistical Institute, Kolkata, India

S. Yasmeen
Department of Botany & Microbiology, Acharya Nagarjuna University, Guntur, India

P.K.D. Mohapatra*
Department of Microbiology, Raiganj University, Raiganj, India

ABSTRACT: The fundamental principle that maintains the stability of homologous proteins, operating under diverse environmental conditions, is not fully clear. Here, we studied Nucleosdie Diphosphate Kinase (NDK) from halophilic (hNDK) and thermophilic (tNDK) archaea in reference to mesophilic (mNDK) bacteria. Results show that relative to mNDK, salt-bridge partners and its energetics is more preferred in hNDK and tNDK. Notably, the intrinsic microenvironment (MIE) features also play a prominent role in the latter cases.

Keywords: Nucleoside Diphosphate Kinase, Salt Bridge Energetics, Microenvironment, Halophilic, Thermophilic, Mesophilic

1 INTRODUCTION

Unlike the normal or mesophilic microbes, halophilic and thermophilic archaea operate their biochemical machinery in high salt (~4.5M NaCl) and in high temperature (~100 °C) respectively (Rothschild & Mancinelli, 2001; Bandyopadhyay & Sonawat, 2000; Bandyopadhyay, 2015; Tokunaga et al. 2006). Specific preference of amino acid residues in sequences (Rao & Argos, 1981) and their distribution in the core and surface of structures (Dym et al. 1995; Britton et al. 1998) of halophiles and thermophiles have been correlated with the extremophilic stability of these proteins. What is the role of these extreme environmental factors? It has been demonstrated in vitro that unlike mesophiles, proteins from these microbes maintain the stability only when the solution conditions are similar as their ecosystems. Removal of high salt and reduction of temperature induce unfolding of halophilic and thermophilic proteins respectively (Wrba et al. 1990; Bandyopadhyay et al. 2001). The observation of molten globule state in the unfolding path of halophilic ferredoxin demonstrated that salt is obligatory for the stability of the protein (Bandyopadhyay et al. 2007).

*Corresponding author: E-mail: akbanerjee@biotech.buruniv.ac.in; pkdmvu@gmail.com

It has been shown that like thermophiles (Kreil & Ouzounis, 2001); halophilic proteins also utilize salt bridges for the maintenance of the stability in high salt (Bandyopadhyay et al. 2019b; Nayek et al. 2014). Poisson and Boltzmann Equation (PBE) and its solver (Baker et al. 2001;

Dolinsky et al. 2004) has been used for the determination of net ($\Delta\Delta G_{net}$) and component energy terms (desolvation: $\Delta\Delta G_{dslv}$, background: $\Delta\Delta G_{bac}$, bridge: $\Delta\Delta G_{brd}$) of salt bridge following Isolated Pair Method (IPM) (Kumar et al. 2000; Hendsch & Tidor, 1994). Recently, Bandyopadhyay et al (2019a) demonstrated that the Network Unit Method (NUM) is essential for the energetics of network salt bridges (Bandyopadhyay et al. 2019a). In this aspect, the contribution of intrinsic microenvironment (MIE) of partners of salt bridge remains to be worked out.

Here, we present detailed analyses on salt bridges of nucleoside diphosphate kinase (NDK) from halophilic (hNDK) and thermophilic (tNDK) archaea in reference to that of mesophilic (mNDK) bacteria. Application of NUM and role of MIE has also been highlighted in this work. Taken together, our comparative studies on energetics and MIE of salt bridges of NDKs seem to have potential applications in structural biology and protein engineering.

1 MATERIALS & METHODS

2.1 *Dataset*

Sequences of hNDK, tNDK and mNDK (n~100) were procured from UniProt database. Representative high-resolution structure of hNDK (2AZ3 of *Halobacterium salinarum*), tNDK (2CWK of *Pyrococcus horikoshii*) and mNDK (2HUR of *Escherichia coli*) were extracted from the Research Collaboratory for Structural Bioinformatics (RCSB), Protein Data Bank (PDB) (Berman et al. 2000).

2.2 *Physicochemical, sequence and substitution properties*

Separate FASTA files (n~100) of hNDK, tNDK and mNDK were subjected for the preparation of BLOCK using ABPT tool of PHYSICO2 (Banerjee et al. 2015). Analysis of sequence either in BLOCK or non-BLOCK format was performed as earlier (Gupta et al.2014a). Using frequency of homo, hetero pairs of BLOCKs, non-conservative to conservative substitution ratio (NCS:CS), most conserve (MCR) and diverse (MDR) residues and dominant hetero-pair (DHP) of BLOCKs were computed using APBEST (Gupta et al. 2017b). Mean relative abundance (MRA) was computed from the mean value of a given property of sequences of halophile/thermophile ($\overline{v}_{h/t}$) relative to mesophile ($\overline{v}_m$).

$$MRA = \frac{\left(\overline{v}_{h/t} - \overline{v}_m\right)}{\overline{v}_m}$$

2.3 *Structure, salt bridge, energetics and microenvironment*

All PDBs were energy minimized prior to any structural analysis using AUTOMINv1.0 if not mentioned otherwise (Islam et al. 2018). Core and surface compositions of the monomeric form of 2AZ3, 2CWK and 2HUR were extracted using COSURIM (Gupta et al. 2017a). Details of salt bridges and their energetics were computed as earlier (Gupta et al. 2014b; Gupta et al. 2015). For network salt bridges, energetics was computed both by IPM (Nayek et al. 2015) and by NUM (Bandyopadhyay et al. 2019a) for comparison purpose. The latter and the MIEs of salt bridge were computed using an in-house automated procedure.

2 RESULTS & DISCUSSION

2.1 *Acidic and basic residues alter the sequence but not the structural properties*

To check the difference of homologous positions of sequence of hNDK and tNDK from that of mNDK, we analyzed sequences in BLOCK format, the result of which are shown in Figure 1 (a-g).

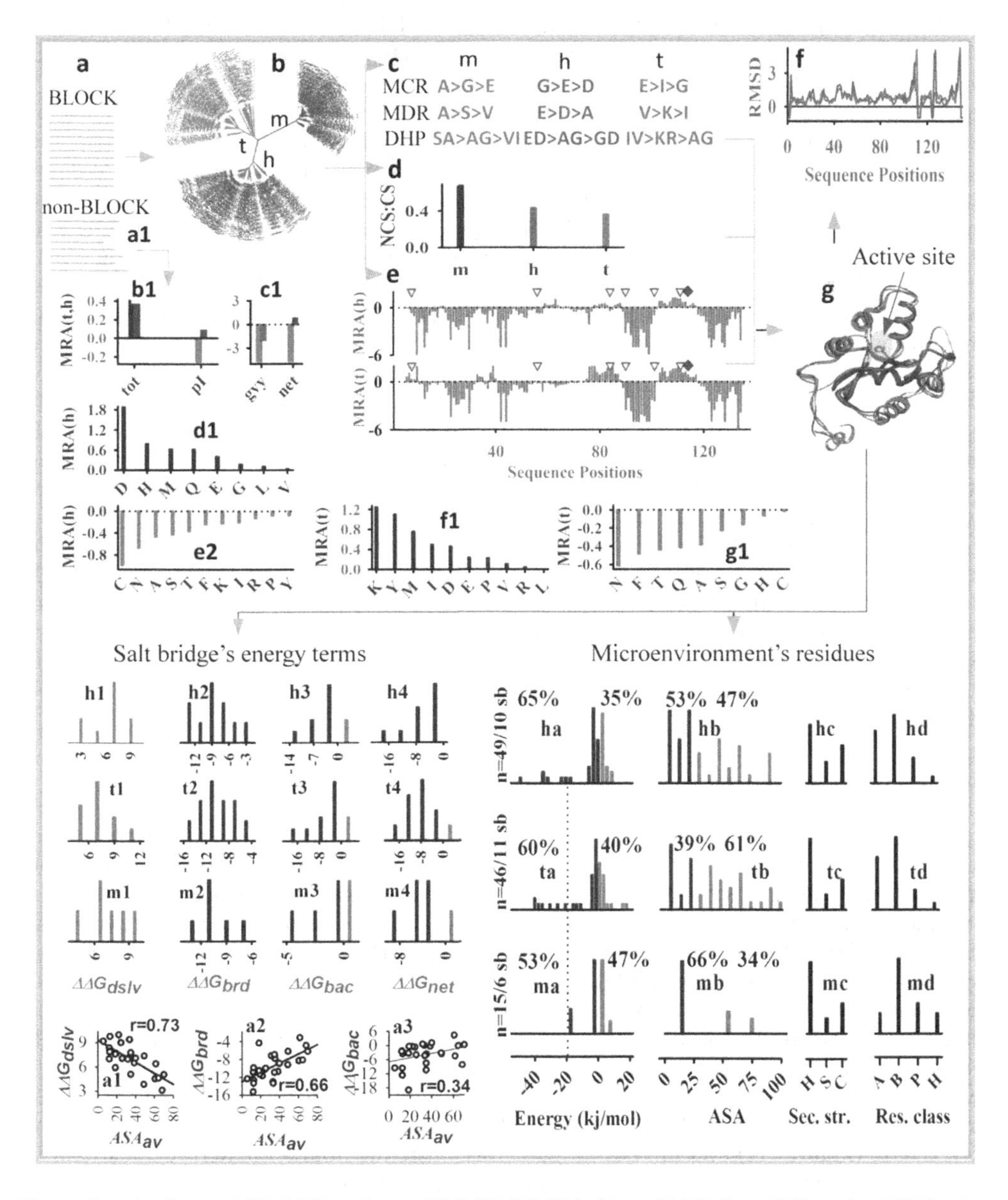

Figure 1. Analyses of BLOCK and non-BLOCK FASTA files of hNDK, tNDK and mNDK using phylogenetic (b), APBEST (c, d), PHYSICO2 (e, b1, c1, d1, e1, f1 and g1) methods. Per-residue-RMSD (f) and structural superposition (g) analysis was done using VMD interface. Distribution of component and net energy terms of salt bridge of hNDK (h1-h4), tNDK (t1-t4) and mNDK (m1-m4) are shown with correlation plots of component energy terms against average accessibility in a1 for $\Delta\Delta G_{dslv}$, a2 for $\Delta\Delta G_{brd}$ and a3 for $\Delta\Delta G_{bac}$ respectively. Distribution of energy of interaction (ha for hNDK, ta for tNDK and ma for mNDK), ASA (hb, tb and mb), secondary structure (Sec. str.; hc, tc and mc) and residue composition (Res. class; hd, td and md) of microenvironment's candidates.

The observation of distinct clades indicates that hNDK and tNDK are not only distinct from mNDK, but also differ from each other (Figure 1b). How these BLOCKs have evolved? While in each case, E (GLU) and G (GLY) have been common as most conserve residue (MCR), the third MCR shows variation (Figure 1c). In hNDK, it is acidic (D) and in tNDK, it is hydrophobic (I). Further, most diverse residue (MDR) in mNDK, hNDK and tNDK is polar/hydrophobic, acidic/hydrophobic and basic/hydrophobic respectively (Figure 1c). The dominant hetero-pair (DHP, i.e. most exchangeable) is largely constituted by acidic/polar, polar/hydrophobic and basic/hydrophobic residues in hNDK, mNDK and tNDK respectively (Figure 1c). It is seen that NCS:CS is far greater in mNDK than that of hNDK and tNDK. The lower ratio in the latter would mean the divergences are restricted (Gupta et al. 2017b), might be due to associated constraints of the extreme environments. Is there a preference for specific amino acid in hNDK and tNDK, relative to mNDK? We see that while acidic and basic residues have positive MRA for hNDK (Figure 1d1) and tNDK (Figure 1f1), polar residues (N, S & T) show negative MRA (Figure 1e2 & g2). As MRA for acidic and basic residues are high, total-charge (tot) and GRAVY (gvy) are also high and low in hNDK and tNDK respectively (b1 & c1). Further and interestingly, while isoelectric value (pI) and net-charge (net) decrease in hNDK due to relatively more preference for acidic over basic residues (see above), these properties show an increase in tNDK (b1 & c1) due to relatively more bias for basic over acidic residues. Nevertheless, higher MRA for both the acidic and selective basic residues in hNDK and tNDK raising the possibility of a concerted role of these amino acids. In fact, we show later that hNDK and tNDK in contrast to mNDK, employ these oppositely charged residues for the formation of highly stabilizing, isolated and network salt bridges. How the amino acid's bias affects the sequence and structure properties of these extreme proteins? Here, we have plotted MRA of Kyte-Doolittle hydrophobicity of hNDK and tNDK (Figure 1e). Interestingly, it is seen that although the topology of hNDK and tNDK is almost identical (Figure 1g) as judged by the per-residue-RMSD plot (Figure 1f), their sequences show an increase in the positional polarity almost identically (Figure 1e). Such an increase of the positional polarity is unlikely to be due to polar residues (i.e. N, S & T) as they have negative MRA (e1 & g1). It, therefore, seems that the acidic and basic residues in hNDK and tNDK are the major player for such an alteration. Is this effect primary or secondary? To check this, we have investigated salt bridge, its energetics and MIEs in details (see below).

2.2 *hNDK and tNDK, unlike mNDK, are stabilized by excess of engineered salt bridges*

In hNDK and tNDK, the frequencies of salt bridges are higher than that of mNDK. Further, the network salt bridge that neutralizes the isolated charge and polar residues especially in the core and thereby imparts stability to the protein, are almost absent in mNDK. Out of 10 (3 in core and 7 in the surface) salt bridges in hNDK (2AZ3_C), 6 pairs are forming 3 network salt bridges. In tNDK (2CWK_B), out of 11 pairs (2 in core and 9 in the surface), 5 pairs are forming 2 network salt bridges. In mNDK (2HUR_A), there are only 6 isolated types of salt bridge. The component ($\Delta\Delta G_{dslv}$, $\Delta\Delta G_{bac}$ and $\Delta\Delta G_{brd}$) and net ($\Delta\Delta G_{net}$) energy terms of these salt bridges are computed using IPM (Kumar et al. 2000; Nayek et al. 2014; Nayek et al. 2015) and NUM (Bandyopadhyay et al. 2019a) for isolated and network salt bridges respectively. The results are presented in Figure 1. Several points are noteworthy from the figure. First, the desolvation term ($\Delta\Delta G_{dslv}$) is seen to be always costly (red color) in hNDK (Figure 1h1), tNDK (Figure 1t1) and mNDK (Figure 1m1). The distribution-pattern of this term is almost similar. Further, it is seen that the $\Delta\Delta G_{dslv}$ is negatively and almost linearly ($r= 0.73$) correlated with the average accessibility (ASA_{av}) of salt bridge (Figure 1a1). This would mean that more the burial of a salt bridge (i.e. less ASA_{av}), higher is the desolvation cost. Such an observation supports the authenticity of the computation (Kumar et al. 2000; Nayek et al. 2014). Second, $\Delta\Delta G_{brd}$, which originates due to direct interaction of positive (base) and negative (acid) partners of a salt bridge, is always contributing as seen in hNDK (Figure 1h2), tNDK (Figure 1t2) and mNDK (Figure 1m2). This energy term is also seen to be positively and almost linearly (r=0.66) correlated with ASA_{av}. This would mean that higher the exposure of a salt bridge (higher ASA_{av}), lesser is the contribution of $\Delta\Delta G_{brd}$ term (Figure 1a2). Third, the $\Delta\Delta G_{bac}$ term that arises due to interaction of either partner of a salt bridge with the side chain of other residues, could either be contributing (blue) or costly (red). In our cases, we see that some fractions of $\Delta\Delta G_{bac}$ are positive in hNDK, tNDK and mNDK (red color; Figure 1h3, t3 & m3), which is much higher for the latter. Oppositely and interestingly, the negativity of $\Delta\Delta G_{bac}$ is much higher in hNDK and tNDK relative to mNDK (Figure 1h3, t3 & m3), might be due to intrinsic composition of MIE (see below). Further, the nonlinear correlation of $\Delta\Delta G_{bac}$ with ASA_{av} (r=0.32; Figure 1a3) support this contention. Forth, remarkably, $\Delta\Delta G_{net}$ is seen to be negative for all salt bridges in hNDK, which is not the case for that of tNDK and mNDK. Finally, since hNDK and tNDK form network salt bridges and since the implementation of IPM in this case would be erroneous (Bandyopadhyay et al. 2019a), we have computed energy terms of network salt bridges using NUM (Bandyopadhyay et al. 2019a). Basic difference of NUM and IPM is shown in the Table 1.

In NUM, the decrease of $\Delta\Delta G_{dslv}$ is due to the common residue, R131. Notably, $\Delta\Delta G_{brd}$ also decreases in NUM due to repulsion of similar charges. Unlike IPM, $\Delta\Delta G_{bac}$ drastically decreases in NUM. Thus, IPM underestimates or overestimates the $\Delta\Delta G_{net}$ of a network salt bridge. The overall contribution of salt bridges to the total stability of hNDK, tNDK and mNDK are seen to be -39.89 kcal/mol/10 SBs, -49.82 kcal/mol/11 SBs and -23.56 kcal/mol/6 SBs respectively. This would mean that although per salt bridge contribution is little lower in hNDK, its overall contributions are much higher in hNDK and tNDK. At extreme of salt or temperature, where dielectric constant is drastically reduced (Kumar et al. 2000; Bandyopadhyay et al. 2019a) that, in turn, affect the hydrophobic interactions (Bandyopadhyay et al. 2019a), these additional stabilities seems to be useful to bypass such constraint of robust environments.

2.3 *Intrinsic microenvironments act favorably to the energetics of salt bridge*

NUM provides the clue for the existence of MIE in the vicinity (≤14 Å) of a salt bridge. Although mNDK, hNDK and tNDK are topologically identical, they are compositionally distinct from each other. How these diversities in composition affect the MIE and energetics of a salt bridge? Notably, the interaction of MIE and partners of a salt bridge are equally likely to be unfavorable or favorable or neutral. The intrinsic composition of hNDK and tNDK is largely hydrophilic (Figure 1e). Which of the above three possibilities is expected in the case of hNDK and tNDK? We thus perform statistical analyses on the MIEs of salt bridges, the result of which is presented in Figure 1. The following observations are noteworthy. First, majority of MIE (≥60%) are making favorable contribution (Figure 1ha and ta) with higher negativity (lower than -20 kj/mol) in the case of hNDK and tNDK. Second, components of MIE are present both in the core and surface of these proteins. Third, surprisingly, larger parts of the MIE are present in the helical (H) rather than in coil (C) or strand (S) segment of NDKs. Forth, although all twenty amino acids are equally likely, only charge (A, acidic; B basic) and polar (P polar) residues are acting as MIE of which the former dominates in

Table 1. Computation of energy terms for a typical pair of salt bridge by IPM and NUM.

Salt bridge	D125-R131 and R131-E55	D125-R131-E55
Energy term	IPM (kcal/mol)	NUM (kcal/mol)
$\Delta\Delta G_{dslv}$	7.99	3.15
$\Delta\Delta G_{brd}$	-14.99	-12.89
$\Delta\Delta G_{bac}$	9.38	-0.25
$\Delta\Delta G_{net}$	-16.38	-9.99

hNDK and tNDK relative to mNDK (Figure 1hd and td). Since acidic and basic residues have positive MRA (Figure 1d1 and f1) in hNDK and tNDK, and since E (and also D in hNDK; Figure 1c) is maintained as MCR, and since we see more favorable interactions in hNDK and tNDK (Figure 1ha and ta) than that of mNDK, it seems that the increase of charge residues in these extreme proteins (i.e. hNDK and tNDK) are acting as intrinsic code for salt bridge's stability. In other words, favorable contribution is preferred over unfavorable or neutral effects of the MIEs in hNDK and tNDK. Here, mention may be made of the fact that the increase of hydrophilicity of the entire sequence of hNDK and tNDK appears to be a secondary effect. The primary purpose is to engineer a favorable charge and polar MIE around salt bridges such that the loss of hydrophobic interaction in high salt or high temperature could be replenished by these specific electrostatic interactions.

3 CONCLUSION

Analysis of sequences of nucleoside diphosphate kinase (NDK) from halophilic (hNDK) and thermophilic (tNDK) archaea in comparison to mesophilic (mNDK) counterpart shows positive MRA for acidic and basic residues that alter sequence but not structural properties. Analysis of salt bridge and its energetics using IPM and NUM show that unlike mNDK, hNDK and tNDK have engineered highly stable isolated and network salt bridges. Implication of NUM further shows that network salt bridge efficiently neutralizes the isolated charge and polar background. Statistical analyses of MIE of the partners of salt bridge show intrinsic basis of the formation of stabilizing salt bridges in extreme NDKs. Enhanced preference for charge residues in hNDK and mNDK induce favourable MIE to make these salt bridges stable in extreme of salt and temperature conditions.

ACKNOWLEDGMENT

We are thankful to the Department of Biotechnology, University of Burdwan for the computational facility laboratory.

REFERENCES

Baker et al. 2001. Electrostatics of nanosystems: application to microtubules and the ribosome. *Proceedings of the National Academy of Sciences*, *98*(18): 10037–10041.

Bandyopadhyay, A.K., 2015. Stability of Halophilic Proteins in Hyper Saline Brine:[2Fe-2S] Ferredoxin as a Paradigm. *Biochemistry and Analytical Biochemistry*, *4*(3): 1–4.

Bandyopadhyay, A.K. & Sonawat, H.M., 2000. Salt dependent stability and unfolding of [Fe2-S2] ferredoxin of Halobacterium salinarum: spectroscopic investigations. *Biophysical journal*, *79*(1): 501–510.

Bandyopadhyay et al. 2001. Structural stabilization of [2Fe-2S] ferredoxin from Halobacterium salinarum. *Biochemistry*, *40*(5): 1284–1292.

Bandyopadhyay et al. 2007. Kinetics of salt-dependent unfolding of [2Fe–2S] ferredoxin of Halobacterium salinarum. *Extremophiles*, *11*(4): 615–625.

Bandyopadhyay et al. 2019a. Stability of buried and networked salt-bridges (BNSB) in thermophilic proteins. *Bioinformation*, *15*(1): 61–67.

Bandyopadhyay et al. 2019b. Insight into the Salt Bridge of Malate Dehydrogenase from Halobacterium salinarum and Escherichia coli. *Bioinformation*, *15*(2): 95–103.

Banerjee et al. 2015. PHYSICO2: an UNIX based standalone procedure for computation of physicochemical, window-dependent and substitution based evolutionary properties of protein sequences along with automated block preparation tool, version 2. *Bioinformation*, *11*(7): 366–368.

Berman et al. 2000. The protein data bank. *Nucleic acids research*, *28*(1): 235–242.

Britton et al. 1998. Insights into the molecular basis of salt tolerance from the study of glutamate dehydrogenase from Halobacterium salinarum. *Journal of Biological Chemistry*, *273*(15): 9023–9030.

Dolinsky et al. 2004. PDB2PQR: an automated pipeline for the setup of Poisson–Boltzmann electrostatics calculations. *Nucleic acids research, 32* (suppl_2):W665–W667.
Dym et al. 1995. Structural features that stabilize halophilic malate dehydrogenase from an archaebacterium. *Science, 267*(5202): 1344–1346.
Gupta et al. 2014a. PHYSICO: An UNIX based Standalone Procedure for Computation of Individual and Group Properties of Protein Sequences. *Bioinformation, 10*(2): 105–107.
Gupta et al. 2014b. SBION: A program for analyses of salt-bridges from multiple structure files. *Bioinformation, 10*(3): 164–166.
Gupta et al. 2015. SBION2: Analyses of Salt Bridges from Multiple Structure Files, Version 2. *Bioinformation, 11*(1): 39–41.
Gupta et al. 2017a, Cosurim: A program for automated computation of composition of core, rim and surface of protein data bank files and its applications. *International Journal of Engineering Science and Technology*, 9(11): 993–999.
Gupta et al. 2017b. Substitutional Analysis of Orthologous Protein Families Using BLOCKS. *Bioinformation, 13*(1): 1–7.
Hendsch, Z.S. & Tidor, B., 1994. Do salt bridges stabilize proteins? A continuum electrostatic analysis. *Protein Science, 3*(2): 211–226.
Islam et al. 2018. AUTOMINv1. 0: an automation for minimization of Protein Data Bank files and its usage. *Bioinformation, 14*(9): 525–529.
Kreil, D.P. and Ouzounis, C.A., 2001. Identification of thermophilic species by the amino acid compositions deduced from their genomes. *Nucleic acids research, 29*(7): 1608–1615.
Kumar et al. 2000. Factors enhancing protein thermostability. *Protein engineering, 13*(3): 179–191.
Nayek et al. 2015. ADSBET: automated determination of salt-bridge energy terms. *International Journal of Institutional Pharmacy and Life Sciences, 5*(3): 28–36.
Nayek et al. 2014. Salt-bridge energetics in halophilic proteins. *Plos one, 9*(4): 1–11.
Nayek et al. 2015. ADSBET2: Automated Determination of Salt-Bridge Energy-Terms version 2. *Bioinformation, 11*(8): 413–415.
Rao, J.M. & Argos, P., 1981. Structural stability of halophilic proteins. *Biochemistry, 20*(23): 6536–6543.
Rothschild, L.J. & Mancinelli, R.L., 2001. Life in extreme environments. *Nature, 409*(6823): 1092–1101.
Tokunaga et al. 2006. Contribution of halophilic nucleoside diphosphate kinase sequence to the heat stability of chimeric molecule. *Protein and peptide letters, 13*(5): 525–530.
Wrba et al. 1990. Extremely thermostable D-glyceraldehyde-3-phosphate dehydrogenase from the eubacterium Thermotoga maritima. *Biochemistry, 29*(33): 7584–7592.

Biotechnology and Biological Sciences – Sen et al. (Eds)
© 2020 Taylor & Francis Group, London, ISBN 978-0-367-43161-7

Advances in biomedical entity identification: A survey

Jainisha Sankhavara & Prasenjit Majumder
Dhirubhai Ambani Institute of Information and Communication Technology, Gandhinagar, Gujarat, India

ABSTRACT: Biomedical entity identification is an essential step for any biomedical natural language processing system. This paper presents a survey on biomedical entity identification with various community challenges. This paper also includes comparison of techniques proposed in the literature for biomedical entity identification. Combined approaches based on neural network and sequential machine learning methods are outperforming other techniques on various biomedical entity recognition datasets.

1 INTROUCTIONS

Biomedical entity identification refers to the task of identifying and classifying biomedical terms into predefined categories. It is a first and very important step to understand the biomedical text. The biomedical entities of major concern are Genes, Proteins, Drugs, temporal expressions, Disease names etc. As the biomedical knowledge grows, it adds new medical terms and entities to the collection of biomedical entities, making the set of biomedical entities incomplete at any point of time. Therefore, the string matching based algorithms which use the dictionary of entities for Named Entity Identification are no longer useful for biomedical domain as it is difficult to have exhaustive dictionary. Biomedical Entity Identification task involves:

- Identifying boundaries of the entities
- Assigning a preferred class to the entity
- Getting the preferred name or concept's unique identifier of the entity

These steps are themselves individual tasks, sometimes referred as biomedical tokenization, biomedical entity classification and biomedical entity normalization respectively.

2 IDENTIFYING BOUNDARIES OF THE ENTITIES

Entity Identification in biomedical domain is a challenging task due to the characteristics of medical entities. Biomedical entities are sometimes very complex like 'nuclear factor kappa-light-chain-enhancer of activated B cells' and 'NF-kB DNA binding with electromobility shift assay'. The average length of biomedical entities is much higher which makes the process of identifying boundaries of entities difficult.

BIO (Rahman 2016) is a popular scheme for identifying boundaries of entities. BIO stands for Beginning, Inside and Outside of the entity. The starting word of the entity is marked as beginning of the entity and subsequent words of the entity are marked as inside of entity. Other non-entity words are marked as outside. BIO tagged representation of the text "nuclear factor kappa-light-chain-enhancer of activated B cells is found in almost all animal cell types" is: "Nuclear/B factor/I kappa-light-chain-enhancer/I of/I activated/I B/I cells/I is/O found/O in/O almost/O all/O animal/O cell/O types/O"

3 CLASSIFICATION OF BIOMEDICAL ENTITIES

After identifying the boundaries of the medical entities, proper class should be assigned to each entity. Biomedical entities are largely classified into following categories: DNA, RNA, Protein, Cell type, Cell line, Chemicals, Genes, Species, Diseases, Treatments etc. Physicians often use ad-hoc abbreviations for biomedical entities. They also use acronyms and abbreviations which are ambiguous like 'PSA' can be 'prostate specific antigen' or 'psoriasis arthritis' or 'poultry science association'. The meaning of all three entities are different but they share a same abbreviation. Thus, the correct expansion of such entity is very important in order to correctly classify the entity.

4 BIOMEDICAL ENTITY NORMALIZATION

Biomedical terminology changes very rapidly which makes it inconsistent. This problem is also known as polysemy. For instance, 'H1N1 influenza', 'swine influenza', 'SI', 'Pig Flu' and 'Swine-Origin Influenza A H1N1 Virus', all refers to the same entity. Such different representations of the same entity should be mapped to a single entity. Due to these reasons, the biomedical entity identification task implicitly needs entity normalization as a subtask which can also address abbreviations, synonymy and polysemy. It requires external resources map the entity to its preferred representation or the concept. Such knowledge sources include ontologies (ex. Gene Ontology (Gene 2018)) and semantic networks (ex. UMLS Metathesaurus (Bodenreider 2004)).

5 COMMUNITY CHALLENGES AND RESOURCES

Community identified challenges in the area of biomedical entity identification and normalization are listed in Table 1.

The common evaluation measures used for Biomedical Entity Identification tasks are Precision, Recall and F-score. Various datasets, for example, GENIA (Kim *et al.* 2003), AIMed (Bunescu *et al.* 2005), JNLPBA04 (Kim *et al.* 2004), NCBI disease corpus (Doğan *et al.* 2014), CHEMDNER corpus (Krallinger *et al.* 2015), GENETAG (Tanabe *et al.* 2005), BioCreative GM(Smith *et al.* 2008), i2b2 (Uzuner *et al.* 2010) etc. are used by the community for Biomedical Entity Identification.

Table 1. Community challenges for biomedical entity recognition over the years 2004-17.

Year	Challenge	Task
2004	BioCreative I	Identification of gene mentions (Yeh *et al.* 2005, Hirschman *et al.* 2005)
2006	BioCreative II	Gene mention tagging (GM) (Smith *et al.* 2008) Gene normalization (GN) (Morgan *et al.* 2008)
2010	BioCreative III	GN: The gene normalization task (Lu & Wilbur 2010, Lu *et al.* 2010)
2012	BioCreative IV	Chemical and Drug Named Entity Recognition (CHEMDNER) (Krallinger *et al.* 2013, Krallinger *et al.* 2015)
2013	CLEF eHealth	Named entity recognition and normalization of disorders (Pradhan *et al.* 2013)
2014	CLEF eHealth	Disease/Disorder Template Filling (Mowery *et al.* 2014)
2015	BioCreative V	CHEMDNER patents (Krallinger *et al.* 2015), Chemical-disease relation (CDR) task (Wei *et al.* 2015)
2015	CLEF eHealth	Clinical Named Entity Recognition (Névéol *et al.* 2015)
2016	CLEF eHealth	Multilingual Information Extraction (Névéol *et al.* 2016)
2017	BioCreative V.5	Chemical Entity Mention recognition (CEMP) (Pérez-Pérez *et al.* 2017), Gene and Protein Related Object recognition (GPRO) (Pérez-Pérez *et al.* 2017)
2017	CLEF eHealth	Multilingual Information Extraction - ICD10 coding (Névéol *et al.* 2017)

6 APPROACHES TO BIOMEDICAL ENTITY IDENTIFICATION

Biomedical Named Entity Recognition approaches mainly characterized into four groups (Ananiadou *et al.* 2006):

1. Dictionary-based approaches that try to find names of nomenclatures in the literature.
2. Rule-based approaches that manually or automatically construct rules and patterns to directly match entities to candidate entities in the literature.
3. Machine learning approaches that employ machine learning techniques, such as SVMs (Cortes *et al.* 1995), CRFs (Lafferty *et al.* 2001), and neural networks to develop statistical models for biomedical entity recognition.
4. Hybrid approaches that merge two or more of the above approaches, mostly in a sequential way, to deal with different aspects of Named Entity Recognition (NER).

Dictionary based approaches uses dictionary as a biomedical resource for the matching of entity occurrences directly. It identifies the biomedical entities form the text using string matching based algorithms. Dictionary based methods utilize comprehensive list of biomedical terms to identify biomedical entities from biomedical text. This approach highly suffers from spelling mistakes, morphological variants of entities and incompleteness of the corresponding biomedical resource. To deal with such situations, spelling variations based algorithms, approximate string matching based algorithms have been proposed.

Rule based approaches uses preestablished rules based on the composition pattern of biomedical entities. These approaches need rules to identify biomedical entities hence we need to define them properly. Rule based approaches give better performance than dictionary-based approaches.

Machine learning based approaches are becoming popular for biomedical entity identification. They use supervised statistical methods to identify entities. These methods are pretrained on tagged dataset and learn to identify medical entities from new data. To train these methods, gold-standard data is used which is created using manual intervention. Machine learning based approaches give better results than dictionary based and rule-based approaches. There are semi-supervised methods and also the methods which create training data with the use of bootstrapping. Classification based approaches like SVM and sequential approaches like Hidden Markov Model (HMM) (Rabiner *et al.* 1993), Maximum Entropy Markov Model (MEMM) (Berger *et al.* 1996), Conditional Random Field (CRF) (Lafferty *et al.* 2001) and Long Short Term Memory (LSTM) (Hochreiter *et al.* 1997) are very much favorable for biomedical entity identification. Sequential methods are even better than classification methods. The state-of-the-art biomedical entity recognition models are based on CRF and LSTM.

Given a word sequence $W = \{w_1, w_2, \ldots, w_n\}$ and its label sequence $L = \{l_1, l_2, \ldots, l_n\}$, the conditional probability of a linear chain CRF is given in Equation 1.

$$P(L/W,\lambda) = \frac{1}{z}\exp\left(\sum_{i=1}^{n}\sum_{j=1}^{m}\lambda_j f_j(l_i, l_{i-1}, w, i)\right) \quad (1)$$

where *fj(li, li-1, w, i)* is a feature function; *li* and *li-1* refer to current and previous state, respectively; *z* is a normalization factor shown in Equation 2.

$$z = \sum_{l}\exp\left(\sum_{i=1}^{n}\sum_{j=1}^{m}\lambda_j f_j(l_i, l_{i-1}, w, i)\right) \quad (2)$$

When any of these dictionary based, rule based and machine learning based approaches are combined for biomedical entity identification, they are known as hybrid approaches.

Table 2. Comparison of biomedical entity identification approaches on JNLPBA04 dataset over the years 2004-2018.

Year	Method	F-measure
2004	HMM + SVM + deep knowledge resources (GuoDong *et al.* 2004)	72.55
2012	Generic classifier ensemble with SVM (Liao *et al.* 2012)	77.85
2012	SVM-CRF (Zhu *et al.* 2012)	92.59
2013	Gimli (Campos *et al.* 2013)	72.23
2013	GA based feature selection for SVM and CRF (Ekbal *et al.* 2013)	75.17
2014	CRF + word representations (Tang *et al.* 2014)	71.39
2014	CRF + rules (Raja *et al.* 2014)	75.77
2015	CRF + MapReduce (Tang *et al.* 2015)	73.31
2015	GA based classifier-ensemble for SVM and CRF (Saha *et al.* 2015)	75.97
2015	Deep neutral network (Yao *et al.* 2015)	71.01
2016	Bidirectional LSTM (character + words) (Rei *et al.* 2016)	72.70
2017	BLSTM + WE + char + dropout + CRF (Gridach *et al.* 2017)	75.87
2018	Bidirectional LM + transfer learning (Sachan *et al.* 2018)	75.03

Various methods proposed in the literature for biomedical entity identification in JNLPBA04 dataset are compared in the Table 2.

GuoDong *et al.* (2004) has explored various deep knowledge resources such as the name alias, the cascaded entity name, dictionary, the alias list LocusLink, abbreviation resolution and POS with SVM and achieved 72.55% F-measure for biomedical NER. Liao *et al.* (2012) has proposed generic classifier ensemble approach using SVM based on the principle that contributing degrees of prediction classes among different classes in the same classifier are different and they also differ among different classifiers. They compared their results with single SVM classifier, vote based SVM-classifier selection, HMM, MEMM and CRF and achieved maximum F-measure of 77.85% on JNLPBA04 dataset.

Zhu *et al.* (2012) has used SVM to separate biological terms from non-biological terms and CRFs to determine the types of biological terms. Their proposed hybrid approach SVM-CRF have surprisingly achieved F-measure of 92.59% on JNLPBA04 data and 97.48% on GENIA data. An open-source tool, Gimli (Campos *et al.* 2013) implements a machine learning technique CRF with a rich set of features which include morphological, orthographic, linguistic and domain knowledge features. It also has a post-processing module which does parentheses correction and abbreviation resolution. Gimli shows 72.23% F-score on JNLPBA04 dataset. Ekbal *et al.* (2013) have used genetic algorithm (GA) in feature selection process for SVM and CRF classifiers with stacked based ensemble to combine the classifiers. On JNLPBA04 dataset and GENETAG dataset, they achieved F-measure values of 75.17% and 94.70%, respectively. Their approach gave ~1%-2% increment over best individual classifier, Majority vote based ensemble and weighted vote based ensemble.

Tang *et al.* (2014) investigated and combined three different types of word representation features for Biomedical Entity Identification, including clustering-based representation, distributional representation, and word embeddings. Their system achieved F-measure 80.96% 71.39% with 3.75% and 1.39% improvement when compared with the systems using baseline features for BioCreAtIvE II GM and JNLPBA04 corpora, respectively. Raja *et al.* (2014) have combined machine learning based approach with rule based approach. Their generated post-processing rules were combined with CRF and achieved F-score of 75.77% on JNLPBA04 dataset.

Tang *et al.* (2015) used parallel optimization framework with CRF for biomedical entity identification, achieving 73.31% F-score with short training time. Saha *et al.* (2015) used single objective optimization based classifier ensemble technique with SVM and CRF gives F-measure values 75.97% and 95.90%, achieving the increments of 1.07% and 0.57% over the individual classifiers for JNLPBA04 and GENETAG dataset, respectively.

Yao *et al.* (2015) used a multilayer neural network to continuously learn the representation of features, achieving 71.01% F-score. Rei *et al.* (2016) has proposed a character level neural model (bidirectional LSTM) in combination with word level model using attention mechanism and achieved 72.70% F-score as compared to F-score 70.75% for word level model. Gridach *et al.* (2017) has achieved F-measure 75.87% using CRF on top of bidirectional LSTM in combination with pretrained word embeddings and character-level embeddings. Sachan *et al.* (2018) has used transfer learning with bidirectional language model for biomedical entity recognition and achieved F-measure 75.03%.

Crichton *et al.* (2017) has studied multi-task learning across 15 biomedical NER datasets using CNN with multiple output layers and observed an average improvement on multi-task learning as compared to single task learning. Ju *et al.* (2018) proposed a neural model to identify nested entities by dynamically stacking NER layers. They used LSTM+CRF as a neural model and this dynamic model achieved F-measure 74.7% and 72.2% on GENIA and ACE2005 dataset, respectively.

Xu *et al.* (2017) has used BiLSTM-CRF model NCBI Disease Corpus and achieved 80.22% F-score for disease named entity identification. Xu *et al.* (2019) has proposed to combine disease dictionary using bidirectional LSTM and CRF with a dictionary attention layer for disease named entity recognition. Zeng *et al.* (2017) showed the effect of bidirectional LSTM and CRF with word embedding and character embedding for drug named entity recognition.

7 CONCLUSION

This paper presents a survey on biomedical entity recognition. Various community identified challenges and tasks for biomedical entity recognition are summarized here. Dictionary based, rule based and machine learning based approaches along with their hybrid approaches are also outlined. Various state-of-the-art machine learning techniques are compared for biomedical entity identification. Combined approaches based on neural network and sequential machine learning methods are outperforming other techniques on most of the biomedical entity recognition datasets.

ACKNOWLEDGEMENT

The present research work is supported by TCS foundation under TCS Research Scholarship Program.

REFERENCES

Ananiadou, S. and McNaught, J., 2006. *Text mining for biology and biomedicine* (pp. 1-12). London: Artech House.

Baum, L.E. and Petrie, T., 1966. Statistical inference for probabilistic functions of finite state Markov chains. *The annals of mathematical statistics*, *37*(6), pp.1554-1563.

Berger, A.L., Pietra, V.J.D. and Pietra, S.A.D., 1996. A maximum entropy approach to natural language processing. *Computational linguistics*, *22*(1), pp.39-71.

Bodenreider, O., 2004. The unified medical language system (UMLS): integrating biomedical terminology. *Nucleic acids research*, *32* (suppl_1),pp. D267-D270.

Bunescu, R., Ge, R., Kate, R.J., Marcotte, E.M., Mooney, R.J., Ramani, A.K. and Wong, Y.W., 2005. Comparative experiments on learning information extractors for proteins and their interactions. *Artificial intelligence in medicine*, *33*(2), pp.139-155.

Campos, D., Matos, S. and Oliveira, J.L., 2013. Gimli: open source and high-performance biomedical name recognition. *BMC bioinformatics*, *14*(1), p.54.

Cortes, C. and Vapnik, V., 1995. Support-vector networks. *Machine learning*, *20*(3), pp.273-297.

Crichton, G., Pyysalo, S., Chiu, B. and Korhonen, A., 2017. A neural network multi-task learning approach to biomedical named entity recognition. *BMC bioinformatics*, *18*(1), p.368.

Doğan, R.I., Leaman, R. and Lu, Z., 2014. NCBI disease corpus: a resource for disease name recognition and concept normalization. *Journal of biomedical informatics*, *47*, pp.1-10.

Ekbal, A. and Saha, S., 2013. Stacked ensemble coupled with feature selection for biomedical entity extraction. *Knowledge-Based Systems*, *46*, pp.22-32.

Gene Ontology Consortium, 2018. The Gene Ontology resource: 20 years and still GOing strong. *Nucleic acids research*, *47*(D1), pp.D330-D338.

Gridach, M., 2017. Character-level neural network for biomedical named entity recognition. *Journal of biomedical informatics*, *70*, pp.85-91.

GuoDong, Z. and Jian, S., 2004, August. Exploring deep knowledge resources in biomedical name recognition. In *Proceedings of the International Joint Workshop on Natural Language Processing in Biomedicine and its Applications* (pp. 96-99). Association for Computational Linguistics.

Hirschman, L., Colosimo, M., Morgan, A. and Yeh, A., 2005. Overview of BioCreAtIvE task 1B: normalized gene lists. *BMC bioinformatics*, *6*(1), p.S11.

Hochreiter, S. and Schmidhuber, J., 1997. Long short-term memory. *Neural computation*, *9*(8), pp.1735-1780.

Ju, M., Miwa, M. and Ananiadou, S., 2018, June. A neural layered model for nested named entity recognition. In *Proceedings of the 2018 Conference of the North American Chapter of the Association for Computational Linguistics: Human Language Technologies, Volume 1 (Long Papers)*, pp. 1446-1459.

Kim, J.D., Ohta, T., Tateisi, Y. and Tsujii, J.I., 2003. GENIA corpus—a semantically annotated corpus for bio-text mining. *Bioinformatics*, *19* (suppl_1),pp. i180-i182.

Kim, J.D., Ohta, T., Tsuruoka, Y., Tateisi, Y. and Collier, N., 2004, August. Introduction to the bio-entity recognition task at JNLPBA. In *Proceedings of the international joint workshop on natural language processing in biomedicine and its applications* (pp. 70-75). Association for Computational Linguistics.

Krallinger, M., Leitner, F., Rabal, O., Vazquez, M., Oyarzabal, J. and Valencia, A., 2013, October. Overview of the chemical compound and drug name recognition (CHEMDNER) task. In *BioCreative challenge evaluation workshop* (Vol. 2, p. 2).

Krallinger, M., Leitner, F., Rabal, O., Vazquez, M., Oyarzabal, J. and Valencia, A., 2015. CHEMDNER: The drugs and chemical names extraction challenge. *Journal of cheminformatics*, *7*(1), p.S1.

Krallinger, M., Rabal, O., Leitner, F., Vazquez, M., Salgado, D., Lu, Z., Leaman, R., Lu, Y., Ji, D., Lowe, D.M. and Sayle, R.A., 2015. The CHEMDNER corpus of chemicals and drugs and its annotation principles. *Journal of cheminformatics*, *7*(1), p.S2.

Krallinger, M., Rabal, O., Lourenço, A., Perez, M.P., Rodriguez, G.P., Vazquez, M., Leitner, F., Oyarzabal, J. and Valencia, A., 2015. Overview of the CHEMDNER patents task. In *Proceedings of the fifth BioCreative challenge evaluation workshop* (pp. 63-75).

Lafferty, J., McCallum, A. and Pereira, F.C., 2001. Conditional random fields: Probabilistic models for segmenting and labeling sequence data.

Liao, Z. and Zhang, Z., 2012, May. A generic classifier-ensemble approach for biomedical named entity recognition. In *Pacific-Asia Conference on Knowledge Discovery and Data Mining* (pp. 86-97). Springer, Berlin, Heidelberg.

Lu, Z., Kao, H.Y., Wei, C.H., Huang, M., Liu, J., Kuo, C.J., Hsu, C.N., Tsai, R.T.H., Dai, H.J., Okazaki, N. and Cho, H.C., 2011. The gene normalization task in BioCreative III. *BMC bioinformatics*, *12*(8), p.S2.

Lu, Z. and Wilbur, W.J., 2010, September. Overview of BioCreative III gene normalization. In Proceedings of the BioCreative III workshop, pp. 24-45.

Morgan, A.A., Lu, Z., Wang, X., Cohen, A.M., Fluck, J., Ruch, P., Divoli, A., Fundel, K., Leaman, R., Hakenberg, J. and Sun, C., 2008. Overview of BioCreative II gene normalization. *Genome biology*, *9* (2), p.S3.

Mowery, D.L., Velupillai, S., South, B.R., Christensen, L., Martinez, D., Kelly, L., Goeuriot, L., Elhadad, N., Pradhan, S., Savova, G. and Chapman, W., 2014, September. Task 2: ShARe/CLEF eHealth evaluation lab 2014. In *Proceedings of CLEF* 2014.

Névéol, A., Cohen, K.B., Grouin, C., Hamon, T., Lavergne, T., Kelly, L., Goeuriot, L., Rey, G., Robert, A., Tannier, X. and Zweigenbaum, P., 2016, September. Clinical information extraction at the CLEF eHealth evaluation lab 2016. In *CEUR workshop proceedings* (Vol. 1609, p. 28). NIH Public Access.

Névéol, A., Grouin, C., Tannier, X., Hamon, T., Kelly, L., Goeuriot, L. and Zweigenbaum, P., 2015, September. CLEF eHealth Evaluation Lab 2015 Task 1b: Clinical Named Entity Recognition. In *CLEF (Working Notes)*.

Névéol, A., Robert, A., Anderson, R., Cohen, K.B., Grouin, C., Lavergne, T., Rey, G., Rondet, C. and Zweigenbaum, P., 2017, September. CLEF eHealth 2017 Multilingual Information Extraction task Overview: ICD10 Coding of Death Certificates in English and French. In *CLEF (Working Notes)*.

Pradhan, S., Elhadad, N., South, B.R., Martinez, D., Christensen, L.M., Vogel, A., Suominen, H., Chapman, W.W. and Savova, G.K., 2013, September. Task 1: ShARe/CLEF eHealth Evaluation Lab 2013. In *CLEF (Working Notes)*.

Pérez-Pérez, M., Rabal, O., Pérez-Rodríguez, G., Vazquez, M., Fdez-Riverola, F., Oyarzabal, J., Valencia, A., Lourenço, A. and Krallinger, M., 2017. Evaluation of chemical and gene/protein entity recognition systems at BioCreative V. 5: the CEMP and GPRO patents tracks.

Rahman, H.U., Chowk, N.G.T.K., Hahn, T. and Segall, R., 2016. Disease named entity recognition using conditional random fields. In *Proceedings of the 7th International Symposium on Semantic Mining in Biomedicine*.

Raja, K., Subramani, S. and Natarajan, J., 2014. A hybrid named entity tagger for tagging human proteins/genes. *Int. J. Data Min. Bioinform*, *10*, pp.315-328.

Rei, M., Crichton, G.K. and Pyysalo, S., 2016. Attending to characters in neural sequence labeling models. *arXiv preprint arXiv:1611.04361*.

Sachan, D. S., Xie, P., Sachan, M., and Xing, E. P. 2018. Effective use of bidirectional language modeling for transfer learning in biomedical named entity recognition. In Machine Learning for Healthcare Conference, pages 383–402.

Saha, S., Ekbal, A. and Sikdar, U.K., 2015. Named entity recognition and classification in biomedical text using classifier ensemble. *International journal of data mining and bioinformatics*, *11*(4), pp.365-391.

Smith, L., Tanabe, L.K., nee Ando, R.J., Kuo, C.J., Chung, I.F., Hsu, C.N., Lin, Y.S., Klinger, R., Friedrich, C.M., Ganchev, K. and Torii, M., 2008. Overview of BioCreative II gene mention recognition. *Genome biology*, *9* (2),p.S2.

Tanabe, L., Xie, N., Thom, L.H., Matten, W. and Wilbur, W.J., 2005. GENETAG: a tagged corpus for gene/protein named entity recognition. *BMC bioinformatics*, *6*(1), p.S3.

Tang, B., Cao, H., Wang, X., Chen, Q. and Xu, H., 2014. Evaluating word representation features in biomedical named entity recognition tasks. *BioMed research international*, 2014.

Tang, Z., Jiang, L., Yang, L., Li, K. and Li, K., 2015. CRFs based parallel biomedical named entity recognition algorithm employing MapReduce framework. *Cluster Computing*, *18*(2), pp.493-505.

Uzuner, Ö., South, B.R., Shen, S. and DuVall, S.L., 2011. 2010 i2b2/VA challenge on concepts, assertions, and relations in clinical text. *Journal of the American Medical Informatics Association*, *18*(5), pp.552-556.

Wei, C.H., Peng, Y., Leaman, R., Davis, A.P., Mattingly, C.J., Li, J., Wiegers, T.C. and Lu, Z., 2015. Overview of the BioCreative V chemical disease relation (CDR) task. In *Proceedings of the fifth BioCreative challenge evaluation workshop*, pp. 154-166.

Xu, K., Yang, Z., Kang, P., Wang, Q. and Liu, W., 2019. Document-level attention-based BiLSTM-CRF incorporating disease dictionary for disease named entity recognition. *Computers in biology and medicine*, *108*, pp.122-132.

Xu, K., Zhou, Z., Hao, T. and Liu, W., 2017, September. A bidirectional LSTM and conditional random fields approach to medical named entity recognition. In *International Conference on Advanced Intelligent Systems and Informatics*, pp. 355-365.

Yao, L., Liu, H., Liu, Y., Li, X. and Anwar, M.W., 2015. Biomedical named entity recognition based on deep neutral network. *Int. J. Hybrid Inf. Technol*, *8*(8), pp.279-288.

Yeh, A., Morgan, A., Colosimo, M. and Hirschman, L., 2005. BioCreAtIvE task 1A: gene mention finding evaluation. *BMC bioinformatics*, *6*(1), p.S2.

Zeng, D., Sun, C., Lin, L. and Liu, B., 2017. LSTM-CRF for drug-named entity recognition. *Entropy*, *19* (6), p.283.

Zhu, F. and Shen, B., 2012. Combined SVM-CRFs for biological named entity recognition with maximal bidirectional squeezing. *PloS one*, *7*(6), p.e39230.

Biotechnology and Biological Sciences – Sen et al. (Eds)
© 2020 Taylor & Francis Group, London, ISBN 978-0-367-43161-7

Study on physicochemical and antioxidant properties of blend of fish skin (*Labeo rohita*) oil and chia seed (*Salvia hispanica*) oil

Nabanita Ghosh*
Research Scholar and Corresponding authour, School of Community Science and Technology, Indian Institute of Engineering Science and Technology, Shibpur, India

Monalisa Roy
Master of Science, School of Community Science and Technology, Indian Institute of Engineering Science and Technology, Shibpur, India

Minakshi Ghosh
Asistant Professor, School of Community Science and Technology, Indian Institute of Engineering Science and Technology, Shibpur, India

D.K Bhattacharyya
Adjunct Professor, School of Community Science and Technology, Indian Institute of Engineering Science and Technology, Shibpur, India

ABSTRACT: The present work deals with enrichment of fish skin oil (FSO) extracted from skins of rohu fish (*Labeo rohita*). Enrichment was conducted by blending chia seed oil (CO) with FSO into the ratios of 1:1(FCO_1), 1:2(FCO_2), 2:1(FCO_3) respectively. The blended oils were subjected to antioxidant activity analysis via total phenoilc content(TPC), Total(TFC), DPPH free radical scavenging activity assay, ABTS assay, FRAP assay, Metal chelatig assay, reducing power assay and are compared with the same of both the crude oils. Results revealed that blending makes the oils more oxidatively stable. For the physicochemical activity acid value, free fatty acid content, peroxide value and p-anisidine value and totox value of both the crude oils and blended oils were determined and from the results it was found that blending makes the oils more stable. These work suggests that blending of FSO and CO improves its quality and stability.

Keywords: FSO, CO, TPC, TFC, DPPH, FRAP, ABTS, Metal Chelating activity, Reducing power, Acid value, Peroxide value, anisidine value

1 INTRODUCTION

This work is a continuation of the authours previous work (Ghosh *et.al* 2019). Fish Skin Oil (FSO) from *Labeo rohita* skin is deficient in certain nutrients. FSO is lack in phytochemicals as it is a animal based oil. On the other side Chia Seed oil (CO) being a plant seed oil contains sufficient phyochemicals including natural antioxidants like tocopherols, phytosterols, carotenoids etc and phenoilc compounds (Ixtaina *et.al.* 2011). Besides having phytonutrients chia also contains essential fatty acid rich oil which have cholesterol loweing effect and properties to reduce obesity (Yingbin Shen *et.al.*2018). Studies revealed that FSO contains moderate amount 35.18% of saturated fatty acids with 27.76% palmitic and stearic acid 5.64% (Ghosh

*Corresponding author: E-mail: naba1990@gmail.com

et. al 2017, ISBN: 978-93-86526-31-1), where as chia seed oil was found to contain a lower content of saturated fatty acids (Cardenas *et.al* 2018). The blending of the two oils will therefore provide a compact nutrient enriched oil with effective health benefits.

2 METHOD AND MATERIALS

2.1 *Extraction of oil*

Fish skin was brought from local market, dried and grinded. Chia seed was brought from Wild Forest brand, grinded to make powder. Both the oils were extracted by sohxlet apparatus with n-hexane. Solution was evaporated and the oil collected was stored at -20°c for further use.

2.2 *Physicochemical property analysis*

FSO and CO were blended in different ratios by weight to obtain 1:1(FCO_1), 1:2(FCO_2), and 2:1(FCO_3). The crude oils and their blended oils were analysed for their physicochemical characteristics according AOCS official methods (AOCS official site)

2.2.1 *Estimation of acid value and free fatty acid*

1.0 gm oil was taken and to these few drops phenolphthalein and 20 mL of ethanol were added. The mixture was titrated with 0.1 M NaOH solution until the development of a pink colour. This value was calculated through Eq. 1.[6]

$$\text{Acid value} = T \times 0.282 \times W \tag{1}$$

Where T is the titre value, 0.0282 is a constant and W indicates the weight of oil used in gram.

Free fatty acid value was calculated utilizing Eq. 2.

$$\text{Free fatty acid} = \frac{T \times M \times 56.1}{W} \tag{2}$$

Where T is the titre value, M represents the molarity of the titrant, W refers to the weight of oil in gram and 56.1 is acid constant.

2.2.2 *Estimation of peroxide value*

It was determined to the AOCS methods, 1990 (AOAC 965.33). 1 gram of oil was taken and to these 1 gram of potassium iodide (KI) and 20 ml of a solvent mixture of glacial acetic acid and chloroform in 2:1 ratio was added. The mixture was heated in a water bath for 10 min. The content was titrated against 0.002 M sodium thiosulfate solution and starch is used as an indicator. This value was calculated with the formula represented in Eq. 3.

$$\text{Peroxide value} = \frac{T \times M \times 100}{W} \tag{3}$$

Where T = titre value, M = molarity of $Na_2S_2O_3$, W = weight of the oil

2.2.3 *Estimation of anisidine value*

It is determined according to the AOCS official method Cd 18-90. 1 gm of oil was dissolved into 25 ml isooctane. 1 ml of anisidine solution was added to the oil and mixed properly. After keeping in dark for 10 min the absorbance was measured at 350 nm against control containing everything except the anisidine solution.. The anisidine value was calculated using Eq. 4.

$$\text{Anicidine Value} = \frac{25 \times (1.2\,Ea - Eb)}{W} \tag{4}$$

Where Ea = absorbance of the fat solution, Eb = absorbance of fat + isooctane + anisidine solution
W= weight of oil sample

2.2.4 *Estimation of totox value*

Totox value indicates the oxidation state of an oil. It is calculated using the equation, Acid Value (AV) + 2 Peroxide Value (PV) (Wan tatt Wai *et al* 2009)

2.3 *Estimation of total phenolic content*

The total phenol content (TPC) of both crude FSO,CO and blended FCO_1, FCO_2, FCO_3 were measured using the Folin-Ciocalteuassay (Slinkard *et.al* 1977). 0.2 mL of oil was taken into test tubes and 0.5 mL Folin-Ciocalteu's reagent was added to it. The solution was kept in dark for 10 min and 1 mL sodium carbonate solution was added to it. The tubes were again kept in the dark for another 30 min. Absorbance at 765 nm was measured with a spectrophotometer (Jasco V-630) and compared to Gallic acid standard curve.

2.4 *Estimation of total flavonoid*

Total Flavonoid content (TFC) was determined to the standard method (Ahmad *et.al* 2015). 125 μL of oils were added to 75 μL of 5% $NaNO_2$ solution. To the mixture 150 μL of 10% aluminium trichloride was added and incubated for 5 min, followed by the addition of 750 μL of NaOH (1M). The final volume of the solution was made up 2500 μL with distilled water and incubated for another 15 min. The absorbance measured at 510 nm in a spectrophotometer (Jasco V-630) and TFC was calculated from the standard curve of catechin.

2.5 *Antioxidativ property analysis*

Antioxidative properties of FSO, CO and FCO_1, FCO_2, FCO_3 were determined involving DPPH, FRAP, Reducing Power, ABTS and Metal chelating capacity assay.

2.5.1 *DPPH free radical scavenging assay*

Free radical scavenging activity of all the oils was measured by DPPH assay (Rebaya et al 2014). 2 mL of methanolic solution of the oil was poured into test tubes and 2 mL, 1 mM DPPH solution was added to it. The tubes were them kept in the dark for 30 min. Absorbance at 517 nm with a spectrophotometer (Jasco V-630) was measured. The percentage inhibition of the DPPH radical was calculated by the Eq. 5.

$$\% \text{ inhibition} = \frac{(Ac - As)}{Ac} \times 100 \tag{5}$$

Where, Ac = absorbance of the control, As = absorbance of the sample

2.5.2 *FRAP assay*

The antioxidant activity by FRAP assay was carried out by the method of the ferric reducing ability of plasma (FRAP) assay of Benzie and Strain 1996 (Gülçin *et al* (2010). 20 μL of oil sample was taken and to that 180 μL of FRAP reagent was added. The absorbance was measured at 593 nm in a UV spectrophotometer (Jasco V-630) at immediate addition of FRAP reagent (0 min) and after 4 min of FRAP reagent addition. Ascorbic Acid is used as standard solution. Value was calculated using the Eq. 6.

$$\text{FRAP activity} = \frac{sample\ changes\ from\ 0\ to\ 4\ min}{standard\ changes\ from\ 0\ to\ 4\ min} \times FRAP\ value\ of\ standard \tag{6}$$

2.5.3 *Reducing power assay*

Reducing Power assay was conducted as per standard method (Irshad *et.al* 2012). In a test tube 2.5 ml oil, 2.5 ml 0.2(M) sodium phosphate buffer and 1%, 2.5 ml Potassium ferricyanide were added, mixed well and incubated at 50°C for 20 min. 2.5 ml 10% trichloroacetic acid was added to it, filtered and the filtrate was collected. To the filtrate 5 ml distilled water and 1 ml ferric chloride was added. The absorbance was measured at 700 nm. Increase in OD value indicates increase in reducing power capacity.

2.5.4 *ABTS decolorization assay*

It was conducted according to standard methods (Rajurkar *et.al* 2011). To 20 μL oil, 2 mL of ABTS was added. The mixture was kept at 37°C for 10 minutes. Absorbance was measured at 734 nm with a spectrophotometer (Jasco V-630). The values were compared with standard BHT. The result was calculated according to the Eq. 5.

$$\%\ \text{Antioxidant activity} = \frac{(A_c - A_s)}{A_c} \times 100 \tag{7}$$

where A_c and A_s are the absorbances of control and sample respectively

2.5.5 *Metal chelating activity*

The ferrous ion chelating activity of all the oils were also measured (Mohan *et.al* (2012). 1 ml $FeSO_4$ was mixed with 20 μL of oil. To this Tris HCl buffer and 2,2′ bipyridyl solution was added and the content volume was made up to 5 ml with the addition of distilled water. The content was incubated for 10 min and the absorbance was checked at 522 nm. Percentage of metal cheating activity was measured according to Eq. 8.

$$\%\ \text{Antioxidant activity} = \frac{(A_c - A_s)}{A_c} \times 100 \tag{8}$$

where A_c and A_s are the absorbances of control and sample respectively

3 STATISTICAL ANALYSIS

All tests were performed in triplets and data is presented as mean ± SD (standard deviation). Statistical significance was performed using ANOVA one way HSD Tukey Test.

4 RESULT AND DISCUSSION

4.1 *Physicochemical property analysis*

Physicochemical properties of FSO, CO, FCO_1, FCO_2, and FCO_3 was included in Table 1. Highest PV was noticed for the FCO_3 blend and it is within the acceptable range of edible oil (**$p < 0.01$, *$p < 0.05$). Both the acid value and p-anisidine value for FCO_3 was also within the required range[15] (**$p<0.01$). Among the 3 oil blends FCO_3 is proved to be the most acceptable one.

4.2 *Determination of totox value*

Totox value of all the oils are mentioned in Figure 1. Lowest totox value is noticed for FCO_3. The lower totox value indicates that the oil is less prone to oxidation and so quality of the oil is higher. From the result FCO_3 shows to be most nutritionally enriched one.

Table 1. Physicochemical Properties of oil.

Indices	FSO	CO	FCO_1	FCO_2	FCO_3
AV(mg NaOH/gm)	3.36 ± 0.03 [a]	7.36 ± 0.01 [b]	11.23 ± 0.01[c]	2.80 ± 0.01 [d]	2.44 ± 0.01 [e]
FFA(%)	1.71 ± 0.03 [a]	1.15 ± 0.03 [b]	2.68 ± 0.02 [c]	1.44 ± 0.04 [d]	1.15 ± 0.03 [b]
PV(mEq O_2/Kg)	5.78 ± 0.03 [a]	3.28 ± 0.03 [b]	5.23 ± 0.03 [c]	10.25 ± 0.01[d]	3.19 ± 0.02 [e]
p-ANISIDINE VALUE	6.77 ± 0.02 [a]	10.28 ± 0.04[b]	9.83 ± 0.03 [c]	2.35 ± 0.04 [d]	3.67 ± 0.02 [e]

All test are performed three times. Data are expressed as mean ± SD. Different letter in the same row defines statistical significance **$p<0.01$ and *$p<0.05$

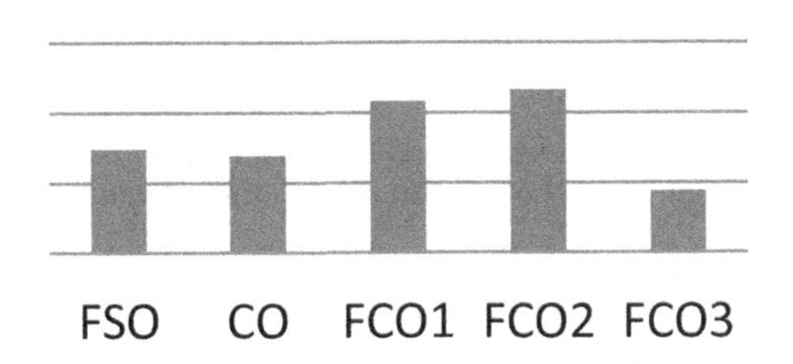

Figure 1. Totox values of oil.

Table 2. Antioxidative activity of oils.

Test Performed	FSO	CO	FCO_1	FCO_2	FCO_3
TPC (µg of GAE/mL)	6.52 ± 0.02 ^{a}	15.87 ±0.02 ^{b}	30.15 ±0.03 ^{c}	61.06 ±0.01 ^{d}	16.45 ±0.02 ^{e}
TFC (µg of RE/mL)	0.0 ^{a}	0.74 ±0.03 ^{b}	0.10 ±0.04 ^{c}	0.75 ±0.04 ^{b}	0.85 ±0.03 ^{d}
DPPH (%)	3.86 ±0.02 ^{a}	35.04 ±0.01 ^{b}	98.47± 0.02 ^{c}	56.18 ±0.02 ^{d}	69.92 ±0.06 ^{e}
FRAP (µg of AAE/mL)	990 ± 1.52 ^{a}	144.66±0.57^{b}	20.24± 0.04 ^{c}	44.45± 0.05 ^{d}	190.33±0.57^{e}
ABTS (%)	3.96 ±0.02 ^{a}	23.12± 0.02 ^{b}	14.17± 0.03 ^{c}	11.23± 0.01 ^{d}	38.75± 0.04 ^{e}
METAL CHELATING ACTIVITY (%)	41.00 ±0.03 ^{a}	19.36± 0.02 ^{b}	86.52± 0.03 ^{c}	80.90± 0.05 ^{d}	39.22± 0.02 ^{e}
REDUCING POWER ASSAY (increase in OD value)	0.7675±0.02^{a}	0.5429±0.02^{b}	0.6304±0.01^{c}	0.6770±0.06^{d}	0.6512±0.03^{d,c}

All test are performed three time. Data is expressed as mean ± SD. Different letter in the same row defines statistical significance $^{**}p<0.01$ and $^{*}p<0.05$.

4.3 *Antioxidative property analysis*

Changes in antioxidant activity of FSO, CO, FCO_1, FCO_2, and FCO_3 along with TPC and TFC value is was included in the Table 2. Changes in antioxidative properties are similar to previous work of similar aspect (Choudhary *et.al* 2013, Ngassapa *et.al* 2012). FSO which was deprived of flavonoid, when blended with CO in different ratios now possess flavonoid and the TFC content of the blended oils are FCO_3 › FCO_2› FCO_1 and the same result was observed in case of FRAP and ABTS assay. Free radical scavenging activity also increases. All the 3 blended oils have higher DPPH value than the crude oils and the increasing range is FCO_1 › FCO_3 › FCO_2. Changes are also noted in metal chelating and reducing power assay. Highest metal chelating activity was given by FCO_1, followed by FCO_2 and FCO_3. In case of reducing power assay highest value was for FCO_2. Total phenoilc content was highest for FCO_2 among the 3 blended oils.

5 CONCLUSION

Dietary fat plays an important and essential role in human health. To avail the nutritional benefit of oils, blending of different oils into appropriate ratio are needed as no single oil can provide all the nutrients. Here, blending increases both the antioxidant and physicochemical properties of the oils From the present study it can be concluded that among the blends of fish skin oil and chia seed oil, the FCO_3 blend is more oxidatively stable due to its lower totox value. Among the seven antioxidant tests that are performed to determine the antioxidative properties of the oils, FCO_3 has given higher values in three tests and it also has lowest totox value too. From this it can be concluded that FCO_3 has higher antioxidative properties than FCO_2 and FCO_1.. there is a correlation This oil blend can be used in food formulation as well as in therapeutic and cooking purpose.

REFERENCES

1. Aheem Rebaya et al (2014). Total phenolic, total flaonoid, tannin and antioxidant capacity of Halimium halimifolium(cistaceae). Jouranl of applied pharmaceutical science. 5(01) 052-057.
2. AOCS official methods.
3. Cardenas M *et.al* (2018). Content of nutrients component and fatty acids in chia seeds(*Salvia hispanica* L) cultivated in Ecuador. Asian J Pharma Clin Res. 11 (02) 387-390.
4. Dildar Ahmad *et.al* (2015). Comparative analysis of phenolics, flavonoids and antioxidant and antibacterial potential of methanolic, hexanic and aquous extracts from *Adiantum caudatum* leaves. Antioxidants. 4:394-409.
5. FN Ngassapa *et.al* (2012). Effects of temperature on the physicochemical properties of traditionally processed vegetable oils and their blends. Tanz. J. Sci. 38 (3).
6. Ixtaina *et.al.* (2011). Characterization of chia seed oil obtained by pressing and sovent extraction. Journal of food composition and analysis. 24:166-174.
7. İlhami Gülçin *et al* (2010). Radical scavenging activity and antioxidant activity of tannic acid. Arabian journal of chemistry. 3(1) 43-53.
8. Md Irshad *et.al* (2012). Comparative analysis of the antioxidant activity of *Cassia fistula* extracts. International journal of Medicinal chemistry.
9. Monika Choudhary *et.al* (2013). Fatty acid composition, oxidative stability and radical scavenging activity of rice bran oil blends. International Journal of Food and Nutritional Sciences. 2(01).
10. Nabanita Ghosh *et.al* (2019). Phytochemical and antioxidative activity of oil extracted from Indian carp fish (*Labeo rohita*). Int Res J Engg Tech (IRJET). 06 (02).
11. Nilima S. Rajurkar *et.al* (2011). Estimation of phytochemical content and antioxidant activity of some selected traditional Indian medicinal plants. Indian J. Pharm Sci. 73(2):146-151.
12. S. Chandra Mohan *et.al* (2012). Metal chelating activity and hydrogen peroxide scavenging activity of medicinal plant *Kalanchoe pinnata*. J. Chem. Pharm. Res. 4 (I):197-202.
13. Slinkard K *et.al* (1977). Total phenol analysis: Automation and comparison with manual methods. Am J Enol Vitic. 28:49-55.
14. UGC Sponsored National seminar- Recents advances in biological sciences, organised by Gurudas College in collaboration with CSIR-IICB,Kolkata and Dumdum Motijhil College. (2017). ISBN:978-93-86526-31-1.
15. Wan tatt. Wai *et al* (2009). Determination of TOTOXvalue in palm oleins using a FI-Potentometric analyzer. Food Chemistry. 113 (1) 285-290.
16. Yingbin Shen *et.al* (2018).Phytochemical and biological characteristics of Mexican chia seed oil. Molecules 23(3219).

Biotechnology and Biological Sciences – Sen et al. (Eds)
© 2020 Taylor & Francis Group, London, ISBN 978-0-367-43161-7

Cadmium induced deterioration of sperm quality: Protective role of coenzyme Q10 in rats

R. Saha & S. Roychoudhury
Department of Life Science & Bioinformatics, Assam University, Silchar, India

A. Varghese
Astra Fertility Group, Mississagua, Canada

P. Nandi
Department of Environmental Science, University of Calcutta, Kolkata, India

K. Kar
Mediland Hospital & Research Centre, Silchar, India

A. Paul Choudhury
Nightingale Nursing Home & Research Centre, Silchar, India

J.C. Kalita
Department of Zoology, Gauhati University, Guwahati, India

N. Lukac, P. Massanyi & A. Kolesarova
Faculty of Biotechnology & Food Sciences, Slovak University of Agriculture in Nitra, Slovakia

ABSTRACT: The study aimed at investigating the protective role of CoQ10 against Cd-induced reprotoxicity in male rats. Wistar rats were exposed to a daily dose of Cd (30 mg/kg bwt; Cd group), Cd+CoQ10 (30 mg/kg bwt + 10 mg CoQ10; Cd-Q10 group) and distilled water (healthy animal control) for 15 days. Semen analysis revealed a significant reduction in sperm concentration, motility, progressive motility, and DNA integrity in both Cd- and Cd-Q10 groups indicating Cd-induced testicular lipid per oxidation (LPO) ($p<0.05$). However, simultaneous co-administration of CoQ10 along with Cd (Cd-Q10 group) was able to improve sperm motility and LPO compared to Cd group ($p<0.05$). Sperm concentration, progressive motility, vitality, and DNA integrity also showed an improving trend although the differences between Cd- and Cd-Q10 groups were statistically not significant. Findings suggest potentiality of CoQ10 in providing moderate protection against reprotoxicity of high dose of Cd by improving sperm motility and reducing oxidative stress.

1 INTRODUCTION

Cadmium (Cd) is a non-essential toxic heavy metal even at low concentrations. Its exposure can cause a number of health issues including deterioration of reproductive health (Kumar & Sharma, 2019). Absorption of Cd is influenced by the routes of exposure: principally through inhalation by means of tobacco smoke, and also through the oral route by means of contaminated water and food (offal and seafood) (ATSDR, 2012). Its toxicity manifests principally through the induction oxidative stress leading to cell damage and subsequent death (Kukongviriyapan et al. 2016, Patra et al. 1999). Coenzyme Q10 (CoQ10) is an antioxidant in plasma membranes and lipoproteins, and is an essential component of the mitochondrial electron transport chain (Hernandez-Camacho et al. 2018). It possesses the ability to scavenge free radicals and prevent the instigation and transmission of lipid per oxidation (LPO) in cellular biomembranes (Cervellati & Grecoa, 2016), thereby stimulating cell growth and inhibiting cell death (Tawfik, 2015). Recently, our research group reported Cd-induced alteration of semen quality in rats at a dose 25 mg/kg bwt Cd as well as its amelioration by 10 mg CoQ10 (Saha et al. 2019). The bjective of the present study was to examine the role of CoQ10 in protecting against reprotoxicity of a relatively high dose of Cd in male Wistar rats *in vivo*.

2 MATERIAL AND METHODS

Male Wistar rats (n=15) of reproductive age were administered cadmium chloride ($CdCl_2$) orally every morning for 15 consecutive days. They were divided into 3 groups: each consisting 5 animals – healthy animal control (received distilled water only), Cd (received Cd at 30 mg/kg bwt), and Cd-Q10 (received Cd + CoQ10 @ 30 mg/kg bwt + 10 mg/kg bwt). On 16th morning, initial and final body weights were noted prior to sacrifice by cervical dislocation. From each animal, pair of testes and *cauda epididymis* were dissected out by laprotomy, gently rinsed in phosphate buffered saline (PBS, pH 7.4) followed by cleaning off the adhering tissues and weighing. *Cauda epididymis* was minced gently in 2ml PBS and incubated at 37°C to await semen analyses (using Labomed-LX 300 phase contrast microscope) using Makler counting chamber (Roychoudhury et al. 2010, 2016), including sperm DNA integrity (using acridine orange dye under fluorescent microscope Olympus-CX 31-TR) (Varghese et al. 2009), and testicular LPO (as amount of malondialdehyde produced) (Saha et al. 2019).

Each experiment was performed thrice. Differences between the experiments were evaluated using one way ANOVA with Scheffe's post hoc comparison of SPSS version 20 software and considered statistically significant at $p<0.05$.

3 RESULTS AND DISCUSSION

Initial body weights were similar in animals of all three groups. In Cd group, there was a significant decline ($p<0.05$) in final body weight in comparison to both healthy animal control as well as Cd-Q10 groups indicating that 10 mg CoQ10 was able to overpower the Cd-induced decline in body weight (Table 1). There was a significant decrease ($p<0.05$) in testicular weight too in both Cd- and Cd-Q10 groups as compared to healthy control. A significant ($p<0.05$) improvement in testicular weight was also noted in Cd-Q10 group as compared to Cd group (Table 1).

Semen analyses revealed that sperm concentration, motility, progressive motility, vitality and DNA integrity were significantly lower ($p<0.05$) in both Cd- and Cd-Q10 groups as compared to healthy animal control confirming the reprotoxicity. Co-application of CoQ10 was

Table 1. Body weight and testicular weight of Wistar rats after cadmium administration and co-application of coenzyme Q10.

Group	Final body weight	Testicular weight
	g	g
Control	170.00 ± 11.52	1.00 ± 0.05
Cd	123.67 ± 0.88^{a}	0.70 ± 0.03^{b}
Cd-Q10	176.67 ± 0.87	$0.84 \pm 0.02^{b,c}$

*Mean ± SEM. a: $p < 0.05$ in comparison to two other groups; b: $p < 0.05$ in Cd and Cd-Q10 groups in comparison to healthy control; c: $p < 0.05$ in Cd-Q10 group in comparison to Cd group

able to recover the toxic effect of Cd on sperm motility as it increased significantly ($p<0.05$) in Cd-Q10 group in comparison to Cd group. Sperm concentration, progressive motility, vitality and DNA integrity also showed an improving trend although the differences between Cd- and Cd-Q10 groups were not statistically significant (Figure 1).

In comparison to healthy control, higher levels of testicular LPO were noted in both Cd- and Cd-Q10 groups ($p<0.05$) in this study. However, the protective role of CoQ10 was evident from the significant decline ($p<0.05$) in LPO in Cd-Q10 group in comparison to Cd group (Figure 2).

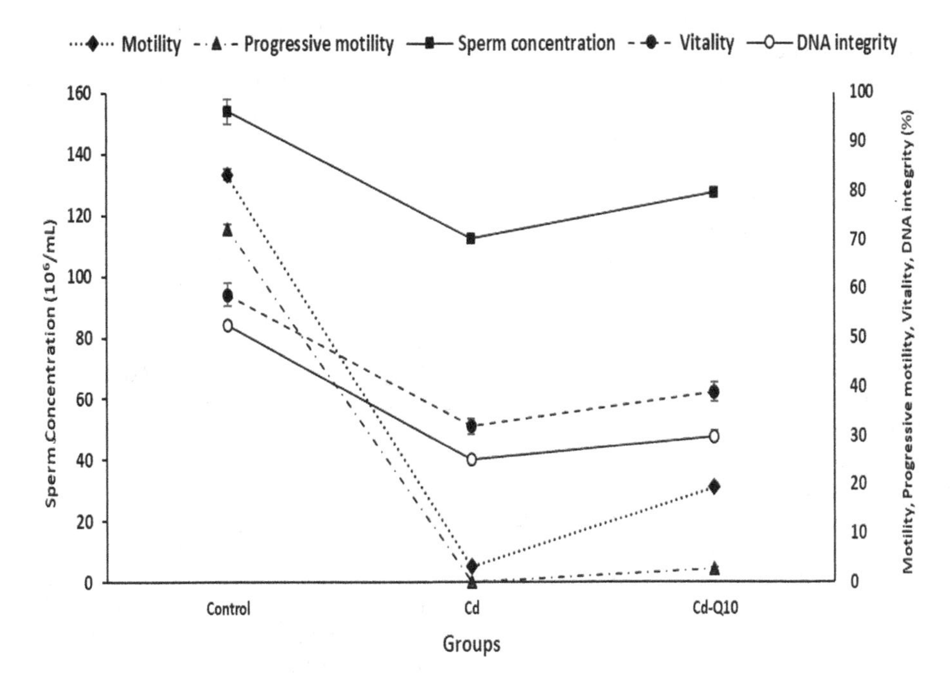

Figure 1. Moderate recovery of reprotoxicity of cadmium by coenzyme Q10 in male rats. Sperm motility, progressive motility, vitality and DNA integrity in healthy animal control, Cd- and Cd-Q10 groups. Mean±SEM.

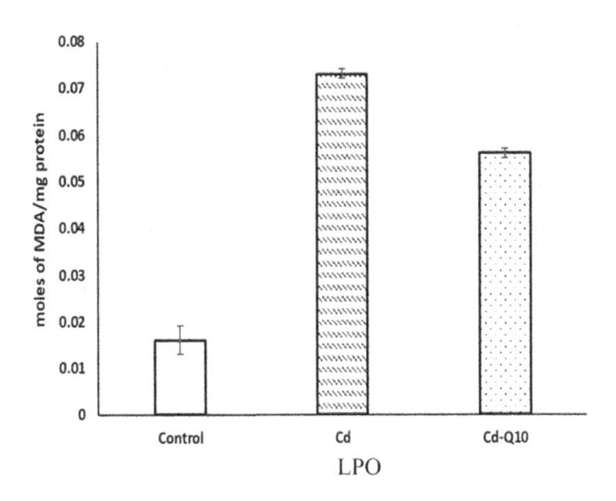

Figure 2. Moderate recovery of reprotoxicity of cadmium by coenzyme Q10 in male rats. Testicular lipid per oxidation (LPO) in healthy animal control, Cd- and Cd-Q10 groups. Mean±SEM.

In biological systems, transition metals are mostly conjugated or bound to proteins (Fraga, 2005). They can interact additively, synergistically or antagonistically and affect each other's absorption, distribution and excretion. Such metals including Cd, chromium, mercury, nickel, zinc, and copper affect the reproductive system (Chandel & Jain, 2014). *In vitro* studies showed the affect of Cd on human, mice (Zhao et al. 2017) and rabbit (Roychoudhury et al. 2010) sperm quality including the concentration dependent decrease in sperm motility. Recently, at a high dose of 25 mg/kg bwt Cd reduced the semen quality in rats and enhanced the testicular LPO, wherein co-application of Q10 at a low dose of 10 mg/kg bwt was found to moderately ameliorate the reprotoxicity (Saha et al. 2019). These findings directly support the results of the present study wherein a higher dose of 30 mg/kg bwt Cd was able to recover the toxic effect of Cd on body weight, sperm motility, and testicular LPO.

Moreover, a previous study on rats reported reduced testicular as well as epididymal weights after 7 and 56 days of $CdCl_2$ administration at a low dose of 1.2 mg/kg bwt, which is in agreement with the present study (de Souza Predes et al. 2010). Cadmium-induced decline in sperm quality parameters as evident in the present study is believed to play putative role in testicular damage, and relates directly to male fertility (Akunna et al. 2017). In fact, reprotoxicity of Cd is manifested through the impairment of testicular function by inducing oxidative stress (Amara et al. 2006, Kukongviriyapan et al. 2016), which is a common pathology in approximately half of all infertile men. Further, it is correlated with pathogenesis of sperm DNA damage (Aitken et al. 2010).

In an earlier study, Cd was found to inhibit sperm motility, morphology and membrane integrity in rabbits even at a low dose (0.62 μg $CdCl_2$/ml) *in vitro* (Roychoudhury et al. 2010). In a human study involving infertile males, a negative correlation was established between seminal Cd levels and sperm concentration and motility. The likelihood of increase in the magnitude of this was associated with advancing age, diet, smoking and tobacco chewing habits (Pant et al. 2014). Another human study also reported negative association between seminal Cd concentrations and sperm progressive motility, vitality, DNA fragmentation and seminal ROS (Taha et al. 2013). Daily administration of Cd at 0.5 mg/kg bwt for 90 days was also found to enhance testicular LPO and inhibit superoxide dismutase (SOD) activity in rats (Patra et al. 1999). A low dose of Cd at 1 mg/kg bwt administered for a period of 5 - 8 weeks showed enhanced LPO and relative depletion of testicular antioxidant levels in rats (Acharya *et al.* 2008). Another rat study also postulated that Cd induces morphological and functional abnormalities on sperm cells by reducing antioxidant status, particularly SOD, glutathione, catalase and malondialdehyde. Single doses of Cd at 5 and 7 mg/kg bwt administered to Wistar rats were found to decrease sperm count, motility and morphology (Akunna et al. 2016). Incorporated in radish bulbs at 1.1 mg Cd/g of diet was capable of decreasing testicular malondialdehyde concentrations and glutathione levels after 12 weeks of treatment, while SOD activity was increased as compared with the control rats (Haouem & El Hani, 2013).

Interestingly, selenium was reported to reduce Cd-induced alterations in testicular LPO in rats when co-administered with Cd (Yiin et al. 1999). In its reduced form ubiquinol, CoQ10 possesses the capacity to neutralize tocopheroxyl radicals (Nagaoka *et al.* 2000). Pre-treatment with *Allium cepa* extract was also found to prevent Cd-induced reprotoxicity by improving testicular weight, sperm quality and testicular LPO status (Ige et al. 2012), whereas in the present study CoQ10 was able to bring about a similar effect. Previous authors also reported the protective effects of CoQ10 in enhancing sperm motility, LPO and DNA fragmentation *in vitro* (Talevi et al. 2013). At the same dose as used in the present study, CoQ10 it was able to recover testicular toxicity of 5 mg/kg bwt Cd in rabbits, particularly by reducing testis histopathological changes, and increasing the levels of testosterone and luteinizing hormones (Abdel-Hady & Abdel Rahman, 2011).

4 CONCLUSIONS

In the present study, moderate increase in body weight and sperm motility was noted along with a decline in testicular LPO in the Cd-Q10 group with 30 mg/kg bwt Cd co-administered along with 10 mg CoQ10 for a period of 15 days in comparison to Cd group with 30 mg/kg bwt Cd-exposure only. This ability of a low dose of CoQ10 to protect against toxic effects of a high dose of Cd probably indicates toward the capability of CoQ10 to replace redox-active metals. This further supports the potential use of CoQ10 in the management of heavy metal induced reprotoxicity and possibly infertility in human males through mechanisms involving enhancement of semen quality and reduction of oxidative stress.

REFERENCES

Abdel-Hady, E.S.K. & Abdel-Rahman, G.H. 2011. Protective effect of coenzyme Q10 on cadmium-induced testicular damage in male rabbits. *American-Eurasian Journal of Toxicological Sciences* 3(3): 153–160.

Acharya, U.R., Misra, M., Patro, J. & Panda, M.K. 2008. Effect of vitamins C and E on spermatogenesisin mice exposed to cadmium. *Reproductive Toxicology* 25(1): 84–88.

Aitken, R.J., de Iuliis, G.N., Finnie, J.M., Hedge, A. & McLachlan, R.I. 2010. Analysis of the relationships between oxidative stress, DNA damage and sperm vitality in a patient population: development of diagnostic criteria. *Human Reproduction* 25(10): 2415–2426.

Akunna, G., Obikili, E., Anyanwu, E. & Esom, E. 2017. Protective and curative role of citrus sinensispeel on cadmium-induced testicular damage: a morphometric and immunihistochemical evaluation using monoclonal antibodies against Ki-67 and proliferating cell nuclear antigen. *European Journal of Anatomy* 21(1): 19–30.

Amara, S., Abdelmelek, H., Garrel, C., Guiraud, P., Douki, T., Ravanat, J.L., Favier, A., Sakly, M. & BenRhouma, K. 2006. Influence of static magnetic field on cadmium toxicity: study of oxidative stress and DNA damage in rat tissue. *Journal of Trace Elements Medicine and Biology* 20(4): 263–269.

ATSDR. U.S. 2012. Toxicological Profile for Cadmium. Department of Health and Human Services, Public Health Service, Centers for Disease Control Atlanta; ATSDR; Atlanta, GA, USA. [assessed on 2 July 2019]. Available online: https://www.atsdr.cdc.gov/toxprofiles/tp.asp?id=48&tid=15.

Cervellati, R. & Grecoa, E. 2016. *In vitro* antioxidant activity of ubiquinone and ubiquinol, compared to vitamin E. *Helvetica Chemica Acta* 99(1): 41–45.

Chandel, M. & Jain, G.C. 2014. Toxic effects of transition metals on male reproductive system: a review. *Journal of Environmental and Occupational Science* 3(4): 204–213.

de Souza Predes, F., Diamante, M.A.S. & Dolder, H. 2010. Testis response to low doses of cadmium in Wistar rats. *International Journal of Experimental Pathology* 91(2): 125–131.

Fraga, C.G. 2005. Relevance, essentiality and toxicity of trace elements in human health. *Molecular Aspects in Medicine* 26(4-5): 235–244.

Haouem, S. & El Hani, A. 2013. Effect of cadmium on lipid peroxidation and on some antioxidants in theliver, kidneys and testes of rats given diet containing cadmium-polluted radish bulbs. *Journal ofToxicology and Pathology* 26(4): 359–364.

Hernandez-Camacho, J.D., Bernier, M., Lopez-Lluch, G. & Navas, P. 2018. Coenzyme Q10 supplementation in aging and disease. *Frontiers in Physiology* 9: 44.

Ige, S.F., Olaleye, S.B., Akhigbe, R.E., Akanbi, T.A., Oyekunle, O.A. & Udoh, U.A.S. 2012. Testiculartoxicity and sperm quality following cadmium exposure in rats: Ameliorative potentials of Allium cepa. Journal of Human Reproductive Sciences 5(1): 37–42.

Kukongviriyapan, U., Apaijit, K. & Kukongviriyapan, V. 2016. Oxidative stress and cardiovascular dysfunction associated with cadmium exposure: Beneficial effects of curcumin and tetrahydrocurcumin. *Tohoku Journal of Experimental Medicine* 239(1): 25–38.

Kumar, S. & Sharma, A. 2019. Cadmium toxicity: effects on human reproduction and fertility. *Reviews on Environmental Health* May 27. Pii:/j/reveh.ahead-of-print/reveh-2019-0016/reveh-2019-0016.xml.

Nagaoka, S., Inoue, M., Nishioka, C., Nishioku, S., Ohguchi, C., Okhara, K., Mukai, K. & Nagashima, U. 2000. Tunneling effect in antioxidant, prooxidant and regeneration reactions of vitamin E. *Journal of Physical Chemistry B* 104(4): 856–862.

Pant, N., Kumar, G., Upadhyay, A.D., Gupta, Y.K. & Chaturvedi, P.K. 2015. Correlation between leadand cadmium concentration and semen quality. *Andrologia* 47(8): 887–891.

Patra, R.C., Swarup, D. & Senapati, S.K. 1999. Effects of cadmium on lipid peroxides and superoxide-dismutase in hepatic, renal and testicular tissue of rats. *Veterinary and Human Toxicology* 41(2): 65–67.

Roychoudhury, S., Massanyi, P., Bulla, J., Choudhury, M.D., Lukac, N., Filipejova, T., Trandzik, J., Toman, R. & Almasiova, V. 2010. Cadmium toxicity at low concentration on rabbit spermatozoa spermatozoa motility, morphology and membrane integrity *in vitro. Journal of Environmental Science & Health A Toxic/Hazardous Substances & Environmental Engineering* 45(11): 1374–1383.

Roychoudhury, S., Sharma, R., Sikka, S. & Agarwal, A. 2016. Diagnostic application of total antioxidant capacity in seminal plasma to assess oxidative stress in male factor infertility. *Journal of Assisted Reproduction and Genetics* 33(5): 627–635.

Saha, R., Roychoudhury, S., Kar, K.K., Varghese, A.C., Nandi, P., Sharma, G.D., Formicki, G., Slama, P. & Kolesarova, A. 2019. Coenzyme Q10 ameliorates cadmium induced reproductive toxicity in male rats. *Physiological Research* 68(1): 141–145.

Taha, E.A., Sayed, S.K., Ghandour, N.M., Mahran, A.M., Saleh, M.A., Amin, M.M. & Shamloui, R. 2013. *Central European Journal of Urology* 66(1): 84–92.

Talevi, R., Barbato, V., Fiorentino, I., Braun, S., Longobardi, S. & Gualtier, R. 2013. Protective effects if*in vitro* treatment with zinc, d-astertate and coenzyme q10 on human sperm motility, lipid peroxidation and DNA fragmentation. *Reproductive Biology and Endocrinology* 11: 81.

Tawfik, M.K. 2015. Combination of coenzyme Q10 with methotrexate suppresses Freund's complete adjuvant induced synovial inflammation with reduced hepatotoxicity in rats: eftfect of oxidative stress and inflammation. *International Immunopharmacology* 24(1): 80–87.

Varghese, A.C., Bragais F.M., Mukhopadhyay, D., Kundu, S., Pal, M., Bhattacharyya, A.K. & Agarwal, A. 2009. Human sperm DNA integrity in normal and abnormal semen samples and its correlation with sperm characteristics. *Andrologia* 41(4): 207–215.

Yiin, S.J., Chern, C.L., Sheu, J.Y. & Lin, T.H. 1999. Cadmium induced lipid peroxidation in rat testesand protection by selenium. *Biometals* 12(4): 353–359.

Zhao L., Ru Y., Liu M., Tang J., Zheng J., Wu B., Gu Y. & Shi, H. 2017. Reproductive effects of cad-mium on sperm function and early embryonic development *in vitro. PLoS ONE* 12: e0186727.

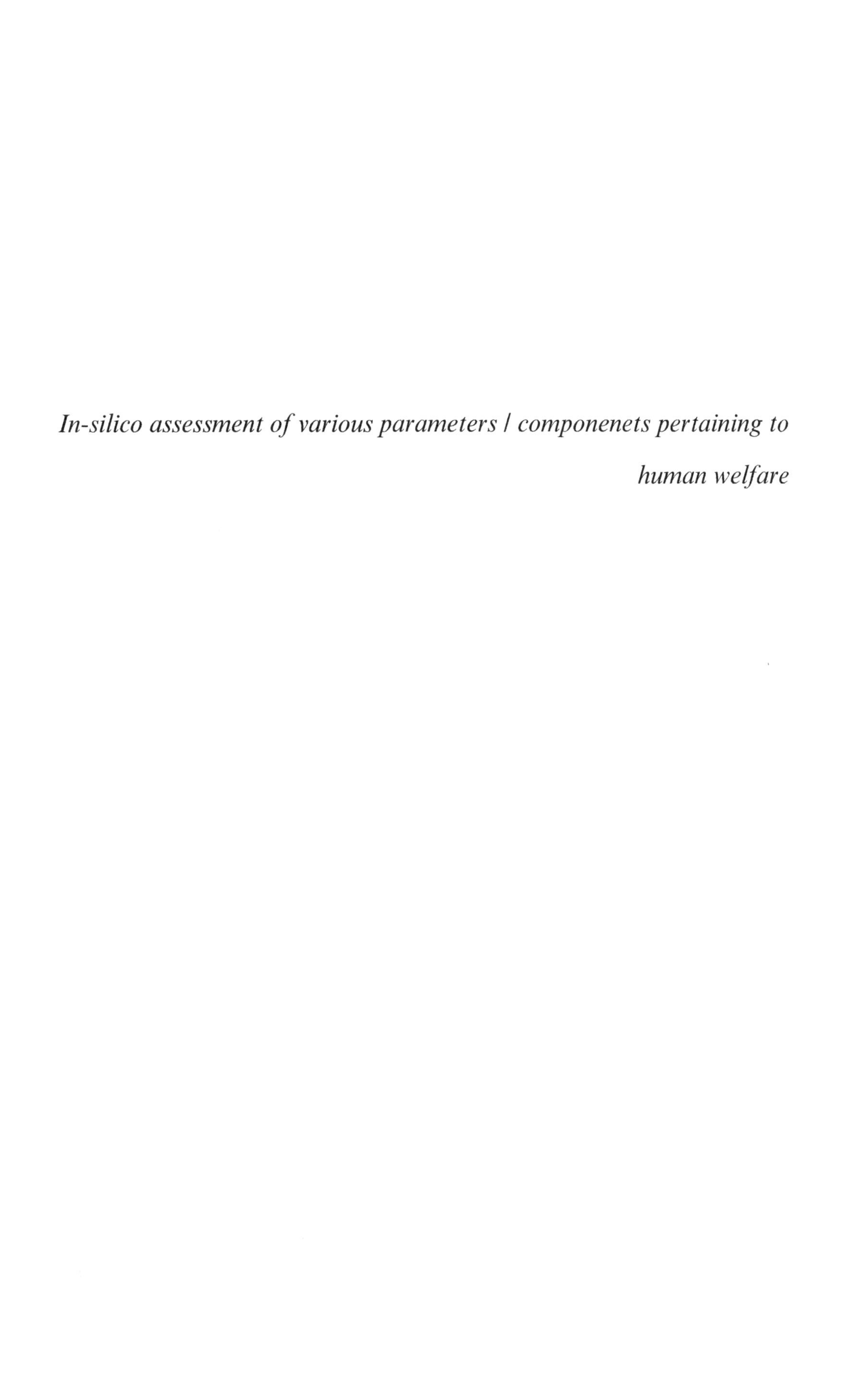

In-silico assessment of various parameters / componenets pertaining to human welfare

Biotechnology and Biological Sciences – Sen et al. (Eds)
© 2020 Taylor & Francis Group, London, ISBN 978-0-367-43161-7

Assessment of anticancer properties of selected medicinal plants

Shreya Vora*
Parul University, Waghodia, Vadodara, Gujarat

ABSTRACT: Cancer is a major public health concern all over the world. Herbal medicines play vital role in the prevention and treatment of cancer. Some medicinal plants have been proved very effective anticancer remedies and have been used since ages. However, the drug formulations were not distinctly clear except after their proper effective assays within the system applied. Active compounds from plants such as *Andrographis paniculata, Aegle marmelos, Glycyrrhiza glabra, Elephantopus scaber, Cistanche tubulosa have* been reported in literature. Dried extracts prepared from these plants and their phytochemical analysis was reported. To check cytotoxic effects of the active compounds in the extracts MTT assay can be formulated. These plant extracts might possibly have effective significance on preventing cancerous tissues.

Keywords: Herbal medicine, Cytotoxicity, phytomedicine, anticancer

1 INTRODUCTION

Cancer is one of the most severe health problems in both developing and developed countries, worldwide. Among the most common (lung, stomach, colorectal, liver, breast) types of cancers, lung cancer has continued to be the most common cancer diagnosed in men and breast cancer is the most common cancer diagnosed in women. The International Agency for Research on Cancer estimates of the incidence of mortality and prevalence from major types of cancer, at national level, for 184 countries of the world revealed that there were 14.1 million new cancer cases, 8.2 million cancer deaths, and 32.6 million people living with cancer (within 5 years of diagnosis) in 2012 worldwide. Therefore, there is a constant demand to develop new, effective, and affordable anticancer drugs. From the dawn of ancient medicine, chemical compounds derived from plants have been used to treat human diseases. Natural products have received increasing attention over the past 30 years for their potential as novel cancer preventive and therapeutic agents. In parallel, there is increasing evidence for the potential of plant-derived compounds as inhibitors of various stages of tumorigenesis and associated inflammatory processes, underlining the importance of these products in cancer prevention and therapy. derived from plants have been used to treat human diseases. The increasing evidence for the potential of plant-derived compounds as inhibitors of various stages of tumorigenesis and associated inflammatory processes, underlining the importance of these products in cancer prevention and therapy. Some medicinal plants such as *Andrographis paniculata, Aegle marmelos, Glycyrrhiza glabra, Elephantopus scaber, Cistanche tubulosa* have been found effective in various types of cancers (Table-1) These medicinal plants maintain the health and vitality of individuals, and also cure various diseases, including cancer without causing toxicity. In this review, these anticancer medicinal plants of natural origin have been presented. These medicinal plants possess good immunomodulatory and antioxidant properties, leading

*Corresponding author: sddave08@gmail.com

Table 1. List of plants having anticancer properties- considered for this study.

Plant name	Family	Parts used	Traditional use
Andrographis paniculata	Acanthaceae	Leaves	Blood purification
Aegle marmelos	Rutaceae	Fruits, leaves	Arthritis, Anaemia, Fractures
Glycyrrhiza glabra	Fabaceae	Roots	Cough, hepatitis
Elephantopus scber	Asteraceae	Whole plant	Astringent, cardiac tonic
Cistanche tubulosa	Orobanchaceae	Stems	Reproductive problems, improve learning ability memory

to anticancer activities. The antioxidant phytochemicals protect the cells from oxidative damage.

2 MATERIALS & METHODS PREPARATION OF PLANT EXTRACTS

Desired parts of selected plants were removed from the plants and then washed under running tap water to remove dust. The plant samples were then air dried for few days and the leaves were crushed into powder and stored in polythene bags for use. The plant powder was taken in a test tube and distilled water was added to it such that plant powder soaked in it and shaken well. The solution then filtered with the help of filter paper and filtered extract of the selected plant samples were taken and used for further phytochemical analysis [23]. Methanolic extract: 10 g of each powdered leaves were placed in conical flask and 100 ml of methanol was added and plugged with cotton. The powder material was extracted with methanol for 24 hours at room temperature with continuous stirring. After 24 hours the supernatant was collected by filtration and the solvent was evaporated to make the crude extract. The residues obtained were stored in airtight bottles in a refrigerator for further use. Plant materials were dried at 37 °C, powdered and extracted in different solvent. The aqueous extract was obtained by boiling dried ground plant material (100 g) for 30 minutes in distilled water (300 ml). All extracts were fine-filtered and freeze dried. For the ethanolic extracts, dried ground plant material (100 g) was percolated with 95% ethanol and concentrated to dryness under reduced pressure. The aqueous extracts were dissolved in sterile water and the ethanolic extracts in Dimethyl Sulfoxide (DMSO) to form stock solutions 20mg/ml which were filter sterilized (0.2 pm) before testing on cell lines.

2.1 *Phytochemical Analysis*

The methanolic extracts of following plants was subjected to different chemical tests for the detection of different phytoconstituents using standard procedures

- Test for Tannins
- Test for Saponins
- Test for Flavonoids
- Test for Steroids:
- Test for Alkaloids
- Test for Terpenoids
- Test for Cardiac glycosides
- Test for carbohydrates:

In vitro assay for cytotoxic activity Cell culture

The cell line under investigation was human breast adenocarcinoma (MCF7). It was purchased from the European Collection of Animal Cell Culture. The cells were cultured in

RPMI 1640 medium supplement with 10% heated foetal bovine serum, 1% of 2 mM l-glutamine, 50 IU/ml penicillin and 50 g/ml streptomycin

MTT Assay: MTT is in vitro cytotoxicity assay, considered one of the most economic, reliable and convenient methods. This is based on its ease of use, accuracy and rapid indication of toxicity as well as their sensitivity and specificity. The assay is in vitro whole cell toxicity assays that employ colorimetric method for determining the number of viable cells based on mitochondrial dehydrogenase activity measurement and differ only in the reagent employed.

2.2 *Calculation*

% viability = (OD of test material/OD of control) X 100
% Inhibition = 100 - (% Viability)

3 RESULTS & DISCUSSION

Table 2. Phytochemicals present in selected medicinal plants.

	A.	A.	G.	E.	C.
Phyto chemicals	Paniculate	Marmelo	Glabr	Scabe	Tubulos
	A	S	A	R	A
Tannins	+	+	+	+	-
Saponins	-	+	+	+	+
Alkaloids	+	+	-	-	+
Terpenoids	+	+	+	+	-
Flavanoids	+	+	+	+	-
Steroids	+	+	-	+	-
Glycosides	+	-	-	-	+
Carbohydrate S	+	+	+	-	-

Graph below shows highest activity of ethanolic extract of *A.Paniculata* at 48 hours:

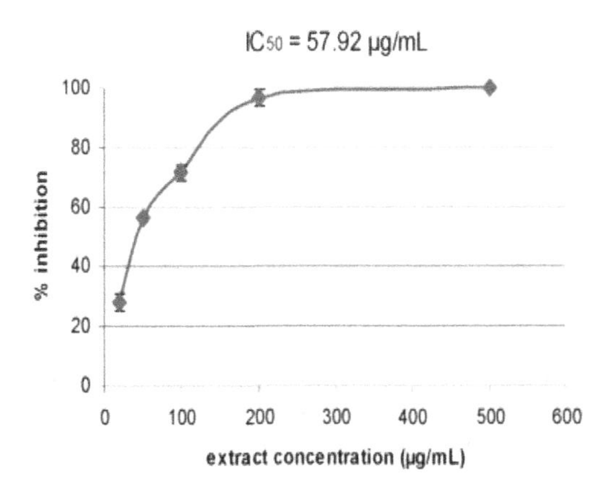

Figure 1. IC_{50} value of *A.paniculata* at 48 hours.

- Ethanol extract shows more inhibition of cells when compared than other, may due to the presence of alkaloids and flavonoids and its main active component andrographolide. Minimum inhibitory concentration was observed based on the percentage of cell viability is 50% at 200 μg/ml for ethanol extracts.
- Graph below shows highest activity of ethanolic extract of *A. Marmelos* at 24 hours:

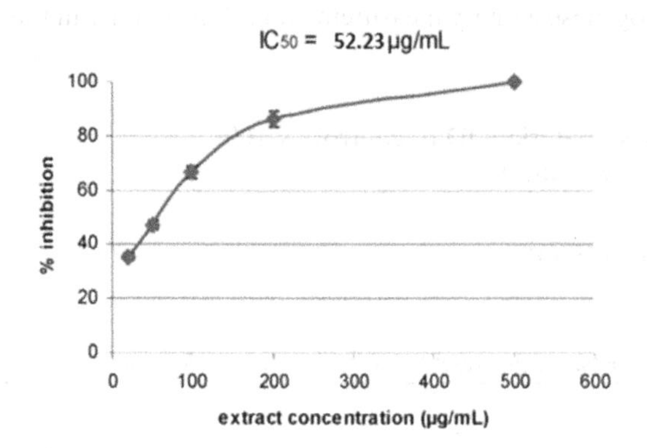

Figure 2. IC_{50} value of A.marmelos at 24 hours.

- The ethanolic extract of the leaves of *A. marmelos* were reported to possess high cytotoxic anticancer effects against human tumor cell line. The results showed the inhibition of *in vitro* proliferation of human tumor cell line at 24 hours is standard for *A. marmelos.*
- Graph below shows highest activity of ethanolic extract of *G. Glabra* at 48 hours:

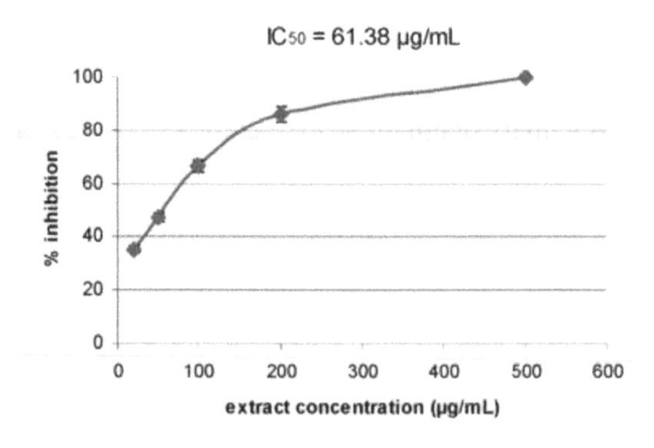

- *G.glabra* ehibit quite potential anticancer activity at 48 hours but its more than standard 50% value. That represents active compounds present in *G.glabra* like glycyrryzin and glycyrric acid shows potential cell cytotoxicity.
- Graph below shows highest activity of ethanolic extract of *E. Scaber* at 48 hours:

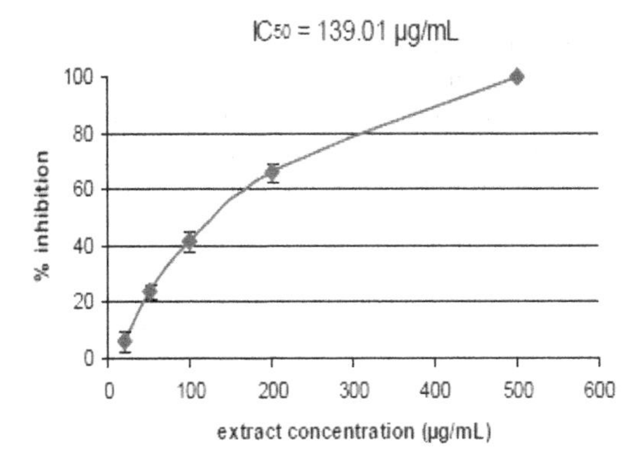

- Graph below shows highest activity of ethanolic extract of *C. Tubulosa* at 48 hours:

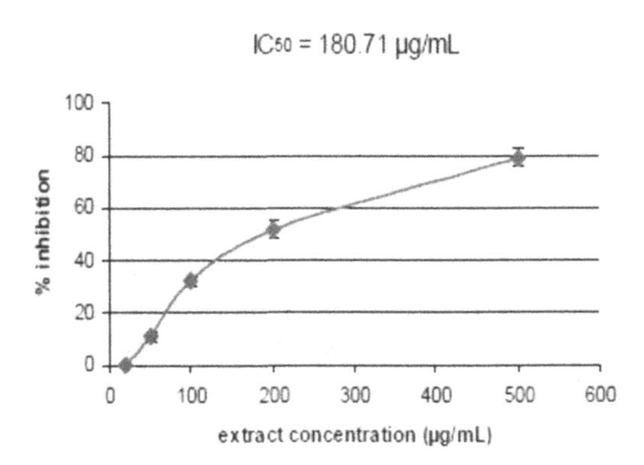

- Antioxidant activity: Ethanol and methanol extracts of the five traditional medicinal plants were tested for their ability to scavenge DPPH radical. The extracts showed dose-dependent DPPH scavenging activities. Methanol extracts generally exhibited higher antioxidant activity compared to ethanol extracts. The DPPH scavenging activity for the methanol extract of *A.paniculata* and *A.marmelos* were comparable to the activity of the synthetic antioxidant. For all extracts, *A.marmelos* showed the highest DPPH scavenging activity, followed by *G.glabra* a, *E.Scaber* and *C.tubulosa* showing the least activity. The concentration of the extracts that was able to scavenge at least 50% of the DPPH dye (I%) was calculated. *The results are summarized in table below:*

Table 3. DPPH Method for Determining the I% of methnolic extracts of plants.

Sr. no.	Volume of sample (200µl)	Percentage (%) of Inhibition (I%)
1.	*A. Paniculata*	49.0
2.	*A. marmelos*	55.0
3.	*G.glabra*	42.5
4.	*E.scaber*	97.2
5.	*C. tubulosa*	80.3

- Highest antioxidant activity was given by *A.marmelos* extract at the concentration of 170µg/ml among all the methanolic leaves which is found to be more than even the ascorbic acid while activity of *A.paniculata* was found to close to the standard. Thus, it is clear that polyphenolic antioxidants in leaves of selected plants play an important role as bioactive principles and the scavenging effect can be attributed to the presence of active phytoconstituents in them.

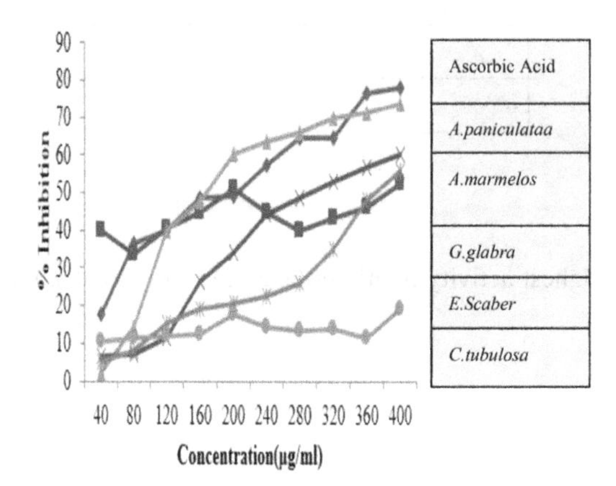

Figure 3. Comparison of Free radical scavenging activity of methanolic plant extrats with standard ascorbic acid.

- All five plants were showing effective anticancer activity but among them also *A. paniculta* and *A.marmelos* are most effective. G. glabra shows higher activity then *E.scaber* and *C. tubulosa* but laser than *A.marmelos*. Anticancer activity is because of antioxidant activity of phyto constituents present in the plant extracts. Therefore, antioxidant properties of selected plants.

4 CONCLUSION

In this study *Andrographis paniculata* and *Aegle marmelos* showed higher antioxidant and anticancer activity compare to *Glycyrria glabra, Elephantopus scaber* and *Cistanche tubulosa*. Though all plants exhibit antioxidant property in lower or higher amount. Therefore all five plants have anticancer properties but *A.marmelos* proven to be most effective and *C.tubulosa* is least explored plant.

Though there are anticancer products of plant origins in clinical testing at the moment, the search for new products is of outmost importance as diseases, including cancer, are constantly developing resistance to existing drugs. Screening plants for other biologically active compounds such as antifungal and antibacterial activities and not just anticancer and antioxidants will also help in the implementation of conservation measures for medicinally useful plants.

REFERENCES

www.who.com

Thun MJ, DeLancey JO, Center MM, Jemal A, Ward EM. The global burden of cancer: priorities for prevention, Carcinogenesis. 2009.31(1):100–110.

Coseri S. Natural products and their analogues as efficient anticancer drugs, Mini-Reviews in Medicinal Chemistry. 2009; 9(5):560–571.

Newman DJ. Natural products as leads to potential drugs: an old process or the new hope for drug discovery? Journal of Medicinal Chemistry. 2008; 51(9):2589–2599.

Newman DJ, Cragg GM, Snader KM. Natural products as sources of new drugs over the period 1981-2002. Journal of Natural Products. 2003; 66(7):1022–1037.
Gordaliza M. Natural products as leads to anticancer drugs, Clinical and Translational Oncology. 2007; 9(12):767–776.
Saklani, Kutty SK. Plant-derived compounds in clinical trials, Drug Discovery Today. 2008; 13 (3-4):161–171.
Gerson-Cwillich R, Serrano-Olvera A, Villalobos-Prieto A. Complementary and alternative medicine (CAM) in Mexican patients with cancer, Clinical and Translational Oncology, 2006; 8(3):200–207.
Tascilar M, de Jong FA, Verweij J, Mathijssen RHJ. Complementary and alternative medicine during cancer treatment: beyond innocence, Oncologist. 2006; 11(7):732–741.
Cassileth BR, Deng G. Complementary and alternative therapies for cancer, Oncologist. 2004; 9 (1):80–89.
Molassiotis, et.al. International Journal of Gynecological Cancer.
Shimoda et al. The American journal of Chinese medicine. 2009;37. 1125–1138.
Fiore C, Salvi M, Palermo M, Sinigagliab G, Armaninia D, Toninello A, 2004. *Biochim Biophys Acta.* On the mechanism of mitochondrial permeability transition induction by glycyrrhetinic acid: 195–201.
Katherine Fallon, 2014. http://fallonillustration.com/#entry-8.
Lee K H, Cowherd C M and Wolo M T., 1975. J Pharm Sci Antitumor agents. XV: deoxyelephantopin, an antitumor principle from Elephantopus carolinianus Willd., 64, 1572–1573.
Leung, H.Y. and K.M. Ko., 2008. Phamaceutical Bio., Herba Cistanche extract enhances mitochnondrial ATP generation in rat hearts and H9c2 cells, 46: 418–424.
McCauley Janice, Ana Zivanovic, Danielle Skropeta, 2013. Research online Bioassays for anticancer activities.
Nadkarni KM., 1976. Indian Materia Medica Mumbai: Popular Prakashan Pvt. Ltd. p.582–584.
R. Ian Freshney. Culture of animal cells.
Shanthy sundaram, satish kumar verma, priyanka dwivedi, 2011. In vitro cytotoxic activity of Indian medicinal plants used traditionally to Treat cancer. Vol. 4, Issue 1.
Song Z.H., S.H. Mo, Y. Chen, P.F. Tu, Y.Y. Zhao and J.H. Zheng, 2008. Studies on chemical constituents of Cistanche tubulosa (Schenk) R. Wight. PMID: 12525059.
Steenkamp V, Gouws MC., 2006. Cytotoxicity of six South African medicinal plant extracts used in the treatment of cancer. S Arf J Bot; 72(4): 630–633.
The Ayurvedic Pharmacopoeia of India Government of India, 1999. Ministry of Health and Family Welfare, Department of Ayush, India, I Part, I Vol, 35–36.
Thun M. J, J. O. DeLancey, M. M. Center, A. Jemal, and E. M. Ward, 2009. Carcinogenesis. The global burden of cancer: priorities for prevention, vol. 31, no. 1, pp. 100–110.
Trivedi, NP and Rawal, UM, 2001. Indian J ExpBiol. Hepatoprotective and antioxidant property of Andrographis paniculata (Nees) in BHC-induced liver damage in mice,; 39: 41–46.

Biotechnology and Biological Sciences – Sen et al. (Eds)
© 2020 Taylor & Francis Group, London, ISBN 978-0-367-43161-7

SSR markers for the assessment of stress and genetic diversity in legumes

Tanushree Agarwal
University of Calcutta, Kolkata, West Bengal, India

ABSTRACT: In recent years there has been an upsurge in the molecular markers based evaluaion of genetic diversity and their use in biotechnological applications. There is a continuous need to find advanced traits from the wild varieties for the integration of these traits into crop species thereby imparting stress resistance. This review focuses on the advent and up rise of SSR marker system which has been abundantly used for studies of genetic diversity, and relatedness between wild and cultivated varieties across leguminous plant species. An approach to highlight advantages of SSR markers have been made, so they could be used to introduce agriculturally important traits for the improvement of crops.

1 SSR MARKERS

Molecular markers have essentially been developed for the detection of DNA polymorphism. Among the various types of markers RFLP, RAPD, AFLP, ISSR and SSR are more commonly used for genetic diversity studies. Simple Sequence Repeat (SSR) markers show higher level of inter and intra-specific polymorphism with ten or more number of tandem repeat units (Queller et al., 1993). SSR markers are codominant, multi-allelic, highly reproducible, easy to use, and display high polymorphism even in closely related species (Souframanien & Gopalakrishna, 2004, and are transferable between populations and across the species. Like other PCR based markers SSRs or Microsatellites are relatively simple, primers restricting its use mostly in the agriculturally important crop plants. In the Rice Genome project 2005, around 18,828 SSR sequences have been detected of which only 10-15 % have been analyzed. This depicts the high potential available for SSR markers. With the rapid advances in molecular biology techniques and development of next generation sequencing technology (NGS) massive amount of information related to nucleotide sequences and SSR markers can be developed and utilized (Churbanov et al., 2012).

1.1 *SSR in genetic diversity of orphan legumes*

Legumes are important crop plants and cultivated in an area of 78.5 million hectares worldwide producing 69 million tons of food grain (FAOstat, 2013). Functional genomics in legumes has been accelerated with studies in *Medicago truncatula* (Town et al., 2006), *Lotus japonicus* (Sato et al., 2005) and *Glycine max* (Sato et al., 2007).

However orphan legumes which are underutilized legumes as compared to the major food crops of the world have received little attention from the scientific community. They remain confined to their place of origin, and are ill documented as to their use and cultivation with low seed productivity. Orphan crops are better adapted to extreme soil and environmental conditions than the major crops with high tolerance to drought constraints and are a repository of other valuable traits which could be used for the improvement of related crops (Cullis et al., 2017). These legumes have not been studied in detail due to the absence of genome sequence information and paucity of molecular markers. Recent advances in sequencing and SSR based marker systems should be exploited for the analysis of such crop species. According to the Kirkhouse Trust, *Arachis hypogea*, *Lathyrus sativus*, *Tylosema esculentum*, *Lablab purpureus*, *Macrotylo mauniflorum* are some of the orphan legumes. Some species of *Vigna* such as *V. subterranean*, *V. aconitifolia* have also been included in this group.

The genus *Vigna* is a large taxa in the leguminosae family comprising of 104 species out of which nine species are cultivated as food crops distributed throughout Asia, South America, Africa and Australia (Lewis et al., 2005). These include *V. subterranean, V. unguiculata, V. vexillata, V. radiata, V. angularis, V. mungo, V. aconitifolia, V. umbellate*, and *V. reflexo-pilosa* (Tomooka *et al.*, 2002). The economically important grown crops of *Vigna* are cowpea, mung-bean and black gram, adzuki bean and rice bean. Genetic diversity analysis of Rice bean or *V. umbellata* gene pool has been successful with the help of SSR markers (Tian et al., 2013). Wang et al. (2015) developed an SSR enriched library of rice bean. In a recent report 44 SSR primers were analyzed successfully to determine the polymorphism among 40 germplasm lines (Thakur et al., 2017).

Black gram (*V. mungo*) crop is grown in India on 3.26 million hectares forming one of the essential crop plants. However, its productivity is low due to lack of genetic variability and susceptibility to pests and diseases. Apart from some genic SSR markers developed from immature pods of black gram, there are no further reports of marker development. (Souframanien and Reddy 2015). SSR markers from other species have been employed for functional analysis (Gupta *et al.* 2008). Around 361 SSR markers derived from adzuki bean, common bean, cowpea, mung bean, chickpea and pigeon pea were used for genetic diversity analysis in 24 genotypes of black gram out of which 39 markers were highly polymorphic (Gupta et al., 2013).

Adzuki bean (*Vigna angularis*) is a nutritional crop grown in Asia, particularly parts of China and Japan, and to some extent in India (Banni et al., 2012). There are reports of only 50 genomic SSR markers and 1429 EST-SSR markers in adzuki bean (Wang et al., 2004). NGS technique has led to the identification of around 7947 EST SSR markers in adzuki bean. These markers would be useful in studying genetic relationships and molecular breeding of such crops with low genomic resources (Chen et al., 2015). Mung bean (*Vigna radiata*) is a tropical/subtropical crop, and cultivated in India as a rich protein source and consumption as Dal in South Asian countries. The yield of this crop is however affected due to its susceptibility to biotic stresses (diseases such as powdery mildew, mung-bean yellow mosaic virus, cercospora leaf spot and leaf curl virus) and abiotic stresses such as drought and salinity and lack of genetic variability and presence of feasible ideotypes. Only a few reports of mung bean SSR markers have been made available (Sangiri et al., 2007; Somata et al., 2009). India contributes to 54% production of mung bean worldwide (Lambrides& Godwin, 2007). However the rate of production is decreasing due to the increase in salinity of cultivable land yet sufficient research has not been performed in *Vigna sp.* for the identification and development of cultivars that are adapted to salt stress conditions (Singh and Singh, 2011). In such case the identification of SSR markers linked to salt tolerance from wild relatives would facilitate molecular breeding and crop improvement strategies. The number of genome wide polymorphic SSR markers is limited for mung bean (Kumar *et al.*, 2004). Gene specific microsatellite markers have been identified and characterized for polymorphism between salt tolerant and susceptible varieties (Sehrawat et al., 2014) In a subsequent report 21 SSR markers from adzuki bean were tested for transferability in mung bean and gave positive results in genetic diversity analysis (Singh et al., 2014). Liu et al. (2016) have identified around 3788 EST-SSRs analyzing transcriptomes of different mung bean genotypes, of which 151 markers were found to be polymorphic thus proving the efficacy of NGS in development of SSR markers. *Vigna unguiculata* or cowpea is cultivated as a crop in Asia and parts of America. There have been reports of around 200 genic and 100 genomic SSR markers from cowpea, out of which 54 markers were found to be polymorphic (Chen et al., 2017). In view of recent analysis SSR markers have emerged as markers of choice for gene identification and molecular breeding to fasten the introduction of new and improved varieties with increased yield and tolerance. There is an urgent need for the genetic analysis of orphan legume crops by development of polymorphic SSR markers to widen the genetic base of such species.

1.2 *SSR markers in biotic and abiotic stress*

Microsatellite markers have found their use in identification and association with disease resistance genes. Majority of research work has been carried out with plant taxa whose genome information is easily available. Legumes being one of the major food crops are affected by several bacterial and fungal diseases such as powdery mildew, rust disease, bacterial blights, as well as several insect pests, such as, legume pod borer, aphids, white fly, leaf hoppers. Plant geneticists propose the use of microsatellites to identify disease resistance genes and make marker assisted selection easier (Maiti et al., 2011). SSR markers have been effectively used with the studies of Cercospora leaf spot in mung bean (Chankaew et al. 2010), resistance to powdery mildew disease in mung bean (Zhang et al., 2008). Cowpea is susceptible to a range of viruses such as Cowpea golden mosaic virus (CPGMV) (Shoyinka, 1974; Taiwo, 2003), Cucumber mosaic virus (CMV), and Cowpea yellow mosaic virus (CYMV). CYMV is responsible for 80-100% of reduction in crop yield (Williams, 1977). Genes which confer resistance to CYMV have been identified with the help of SSR markers. Around 4 SSR markers were identified which were able to detect around 17 resistant plants among 20 genotypes of cowpea showing polymorphism between

Table 1. Comprehensive overview of the use of SSR markers for disease resistance analysis.

Scientific Name	Target of Study	No. of SSR markers analyzed linked to disease resistance	Reference
Vigna unguiculata	CYMV resistance	4	Gioi et al., 2012
	Rust resistance	3	Uma et al., 2015
Vigna mungo	Stable resistance against MYMV	21	Singh et al., 2013
	Genetic diversity pertaining to MYMIV resistance and susceptible	29	Hari et al., 2017
	MYMIV resistance	12	Naik et al., 2017
Vigna radiata	Genetic diversity among YMV resistant and susceptible varieties	29	Hari et al., 2017
	Powdery Mildew resistance	10	Kumar et al., 2017
Vigna umbellata	Interspecific diversity between *V. umbellata* and *V. nakashimae* for bruchid resistance	101	Somta et al., 2006
Glycine max	YMV disease resistance	97	Kumar et al., 2014
	SMV resistance	12	Gupta et al., 2016
	Powdery Mildew resistance	23	Tuong et al., 2016
Medicago truncatula	Aphanomyces root rot	3	Nayel et al., 2009
	Bacterial wilt resistance	3	Ben et al., 2013
Cicer arietinum	Fusarium wilt resistance	57	Sabbavarapu et al., 2013
	Dry root rot resistance	2	Talekar et al., 2017
	Fusarium wilt resistance	3	Jekishandas et al., 2017
Cajanus cajan	Fusarium wilt resistance	5	Khalekar et al., 2013
	Fusarium wilt resistance	3	Singh et al., 2016
Arachis hypogea	Multiple resistant traits such as Early Leaf spot, Groundnut Rosette Disease, Rust	376	Kanyika et al., 2015
	Early leaf spot disease resistance	4	Zongo et al., 2017
	Late leaf spot and rust resistance	1	Divyadharsini et al., 2017
Lathyrus sativus	Transcriptome resistance response to *Ascochyta lathyri*	1	Almeida et al., 2015

resistant and susceptible lines (Gioi et al., 2012). Bean Common Mosaic Virus (BCMV), a member of potyvirus transmitted by aphids, is one of the major problems of cowpea especially crops grown in Karnataka, India. In a recent report 4 polymorphic SSR markers were identified while characterizing BCMV resistance gene (Manjunatha et al., 2017). Mungbean Yellow Mosaic Virus (MYMV), a group of Begomovirus, transmitted by whitefly (*Bemisia tabaci*) is the major cause of yield loss of pulses in India. The rapid introduction of mutants of MYMV and its complex disease resistance mechanism makes it difficult to control the disease by conventional methods (Karthikeyan et al., 2012) In a recent report 12 SSR markers were analyzed and VR9, an SSR marker was identified to be closely related to YMV (Yellow Mosaic Virus) in black gram and could be effectively utilized in plant breeding programme (Naik et al., 2017).

Despite having bruchid resistance (Somta *et al.*, 2006); as well as resistance to yellow mosaic virus (Borah *et al.*, 2001) and bacterial leaf spot (Arora *et al.*, 1980), rice bean (*Vigna umbellata*) has remained scientifically ignored much like its close relatives.

In comparison to the use of SSR markers for analysis of disease resistance, there is very limited report for its application in abiotic stress. A comprehensive overview has been represented below in Table 1 of the specific use of SSR markers in analysis of disease resistance across various legume crop plants specifically *Vigna sp.* indicating the future potential of SSR markers in disease resistance and stress.

2 CONCLUSIONS

The current review focuses on the fact that SSR markers have clearly emerged as an efficient molecular marker method for the assessment of genetic diversity and disease resistance genes and continues to be the most lucrative approach to be adopted in molecular analysis. In order to meet the growing global demand and food supply, resistant cultivars have to be developed and released. The legumes showing susceptibility against stresses are to be replaced with the tolerant varieties which are available in the germplasm banks or remain unexplored in the wild or confined to small cultivated lands. SSR markers provide a cost effective and time saving approach to analyze available germplasms for the identification of immune responsive genes. The introduction of advanced technologies such as next generation sequencing have made the identification of SSR markers almost effortless.

There is an urgent requirement for the analysis of polymorphic SSR markers in underutilized plants, which carry the resistance genes, for their introgression into the major crops to develop stress resilience and resistance. A comprehensive knowledge of SSR markers and its potential may accelerate its identification and utility in major crop plants, which is much needed for effective crop management strategies.

ACKNOWLEDGEMENT

The present research work has been carried out as part of honorary work after completion of Ph.D. at University of Calcutta. I acknowledge my teachers who have helped me to bring this review to completion. I would like to thank my friend Anamika Pal, currently at TERI, Delhi for her constant help.

REFERENCES

Churbanov, A., Ryan, R., Hasan, N., Bailey, D., Chen, H., Milligan, B., & Houde, P. 2012. High SSR: high-throughput SSR characterization and locus development from next-generation sequencing data. Bioinformatics (Oxford, England) 28(21): 2797–2803.

Cullis, C. & Kunert, K.J. 2017. Unlocking the potential of orphan legumes. Journal of experimental botany. 68(8):1895–1903.

Lewis G., Schrire B., Mackind B., Lock M. 2005. Legumes of the World. Royal Botanic Gardens, Kew, UK.

Queller, D.C., Strassmann, J.E., Hughes, C.R. 1993. Microsatellites and kinship. Trends Ecol Evol 8: 285–288.

Sato, S., Nakamura, Y., Asamizu, E., Isobe, S., & Tabata, S. 2007. Genome sequencing and genome resources in model legumes. Plant physiology 144(2): 588–593.
Souframanien, J. & Gopalakrishna, T. 2004. A comparative analysis of genetic diversity in black gram genotypes using RAPD and ISSR markers. Theor. Appl. Genet. 109: 1687–1693.
Tomooka N., Maxted N., Thavarasook C., Jayasuriya A.H.M. 2002. Two new species, new species combinations and sectional designations in *Vigna* subgenus Ceratotropis (Piper) Verdcourt (Leguminosae, Phaseoleae). Kew Bulletin 57: 613–624.
Town, C.D. 2006. Annotating the genome of *Medicago truncatula*. Curr. Opin.Plant. Biol. 9:122–127.
Tian J., Isemura T., Kaga A., Vaughan D.A., Tomooka N. 2013. Genetic diversity of the rice bean (*Vigna umbellata*) genepool as assessed by SSR markers. Genome. 56(12): 717–727.
Wang, L.X., Chen, H.L., Bai, P., Wu, J.X., Wang, S.H., Blair, M.W. and Cheng, X.Z., 2015. The transferability and polymorphism of mung bean SSR markers in rice bean germplasm. Molecular breeding 35(2): 77.

Biotechnology and Biological Sciences – Sen et al. (Eds)
 ISBN 978-0-367-43161-7

Improvement of drug by descriptors and docking criteria: Benzoazetinone-P450 system as a paradigm

S. Banerjee & A. Goswami
Department of Biological Sciences, Indian Statistical Institute, Kolkata, West Bengal, India

D. Mitra, S. Bhattacharyya & A.K. Bandyopadhyay*
Department of Biotechnology, University of Burdwan, Burdwan, West Bengal, India

S. Yasmeen
Department of Botany & Microbiology, Acharya Nagarjuna University, Guntur, Andhra Pradesh, India

R.N.U. Islam
Department of Zoology, University of Burdwan, Burdwan, West Bengal, India

I. Ansary
Department of Chemistry, University of Burdwan, Burdwan, West Bengal, India

ABSTRACT: Rational drug-design has been a successful procedure for the discovery of drugs. Screening of molecular diversity in natural and synthetic compounds has achieved quantum leap in medicinal chemistry. However, the fact of withdrawal of approved drugs from the market signals more rigorous practices of these cost effective and time-consuming procedures. Here, we apply a new protocol over the reported antifungal benzoazetinone to screen out the best drug of the system. The correlation plot of binding energy vs polarizability and screening of empirical and quantum mechanical descriptors allowed us to discover a novel antifungal agent over the reported one. The protocol is applicable to other similar systems.

Keywords: New lead, Descriptors, Empirical, Quantum Mechanical, Docking, Virtual Screening

1 INTRODUCTION

Drug discovery needs careful approaches as it involves high cost and time with higher attrition rate for the discovery (Šunjić & Parnham, 2011). Small organic molecule, which in recent time has been the source for the discovery of new drug (Wang et al. 2000), has the possibility to be a toxin or mutagen or metabolite or other non-drug compounds. Pharmacokinetic optimization of molecular descriptors based on the physicochemical properties, hetero and homo cyclic ring-systems, and functional groups has been the procedure for screening of drug candidates (Khanna & Ranganathan, 2009; Cruciani et al. 2000). Creation of diverse library of small organic molecues using combinatorial procedure and short-listing these by descriptors and virtual screening criteria has been a regular practice to achieving the exhaustiveness in the discovery process (Šunjić & Parnham, 2011; Gordon et al. 1994; Anderson, 2003; Furberg & Pitt, 2001). However, the report of withdrawal of approved drugs from the market (under phase-IV

*Corresponding author: Email: akbanerjee@biotech.buruniversityac.in

monitoring) indicates that the pharmacokinetic characterization and screening of effective leads require approaches that are extremely through and careful (Furberg & Pitt, 2001). However, in an earlier study, it has been demonstrated that CYP53A15, a cytochrome-P450 isozyme, is a promising antifungal target against the benzoazetinones. Notably, the isozyme, CYP53A15 is especially abundant in the genome of pathogenic fungi (Ansary et al. 2017).

Here, we present a protocol for the improvement of drug-properties of an existing drug by using empirical and quantum mechanical descriptors along with virtual screening and strict short-listing procedure. The work also highlights the correlation of binding energy and pharmacokinetic properties. Taken together, the study provides new insight, which we hope would have potential application in the drug discovery.

2 MATERIALS & METHODS

2.1 *Dataset*

We used the original antifungal agent (3i) and target (CYP53A15) (Ansary et al. 2017) for the study. CYP53A15, which is abundant in the genome of pathogenic fungi, is belonging to 4-monooxygenase class of enzyme of cytochrome-P450 super family. The original lead was modified *in silico* by using various in-built functional groups followed by the optimization using semi-empirical/PM3 and or DFT/6-31G (d,p) level of theory of Gaussian software. The latter optimization is done only for a selective population (~38) of the total (112) drug candidates.

2.2 *Empirical and quantum mechanical descriptors*

Empirical (EMD) and quantum (QMD) mechanical descriptors for 112 and 38 optimized compounds were computed using the reported procedures respectively (Ansary et al. 2017; Ansary et al. 2019).

2.3 *Docking and virtual screening*

The rigid docking of each of these compounds was performed by the process of virtual screening at the active site (i.e. heme-binding site) of the modeled target, i.e. CYP53A15 (Ansary et al. 2017). The docking parameters were kept identical as it was for the original compound (3i) (Ansary et al. 2017). AUTODOCK and its associated tools (Morris et al. 2009) were used for the purpose if not mentioned otherwise.

3 RESULTS & DISCUSSION

Effective therapeutic agents need to well qualify the EMD and QMD properties (Khanna & Ranganathan, 2009; Matuszek & Reynisson, 2016). Post-screened lead that possesses highest binding energy against the target is considered the best. What determines the binding energy of a drug molecule? To check this, we have employed a new protocol that follows the generation and optimization of molecular diversity, obtaining binding energy by virtual screening and short-listing by EMD and QMD properties.

3.1 *Molecular volume is an alternate parameter of the polarizability*

Molecular volume, which is used for QSAR analysis, is related to the interactivity and absorptivity of drug molecules (Zhao et al. 2003). We see that molecular volume and polarizability (α) are highly correlated (r=0.96) to each other (Figure 1A). This observation allows us to use the former as the primary screening parameter rather than the latter. While volume is empirically determined, non-linear optical procedure is required for accurate determination of the

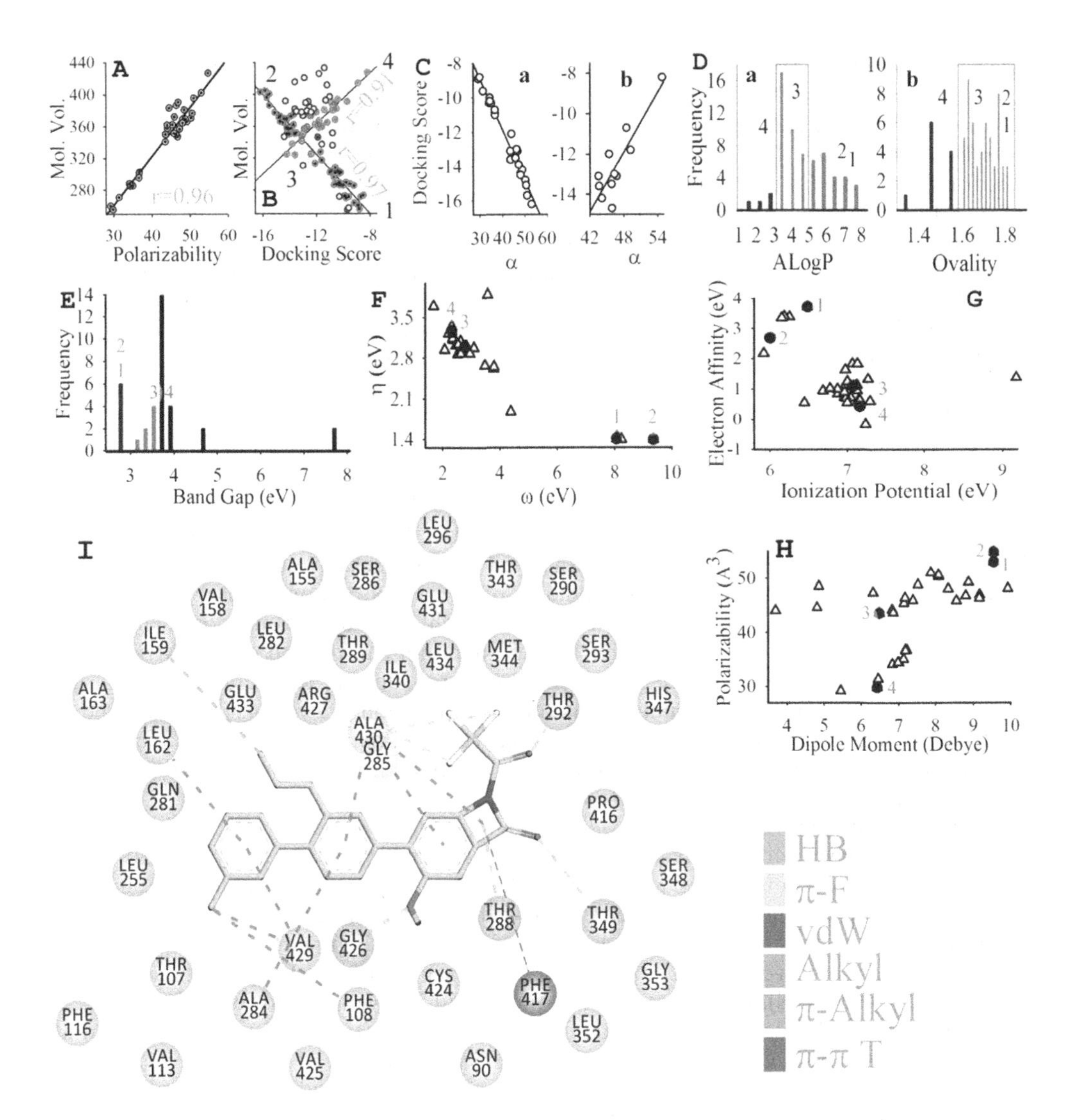

Figure 1. The plot shows various bivariate distributions and drug-target (CYP53A15) complex. Correlation graphs for volume vs polarizability (A), volume vs docking score (B) and docking score vs α (direct: Ca and inverse: Cb). Distribution graph for ALogP vs frequency (Da), Ovality vs frequency (Db), band-gap vs frequency (E), ω (electrophilicity index) vs η (chemical hardness) (F), adiabatic ionization potential vs adiabatic electron affinity (G) and static dipole moment vs static polarizability (H). The complex of the selected-lead and the target (I) is shown with different types of interactions.

polarizability. The use of volume, however, helps to bypass the costly computational steps for polarizability. It is more so when sample size is large and that is populated with complex structures.

3.2 *Interactivity of drugs by volume and polarizability (α) parameters*

Intermolecular interaction, cellular permeability, and absorptivity are largely determined by both the polarizability (α) and volume parameters. The former is a QMD and the latter is an EMD. The binding energy for 112 variants of the original benzoazetinone (3i) is plotted against the volume parameter (Figure 1B). Interestingly, we see that our database has directly (1 to 2; red circles; r=0.97), inversely (3 to 4; green circles; r=0.91) and non-linearly (blue circles) related types of compound. As α is computationally expansive, it was determined only for selected structures. The plot of α vs docking score gives the similar observations (Figure 1Ca for direct and Cb for inverse relations). α originates due to intermolecular dispersion force. While non-polar component induces α, molecular size, shape and polar component in a molecule also influence this parameter. Although, at positions 4 and 2 (Figure 1B) the α values are high, binding energies at these positions are seen to be lowest and highest respectively. Again, although α, at position 1 is lowest, its binding energy exceeds than that at position 4. It, thus, indicates that the interaction of drug with its target depends on complex structural factors. Which binding energy of these four extreme positions (Figure 1B) is the best? To check this, we have analyzed drug likeness properties of compounds of our database using EMD and QMD criteria.

3.3 *Optimum binding energy is obtained at the intersection region*

The linearly fitted lines intersect at position 3 (Figure 1B). While EMDs are computed for all 112 compounds, cost effective QMDs are analyzed only for selective (~38) compounds. We were interested to obtain the best drug candidate over the reported one (Ansary et al. 2017). Thus, we tested Lipinski's rule of five (molecular weight, hydrogen bond donor, hydrogen bond acceptor, lipophilicity and molar refractivity; data not shown) and other EMD parameters (such as volume, polar surface area; ovality) of compounds of our database. By ALogP criteria (Figure 1Da; green bars), unlike compounds at position 1, 2 and 4; the compound at position 3 is most promising as it has a balance between the lipophilic and lipophobic properties. Similarly, as far as mutagenicity is concerned, compounds a position 3, 1 and 2 are seen to be in the non-mutagenic region (Figure 1Db; green bars).

Since drugs are ultimately applied on human or animal subjects, and since empirical descriptors may have indifferent results (Matuszek & Reynisson, 2016; Ansary et al. 2019), we also computed and compared crucial QMDs (Electronic, thermochemical and non-linear optical) for our compounds. Intermediate band-gap is indicative of a balance between the chemical stability and reactivity, which is seen at position 3 (Figure 1E; green bars). While lower band-gap is observed for position 1 and 2 (blue bar), higher band-gap is seen at position 4 (Figure 1E; black bars). Compounds at position 1 and 2 have much lower global chemical hardness (i.e. η) and much higher electrophilicity power (i.e. ω). In this sense, compound at position 3 and 4, brings a balance between the chemical hardness and the global electrophilicity index (Figure 1F). Comparison of adiabatic ionization potential and adiabatic electron affinity (Figure 1G) shows, unlike compounds at position 1 and 2, compounds of position 3 and 4 occupy the most preferred region of clinically approved drugs (Matuszek & Reynisson, 2016; Ansary et al. 2019). Similarly, compounds at position 3 and 4, unlike that for position 1 and 2, occupy a region in the α vs dipole moment graph (Figure 1H), which is most compatible for clinically approved drugs (Matuszek & Reynisson, 2016).

Taken together and remarkably, the compound(s) from the intersection region (~position 3) qualify all the EMD and QMD criteria. Further and notably, the compounds at position 1, which includes the reported drug (3i), exceed the prescribed range of some of these descriptor properties (Ansary et al. 2019). We, therefore, selected a compound from the intersection region whose binding energy has been -14.7 kcal/mol. Although, this binding energy (-14.7 kcal/mol) is lower than the best representative molecule at position 2 (-16.2 kcal/mol), it is higher by -5.9 kcal/mol than the reported drug and further, it has the best drug likeness properties than any other compound of our database.

3.4 *Ligand-target interactions are improved by -6.0 kcal/mol*

The selected ligand-target complex is shown (Figure 1F) with different types of interaction. It is seen that ligand establishes hydrogen bond (HB), π-F, van der Waals (vdW), alkyl-alkyl

(alkyl), π-alkyl and π-π T-shaped (π-π T) types of interaction at the active site of the target (CYP53A15). Compared to the reported drug molecule (3i), the interactions have been much more extensive for the selected compound, such that its binding energy has been more favorable by -5.9 kcal/mol. Taken together; it seems that the selected drug candidate makes optimum interaction with the target and thus, considered the best of our database.

4 CONCLUSION

We generated and optimized diverse types of compound by introducing available functional groups in the core structure of the reported drug molecule. There exist three types of lead in our database that are directly, inversely and non-linearly related with the molecular volume or polarizability (α). Comparison over a wide range of EMD and QMD properties demonstrate that the drug candidates at the interaction region not only qualify most of the drug likeness properties but also improve the binding energy. Although higher binding energy than the selected compound is available in our database, they suffer from certain crucial descriptor properties. Thus, the score of the selected drug candidate is taken as optimum. The method seems to be applicable to other similar systems.

ACKNOWLEDGMENT

We are thankful to the Department of Biotechnology, University of Burdwan for the computational facility laboratory.

REFERENCES

Šunjić, V. & Parnham, M.J. 2011. *Signposts to Chiral Drugs: Organic Synthesis in Action*. Springer.

Wang et al. 2000. Structure-based discovery of an organic compound that binds Bcl-2 protein and induces apoptosis of tumor cells. *Proceedings of the National Academy of Sciences*. 97(13):7124–9.

Khanna, V. & Ranganathan, S. 2009. Physicochemical property space distribution among human metabolites, drugs and toxins. *BMC bioinformatics*. 10(15):S10.

Cruciani et al. 2000. VolSurf: a new tool for the pharmacokinetic optimization of lead compounds. *European Journal of Pharmaceutical Sciences*. 11:S29–39.

Gordon et al. 1994. Applications of combinatorial technologies to drug discovery. 2. Combinatorial organic synthesis, library screening strategies, and future directions. *Journal of medicinal chemistry*. 37 (10):1385–401.

Anderson, A.C. 2003. The process of structure-based drug design. *Chemistry & biology*. 10(9):787–97.

Furberg, C.D. & Pitt, B. 2001. Withdrawal of cerivastatin from the world market. *Trials*. 2(5):205.

Ansary et al. 2017. Synthesis, molecular modeling of N-acyl benzoazetinones and their docking simulation on fungal modeled target. *Synthetic Communications*. 47(15):1375–86.

Ansary et al. 2019. Regioselective Synthesis, Molecular Descriptors of (1, 5-Disubstituted 1, 2, 3-Triazolyl) Coumarin/Quinolone Derivatives and Their Docking Studies against Cancer Targets. *ChemistrySelect*. 4(12):3486–94.

Morris et al. 2009. AutoDock4 and AutoDockTools4: Automated docking with selective receptor flexibility. *Journal of computational chemistry*. 30(16):2785–91.

Matuszek, A.M. & Reynisson, J. 2016. Defining known drug space using DFT. *Molecular informatics*. 35 (2):46–53.

Zhao et al. 2003. Fast calculation of van der Waals volume as a sum of atomic and bond contributions and its application to drug compounds. *The Journal of organic chemistry*. 68(19):7368–73.

Biotechnology and Biological Sciences – Sen et al. (Eds)
© 2020 Taylor & Francis Group, London, ISBN 978-0-367-43161-7

Salt-bridge, aromatic-aromatic, cation-π and anion-π interactions in proteins from different domains of life

R.N.U. Islam
Department of Zoology, University of Burdwan, Burdwan, West Bengal, India

C. Roy, D. Mitra & A.K. Bandyopadhyay*
Department of Biotechnology, University of Burdwan, Burdwan, West Bengal, India

S. Yasmeen
Department of Botany & Microbiology, Acharya Nagarjuna University, Guntur, Andhra Pradesh, India

S. Banerjee
Department of Biological Sciences, Indian Statistical Institute, Kolkata, West Bengal, India

ABSTRACT: Cellular proteins from different domains of life are operating in their respective ecosystems. Weak forces form the tertiary structure. Is there a modulation of these interactions for different domains of life? Using a large database of X-ray structures of protein, we demonstrate that unlike eukarya and bacteria, archaeal proteins show an increase in salt-bridge, a decrease in π-π (hydrophobic) and the maintained level of cation-π (also anion-π) types of interactions. These observations clearly suggest environment specific modulation of weak forces. It, thus, seems that the intrinsic composition of amino acid in archaeal proteins and the extreme environment together maintain the stability of these proteins.

Keywords: PROTEINS, ARCHAEA-BACTERIA-EUKARYA, SALT-BRIDGE, Π-Π, CATION-Π, ANION-Π

1 INTRODUCTION

Weak forces maintain an intricate balance between the rigidity and the flexibility in proteins (Dill, 1990; Frieden, 1975). Unlike mesophiles, archaeal proteins operate in extreme environments. Here, the dielectric property of solvent is ~50 that favor the electrostatic over the hydrophobic interactions. The interplay of the intrinsic composition of proteins and the extrinsic solvent conditions (Anfinsen, 1973; Bandyopadhyay & Sonawat, 2000; Bandyopadhyay et al. 2001) needs thorough investigations. Understanding the balance of weak forces is vital for the stability and interactivity of these proteins (Dill, 1990; Frieden, 1975). While salt-bridge (Bandyopadhyay et al. 2019b; Kumar et al. 2000a; Nayek et al. 2014), cation-π (Biot et al. 2002) and anion-π (Frontera et al. 2011) are electrostatic, π-π interaction (Roy & Dutta, 2018) is hydrophobic in nature (Burley & Petsko, 1985; McGaughey et al. 1998). Notably, these interactions follow defined geometric and stereochemical criteria (Kumar et al. 2000a; Kumar et al. 2000b; Nayek et al. 2014; Biot et al. 2002; McGaughey et al. 1998; Frontera et al. 2011). X-ray structures of protein and their complexes have been used to explore these interactions (Biot et al. 2002; McGaughey et al. 1998; Frontera et al. 2011) to understand

*Corresponding author: akbanerjee@biotech.buruniversityac.in

their stability and biological specificity. Nonetheless, more work remains to gain insight into the balance of these weak-forces in proteins from these domains of life.

Here, we involve a large database of eukaryal, bacterial and archaeal proteins to compare salt-bridge, π-π, cation-π and anion-π types of interactions. The work also highlights the differential strategies in terms of these forces. Taken together, the study seems to have potential application in protein engineering and structural biology.

2 MATERIALS & METHODS

2.1 *Dataset*

We have employed strict screening criteria (domains of life, resolution, chain length etc) to extract crystallographic structures of protein of archaea, bacteria and eukarya from RCSB, PDB database (Berman et al. 2000).

2.2 *Analysis of salt-bridge, π-π, cation-π and anion-π interactions*

Salt-bridges were extracted and analyzed using domain-specific crystal structures of protein as input following earlier procedures (Gupta et al. 2014; Gupta et al. 2015). Further, an in-house procedure was developed for the analysis of π-π, cation-π and anion-π interactions. The rules for these interactions such as distances, angles ($\theta 1$, $\theta 2$ and γ) and others were followed as defined (Biot et al. 2002; McGaughey et al. 1998; Frontera et al. 2011). Post-run analyses were carried out using MS Excel and Sigma Plot v12.0 programs.

3 RESULTS & DISCUSSION

Archaea, unlike mesophiles, thrive in belligerent environments that alter the solvent properties (Bandyopadhyay et al. 2019b; Bandyopadhyay et al. 2019c). Proteins in the cytoplasm are also operating under similar conditions as the environments, which is detrimental for mesophilic counterparts. How then archaea and its proteins are willful in their robust ecosystems? Although, these proteins possess abundance of salt-bridges (Kumar et al. 2000b; Nayek et al. 2014), the question as to, be there a modulation of weak interactions, remains unanswered.

3.1 *Salt-bride forming residues (SBFR) are abundant in archaea*

Table 1 shows the mean of salt-bridge forming residues (SBFR i.e. acidic: D, E and basic: K, H, R) that are extracted from eukaryal (706), bacterial (759) and archaeal (441) proteins.

It is interesting to note that with respect to (w.r.t) bacteria (n=759) and eukarya (n=706), archaeal proteins (n=441) recruit higher proportion of acidic (D & E) and basic (K & R) residues for the formation of salt-bridge. H is more and less in archaea than eukarya and bacteria respectively. How are SBFR distributed in the surface and core of archaeal proteins? Except few cases (E in core w.r.t eukarya; H in surface and core w.r.t bacteria and R in core w.r.t eukarya), SBFR in archaea are abundant both in the surface and core relative to the eukaryal and bacterial counterparts (Table 1). Notably, SBFR increase the mean hydrophilicity of archaeal proteins (Bandyopadhyay et al. 2019b). Thus, the increase of salt-bridge and polarity of archaeal protein seems to the primary and secondary effects respectively.

3.2 *Binary items of salt-bridge increase in archaeal proteins*

The distribution (Figure 1, a for eukarya; e for bacteria and i for archaea) and the binary items (Table 2) of salt-bridges for 706, 759 and 441 non-redundant, high resolution and minimized X-ray structures of proteins from eukaryal (eu), bacterial (ba) and archaeal (ar) domains of life are compared. To the best of our knowledge, analyses of salt-bridges and

Table 1. Details of mean and normalized (per 300 residues protein) content of acidic (D & E) and basic (H, R & K) residues in protein (P) and in salt-bridge (SB). All these structures were minimized by AUTOMINv1.0 (Islam et al. 2018) prior analyses if not mentioned otherwise.

			P surface		P core		In SB		SB surface		SB core	
Domains	SBFR	InP	abs	%	abs	%	abs	%	abs	%	abs	%
EUKARYA (n=706)	D	18.0	12.5	69.4	5.5	30.6	9.4	52.4	5.2	29.0	4.2	23.4
	E	19.1	14.4	75.6	4.7	24.4	10.5	55.1	6.6	34.7	3.9	20.5
	K	19.1	15.8	82.9	3.3	17.1	8.4	44.1	6.1	32.2	2.3	11.9
	H	7.1	3.8	53.4	3.3	46.6	1.8	25.5	0.7	10.1	1.1	15.4
	R	14.2	9.8	68.8	4.4	31.2	9.7	68.6	5.7	40.0	4.1	28.6
ARCHAEA (n=441)	D	17.1	11.9	69.2	5.3	30.8	10.3	60.0	5.8	33.9	4.5	26.0
	E	27.0	21.5	79.8	5.5	20.2	17.2	63.5	11.8	43.5	5.4	20.0
	K	23.0	19.5	84.7	3.5	15.3	12.4	53.6	9.2	40.0	3.1	13.7
	H	5.0	2.4	47.9	2.6	52.1	1.4	27.6	0.5	10.2	0.9	17.4
	R	17.3	12.8	73.7	4.5	26.3	13.7	79.3	8.8	50.7	4.9	28.5
BACTERIA (n=759)	D	19.6	13.5	68.9	6.1	31.1	10.1	51.6	5.6	28.4	4.6	23.2
	E	20.3	16.0	78.6	4.3	21.4	11.0	54.3	7.2	35.6	3.8	18.7
	K	16.9	14.3	84.7	2.6	15.3	8.0	47.2	5.9	35.0	2.1	12.2
	H	7.2	3.9	54.6	3.3	45.4	2.1	29.2	0.8	11.5	1.3	17.7
	R	15.6	11.0	71.0	4.5	29.0	11.1	71.0	6.8	43.6	4.3	27.4

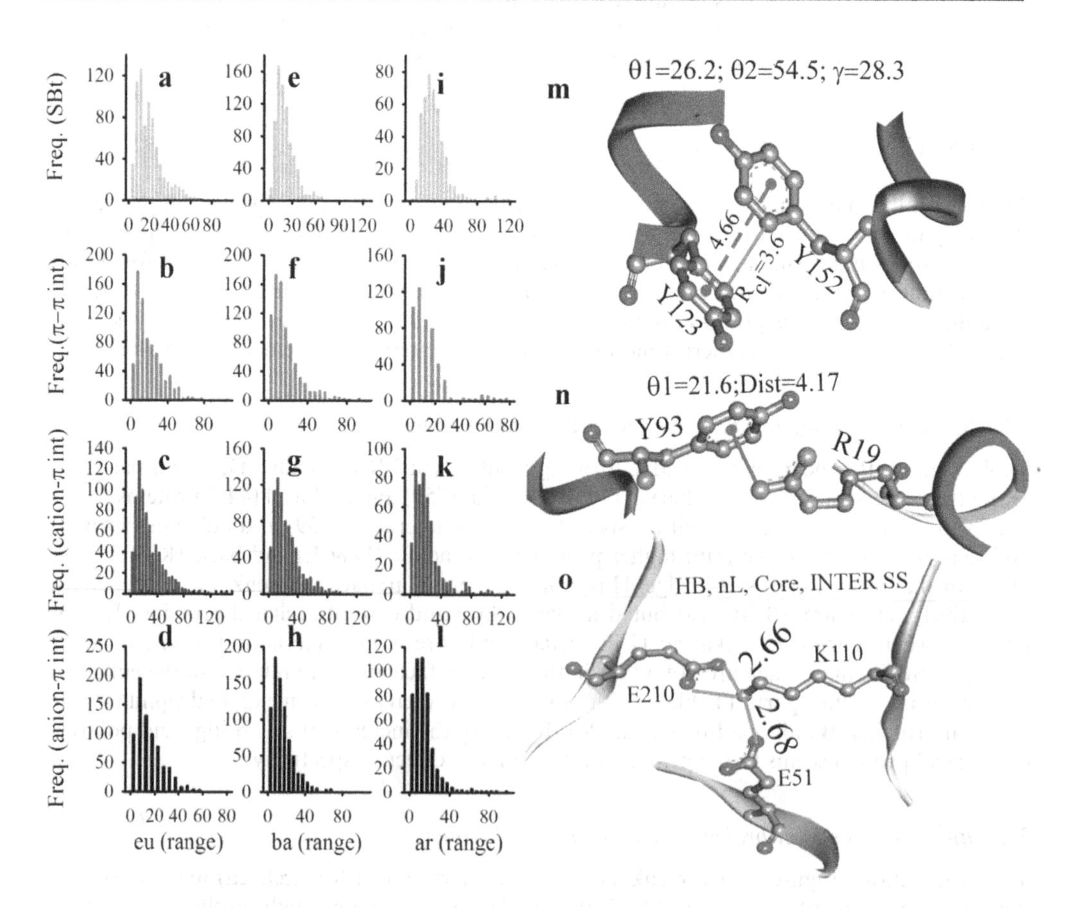

Figure 1. Distributions of salt-bridge, π-π, cation-π and anion-π types of integration of proteins from eukarya (a-d), bacteria (e-h) and archaea (i-l). Typical example of π-π (m), cation-π (n) and salt-bridge (o) interactions are also shown.

Table 2. Binary items of salt-bridges from eukaryal, bacterial and archaeal proteins. Each entry in the table is the mean and normalized value. The cut-off of the accessibility was set to 22.0 Å^2. The local (L) and non-local (nL) boundary was set to a separation of 9 residues. Secondary structure (i.e. Sst) includes both helices and strands. ND: not determined; co: core; su: surface; iso: isolated; net: network; tot: total.

Binary items	Binary types	Eukarya (n=706)			Bacteria (n=759)			Archaea (n=441)		
		co	su	tot	co	su	tot	co	su	tot
Freq	Freq-iso	2.7	7.6	10.3	2.2	7.9	10.2	2.6	10.1	12.8
	Freq-net	4.7	5.2	9.9	4.5	5.8	10.3	6.1	9.4	15.5
HB	HB -iso	2.5	7.1	9.6	2.1	7.4	9.5	2.5	9.5	12.0
	HB -net	4.3	4.8	9.1	4.1	5.4	9.4	5.7	8.7	14.3
nHB	nHB -iso	0.2	0.5	0.7	0.1	0.5	0.6	0.2	0.6	0.8
	nHB -net	0.4	0.4	0.9	0.4	0.4	0.9	0.4	0.7	1.1
L	L-iso	0.7	3.6	4.3	0.6	3.8	4.4	0.8	5.2	6.0
	L-net	1.2	2.1	3.3	1.0	2.4	3.3	1.7	4.2	5.9
nL	nL-iso	2.0	4.0	6.0	1.7	4.2	5.8	1.9	4.9	6.8
	nL-net	3.6	3.1	6.6	3.5	3.4	6.9	4.4	5.2	9.6
Sst	Sst-iso	ND	ND	6.6	ND	ND	6.7	ND	ND	8.7
	Sst-net	ND	ND	6.5	ND	ND	6.8	ND	ND	11.0
coil	Coil-iso	ND	ND	3.7	ND	ND	3.5	ND	ND	4.1
	Coil-net	ND	ND	3.5	ND	ND	3.5	ND	ND	4.5

other weak forces on such a high sample size (n=1906) are rare. The frequency distribution of salt-bridges (of structures from these domains) has been typical of Gaussian type, whose mean is seen to be higher in archaea (Figure 1i) than that of the eukarya (Figure 1a) and the bacteria (Figure 1e).

The normalized total mean frequency of salt-bridges is subdivided into different binary items such as isolated (Freq-iso) vs network (Freq-net), hydrogen bonded (HB-iso & HB-net) vs non-hydrogen bonded (nHB-iso & nHB-net), local (L-iso & L-net) vs non-local (nL-iso & nL-net) and secondary structure (Sst-iso & Sst-net) vs coil (coil-iso & coil-net) types. Based on the average accessibility of partners of salt-bridge, these binary items are further subdivided into core (co) and surface (su) types. These itemized results are then compared among different domains (Table 2). Several points are noteworthy from the table. First, the greater abundance of all the binary items of salt-bridge in archaeal proteins than that of eukarya and bacteria may indicate that salt-bridge is the dominant weak force in the former. Second, the higher frequency of the network over the isolated types of salt-bridge in archaea than that of eukarya and bacteria implying that the former has acquired additional stability to counteract the adverse effect of the environmental stress (high temperature, high salt, etc). At this point, it is worth noting that the network salt-bridges (Figure 1o) are more efficient to eliminate isolated charge and polar component especially from the core of proteins and thereby contribute more to the stability than the isolated types of salt-bridge (Bandyopadhyay et al. 2019a; Kumar & Nussinov, 1999).

Third, interestingly, unlike eukarya and bacteria, in archaea, it is seen that the overall frequency of HB and Sst types of salt-bridge dominates over nHB and coil types respectively. While HB types of salt-bridge confer additional stability than the nHB types, Sst types act as determinant of the topology of proteins (Bandyopadhyay et al. 2019b). Thus, the above mentioned preferences of binary items in archaeal proteins seem to be useful to vanquish the harsh effect of the extreme environment. Taken together, it is seen that archaeal proteins ramped up salt-bridges at all levels of its binary items. Why it is so? Intuitively, it appears to be a similar case of the comparison of the stability of GC-rich vs an AT-rich DNA segments. The former is more stable than the latter. Similarly, in archaeal proteins, the higher frequency of salt-bridges at all level of binary items seems to be a strategy to resist the pernicious effects of the extreme environments.

3.3 *Hydrophobic (π-π) force is reduced in archaeal proteins*

3D structure of protein is thermodynamically compromised (Bandyopadhyay et al. 2019c; Pace & Herman, 1975). It, thus, maintains an intricate balance between the rigidity and flexibility to attain biological specificity (Dill, 1990; Jaenicke & Böhm, 1998). In the above section, we have seen, unlike eukaryal and bacterial cases, in archaeal proteins, salt-bridges are abundant at all level of the binary items. What about the π-π (hydrophobic) interactions? To check this, we have analyzed π-π interactions of proteins of our database (n=1906), in domains of life specific manner, along with its binary items; the result of which is presented in Table 3. Following points are noteworthy from the table.

First, the distributions of π-π interactions for eukarya, bacteria and archaea are shown in Figure 1 (b, eukarya; f, bacteria and j, archaea). These are typical of normal distributions. A typical example of π-π interaction is also presented in Figure 1m for clarity, where two distantly spaced (30 residues apart) helix containing tyrosine residues are interacting through a centroid-centroid distance of 4.66 Å. The closest distance (R_{close} limit ≤ 4.5 Å) between the atoms of interacting rings is 3.6 Å. θ1, θ2 and γ are seen to be within the define ranges as 26.2°, 54.5° and 28.3° respectively. Notably, the cut-off limit for these angles for T-shaped and stacked π-π interactions are almost alternatively similar (McGaughey et al. 1998). Second, interestingly, the computed mean and normalized level of π-π interactions in archaea is far less than that of the eukarya and the bacteria (Table 3). Notably, in eukarya and bacteria, the levels of such interactions are almost comparable. Third, in all these domains, bb-type of π-π interactions dominates over the be-type and ee-type. Forth, inter-helix (hh) and helix-coil (hc)

Table 3. π-π (hydrophobic), cation-π (electrostatic) and anion-π (electrostatic) interactions (normalized and mean) for eukaryal, bacterial and archaeal proteins. The cut-off for π-π was set to 7.5 Å. The cut-off for cation-π and anion-π were set to 6.0 Å. Other criteria were kept similar as earlier (McGaughey et al. 1998). Interacting residue pairs are shown by amino acid's single letter code. Accessibility cut-off is 22.0. bb: both residues are buried; be: one is buried and other is exposed; ee: both residues are exposed; hh: both are in helix; hs: one in helix and other in strand; hc: one in helix and other in coil; ss: both in strand; sc: one in strand and other in coil: cc both in coil; e: eukarya; b: bacteria and a: archaea; na: not available.

π-π				Cation-π				Anion-π			
Int. type	e	b	a	Int. type	e	b	a	Int. type	e	b	a
Total	19.2	17.3	13.6	total	13.9	13.7	14.4	total	10.9	11.2	10.9
YY	1.9	1.9	1.8	RY	4.6	4.9	5.6	DY	1.9	1.9	1.8
YF	2.2	2.1	2.1	RF	4.3	4.2	4.0	DF	2.3	2.3	2.0
YW	1.4	1.3	1.0	RW	2.2	2.4	1.7	DW	0.9	1.0	0.7
YH	1.5	1.2	0.9	KY	1.5	1.1	1.8	EY	2.0	2.0	2.5
FF	4.8	3.9	3.5	KF	0.9	0.7	1.0	EF	2.6	2.5	2.9
FW	2.3	1.8	1.1	KW	0.4	0.4	0.4	EW	1.2	1.4	1.1
FH	2.0	1.5	1.0	na	na	na	na	na	na	na	na
WW	0.6	0.6	0.3	na	na	na	na	na	na	na	na
HW	0.5	0.5	0.2	na	na	na	na	na	na	na	na
HH	2.3	2.4	1.8	na	na	na	na	na	na	na	na
bb	12.4	10.8	8.6	bb	6.1	5.3	5.7	bb	4.6	4.5	4.2
be	5.3	5.0	3.8	be	5.3	5.7	6.1	be	4.4	4.7	4.7
ee	1.5	1.5	1.1	ee	2.5	2.7	2.7	ee	1.9	2.1	2.0
hh	4.1	4.0	3.6	hh	3.2	3.4	4.1	hh	2.4	2.5	2.9
hs	3.0	2.6	1.8	hs	1.5	1.5	1.5	hs	1.2	1.5	1.3
hc	5.2	4.2	3.5	hc	3.6	3.7	3.6	hc	3.0	2.9	3.0
ss	1.7	1.6	1.3	ss	1.1	1.1	1.2	ss	0.7	0.8	0.8
sc	2.3	2.2	1.5	sc	1.7	1.5	1.6	sc	1.3	1.4	1.3
cc	3.0	2.6	1.9	cc	2.9	2.5	2.4	cc	2.2	2.1	1.6
L	4.0	3.7	3.5	L	4.1	4.1	4.8	L	2.6	2.6	2.7
nL	15.2	13.6	10.1	nL	9.8	9.6	9.6	nL	8.4	8.6	8.2

types of π-π interactions dominate over other types (Table 3), which confer more stability to protein. Fifth, irrespective of domains, the non-local (nL) types of π-π interaction dominate over the local (L) types, which help proteins to adapt globular shape.

Taken together, the lower level of π-π (hydrophobic) interactions (Table 3) and higher level of salt-bridge (Table 2) in archaeal proteins is a clear indication of the modulation of weak forces. In extreme (high temperature and high salt) environments, the dielectric constant is ~50 (Bandyopadhyay et al. 2019a), which severely affect the hydrophobic force. At the same time, under these conditions, salt-bridge interactions remain largely unaffected (Dill, 1990). This could be the reason for the alteration of intrinsic composition of SBFR (Table 1) and hydrophobic residues (data not shown) in archaeal proteins relative to its mesophilic counterparts (i.e. eukaryal and bacterial proteins).

3.4 *Cation-π and anion-π interactions are maintained in archaea*

The cation-π interaction originates due to the proximity of the side-chain of ARG or LYS (always positive at cellular pH ~7.0) to a delocalized π-ring (TYR/PHE/TRP) system. The angle (between the centroid-centroid vector and the normal vector the ring) cut-off is ±40° (Biot et al. 2002). Similarly, an anion-π interaction, which is due to the proximity of the side chains of ASP or GLU (always negative at cellular pH~7.0) to the aromatic ring system, is calculated as cation-π interaction (Frontera et al. 2011). Notably, these electrostatic interactions occur beyond the distance limit of salt-bridge interaction (i.e. ≤4.0Å). The distributions of cation-π and anion-π interaction for eukarya (c & d), bacteria (g & h) and archaea (k & l) respectively are shown in Figure 1. Typical example of cation-π interaction is also presented in Figure 1n for clarity. Notably, the overall mean frequencies for cation-π and anion-π types of interaction are seen to be similar for proteins from all three domains (Table 3). Further, unlike salt-bridge and π-π types of interactions, the frequency of the individual interacting pair (for cation-π & anion-π) in archaea follows the similar level and pattern as that of the eukarya and the bacteria, clearly indicating maintenance of these forces. As far as binary items are concerned, in all three domains, accessibility (bb, be and ee), secondary structure (hh, hs, hc, ss, sc and cc) and residue separation (L and nL) based items of interaction also follow similar patterns. It is worth noting here that although in archaeal proteins, the frequency of acidic (D & E) and basic (K & R) residues are higher (Table 1), aromatic (Y, F & W) residues are lower (data not shown) than that of eukarya and bacteria. Such intrinsic differential distributions of charge and aromatic residues in archaeal proteins seem to be responsible for the higher, lower and preserved level of salt-bridge, π-π and cation-π (also anion-π) types of interaction respectively.

4 CONCLUSION

Using a large database of proteins from eukarya, bacteria and archaea domains of life, we demonstrate that the latter has greater abundance of acidic and basic residues. These excess of oppositely charged residues in archaea increase almost all binary items of salt-bridge to contribute favourably to the overall stability of its proteins. At the same time, unlike eukarya and bacteria, due to reduction of hydrophobic residues in archaeal proteins, the π-π (i.e. hydrophobic) interactions are lowered. Such increase of salt-bridge and decrease of hydrophobic forces are the clear indication of modulation of weak interactions. The intrinsic differential composition of charged and hydrophobic residues in the archaeal proteins (relative to that of eukarya and bacteria) seems to be responsible for the increase, decrease and preserved level of salt-bridge, π-π and cation-π (also anion-π) types of interaction respectively.

ACKNOWLEDGMENT

We are thankful to the Department of Biotechnology, University of Burdwan for the computational facility laboratory.

REFERENCES

Anfinsen, C.B. 1973. Principles that govern the folding of protein chains. *Science*, 181(4096): 223–30.

Bandyopadhyay, A.K. & Sonawat, H.M. 2000. Salt dependent stability and unfolding of [Fe2-S2] ferredoxin of Halobacterium salinarum: spectroscopic investigations. *Biophysical journal*, 79(1): 501–510.

Bandyopadhyay et al. 2001. Structural stabilization of [2Fe-2S] ferredoxin from Halobacterium salinarum. *Biochemistry*, 40(5): 1284–92.

Bandyopadhyay et al. 2007. Kinetics of salt-dependent unfolding of [2Fe–2S] ferredoxin of Halobacterium salinarum. *Extremophiles*, 11(4): 615–25.

Bandyopadhyay et al. 2019a. Stability of buried and networked salt-bridges (BNSB) in thermophilic proteins. *Bioinformation*, 15(1): 61–67.

Bandyopadhyay et al. 2019b. Analysis of salt-bridges in prolyl oligopeptidase from Pyrococcus furiosus and Homo sapiens. *Bioinformation*, 15(3): 214–25.

Bandyopadhyay et al. 2019c. Insights from the salt bridge analysis of malate dehydrogenase from H. salinarum and E. coli. *Bioinformation*, 15(2): 95–103.

Biot et al. 2002. Probing the energetic and structural role of amino acid/nucleobase cation-pi interactions in protein-ligand complexes. *Journal of Biological Chemistry*, 277(43): 40816–40822.

Burley, S.K. & Petsko, G.A. 1985. Aromatic-aromatic interaction: a mechanism of protein structure stabilization. *Science*, 229(4708): 23–28.

Dill, K.A. 1990. Dominant forces in protein folding. *Biochemistry*, 29(31): 133–55.

Frieden, E. 1975. Non-covalent interactions: key to biological flexibility and specificity. *Journal of chemical education*, 52(12): 754–759.

Frontera et al. 2011. Putting anion-pi interactions into perspective. *Angewandte Chemie (International ed. in English)*, 50(41): 9564–9583.

Gupta et al. 2014. SBION: A program for analyses of salt-bridges from multiple structure files. *Bioinformation*, 10(3): 164–166.

Gupta et al. 2015. SBION2: Analyses of Salt Bridges from Multiple Structure Files, Version 2. *Bioinformation*, 11(1): 39–42.

Islam et al. 2018. AUTOMINv1. 0: an automation for minimization of Protein Data Bank files and its usage. *Bioinformation*, 14(9): 525–529.

Kumar, S. & Nussinov, R. 1999. Salt bridge stability in monomeric proteins. *Journal of molecular biology*, 293(5): 1241–55.

Kumar et al. 2000a. Electrostatic strengths of salt bridges in thermophilic and mesophilic glutamate dehydrogenase monomers. *Proteins: Structure, Function, and Bioinformatics*, 38(4): 368–83.

Kumar et al. 2000b. Factors enhancing protein thermostability. *Protein engineering*, 13(3): 179–91.

McGaughey et al. 1998. π-Stacking Interactions Alive and Well in Proteins. *Journal of Biological Chemistry*, 273(25): 15458–15463.

Nayek et al. 2014. Salt-bridge energetics in halophilic proteins. *Plos one*, 9(4): e93862.

Pace, C.N. & Hermans, J. 1975. The stability of globular protein. *CRC critical reviews in biochemistry*, 3 (1): 1–43.

Roy, C. & Dutta, S. 2018. ASBAAC: Automated Salt-Bridge and Aromatic-Aromatic Calculator. *Bioinformation*, 14(4): 164–166.

Biotechnology and Biological Sciences – Sen et al. (Eds)
© 2020 Taylor & Francis Group, London, ISBN 978-0-367-43161-7

Integrated transcriptional meta-analysis of pigmentation disorders

Rajkumar Chakraborty, Shreeya Kedia & Yasha Hasija*
Delhi Technological University, Delhi, India

ABSTRACT: Over the years, there has been a sharp increase in the number of pigmentation disorders. Pigmentation disorders can broadly be classified into hyperpigmentation and hypopigmentation disorders. Hypopigmentation disorders are categorized by loss of skin pigment melanin either due to trauma or disease. Hyperpigmentation disorders, on the other hand, involve an increase in skin pigment leading to dark patches on the skin. The meta-analysis involved analyzing the transcriptome for two hypopigmentation (Vitiligo and Hermansky-Pudlak syndrome) and two hyperpigmentation disorders (Melasma and Epidermolysis Bullosa Simplex with mottled pigmentation) each. The main aim of the analysis was to identify differentially expressed genes for each disease using Bioconductor, functional analysis using DAVID, EnrichR, and Genemania as well as creating a network for the DE's and obtaining hub genes using Cytoscape. This study identify common pathways involved in hypopigmentation and hyperpigmentation disorders as well as identify specific mutated function in each disease.

1 INTRODUCTION

Skin, the largest organ in the body, is a complex barrier organ providing defenses against injury and pathogen [1,2]. The skin is majorly composed of three layers: the outermost layer of the skin is called epidermis and forms a waterproof barrier. It is avascular and consists of cells like keratinocytes, Merkel cells, melanocytes and the Langerhans cells [3]. Present beneath the epidermis is the dermis and consists of connective tissues and also confers protection against stress and strain. It is tightly connected to the basal membrane and consists of hair follicles, sweat glands as well as nerve endings that confer a sense of pain and heat, Beneath the dermis is the subcutaneous tissue, which connects the skin to the bone and muscle and is primarily composed of loose connective tissue, elastin and adipose tissue. Cells like macrophages, fibroblasts and adipocytes are present in subcutaneous region.

The intricate properties of the skin are disturbed in genodermatoses, a broad category comprising of genetically inherited diseases where abnormalities are attributed to genetic mutations [4].

The conventional method of gene-phenotype analysis includes studying each independent disease and its associated mutations. The method is a patient-specific approach and would require the study of each patient and phenotype independently, making it time-consuming and expensive. Common themes like mutation on signalling pathways, pigmentation defects, growth retardation, cell cytoskeleton mutations and DNA damage repair mechanisms are some of the most popular themes associated with a majority of dermatological disorders [4–7].

The process of deposition of melanin in skin, eyes and hair in mammals is called pigmentation. The type, amount and location of melanin pigmentation differs amongst organisms as a result of genetic heterogenicity of molecular pathways. In mammals, the process of melanin productions by cells called melanocytes is known as melanogenesis and can be divided into

*Corresponding author: E-mail: yashahasija06@gmail.com

reddish-hued photomelanin and brown- black colored eumelanin. Melanin, besides providing pigmentation, is also involved in photoprotection from UVR radiation and concordant with these functions, most melanocytes are therefore located in epidermis of skin or hair follicles [8].

Pigmentation disorders can be classified into Hypopigmentation and Hyperpigmentation disorders. The study were carried out on four pigmentation disorders two of each category mentioned above pigmentation disorders, namely:

Vitiligo, Hermansky-Pudlak Syndrome, Melasma, Epidermolysis Bullosa Simplex- Mottled Pigmentation.

The main aim of this work is to conduct a meta-analysis of the transcriptome data from already published literature to identify conserved features in these dermatological disorders, primarily focusing on pigmentation disorders. Databases like the NCBI (National Centre for Biotechnology Information) Gene Expression Omnibus (GEO) and EBI Array Express can be used to obtain already published transcriptome analysis data.

2 MATERIALS & METHODS

2.1 *Collection of data and pre-processing*

The aim is to obtain transcriptional profile data from public repositories like National Centre for Biotechnology Information (NCBI) Gene Expression Omnibus (GEO) and ArrayExpress. The data for different dermatological diseases corresponding to genome wide microarray studies will be identified and collected. Inorder to conduct a meta-analysis two type of cohorts will be identified namely the Discovery cohorts and the Validation cohorts.

2.2 *Transcriptome analysis*

The aim is to conduct a meta-analysis to identify differentially expressed genes in each dermatological disorder. Microarray studies were selected for each disorder and Biconductor Affy Matrics package was used for their analysis.

2.3 *Analysis of Gene Ontology, pathway and network*

Gene Ontology provides hierarchical controlled vocabulary divided into three categories: Biological process, Cellular component, Molecular function. Analyzing the pathways and networks would help determine related genes, presence of any compensatory gene and the phenotype conferred by each gene. This was done using DAVID (Database for Annotation, Visualization and Integrated Discovery) and EnrichR.

3 RESULTS & DISCUSSION

GEO Accession Number GSE90880, GSE83920, GSE72140, GSE72140 were found to be associated with vitiligo, Hermansky Pudlak Syndrome, Melasma and Epidermolysis bullosa simplex mottled pigmentation respectively from GEO database.

Differentially expressed genes (DEGs) were filtered out after careful analysis using Bioconductor Affy Matrics package. Background noise and data were normalize using RMA method. Moreover, statistical significance of difference between average means of control and test were calculated using t-test. Found DEGs for each of the four dieases are shown in the below table no. 1.

Pathway analysis of retrieve genes were carried out, and results are described below.

Table 1. Table of differentially expressed genes for respective pigmentation disorders.

Diseases	Number of Genes	Genes Symbols
Vilitigo	35	CD44, CLNK, MC1R, C1QTNF6, CASP7, BTNL2, LTA, TICAM1, TNF, ZMIM1, NLRP1, SLA, TLR2, and, TLR4, SOD2, and, SOD3, CCR6, CAT, BACH2, IL2RA, GZMB, PTPN22, IKZF4, SH2B3, HLA-DQB1, and, HLA-DRB1, CD80, MTHFR, IFIH1, HLA-A,, HLA-B, and, HLA-C, IFNAR1, GSTP1, FOXP3, FGFR1OP, RERE, UBASH3A, CTLA4, CXCR5
Hermanski-Pudlak Syndrome	4	BHLHE22, BLOC1S3, LGALS3, VPS33A, HSPB3, RAB38, PRDX5, SOS1, VPS39, CD1B, NXF1, DTNBP1, PDIA2, PADI1, PWAR4, SLC7A11, LYST, PI4K2A, F2R, PAWR, BLOC1S5, GDI1, HPS1, TFF1, HPS3, CHM, MAP1LC3B, RABGGTA, PARKIN, CHI3L1, HPS4, ELANE, AP3D1, HPS6, KXD1, RAB32, AP3B1, HPS5, FBXW7, F2RL3, VWF, SLC35D3, ABCC4, P4HB, RASGRF1, BLOC1S6, BLOC1S4
Melasma	21	PTGS1, PPARGC1A, SULT1E1, HMGCS2, DGAT2L3, HPGD, PPARA, ALOX15B, ACADM, DCT, HACL1, PECR, RDH11, SILV, SLC27A2, PTGIS, DHRS9, ACSS2, MVD, HMGCS1, HSD11B1
EBS-MP	37	FMO2, TSPY1, CCL5, PRKAR2B, TIMP4, PCDHA10, PFKFB1, G0S2, CTSW, POLR23, LIPE, CYP1A1, LPL, CCDC23, TUSC5, LEP, CCL22, SCD, PPP1R1A, RPS23, BDP1, GLYAT, PSG4, MMP12, PCK1, SPRR2B, RPL7L1, GPAM, PDE8B, ACVR1C, ZNF257, CIDEC, NLRP2, PDE3B, ABCD2, PAPPA2, LGALS12

3.1 *Vitiligo*

Vitiligo is an acquired, auto-immune disorder characterized by loss of melanocytes resulting in white, hypopigmented patches in the face, skin and hair. Functional analysis using DAVID (Database for Annotation, Visualization and Integrated Discovery) revealed that a majority of the genes identified were involved in immune responses, including antiviral defense, type 1 interferon signaling, innate immunity and host- viral infection.

3.2 *Hermansky-Pudlak syndrome (HPS)*

Hermansky-Pudlak syndrome is generally characterized by mutation is alveolar epithelial cell type II (AECII) and is shown to cause pulmonary fibrosis. Studies reveal that mitochondrial homeostasis plays an important role in alveolar epithelium dysfunction, which eventually leads to pulmonary fibrosis. An analysis of mitochondrial biogenesis and turnover was made based ATP levels, mitochondrial membrane potential and the levels of reactive oxygen species (ROS). Observing the data obtained above, genes involved in oxidation- reduction processes, ATP synthesis, oxidoreductase and dioxygenase have been significantly downregulated, indicating a corresponding decrease in mitochondrial biosynthesis. Additional studies reveal significant reduction in expression of mitochondrial biogenesis protein PGC-1α and it's downstream targets NRF-1, NRF-2 and TFAM. A further reduction in PARKIN levels reveals further impairment in not just mitochondrial biogenesis but also mitochondrial turnover processes, suggesting that mitochondrial homeostasis plays an important role in Hermansky-Pudlak syndrome symptoms like interstitial pneumonitis.

3.3 *Melasma*

Melasma is generally referred to as dark skin coloration and is characterized by appearance of dark colored patches on the jawline and face. Common causes are increased sun exposure, hormonal changes and genetic predisposition. Functional analysis using revealed that a majority of the genes identified were involved in cell proliferation regulation, peptidase activity regulation, lipid metabolism and response to xenobiotic stimulus. Also upregulated were genes in fibroblast and keratinocyte proliferation.

3.4 *Epidermolysis Bullosa Simplex with Mottled Pigmentation*

Epidermolysis Bullosa Simplex with Mottled Pigmentation is an autosomal dominant disorder characterized by blistering of epidermis, hyperkeratosis and pigmentation defects. Functional analysis using DAVID revealed that a majority of the genes identified were involved in retinoid acid metabolism and lipid metabolism. Other affected pathways included proximal and distal pattern formation and embryonic limb morphogenesis.

Further Pathway analysis shows that major pathways associated with Hypopigmentation are Innate immunity mechanisms, Interferon gamma-mediated response and Mitochondrial Biogenesis and major pathways associated with Hypopigmentation are Lipid Metabolism, UV light on melanocytes and Retinoic acid production.

Also, Table 2 and Figure 1 shows TYR is common between all four diseases, further one gene i.e. TYRP1 is common between HPS and Melasma. PMEL and OCA2 are related to HPS and Vitiligo. All the four genes are associated with melanin synthesis, which shows their involvement in pigmentation disorder.

Table 2. Tables shows the found common genes with in the diseases.

Diseases	Numbers	Genes	Gene Title
EBS-MP, Hermanski-Pudlak Syndrome, Melasma, Vilitigo	1	TYR	Tyrosinase
Hermanski-Pudlak Syndrome, Melasma	1	TYRP1	Tyrosinase-related protein 1
Hermanski-Pudlak Syndrome, Vilitigo	2	PMEL, OCA2	Premelanosome, Protein Melanocyte-Specific Transporter Protein

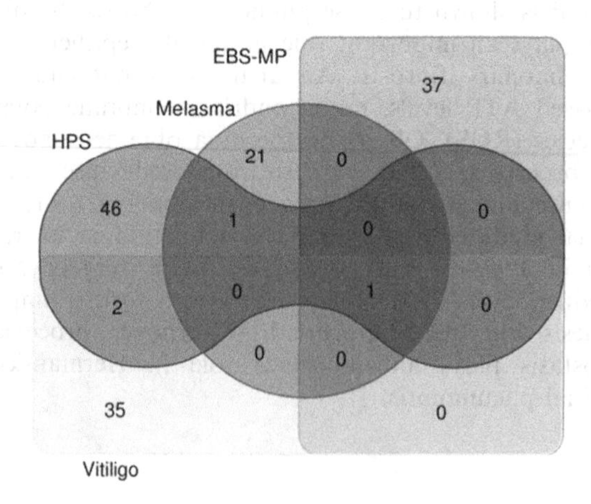

Figure 1. Venn diagram showing common genes among selected diseases.

4 CONCLUSION

The focus of this study was to conduct a transcriptional meta-analysis of pigmentation disorders. Over the years, there has been a sharp rise in the incidence of pigmentation disorders. Broadly, pigmentation disorders can be classified into hyperpigmentation and hypopigmentation disorders. In my study, I analyzed microarray data from two diseases in each category: Hypopigmentation (Vitiligo and Hermansky-Pudlak syndrome) and Hyperpigmentation (Melasma and Epidermolysis Bullosa Simplex with Mottled Pigmentation). The aim was to identify individual pathways as well as to simultaneously identify common pathways in each of the categories.

- Hypopigmentation disorders: These disorders are categorized by loss of pigmentation in skin and hair. This can be either due to loss of melanocytes or due to their inability to produce melanin.

  The hypopigmentation disorder vitiligo is characterized by an auto-immune loss of melanocytes. The functional annotation revealed mutation in viral defense pathways. Study of animal models revealed that long term viral infection can lead to autoimmunity and this observation was consistent with the fact that viruses like Epstein bar and herpes virus have been isolated form vitiligo skin. Also, an increasing incidence of vitiligo has been observed in HIV positive patients. Other mutations included alterations in innate immunity pathways, wherein inflammation is initiated due to the presence of reactive oxygen species. The interferon mediated response was also found to be altered leading to autoimmune loss of melanocytes.

  Hermansky-Pudlak is a genetic disorder involving multiple body symptoms and leads to blood platelet dysfunction, impairment in vision and loss of pigment of skin and hair (oculocutaneous albinism). Functional annotation revealed mutations in mitochondrial genes, particularly genes involved in ATP synthesis. Studies reveal that mitochondrial homeostasis plays an important role in alveolar epithelium dysfunction, which eventually leads to pulmonary fibrosis. Mutations were also found in gene expression controls especially at the translational level.
- Hyperpigmentation disorders: These are classified by an increase in melanin synthesis resulting in dark patches on the skin.

  An analysis of melasma and Epidermolysis Bullosa Simplex with Mottled Pigmentation revealed that both diseases presented mutations in pathways associated with retinoid acid metabolism, lipid metabolism, Bone morphogenesis pathways and response xenobiotic stimulus (drug metabolism). One of the important functions of the skin includes prevention of loss of water. Decrease in lipid metabolism in hyperpigmentation disorders results in disruption of stratum corneum and resulting trans-epidermal water loss. Retinoic acid acts as a transcription factor resulting in upregulation of certain keratinocytes and downregulation of others. The disease is characterized by loss of retinoid acid metabolism, leading to a simultaneous increase in keratinocytes, characteristic with the disease. Mutations in bone morphogenesis proteins also result in abnormal keratinocyte differentiation. Other mutations observed included alterations in Wnt pathway as well as the drug metabolizing capacity of cells (particularly cytochrome P450 family).

ACKNOWLEDGMENT

This study has been supported by Department of Biotechnology, Government of India, project file no., No.BT/PR5402/BID/7/408/2012.

REFERENCES

[1] DeStefano GM, Christiano AM. The Genetics of Human Skin Disease. Cold Spring Harb. Perspect. Med. [Internet]. 2014 [cited 2019 Jun 29];4:a015172–a015172. Available from: http://www.ncbi.nlm.nih.gov/pubmed/25274756.

[2] Dréno B, Araviiskaia E, Berardesca E, et al. Microbiome in healthy skin, update for dermatologists. J. Eur. Acad. Dermatology Venereol. 2016.
[3] Seneschal J, Clark RA, Gehad A, et al. Human epidermal Langerhans cells maintain immune homeostasis in skin by activating skin resident regulatory T cells. Immunity. 2012;
[4] Sadreyev RI, Feramisco JD, Tsao H, et al. Phenotypic categorization of genetic skin diseases reveals new relations between phenotypes, genes and pathways. Bioinformatics. 2009;
[5] Bhattacharyya M, Feuerbach L, Bhadra T, et al. MicroRNA transcription start site prediction with multi-objective feature selection. Stat. Appl. Genet. Mol. Biol. 2012;
[6] Chamcheu JC, Siddiqui IA, Syed DN, et al. Keratin gene mutations in disorders of human skin and its appendages. Arch. Biochem. Biophys. 2011.
[7] Kashiyama K, Nakazawa Y, Pilz DT, et al. Malfunction of nuclease ERCC1-XPF results in diverse clinical manifestations and causes Cockayne syndrome, xeroderma pigmentosum, and Fanconi anemia. Am. J. Hum. Genet. 2013;
[8] Ainger SA, Jagirdar K, Lee KJ, et al. Skin Pigmentation Genetics for the Clinic. Dermatology. 2017.

Biotechnology and Biological Sciences – Sen et al. (Eds)
© 2020 Taylor & Francis Group, London, ISBN 978-0-367-43161-7

Systems biology approach to predict novel microbial enzymes towards functionalization of piperine biosynthetic pathway in microbial host

Dipankar Ghosh*, Indranil Banerjee, Sudipta Sarkar, Spandan Saha, Ayantika Chakraborty & Bidisha Sarkar
Microbial Engineering Laboratory, Department of Biotechnology JIS University Agarpara Kolkata, India

ABSTRACT: Piperine is an alkaloid biomolecule. Piperine is found in black pepper seeds (*Piper nigrum*) from plant regime. Piperine has shown broad spectrum therapeutic activities. Piperine productivity suffers due to dependence on plant regimes considering longer generation time of plant. However, extraction of piperine from plant is another big hurdle. These facts limit the productivity of piperine towards commercial implementation. Piperine biosynthetic pathway in plant is well known. However, exact enzymatic reaction mechanisms are not yet known completely. Moreover, plant based enzymes suffer due to its post translational modification and stability. Based on this current scenario, current study focuses on systems biology approaches to predict novel microbial enzymes (as alternative to plant based enzymes) towards functionalization of piperine biosynthetic pathway in microbial host. It will not only solve the issue related to plant based enzymes but it could also improve the productivity using microbial platform in near future.

1 INTRODUCTION

Piperine is an alkaloid biomolecule. Piperine is usually found in the seeds of black pepper (*Piper nigrum*) from plant regime. Piperine paves the pharmaceutical sciences as a scaffold biomolecule for developing diverse derivatives of bioactive compounds towards therapeutic implementation in broad range of human diseases. Precisely, piperine and its derivatives have shown broad spectrum therapeutic activities i.e. antioxidant activity, anti-inflammatory activity, bio-enhancing and anti neurological disorder activity. Moreover, piperine harnesses drug resistance phenomenon to ameliorate clinical efficiency of antibiotics and anticancer drugs (Gorgani et al. 2017). Even usage of piperine as a scaffold for bioactive biomolecules remains still in its infancy but the steady progression of exploration of its molecular structure may lead to enormous encourages in novel drug discovery. Besides, Piperine productivity suffers due to dependence on plant regimes. However, extraction of piperine is another big hurdle. These facts and figures limit the productivity and yield of piperine towards commercial implementation as therapeutic compound. Piperine biosynthetic pathway in plant is well known. Plant secondary metabolite piperine biosynthesis is probably initiated by sequence of reactions i.e. condensation, decarboxylation, oxidative deamination and cyclization (Chopra et al. 2016). However, exact enzymatic reactions mechanisms are not yet known completely. Moreover, enzymes involved in each reaction steps in plant are not completely known and characterized yet. To this end, few enzymatic studies have been done considering biosynthetic pathway of analogous biomolecules likely coumaroylagmatine and feruloyltyramine

*Corresponding author: dghosh.jisuniversity@gmail.com or d.ghosh@jisuniversity.ac.in

(Bird & Smith 1983, Negrel & Jeandet 1987). Moreover, plant based enzymes suffer due to its post translational modification and stability. Based on this current scenario, current study focuses on systems biology approaches to predict novel microbial enzymes towards functionalization of Piperine Biosynthetic Pathway in Microbial Host.

2 MATERIALS AND METHODS

In this current study, a novel combinatorial computational framework has been optimized and applied including Atlas of Biochemistry database to find out novel pathways (with free energy change) along with its corresponding predicted novel pathway enzymes (Hadadi et al. 2016, Ghosh & Debnath 2019). The Neighbor-Joining method has also been applied for reconstructing phylogenetic relationship considering predicted novel enzyme classes (Saitou & Nei M 1987).

3 RESULTS & DISCUSSION

In this current study, we have successfully carried out computational combinatorial framework algorithms to predict novel synthetic metabolic pathways and its corresponding probable enzymes to functionalize in microbial cell factories for piperine generations (Figure 1). Where, query input database are the choice of substrates i.e. lysine for target biomolecule piperine generation. Predicted enzyme classes include lysine decarboxylase (EC 4.1.1.18; EC 4.1.1.-; EC 4.2.1.-), diamine Oxidase or copper amine oxidase (EC 1.4.3.22), amino acid transaminase (EC 2.6.1.-), amino acid oxidase (EC 1.4.3.-), and N-acetyl transferase (EC 2.3.1.-). In this study, Cadaverine, 5 Amino Pentanal, and Piperidine are metabolic intermediates. Though few enzymatic reaction steps show predicted free energy change positive, however, those reaction steps have be optimized and validated for actual free energy change calculations through laboratory based experimentations in future in microbial host. On the other phylogenetic analysis has been perform to find out the closer enzyme classes towards expression in microbial host as first choice to functionalize each and every enzymatic steps for predicted piperine biosynthetic pathway (Figure 2).

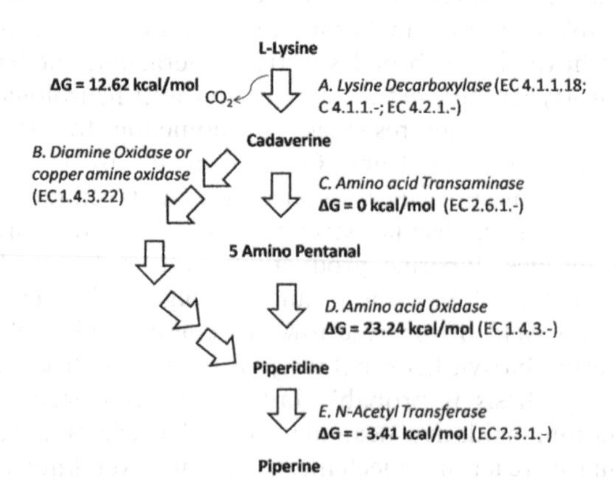

Figure 1. Designed and optimized biosynthetic pathway for piperine production.

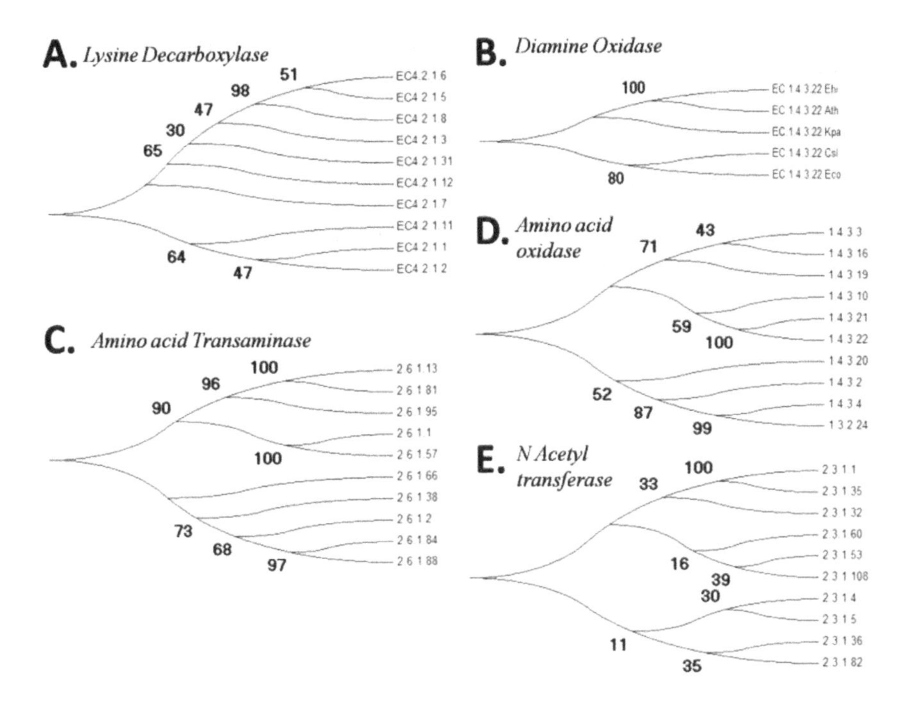

Figure 2. Phylogenetic relationship amongst the predicted pathway enzyme classes with boot strapping value of 500 (where A. lysine decarboxylase; B. diamine oxidase; C. amino acid transaminase; D. amino acid oxidase; and E. N-acetyl transferase enzyme classes).

4 CONCLUSION

The current combinatorial computational approach seems very promising avenue to predict synthetic metabolic pathways and corresponding enzyme classes for piperine generation in microbial platform however it exits in its infancy. In future an extensive research requires characterizing most potent enzyme from these enzyme classes to functionalize this predicted piperine biosynthetic pathway. This approach will not only reduce the time of piperine production and purification cost as well as improve the piperine productivity using fast growing suitable microbial host instead of dependent on slow growing plant platform.

ACKNOWLEDGEMENT

Authors would like to thank JIS University and JIS Group of Educational Initiatives for financial support and help.

REFERENCES

Bird, C.R. & Smith, T.A. 1983. Agmatine coumaroyltransferase from barley seedlings. *Phytochemistry* 22: 2401-2403.

Chopra, B., Dhingra, A.K., Kapoor, R.P. & Prasad D.N. 2016. Piperine and Its Various Physicochemical and Biological Aspects: A Review. *Open Chemical Journal* 3: 75-96.

Ghosh, D. & Debnath, S. 2019. Systems biology and synthetic biology approaches towards anti-hairfall compound biosynthetic pathways design and pathway enzymes findings. *International Journal of Innovation Knowledge Concepts* 7 (1): 166-169.

Gorgani, L., Mohammadi, M., Najafpour, G.D. & Nikzad, M. 2017. Piperine-The Bioactive Compound of Black Pepper: From Isolation to Medicinal Formulations. *Comprehensive Reviews in Food Science and Food Safety* 6: 124-140.
Hadadi, N., Hafner, J., Shajkofci, A., Zisaki, A. & Hatzimanikatis V. 2016. ATLAS of Biochemistry: A Repository of All Possible Biochemical. *ACS Synthethic Biology* 5: 1155-1166.
Negrel, J. & Jeandet, P. 1987. Metabolism of tyramine and feruloyltyramine in TMV inoculated leaves of Nicotiana tabacum. *Phytochemistry* 26: 2185-2190.
Saitou, N. & Nei, M. 1987. The Neighbor-joining method: a new method for reconstructing phylogenetic trees. *Molecular Biology Evolution* 4(4): 406-425.

Biotechnology and Biological Sciences – Sen et al. (Eds)
© 2020 Taylor & Francis Group, London, ISBN 978-0-367-43161-7

BmNPV late expression factor (lef-1) a potent target for inducing virus resistance against Grasserie infected *Bombyx mori* by RNA interference technology

Sahar Ismail, K.S.Tulsi Naik, K.M. Ponnuvel & R.K. Mishra
Seri Biotech Research Laboratory, Kodathi, Bangalore, Karnataka

M.V. Rajam
Department of Genetics, University of Delhi, South Campus, New Delhi

ABSTRACT: RNA mediated silencing technology (RNAi) has become the tool of choice for induction of virus resistance in many organisms. A significant feature of this technology is the double stranded RNA (dsRNA), the potent triggers of RNAi. In this study, *E. coli* are engineered to produce dsRNA of the cloned genes from a plasmid (L4440) containing gene of interest under the control of double T7 promoter which efficiently produces dsRNA once it is transformed into *E.coli* HT115 host strain and upon induction with IPTG. In this study, lef-1 (late expression factor-1) involved in BmNPV DNA replication was targeted against which dsRNA was produced. The RT-PCR analysis revealed that larva fed with 30μg of *E. coli* expressing lef-1 dsRNA showed significant decrease in the expression of viral genes involved in the BmNPV multiplication and showed in-crease in survivability of infected silkworms upto 50%. These results prove the successful use of *E.coli* expressing dsRNA as an efficient way of delivering dsRNA in silkworms as well as inducing resistance against BmNPV.

1 INTRODUCTION

Bombyx mori L. (Lepidoptera: Bombycidae) nucleo polyhedrosis virus (BmNPV) is a highly pathogenic virus that causes Grasserie infection in silkworms, effective management of the virus has been a challenge because of its rigid nature and the lack of effective control/preventive measures. The disease often breaks out in tropical sericultural countries where there are high humidity and temperature fluctuations throughout the year. As there are no preventive measures to stop the spread of infection, the only way out to avoid the spread of the disease is by burning of infected silkworm lots.

In 1998, Fire and his coworkers announced the discovery of RNA interfernce in *Caenorhabditis elegans*. And as a result of their discovery over the years, RNAi has become a powerful tool for functional genomics and widely used in insect genetic research. Though RNAi technology holds promising future in the field of new gene directed therapies, the difficulties involved in the efficient delivery of siRNA to the target cells has slowed down the rapid expansion of RNAi-based technologies,. Therefore, most of the current studies aim at improving existing methodologies and adopting innovative technologies to increase the efficiency of RNAi. Studies conducted by Timmons & Fire, 1998 and Timmons *et al.*, 2001 have shown that *E.coli* engineered to produce interfering RNAs was able to efficiently produce mutants of C.elegans.

RNAI experiments were carried out successfully in lepidopteran especially in *Bombyx mori* mostly to study the physiology related to Embryonic development, postembryonic development (ecdysis, larval-pupal molt, metamorphosis, adult wing and leg formation), cocoon colour, sex pheromone synthesis (Tabunoki *et al.*,2004, Liu *et al.*, 2008, Ohnishi *et al.*,2006, Masumoto *et al.*, 2009). Transgenic *Bombyx mori* was developed to express hairpin RNAs against viral gene ie-1 and was successful in using RNAi technology for preventing BmNPV infection in silkworms. And also knock down of ecdysis triggering hormone using RNAi was efficient in preventing ecdysis in silkworms.(Kanginakudru *et al.,* 2007; Dai *et al.,*2008). Silkworms expressing BmNPV lef-1 dsRNA was developed by combining RNAi and *B.mori* germline transformationthus study was able to show moderate inhibition of lef-1 gene expression in *Bombyx mori* (Tamura *et al.,* 2000 and Isobe *et al.,* 2004).

An extensive study of biology of nucleopolyhedrosis virus (NPV) was done in AcNPV *Autographa californica* (the alfalfa looper moth) multiple nucleopolyhedrovirus. This study led to identification of many genes essential for viral DNA replication including immediate-early-1 (*ie-1*), late expression factor-1 (*lef-1*), *lef-2*, *lef-3*, *dnapol* and *p143* (*dnahel*). It is also well known that approximately 80-90% of genome similarity is found between AcNPV and BmNPV. Therefore it is also confirmed that among the viral genes needed for baculovirus propagation, two genes are known to contribute majorly for baculovirus infection, *ie-1* encoding IE-1 (an essential regulatory protein for viral gene expression and DNA replication) and *lef-1* encoding the DNA primase essential for baculovirus DNA replication. Thus, *ie-1* and *lef-1* are appropriate candidates for RNAi-mediated gene silencing (Asha et al., 2002) (Zhang et al., 2014). Therefore, in this study we have targeted lef-1gene for inducing virus resistance against Grasserie disease caused by BmNPV by RNA interference technology by engineering *E. coli* to produce dsRNA against targeted gene.

2 MATERIALS AND METHODS

2.1 *Collection of infected larvae and purification of BmNPV occlusion bodies (polyhedra)*

The BmNPV infected 5th instar silkworm larvae were collected from the rearing house. The infected hemolymph (a white milky fluid/polyhedra) was collected by puncturing the prolegs of the infected larvae. The genomic DNA was isolated from the purified polyhedra. Approximatel 30ng of isolated genomic DNA was used to amplify the the Late expression factor (lef-1)gene (Accession Number L33180) (Subbaiah et al., 2011) (Table 1).The PCR amplification was performed using the following conditions: initial denaturation of 5 min at 94°C; 30 cycles of 30 s at 94°C; 30 s at 52°C; and 1 min at 72°C; and a final extension of 10 min at 72°C. The amplified PCR products was analyzed on 1.2% agarose gel along with 100 bp DNA ladder. The amplified lef-1 PCR product was sequenced to confirm the sequence similarity with the published lef-1 sequence in NCBI database(Khurad et al., 2004).

2.2 *Vector construction and dsRNA expression*

The lef-1, gene of ~310 bp (Cao et al., 2016)PCR amplified using BmNPV viral DNA was then cloned into L4440 plasmid possessing two T7 promoters in inverted orientation flanking the multiple cloning sites. The resulting clones L4440-lef1 and L4440 vector alone (without the insert) was transformed into *E.coli* HT115 (DE3) competent cells lacking RNase III activity. The T7 polymerase was induced by addition of 0.25mM IPTG. To confirm the expressed dsRNA, cDNA was synthesized and amplified using gene-specific primers to confirm the presence of the gene and also induction.

Table 1. List of primers selected for the analysis of viral gene expression.

Primer Name	Primer sequences `5-3`
Lef-1	FP-CGCGGATCCGGTACAAGGCCTTCGAAA
	RP-CCGTCCGGACCGGACCACATCCACCAA TTCCAT
Lef 3	FP-ACGCGTCGACTTGTTTTTGAAGTCGCG
	RP-TGCTCTAGACTTGCGTTTGTGCAATTTTGCA
GP41	FP-CGTAGTAGTAGTAATCGCCG
	RP- AGTCGAGTCGCGTCGCTTT

2.3 *Bioassay to validate the effect of DsRNA in BmNPV infected silkworm.*

The 5th instar *B. mori* larvae were fed individually with fresh mulberry leaves spread uniformly with Fifty µl (50 µl) of bacteria expressed dsRNA and NPV polyhedra(4000OBs). For each set, 50 larvae were used and the whole experiment was done in triplicates. After the complete leaves were fed by larvae fresh leaves were fed everyday twice. Larvae were collected after 48 hrs of feeding. Total DNA was extracted from the midguts to analyze viral copy number using Gp41 gene specific primers and total RNA was isolated and cDNA was synthesized to perform Real time PCR analysis to check the differentially expressed viral genes in the midguts of dsRNA treated and non-treated silkworms. Except for the larvas taken for the dis-section for the analysis of expression of viral genes, remaining larva were continued for another 4-5 days till they spun cocoocn, to check the survivability of the dsRNA fed infected silk-worms and percentage of survival was also calculated.

3 RESULTS AND DISCUSSIONS

3.1 *Amplification of lef-1 essential viral gene(s) involved in BmNPV multiplication from the infected silkworms.*

The Grasserie infected larvae look restless, with swollen and swelled integuments. At an advanced stage of infection, the body wall ruptures and white turbid hemolymph containing a large number of OBs/polyhedra oozes out. The heavely infected larva look white at the time of death and became dark within 3–5 hrs as they putrefied (**Figure 1A**). PCR amplification using lef-1 primers with genomic DNA isolated from the polyhedra yielded 310-bp as expected, (**Figure 1B**). Further, the sequencing of the and lef-1 PCR product confirmed that the amplified product was of BmNPV origin. No differences at the nucleotide sequence level in lef-1 against the NCBI accession number L33180.1. Therefore, the lef-1 gene of BmNPV was successfully amplified and sequence similarity also confirmed the lef-1 gene.

3.2 *Cloning and expression analysis of dsRNA lef-1 in bacteria.*

The lef-1 gene BmNPV was cloned into L4440 vector, the clones were confirmed by PCR using gene-specific and vector specific primer and restriction digestion. The confirmed clones were transformed into HT115 DE3 bacterial host strain that lacks RNase III activity for expression/

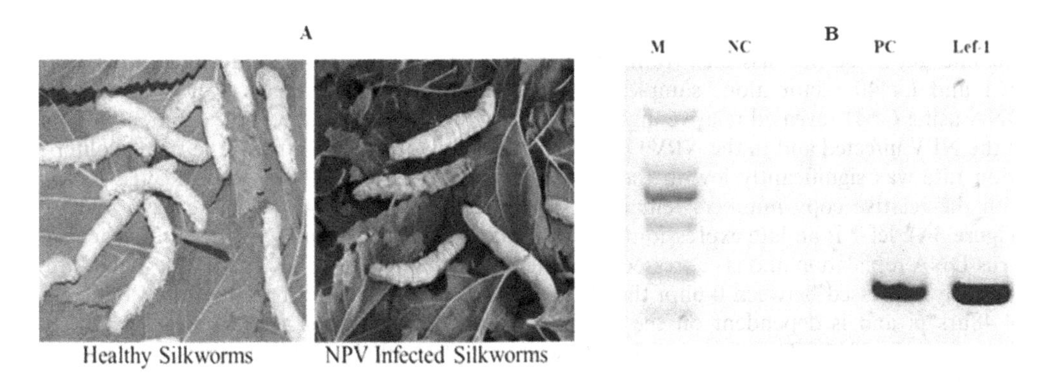

Figure 1. **Analysis of viral genes in silkworms infected with *Bombyx mori* Nucleopolyhedrosis virus-(BmNPV) (A):** An 5th instar BmNPV infected silkworms with swollen integuments, putrefied BmNPV infected larvae (white milky fluid turned brown due to oxidation) **(B):** Viral gene *lef-1* amplified from the genomic DNA isolated from the infected silkworm for cloning into dsRNA producing L4440 vector. M: 100bp DNA ladder, NC: Negative control, PC positive control, Lef-1: lef-1 gene amplified from the purified BmNPV spore DNA.

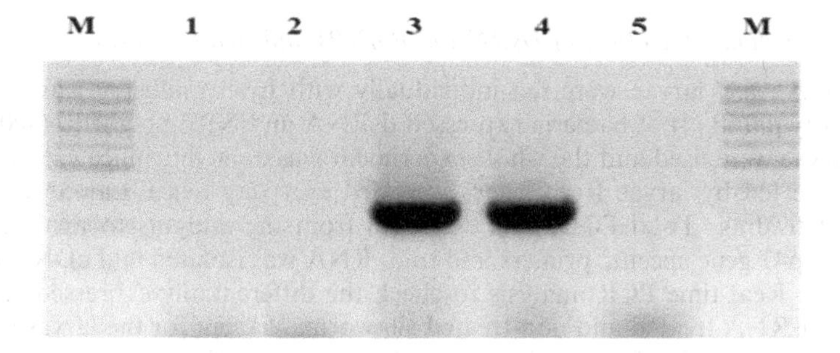

Figure 2. **Cloning and expression analysis of dsRNA produced in bacteria** In this study the dsRNA produced in the bacterial system for stable expression of dsRNA and these bacterial cells are fed to silkworms orally to elicit RNA interference against the viral genes.: A Semi-quantitative analysis to confirm the induced lef-1 (I-lef-1) and Induced vector L4440 without insert (I-L4440 vector) dsRNA in bacteria before feeding to the silkworms.M: DNA Marker, 1: Negative control 2: Induced L4440 vector without gene of interest 3: Induced l4440 vector with lef-1 gene of interest 5: positive control.

production of dsRNA. The transformed cells were induced with 0.25mM of IPTG and continued the induction for 4 hours. The induction of dsRNA was confirmed by semi-quantitative expression analysis of lef-1 genes in induced L4440 with cloned lef-1 genes while there was no such amplification observed in the induced L4440 vector alone. This induction indicated the expression of dsRNA from cloned lef-1 gene of BmNPV in bacteria (**Figure 2**). Many of the studies involved in identifying an efficient delivery system of RNAi into living systems were able to succeffully show that dsRNAs can expressed inside bacteria, or they can also be synthetically made in vitro, and then can be fed to insects through artificial diet or through their natural feed like leaves in case of silkworms. (Turner et al., 2006; Baum et al., 2007; Tian et al., 2009; Surakasi et al., 2011; Yu et al., 2013). There are many methods of introducing dsRNA into the system like use of nanoparticle, microinjection etc, but among all the available methods for delivering dsRNA into the host, oral delivery/ingestion is a less invasive, less labor intensive, cost-effective and is practically more feasible method for RNAi delivery (Yang *et.al.*, 2014) (Piot *et a*l., 2015). And is it is also more feasible if it will be utilized for field applications for RNAi-mediated pest control. And also for insect that are very small and minute and are intolerant of injection (Yu *et al.*, 2013) (Yogindran and Rajam, 2015).

3.3 *Analysis of viral gene copy number and differential expression of viral genes in dsRNA fed silkworms.*

The mid-gut tissue was dissected from infected silkworms after 48 hours of feeding of dsRNA lef-1 and L4440 vector alone samples. The quantitative real-time analysis of the genomic DNA using Gp41 revealed relative high copy numbers of approximately1×10^5 copies at 48h pi in the NPV infected and in the NPV+L4440 treated silk-worms. In contrast, the viral proliferation rate was significantly low in the larval mid-gut of the dsRNA lef-1fed infected larva, with the relative copy numbers remaining below 100 copies until 48 hpi (Bao et al., 2009) (**Figure 3A**). lef-1 is an late expression factor encoding the DNA primase essential for baculovirus DNA replication and is expressed after the expression of earl genes like ie-1, ie-2 that are generally expressed between 0-6hpi therefore the lef-1 proteins could only be found between 24-48hrs pi and is dependent on the time of expression of early genes (Asha et al., 2002) (Zhang et al., 2014). These results also showed that BmNPV invaded the mid-gut tissue of both NPV infected as well as dsRNA lef-1 fed silkworms, but the viral proliferation in the mid-gut of the dsRNA lef-1 fed silkworms was significantly inhibited by RNAi mechanisms triggered by dsRNA ingested by the silkworms.

The real time analysis of viral genes also showed differential expression between NPV infected and dsRNA treated infected silkworms. In this study it was observed that there was approximately 6-fold decrease in the late gene expression in dsRNA fed infected silk-worms

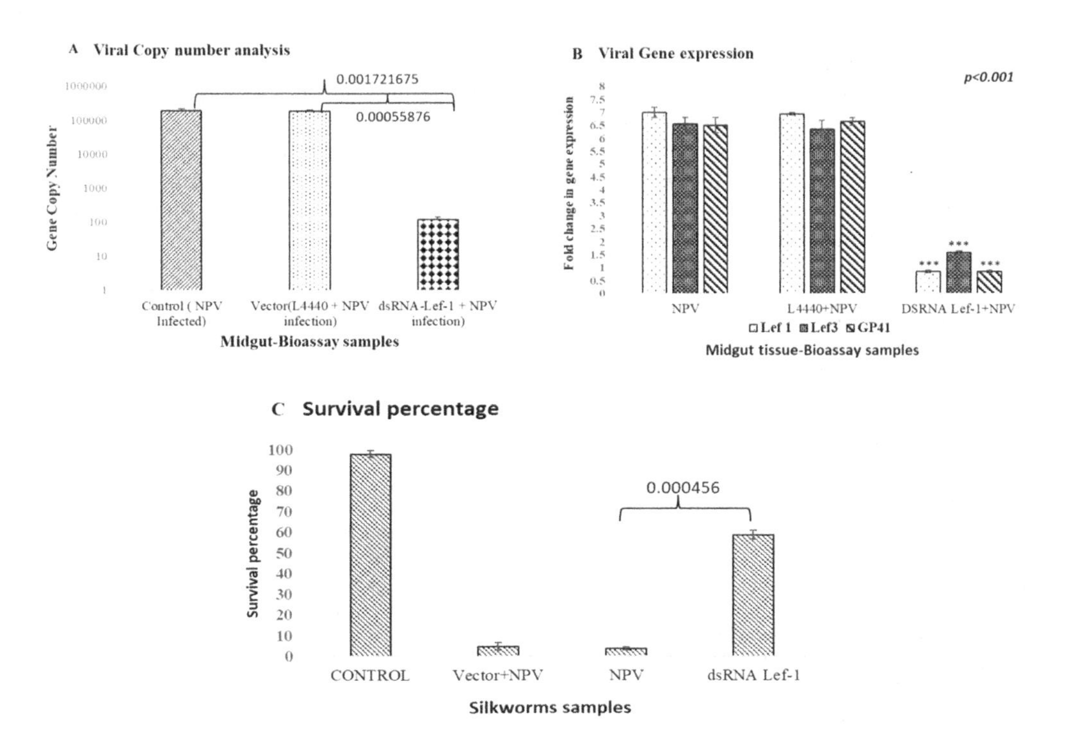

Figure 3. **Differential gene expression analysis of genes involved in NPV multiplication.** The 5th instar larvae were collected after feeding with dsRNA ie-1 and dsRNA lef-1 expressing bacteria total RNA was isolated followed by cDNA synthesis the gene expression analysis was carried out using Real time PCR. **(A)** Copy Number estimation using Gp41 specific primers in bioassay samples dissected from NPV infected and dsRNA fed infected silkworms showing significant pvalue of 0.00172 between NPV infected and dsRNA treted samples and 0.000558 between vector +NPV infection and dsRNA treated **B**). QPCR analysis of BmNPV viral gene expression in midgut samples of dsRNA fed infected silkworms showing significant (***p<0.001) decrease in the viral gene expression compared to infected silkworms. **(C):** Comparison of effect of dsRNA lef-1 on the of Survival percentage of infected silkworms.

compared to infected silkworms (**Figure 3B**). Therefore, the results are clearly evident that the relative decrease in the expression of the viral genes especially Lef-1, Lef-3 and Gp41mostly belong to late genes hence the decrease in the expression levels may be attributed to the effect of RNAi induced by lef-1 dsRNA on the lef-1 genes.

To compare the survival rate of NPV infected and the dsRNA fed NPV infected silkworms, number of dead larvae against the survived larvae in all the experimental group was calculated till all the larvae started spinning cocoons. From this study it was observed that the lef-1dsRNA fed BmNPV infected larvae showed 53.6% survivability and the results obtained was significant with p value (p<0.001) and the larvae fed with NPV and NPV + L4440 vector showed only 1-2% survivability and all the remaining larvae showed the disease symptoms.. The survivability may be attributed to the RNAi mechanism triggered by the siRNA in the mid-gut tissue upon virus infection. (**Figure 3C**).

4 CONCLUSION

The study has shown that larvae fed with *E.coli* expressing *lef-1* dsRNA showed a significant reduction in the expression of targeted viral genes involved in the *Bm*NPV multiplication and showed substantial increase in the survivability of infected silkworms. Therefore, the study could clearly show that dsRNA can be produced successfully in bacteria and effective dosage of dsRNA to elicit RNAi in silkworms can be delivered efficiently into silkworms orally through mulberry leaves. And feeding 50µl of *lef* dsRNA (~30ug dsRNA) at 5th instar to silkworms was effective in silencing the *late gene expression that is shown to be involved in viral DNA multiplication* during BmNPV infection in silkworm. The dsRNA fed infected silkworms not only survived the infection but also spun cocoon indicating that the dsRNA lef-1 was able to control the virus multiplication if the silkworms are fed the dsRNA at the early stage and those silkworms that survived the infection was able to spin good quality cocoons.

ACKNOWLEDGEMENT

This research was funded by Central Silk Board, Bangalore, Karnataka

REFERENCES

Asha acharya, Satyanarayana sri ram, Seema sehrawat, Masudur rahman, Deepak Sehgal and K. P. Gopinathan (2002) *Bombyx mori* nuclopolyhedrovirus: Molecular biology and biotechnological applications for large scale synthesis of recombinant protein, *Current science* 83, 4.

Baum, J.A., Bogaert, T., Clinton, W., Heck, G.R., Feldmann, P., Ilagan, O., Johnson, S., Plaetinck, G., Munyikwa, T., Pleau, M., Vaughn, T. and Roberts, J. (2007) Control of coleopteran insect pests through RNA interference. *Nature Biotechnology*, 25, 1322–1326.

Bao Y.Y., Tang X.D., Lv Z.Y., Wang X.Y., Tian C.H., Xu Y.P., et al. (2009). Gene expression profiling of resistant and susceptible *Bombyx mori* strains reveals nucleopolyhedrovirus-associated variations in host gene transcript levels. *Genomics* 94 138–145.

Cao, Ming-ya & Kuang, Xiu-xiu & Li, Hai-qing & Lei, Xue-jiao & Xiao, Wen-fu & Dong, Zhanqi & Zhang, Jun & Hu, Nan & Chen, Ting-Ting & Lu, Cheng & Pan, Min-Hui. (2016). Screening and optimization of an efficient Bombyx mori nucleopolyhedrovirus inducible promoter. Journal of Biotechnology. 231. 10.1016.

Dai, H., Ma, L., Wang, J., Jiang, R., Wang, Z. and Fei, J. (2008) Knockdown of ecdysis-triggering hormone gene with abinary UAS/GAL4 RNA interference system leads to lethal ecdysis deficiency in silkworm. Acta Biochimicaet Biophysica Sinica, 40, 790–795.

Fire, A., Xu, S., Montgomery, M., Kostas, S., Driver, S. and Mello, C. (1998) Potent and specific genetic interference by double stranded RNA in Caenorhabditis elegans. Nature 391: 806–811.

Isobe R., Kojima K., Matsuyama T., Quan GX., Kanda T., Tamura T., Sahara K., Asano SI., Bando H. (2004) Use of RNAi technology to confer enhanced resistance to BmNPV on transgenic silkworms. Arch. Virol. 149:1931–1940.

Kanginakudru, S.; Royer, C.; Edupalli, S.V.; Jalabert, A.; Mauchamp, B.; Chandrashekaraiah; Prasad, S.V.; Chavancy, G.; Couble, P. & Nagaraju, J (2009). Targeting ie-1 gene by RNAi induces baculoviral resistance in lepidopteran cell lines and in transgenic silkworms. Insect molecular biology, 16:5,635–644.

Khurad A.M., Mahulikar A, Rathod M.K., Rai M.M, Kanginakudru S., and Nagaraju J (2004) Vertical transmission of nucleopolyhedrovirus in the silkworm, *Bombyx mori L* Journal of Invertebrate Pathology 87, 8–15.

Masumoto M., Yaginuma T. & Niimi T. (2009).Functional analysis of Ultrabithorax in the silkworm, *Bombyx mori*, using RNAi. Development genes and evolution, 219, 437–444.

Ohnishi, A.; Hull, J.J. & Matsumoto, S. Targeted disruption of genes in the *Bombyx mori* sex pheromone biosynthetic pathway. *Proceedings of the National Academy of Sciences of the United States of America*, (2006) 103, 4398–4403.

Subbiah E V., Corinne Royer, Sriramana Kanginakudru, Vallari V Satyavathi, Adari Sobhan Babu, Vankadara sivaprasad *et al.*, (2013) Engineering Silkworms for resistance to Baculovirus through Multigene RNA interference: Genetics: 193: 63–75.

Surakasi, V.P., Mohamed, A.A.M. and Kim, Y. (2011) RNA interference of beta 1 integrin subunit impairs development and immune responses of the beet armyworm, *Spodoptera exigua. Journal of Insect Physiology*, 57, 1537–1544.

Tamura, T., Thibert, C., Royer, C., Kanda, T., Abraham, E., Kamba, M., *et al.*, (2000) Germline transformation of the silkworm *Bombyx mori* L. using a piggyBac transposon-derived vector. Nat Biotechnol 18: 81–84.

Tabunoki, H.; Higurashi, S.; Ninagi, O.; Fujii, H.; Banno, Y. & Nozaki, M. A carotenoid-binding protein (CBP) plays a crucial role in cocoon pigmentation of silkworm (*Bombyx mori*) larvae. *FEBS letters*, (2004). 567 175–178.

Tian, H., Peng, H., Yao, Q., Chen, H., Xie, Q., Tang, B. and Zhang, W. (2009) Developmental control of a Lepidopteran pest *Spodoptera exigua* by ingestion of bacterial expressing dsRNA of a non-midgut gene. *PLoS ONE*, 4, e6225.

Tian, H., Peng, H., Yao, Q., Chen, H., Xie, Q., Tang, B. and Zhang, W. (2009) Developmental control of a Lepidopteran pest *Spodoptera exigua* by ingestion of bacterial expressing dsRNA of a non-midgut gene. *PLoS ONE*, 4, e6225.

Timmon, L., Court, D.L. and Fire, A. (2001) Ingestion of bacterially expressed dsRNAs can produce specific and potent genetic interference in *Caenorhabditis elegans*. *Gene*, 263, 103–112.

Timmon, L. and Fire, A. (1998) Specific interference by ingested dsRNA. *Nature*, 395, 854.

Katsuma S, Mita K, Shimada T (2007) ERK- and JNK-dependent signaling pathways contribute to *Bombyx mori* nucleopolyhedrovirus infection. J Virol 81(24),13700–13709.

Turner, C.T., Davy, M.W., MacDiarmid, R.M., Plummer, K.M., Birch, N.P. and Newcomb, R.D. (2006) RNA interference in the light brown apple moth, *Epiphyas postvittana* (Walker) induced by double-stranded RNA feeding. *Insect Molecular Biology*, 15, 383–391.

Yogindran, Sneha and Manchikatla Venkat Rajam. (2016) "Artificial miRNA-mediated silencing of ecdysone receptor (EcR) affects larval development and oogenesis in Helicoverpa armigera." *Insect biochemistry and molecular biology* 77: 21–30.

Yu N, O Christiaens, J Liu, J Niu, K Cappelle, S Caccia, H Huvenne, G Smagghe (2013) Delivery of dsRNA for RNAi in insects: an overview and future directions Insect Science, 20, pp. 4–14.

Zhang, P & Wang, J & Lu, Y & Hu, Y & Xue, Renyu & Cao, Guangli & Chengliang, Gong. (2014). Resistance of transgenic silkworm to BmNPV could be improved by silencing ie-1 and lef-1 genes. *Gene therapy*. 21. 10.1038/gt.2013.60.

Biotechnology and Biological Sciences – Sen et al. (Eds)
© 2020 Taylor & Francis Group, London, ISBN 978-0-367-43161-7

Cell-free mature miR-100-5p expression in plasma and its correlation with the histopathological markers in oral squamous cell carcinoma

Ruma Dey Ghosh*, Akash Bararia & Sudhriti Guha Majumder
Department of Head and Neck Oncology, Tata Translational Cancer Research Centre

Ajay Manickam, Tsewang Yougyal Bhutia, Prateek Jain, Kapila Manikantan, Rajeev Sharan & Pattatheyil Arun
Head and Neck Surgical Unit, Tata Medical Centre, Kolkata, India

ABSTRACT: In Oral Squamous Cell Carcinoma (OSCC), the risk assessment, probability of developing locoregional recurrence and metastatic deposit, and the tailor-made post-operative treatment strategies for each patient are recommended by the multidisciplinary team (MDT) mainly on the basis of TNM staging and histopathological report. So far, there is no molecular biomarker to classify subgroups more accurately in OSCC. Earlier studies revealed that the level of mature miR-100 expression is significantly altered and frequently associated with poor prognosis suggesting that miR-100 can be a potential biomarker candidate in OSCC. In the present study, total 40 patients with newly diagnosed OSCC were selected and EDTA whole blood (onset of surgery), histopathology, follow-up outcome data were collected. We measured cell-free mature miR-100-5p expression through qRT-PCR using plasma samples from each patient. Here, we explored and tried to correlate their functional relationship between the pathological conditions and the level of cell-free mature miR-100-5p expression in plasma with respect to disease recurrence and metastasis in patients with OSCC.

Keywords: Oral cancer, OSCC, microRNA, cell-free-miRNA, biomarker, miR-100-5p, prognosis

1 INTRODUCTION

In India 16.1% of all cancer types is oral cancer, with number of New cases registered being 1,19,992 amongst which India men's tops the position in terms of OSCC disposition.[1, 2, 3] In oral squamous cell carcinoma (OSCC), for early-stage disease, surgery is the mainstay of multimodality treatment procedure. The lymphovascular invasion (LVI), perinural invasion (PNI), exptranodal extension (ENE) are three major histopathological markers. The positivity of any of these markers are associated with poorer prognosis and those patients are considered as high-risk patients.[4, 5, 6] Earlier studies revealed that the dysregulation of miR-100 is correlated with the induced cellular migration, induced epithelial-mesenchymal-transition (EMT), cell-differentiation and frequently associated with poor prognosis suggesting that miR-100 can be a potential biomarker candidate to predict prognosis more accurately in OSCC.[7, 8, 9, 10] However, the cell-free mature miR-100 expression in plasma profile to predict the patient-outcome has not yet been studied in patients with OSCC. In the present study, we studied the functional relationship between the pathological conditions and the level of

*Corresponding author: E-mail: deyrumai@yahoo.co.in

cell-free mature miR-100-5p expression in plasma and analyse to correlate with the patient-outcome mainly disease-recurrence and metastasis in OSCC-patients.

2 MATERIALS & METHODS

The proposed work was approved by our Institutional Review Board (IRB).

Patient selection and Blood sample collection: Informed consents were taken from all patients before sample collection. The patients with newly diagnosed OSCC were recruited for the study by the Head and Neck Cancer Unit at Tata Medical Center, Kolkata in compliance to the inclusion and exclusion criteria. Plasma samples were stored in aliquots after centrifugation in our institutional bio-bank facilities of the Tata Medical Centre.

Patient's data collection: Detail-clinical information for each patient is stored in our online module system known as HMS (Hospital Management System) at TMC. The respective histopathological data, outcome data and other information for all selected patients were collected through our HMS. As per IRB regulations, patient's personal information was kept as anonymous while carrying out the work.

Quantitative real-time PCR (qRT-PCR) and RT-PCR: Total RNA was isolated from each patient's plasma sample using Trizol LS reagent according to manufacturer's protocol. To analyse the mature miR-100-5p expression, 100 ng total RNA was used for cDNA synthesis by using the 5x miScript HiSpec buffer in miScript II RT kit (Qiagen) according to manufacturer's instructions. Then PCR product was diluted and 1 ng cDNA was used per reaction for quantitative real-time PCR by using specific miScript primer assay and miScript Universal primer provided in the miScript SYBR Green PCR kit (Qiagen) following the manufactures protocol. RNU6B (U6) served as an endogenous control for miRNA expression analysis.

Data analyses: All the results were statistically analysed using available commercial software's and packages.

3 RESULTS & DISCUSSION

In the present investigation, a total of 40 newly diagnosed OSCC-patients were selected along with all necessary information and patient outcome data. The average age was 50.475 and within the age range between 27 to 84. The male and female ratio was 33:7 (true male number): (true female number). OSCC samples were collected from different primary subsites of the tumor like, gingivobuccal mucosa, anterior tongue, floor of the mouth, hard palate, anterior alveolar follicle etc., covering the entire oral cavity. On the basis of final histopathology report for each patient, all patients were classified with true Node negative and true Node positive. We have also documented the information regarding demographic data other histopathological information mainly, cellular differentiation, LVI, PNI, ENE, bone involvement, worst pattern invasion. During follow-up (18 months after surgery) data collection we have documented the information regarding the patient outcome.

We performed quantitative real time expression for cell-free mature miR-100-5p in 40 plasma samples. With respect to the above classifications the qPCR data was correlated to predict patient prognosis. The patient analysis was broadly classified based on disease recurrence and metastasis of the patient pool.

4 CONCLUSION

The tumour staging and histopathological markers for the assessment of the tumour invasiveness as well as the level of differentiation are the most important factors for risk assessment and to predict prognostication in OSCC. The present study also suggested that cell-free mature miR-100-5p expression in plasma could be additional insight to predict OSCC-prognostication. Further studies will be required to understand the molecular biology and pathogenesis of OSCC to confirm cell-free miR-100-5p expression for the prediction of OSCC patient-prognosis accurately.

ACKNOWLEDGMENT

This work was supported by HRD WS research grant (V.25011/452-HRD/2016-HR) awarded to Dr. Ruma Dey Ghosh from Department of Health Research (DHR, Govt. of India), New Delhi, India.

REFERENCES

1. Bray F, Ferlay J, Soerjomataram I, et al. Global cancer statistics 2018: GLOBOCAN estimates of incidence and mortality worldwide for 36 cancers in 185 countries. CA Cancer J Clin. 2018 Nov;68 (6):394-424. doi: 10.3322/caac.21492. PubMed PMID: 30207593.
2. Ferlay J, Colombet M, Soerjomataram I, et al. Estimating the global cancer incidence and mortality in 2018: GLOBOCAN sources and methods. Int J Cancer. 2019 Apr 15;144(8):1941-1953. doi: 10.1002/ijc.31937. PubMed PMID: 30350310.
3. International agency for cancer research WHO. 356-india-fact-sheet [Fact Sheet]. The Global Cancer Observatory. 2019 May, 2019.
4. Ho AS, Kraus DH,Ganly I, et al. Decision making in the management of recurrent head and neck cancer. Head Neck. 2014 Jan;36(1):144-51. doi: 10.1002/hed.23227. PubMed PMID: 23471843.
5. Koyfman SA, Ismaila N, Crook D, et al. Management of the Neck in Squamous Cell Carcinoma of the Oral Cavity and Oropharynx: ASCO Clinical Practice Guideline. J Clin Oncol. 2019 Feb 27: JCO1801921. doi: 10.1200/JCO.18.01921. PubMed PMID: 30811281.
6. Omura K. Current status of oral cancer treatment strategies: surgical treatments for oral squamous cell carcinoma. Int J Clin Oncol. 2014;19(3):423-30. doi: 10.1007/s10147-014-0689-z. PubMed PMID: 24682763.
7. Cancer Genome Atlas N. Comprehensive genomic characterization of head and neck squamous cell carcinomas. Nature. 2015 Jan 29;517(7536):576-82. doi: 10.1038/nature14129. PubMed PMID: 25631445; PubMed Central PMCID: PMCPMC4311405.
8. Henson BJ, Bhattacharjee S,O'Dee DM, et al. Decreased expression of miR-125b and miR-100 in oral cancer cells contributes to malignancy. Genes Chromosomes Cancer. 2009 Jul;48(7):569-82. doi: 52 Col:2010.1002/gcc.20666. PubMed PMID: 19396866; PubMed Central PMCID: PMCPMC2726991.
9. Motawi TK, Rizk SM, Ibrahim TM, et al. Circulating microRNAs, miR-92a, miR-100 and miR-143, as non-invasive biomarkers for bladder cancer diagnosis. Cell Biochem Funct. 2016 Apr;34(3):142-8. doi: 10.1002/cbf.3171. PubMed PMID: 26916216.
10. Qin C, Huang RY, Wang ZX. Potential role of miR-100 in cancer diagnosis, prognosis, and therapy. Tumour Biol. 2015 Mar;36(3):1403-9. doi: 10.1007/s13277-015-3267-8. PubMed PMID: 25740059.

Biotechnology and Biological Sciences – Sen et al. (Eds)
© 2020 Taylor & Francis Group, London, ISBN 978-0-367-43161-7

Structure prediction and functional characterization of uncharacterized protein Rv1708 of *Mycobacterium tuberculosis* (Strain ATCC 25618/H37Rv)

Shamrat Kumar Paul, Tasnin Al Hasib, Pranta Ray, Lutful Kabir & Abul Bashar Ripon Khalipha*
Bangabandhu Sheikh Mujibur Rahman Science and Technology University, Bangladesh

ABSTRACT: Tuberculosis (TB) is one of the most known ancient human disease which is caused by *Mycobacterium Tuberculosis* (MTB). Surprisingly, at present it is one of the major causes of death of human around the globe. Complete genome sequencing of MTB provides us a storehouse of genomic information to study about MTB's complex pathogenicity. Among 3924 Open Reading frame of MTB, the Uncharacterized Protein Rv1708 is encoded by Rv1708 gene and it is a inferred to be a cell cycle regulatory protein. It is anticipated that during cell division the Uncharacterized Protein Rv1708 play a critical role in septum formation of MTB. Thus, it becomes the mediator of cell cycle progression and cell division. Inhibition of bacterial cell division by blocking its associated protein is known to be therapeutic target for defeating disease. But the Uncharacterized Protein Rv1708 of the MTB remained unexplored. So, our aim is to propose the structural and functional features in addition to reveal the physicochemical properties. Homology modelling of the Uncharacterized Protein Rv1708 was generated by using Phyre2 and Swiss Model. The in-stability index generated by using the ExPasy's ProtParam tool shows that the Uncharacterized Protein Rv1708 is stable and its nature is acidic. Ramachandran map analysis by MolProbity reveals 95.5% of all residues were in allowed regions and 87.7% of all residues were in favored regions; which is indicating strong evidence of good quality of protein structure. This in-silico process will unveil the role of unexplored Uncharacterized Protein Rv1708 of MTB, and so it can pave the way for enriching our knowledge of the pathogenesis and drug-targeting approach for MTB.

1 INTRODUCTION

Mycobacterium tuberculosis (MTB), the causative agent of tuberculosis (TB), is one of the ancient bacterial species that still have a staggering impact on mortality rate since two million people die each year globally despite the global use of live, attenuated vaccine and antibiotics (Gengenbacher & Kaufmann, 2012). TB is caused by breathing in air contaminated with microscopic droplets that contain the untreated, active form of MTB. These microscopic droplets spread in the air when someone with an untreated, active form of TB speaks, coughs, sneezes, sighs, laughs, or spits (Malenki, 2018). Inhaling only a few droplets of such bacteria can lead to infected with TB and poses the lifetime risk of falling ill with it of 10 percent. In addition, people living with malnutrition, diabetes, HIV or people who use tobacco, possess a much higher risk of falling ill. Genome sequencing of MTB reveals that it is the second-largest bacterial genome sequence ever found which is rich in repetitive DNA, especially insertion sequences. It has also duplicated housekeeping genes and multigene families (Okou et al., 2007). Among 3,924 open reading frames of MTB, 91% of them have the potential of coding

*Corresponding author: khalipha1982@gmail.com

capacity. Some of them are belonging in-frame stop codons or frameshift mutations. This play crucial role in frameshifting during translation in association with pseudogenes (Cole et al., 1998). The genomic context of Uncharacterized Protein Rv1708 is observed and cross-checked by accession NC000962.3 from National Centre for Biotechnology Information (NCBI). Literature review showed that, a septal ring mediates the cell division of MTB, in association with a dozen of known proteins gather to the division site. In MTB, septal ring is a polymer of tubulin like Filamenting temperature-sensitive mutant Z (FtsZ) Protein. Septum formation and FtsZ polymerization are mediator of transcription and are clambered by protein interactions. A recent study shows that the Uncharacterized Protein Rv1708; a cell cycle regulatory protein that steers cell cycle progression and consequently cell division (Misra, Maurya, Chaudhary, & Misra, 2018). However, the structure of this Uncharacterized Protein Rv1708 is not reported yet. Also, the detailed physicochemical characterization and promising structure is not elucidated, so we have proposed a computer-aided homology modelled structure prediction of Uncharacterized Protein Rv1708 of MTB.

2 METHODOLOGY

2.1 *Retrieval of target amino acid sequence*

The amino acid sequence of Uncharacterized Protein Rv1708 of MTB (strain ATCC25618/H37Rv) was retrieved from UniportKB database with the ID P9WLT1. Due to unavailability of 3-D structure in Research Collaboratory for Structural Bioinformatics (RCSB) protein data bank (PDB), modelling of this uncharacterized protein was undertaken utilizing 318 amino acid long sequence of Uncharacterized Protein Rv1708 of MTB (Consortium, 2019).

2.2 *Physicochemical characterization*

Physicochemical properties of the retrieved sequence were determined using two web-based servers, ExPasy's ProtParam tool used for the calculation and interpretation of amino acid composition, instability and aliphatic index, extinction coefficients and grand average of hydropathicity (GRAVY) (Gasteiger et al., 2009). Theoretical isoelectric point (pI) was also calculated using Sequence Manipulation Suite (SMS) Version2.

2.3 *Secondary structure prediction*

The self-optimized prediction method with alignment (SOPMA) (Geourjon & Deléage, 1995) and PSIPRED program was used to predict the secondary structure of Uncharacterized Protein Rv1708. Disorder prediction was performed using DISOPRED tool (Deng, Gumm, Karki, Eickholt, & Cheng, 2015).

2.4 *Homology modeling and validation*

There is no experimentally elucidation of 3D structure available for Uncharacterized Protein Rv1708 of MTB in RCSB protein data bank (PDB), therefore homology modelling of the protein of Uncharacterized Protein Rv1708 was done using two program Swiss Model (Schwede, Kopp, Guex, & Peitsch, 2003) and Phyre2 (Kelley, Mezulis, Yates, Wass, & Sternberg, 2015). Secondary structure has also been predicted using Phyre2. 3D model of Uncharacterized Protein Rv1708 generated from Swiss Model and Phyre2 was compared and only the most suiTable 3D model was selected for final validation. The final modelled structure was validated using Ramachandran plot analysis by MolProbity for stereo-chemical property (Chen et al., 2010).

3 RESULT AND DISCUSSION

3.1 *Physicochemical characterization*

The amino acid sequence of Uncharacterized Protein Rv1708 of MTB was retrieved in FASTA format and used as query sequence for determination of physicochemical parameters. The instability index of Uncharacterized Protein Rv1708 of MTB is 31.61 (<40) indicated the stable nature of the protein (Guruprasad, Reddy, & Pandit, 1990). The molecular weight of protein is 34380.48Da and is acidic in nature (pI 6.00, 6.40*). (*pI determined by SMS Version2) High extinction coefficient values 16180 M^{-1} cm^{-1} indicate the presence of Cys, Trp and Tyr residues when all pairs of Cys residues are assumed to form cystines and the value is 15930 M^{-1} cm^{-1} when all Cys residues are reduced. The aliphatic index value (99.31) is pretty higher which suggested as a positive factor for increased thermos-stability in a wide range of temperature The nature of protein found hydrophilic which indicate the possibility of better interaction with water (Ikai, 1980) and it was calculated by the lower grand average of hydropathicity (GRAVY) indices value (-0.070) as shown in Table 1.

3.2 *Secondary structure prediction*

For the prediction of secondary structure by SOPMA the parameters (window width: 17; similarity threshold: 8; division factor: 4) were considered as default. Analyzing 511 proteins (subdatabase) and 33 aligned proteins, SOPMA predicted 39.62 percent of residues as random coils in comparison to alpha helix (37.74 percent), extended strand (17.92 percent) and Beta turn (4.72 percent) as shown in Table 2. PSIPRED shows the higher confidence of prediction of helix, strand and coil. Intrinsic disorder profile was computed using DISOPRED and <87 percent of the amino acid are below the confidence score of 0.5 for disordered condition, suggested the lowest possibility of distortion and conferred the high stability to the predicted protein (Figure 2 Left).

3.3 *Homology modelling and structural validation*

FASTA format sequence of Uncharacterized Protein Rv1708 was inserted as target sequence in Swiss-Model workspace. Both HHBlits and BLAST parallelly was used to search the Swiss Model Template Library (SMTL). Total 6327 templates were found to match the target sequence resulted in total 50 templates, among the five most favorable template showed in Table 3 (2bek.1.A, 2bek.1.A, 2bej.1.A, 2bej.1.A, 1wcv.1.A) are those who have GMQE value greater than 0.50. Target sequence was selected based on the Qualitative Model Energy Analysis

Table 1. Physicochemical parameters computed using Expasy's Prot-Param and SMS tool.

Physio-chemical parameters	Values
No. of amino acid (aa)	318
Molecular weight (MW)	34380.48
Theoretical Isoelectric point (pI)	6.00, 6.40*
Aliphatic Index	99.31
Instability Index	31.61
Extinction Coefficient (All Cys form Cysteine)	16180
Extinction Coefficient (All Cys reduced)	15930
Total no. of negatively charged residues (Asp + Glu)	38
Total no. of positively charged residues (Arg + Lys)	34
GRAVY (Grand average of hydropathicity)	-0.070

* pI determined by SMS Version2

Table 2. Secondary structure elements prediction by SOPMA.

Secondary structure elements	Values (%)
Alpha helix (Hh)	37.74%
3_{10} helix (Gg)	0.00%
Pi helix (Ii)	0.00%
Beta bridge (Bb)	0.00%
Extended strand (Ee)	17.92%
Beta turn (Tt)	4.72%
Bend region (Ss)	0.00%
Random coil (Cc)	39.62%
Ambiguous states (?)	0.00%

Table 3. Alignment of selected template.

Name	Title	GMQE	QSQE	Identity (%)
2bek.1.A	Segregation protein	0.58	0.53	43.21
2bek.1.A	Segregation protein	0.60	0.51	46.61
2bej.1.A	Segregation protein	0.57	0.35	43.62
2bej.1.A	Segregation protein	0.59	0.33	47.03
1wcv.1.A	Segregation protein	0.57	0.34	43.62

(QMEAN) score (-1.85), Global model quality estimate (GMQE) 0.58, percentage of sequence identity (43.21 percent), similarity (0.40) and coverage (0.76). Model was based on target-template alignment using Swiss model workbench where insertion, deletions remodeled and side chains were then rebuilt. Our model showed resemblance with 2bek.1.A; which is the bacterial chromosome segregation protein Soj of *Thermus thermophilus* and identified as the template for Uncharacterized Protein Rv1708 homology modeled structure. So, the model generated was by using the template 2bek.1.A of Segregation Protein considering the and saved in PDB format and visualized by Discovery Studio 4.5 Visualizer (Figure 1 left). The respective values Z-scores of CBeta interaction energy, torsion angle energy, solvation energy, secondary structure in case of Uncharacterized Protein Rv1708 are -1.94, -1.49, 0.18 and -0.56. The overall QMEAN score for Uncharacterized Protein Rv1708 is -1.85. By QMEAN generated results of Uncharacterized Protein Rv1708, it is conferred as a qualified model for drug target scopes. Similarly, the

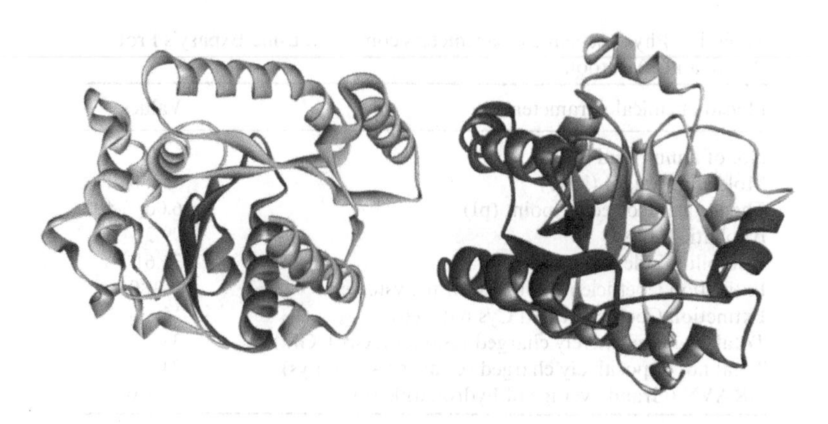

Figure 1. Uncharacterized Protein Rv1708 Structure by Swiss Model (Left) & Phyre2 (Right).

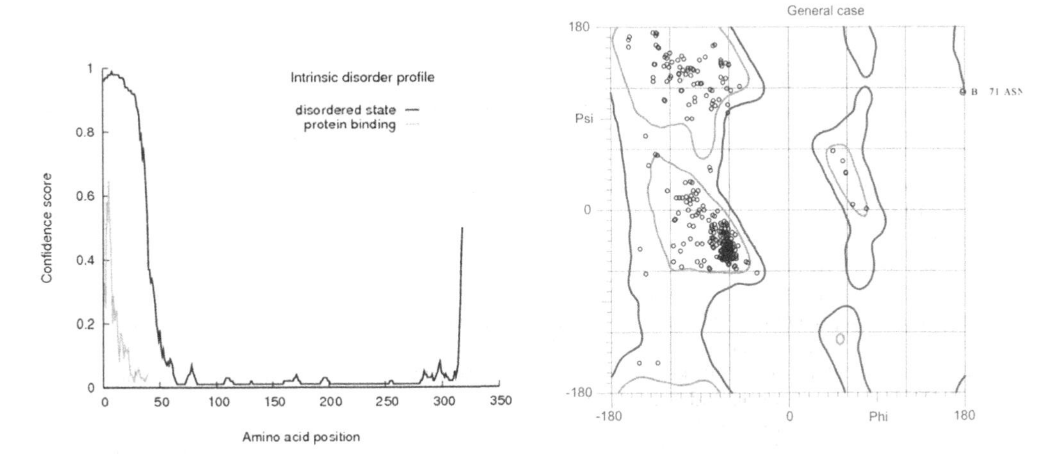

Figure 2. Intrinsic Disorder Profile (Left), Ramachandran Plot using MolProbity (Right).

Table 4. Ramachandran plot analysis by MolProbity.

Ramachandran plot statistics	Residue	%
Residues in the favored regions	481	95.4
Residues in the allowed regions	502	99.6
Residues in the outliers (phi, psi)	02	0.40
Total number of residues	504	-

homology modelling of Uncharacterized Protein Rv1708 was performed by Phyre2. Based on the 6 templates (c3ez6B, c2ozeA, d2afhe1, c2bekB, crpfsA, c5aorA and c3q91B), protein model was generated with 85 percent of the residues modelled at 100 percent confidence (Figure 1-right). Both predicted structural models were evaluated by MolProbity for stereochemical property. Between two structures, Ramachandran Plot data (Figure 2 right) from Swiss Model is promising and that's why we have considered structure from Swiss Model for the result analysis. Here 95.4% of all residues were in favored (98%) regions and 99.6% of all residues were in allowed (>99.8%) regions. There were 2 outliers (phi, psi): Asn71 Val131 and these data of predicted structure from Swiss Model ensures the good quality of the protein structure; data showed in Table 4.

4 CONCLUSION

In this study, we have concluded the structural model of Uncharacterized Protein Rv1708 (strain ATCC 25618/H37Rv) of MTB through in-silico approach. The physicochemical parameters prediction and functional annotation are useful for understanding the action of this proteins' activity. Our Homology-Modelled protein provides insights into the functional role of Uncharacterized Protein Rv1708 (strain ATCC 25618/H37Rv) in pathogenesis, which will help to design potential therapeutic drug against this protein for inhibition of septum formation of MTB.

ACKNOWLEDGMENT

We are grateful to the Department of Biochemistry and Molecular Biology of Bangabandhu Sheikh Mujibur Rahman Science and Technology University, Gopalganj-8100, Bangladesh for giving us a computational platform for completing this project.

REFERENCES

Chen, V.B., Arendall, W.B., Headd, J.J., Keedy, D.A., Immormino, R.M., Kapral, G.J., Richardson, D.C. (2010). MolProbity : all-atom structure validation for macromolecular crystallography. *Acta Crystallographica Section D Biological Crystallography*. https://doi.org/10.1107/s0907444909042073

Cole, S.T., Brosch, R., Parkhill, J., Garnier, T., Churcher, C., Harris, D., Barrell, B.G. (1998). Deciphering the biology of mycobacterium tuberculosis from the complete genome sequence. *Nature*. https://doi.org/10.1038/31159

Consortium, T. U. (2019). UniProt: a worldwide hub of protein knowledge. *Nucleic Acids Research*. https://doi.org/10.1093/nar/gky1049

Deng, X., Gumm, J., Karki, S., Eickholt, J., & Cheng, J. (2015). An overview of practical applications of protein disorder prediction and drive for faster, more accurate predictions. *International Journal of Molecular Sciences*. https://doi.org/10.3390/ijms160715384

Gasteiger, E., Hoogland, C., Gattiker, A., Duvaud, S., Wilkins, M.R., Appel, R.D., & Bairoch, A. (2009). Protein Identification and Analysis Tools on the ExPASy Server. In *The Proteomics Protocols Handbook*. https://doi.org/10.1385/1-59259-890-0:571

Gengenbacher, M., & Kaufmann, S.H.E. (2012). Mycobacterium tuberculosis: Success through dormancy. *FEMS Microbiology Reviews*. https://doi.org/10.1111/j.1574-6976.2012.00331.x

Geourjon, C., & Deléage, G. (1995). Sopma: Significant improvements in protein secondary structure prediction by consensus prediction from multiple alignments. *Bioinformatics*. https://doi.org/10.1093/bioinformatics/11.6.681

Guruprasad, K., Reddy, B.V.B., & Pandit, M.W. (1990). Correlation between stability of a protein and its dipeptide composition: A novel approach for predicting in vivo stability of a protein from its primary sequence. *Protein Engineering, Design and Selection*, *4*(2), 155–161. https://doi.org/10.1093/protein/4.2.155

Ikai, A. (1980). Thermostability and aliphatic index of globular proteins. *Journal of Biochemistry*.

Kelley, L.A., Mezulis, S., Yates, C.M., Wass, M.N., & Sternberg, M.J.E. (2015). The Phyre2 web portal for protein modeling, prediction and analysis. *Nature Protocols*. https://doi.org/10.1038/nprot.2015.053

Malenki, G. (2018). The use of isoniazid as prophylaxis in patients co-infected with tuberculosis and HIV. *Canadian Journal of Respiratory Therapy*.

Misra, H.S., Maurya, G.K., Chaudhary, R., & Misra, C.S. (2018). Interdependence of bacterial cell division and genome segregation and its potential in drug development. *Microbiological Research*. https://doi.org/10.1016/j.micres.2017.12.013

Okou, D.T., Steinberg, K.M., Middle, C., Cutler, D.J., Albert, T.J., & Zwick, M. E. (2007). Microarray-based genomic selection for high-throughput resequencing. *Nature Methods*. https://doi.org/10.1038/nmeth1109

Schwede, T., Kopp, J., Guex, N., & Peitsch, M.C. (2003). SWISS-MODEL: An automated protein homology-modeling server. *Nucleic Acids Research*.

Biotechnology and Biological Sciences – Sen et al. (Eds)
© 2020 Taylor & Francis Group, London, ISBN 978-0-367-43161-7

Exploration of molecular docking and dynamic simulation studies to identify the potential of CID 50906864 as a potential anti-angiogenic molecule

Faizan Ahmad & P.T.V Lakshmi
Centre for Bioinformatics Pondicherry University, Pondicherry, India

ABSTRACT: Mono-carbonyl analogue of Curcumin (CID 50906864) which causes ER-stress induced apoptosis in Non-Small Cell Lung Cancer (NSCLC) was suspected to be a multi-target small molecule, however its target as angiogenin was unknown. Since it targets angiogenin by binding with the angiogenin P1 catalytic site residues of His13 and Lys40, was used as molecule 1 and as reference for comparing the study with the tyrosine and serine amino acid conjugated with 3-amino uridine nucleoside as molecules 2 and 3 respectively that were designed. The docking revealed the scores of -8.16 Kcal/mole, -7.76 Kcal/mole and -5.57 kcal/mole for the Molecules 3, 1 & 2 respectively, which further stabilized throughout the simulation from 10 to 20 ns. Therefore molecule1 (CID 50906864) could be considered as an effective anti-angiogenic multi-target small organic molecule.

Keywords: Mono-carbonyl Analogue of curcumin, Anti-Angiogenin, ER stress, Molecular docking and Molecular dynamics simulation

1 INTRODUCTION

Angiogenin by itself Angiogenin by itself plays a central role in angiogenesis and neovascularization through endothelial cell proliferation and is also necessary for bringing angio-genesis induced by other angiogenic factors [*Macchiarini, P., et al,* (1992) *& Kishimoto, Koji, et al* (2005)]. Angiogenin, under stress conditions act as a stress granules in the nucleus, where it produces tiRNA to reprogram translational process which helps in growth, survival and delayed apoptosis [*Li, Shuping, and Guo-Fu Hu* (2012)]. Hence, angiogenin be-came a crucial target to suppress angiogenin induced tumor, cancer, and metastasis. A comparative evaluation of mono-carbonyl analogue of curcumin against NSCLC cell lines have been studied to evaluate their therapeutic potential and a molecule (CID 50906864) was found to be most effective against NSCLC cell lines by inducing ER stress apoptosis, however their role as an antiangiogenic molecule in non-small lung cancer is unknown [*Debnath et al.,* (2014).]. Therefore, molecular Docking, MD simulation based comparative analysis of energies by modified molecules (3′-amino functionalize Uridine nucleoside analogue of serine as molecule 2 and with tyrosine as molecule 3) have been designed (Figure 1) by substitution of the thymidine to uridine base and investigated in the study to enhance their competitive inhibition with the substrate (RNA) to bind with angiogenin (1B1I) and compared with Molecule 1.

2 COMPUTATIONAL METHODS AND TOOLS

Ligand structure of the first was obtained from literature, while the other two were modified by base substitution of thymidine to uracil and sketched using Marvin Sketch [MarvinSketch (16.8.29.0) (2016)] prepared and docked to the prepared receptor angiogenin (1B1I) using

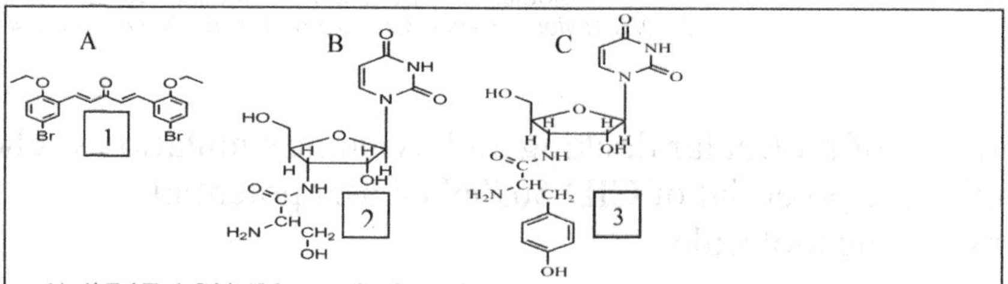

A) (1*E*,4*E*)-1,5-bis(5-bromo-2-ethoxyphenyl)penta-1,4-dien-3-one

B) 2-amino-*N*-((2*S*,3*S*,4*R*,5*R*)-5-(2,4-dioxo-3,4-dihydropyrimidin-1(2*H*)-yl)-4-hydroxy-2-(hydroxymethyl)tetrahydrofuran-3-yl)-3-hydroxypropanamide

C) 2-amino-*N*-((2*S*,3*S*,4*R*,5*R*)-5-(2,4-dioxo-3,4-dihydropyrimidin-1(2*H*)-yl)-4-hydroxy-2-(hydroxymethyl)tetrahydrofuran-3-yl)-3-(4hydroxyphenyl)propanamide

Figure 1. 2D Structure & IUPAC names of the Ligands used in Docking.

Flexible docking mode through Autodock. The top complexes was selected for Molecular dynamic simulation for 20ns using Gromacs 5.1.2. RMSD & RMSF trajectory and average number of hydrogen bonds were analyzed.

3 RESULT AND DISCUSSION

The binding energy of molecule 1, 2 and 3 were recorded as -7.76 kcal/mole, -7.48 kcal/mole and -8.74 kcal/mole respectively, which ranked in the order of 3>1>2 (Table 1) and it was in agreement with the energy obtained from the reported thymidine conjugate of tyrosine (-6.25kcal/mole) and serine (8.75 kcal/mole) [*Debnath et al.,* (2014)]. Interestingly, the ligand inhibition constant seemed to reflect the pattern as it showed lowest values with highest binding energies, which again ranked in the order of 2<1<3 as same as the binding energy

Table 1. Docking analysis of molecule 1, 2 and 3 with protein.

Sr. No.	H bond Donor	H bond Acceptor	Distance (Å)	Binding energy (Kcal/Mole)	Inhibition constant
1	HIS13:HE2	UNK0: O	2.1	-7.76	2.05 µm
	LYS40:HZ1	UNK0: O	1.7		
	LEU115:HN	UNK0: O	1.7		
2	ARG5:HH11	UNK0: O	2.1	-7.48	3.28 µm
	GLN12:HE21	UNK0: O	2.2		
	HIS13:HE2	UNK0: O	1.9		
	HIS114:HD1	UNK0: O	2.0		
	UNK0:H	VAL113: O	2.0		
	UNK0:H	GLN12:OE1	1.9		
3	HIS8:HD1	UNK0: O	2.2	-8.16	1.05 µm
	HIS13:HE2	UNK0: O	2.4		
	HIS13:HE2	UNK0:N	2.3		
	LYS40:HZ1	UNK0: O	2.2		
	UNK0:H	UNK0:N	2.1		
	ARG121: HH22	UNK0: O	2.0		
	UNK0:H	ARG5: O	2.2		
	UNK0:H	LEU115:O	2.0		
	LEU115:HN	ASP41:OD1	2.4		

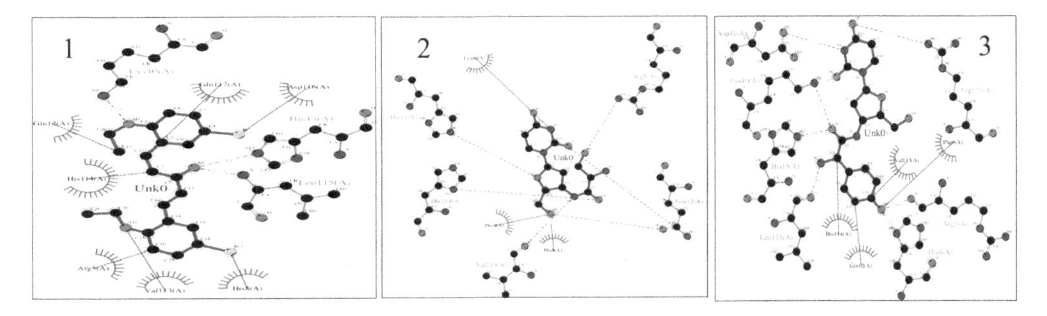

Figure 2. Two dimensional interaction diagram of complex of molecule 1, molecule 2 and molecule 3.

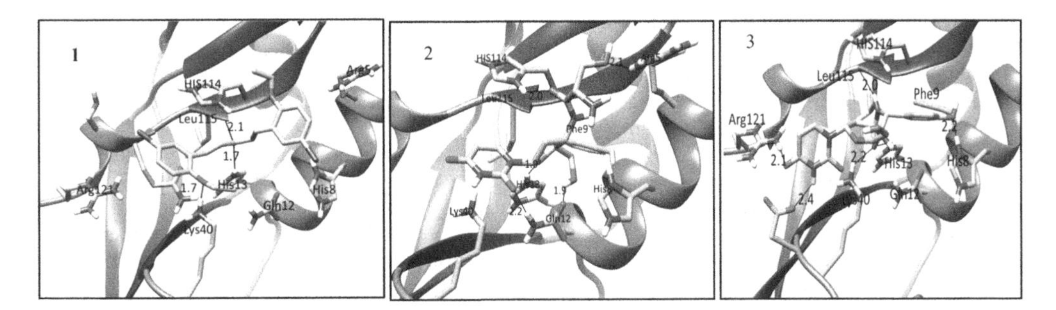

Figure 3. Three dimensional interaction of molecule 1, 2 and 3 with angiogenin protein pocket residue.

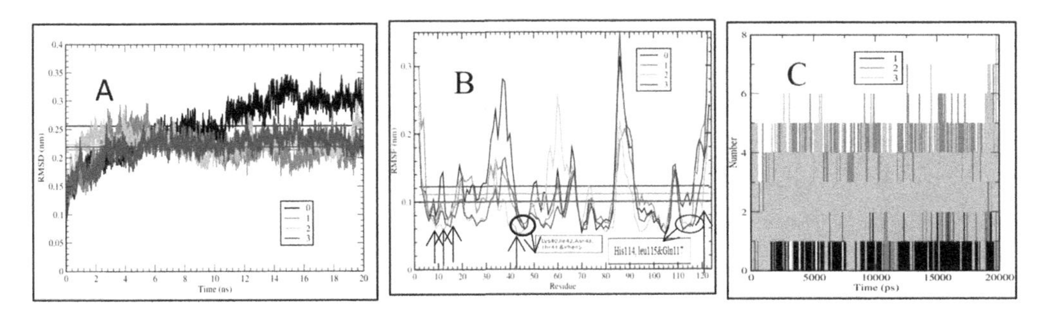

Figure 4. (A) shows the RMSD of protein backbone (B) shows RMSF of Backbone [where 0 (Black Line) indicates Apo-protein, 1 (red line), 2 (Blue) and 3 (green) indicated protein complex with molecule 1, 2 and 3 respectively (C) shows the average no of hydrogen bond formed between molecule 1 (black line), molecule 2 (red line) and molecule 3 (blue line).

(Table 1). The residue His13, Lys40 & His114 constituted the catalytic triad or P1 catalytic site (Figure 2) and it exhibited good interaction with all the compounds, which best fitted in the pocket required for angiogenesis through t-RNA catalysis (Figure 3). Hence these poses which were considered to interact with P1 catalytic site required for ribonucleolytic inhibition of angiogenin-induced angiogenesis obtained as the first cluster with highest binding energy was extended for simulation studies. The RMSD generated for the complexes was in accordance to that of [*Varma et. al.*(2016)] which showed less than 0.25nm exhibiting stability from 8 ns onwards against the Apo-protein, which probably took longer duration (11ns)

(Figure 4A). Likewise, the P1 catalytic site residues of His13, Lys40 and His114 showed less fluctuation than the Apo-protein but similar to the inhibitor molecule 2 and 3 (Figure 4B). Molecule 3, Molecule 1 and molecule 2 showed six, three and six average no of hydrogen bond respectively (Figure 4C).

4 CONCLUSION

Hence, molecule 1 which although shows negative binding energy towards angiogenin inhibition and seemed to be more compared to the conjugate of serine amino acid, still could be considered as the potent molecule that could act as anticancer agent by inducing ER stress.

REFERENCES

Debnath, J., Dasgupta, S., & Pathak, T. (2014). Amino and carboxy functionalized modified nucleosides: a potential class of inhibitors for angiogenin. Bioorganic chemistry, 52, 56–61.

Kishimoto, Koji, et al. "Endogenous angiogenin in endothelial cells is a general requirement for cell prolif-eration and angiogenesis." Oncogene 24.3 (2005): 445–456.

Li, Shuping, and Guo-Fu Hu. "Emerging role of angiogenin in stress response and cell survival under ad-verse conditions." Journal of cellular physiology 227.7 (2012): 2822–2826.

Macchiarini, P., et al. "Relation of neovascularisation to metastasis of non-small-cell lung cancer." The Lancet 340.8812 (1992): 145–146.

MarvinSketch (16.8.29.0) ChemAxon, http://www.chemaxon.com, 2016.

Varma S Grover S et. al. Hydrophobic Interactions Are a Key to MDM2 Inhibition by Polyphenols as Revealed by Molecular Dynamics Simulations and MM/PBSA Free Energy Calculations PLOS ONE 2016 vol: 11 (2) pp: e0149014.

Biotechnology and Biological Sciences – Sen et al. (Eds)

 ISBN 978-0-367-43161-7

Statistical design in accelerated downstream processing

Arshad Jawed & Mohd Wahid
Research & Scientific Studies Unit, College of Nursing, Jazan University, Jazan, KSA

Payel Chaudhury
Department of Microbiology, University of Burdwan, Burdwan, India

Shafiul Haque & Sajad A.Dar
Research & Scientific Studies Unit, College of Nursing, Jazan University, Jazan, KSA

Shikha Joon
Department of Biotechnology, JNU, New Delhi, India

ABSTRACT: Bioprocess development is an unending process that takes years to commercialize and is attempted by many groups in parallel all over the world. Ironically, it's soon superseded by improved, more efficient ones. Classical optimization methods e.g. One-factor-at-a-time (OFAT) approach, borrowing, sequential removal, content swapping, etc., are not efficient, erroneous and painfully slow. Modern statistical methods prove to be quicker, involve less number of experiments and are highly efficient with better reproducibility. Statistical optimizations methods such as Response Surface Methodology (RSM), Central Composite Design (CCD), and Artificial Neural Networks (ANN) provide real-optima with results closer to the desired target recoveries. In spite of the huge potential, the adaptation of statistical methods is slow and sluggish due to the mathematics involved. In this article, we attempt Cholesterol Oxidase (COD) recovery as a model process using batch bead-milling for downstream processing optimizations using RSM.

Keywords: Screening, Optimization, OFAT, RSM, CCD, bead-mill

1 INTRODUCTION

Statistical methods have accelerated the development of bioprocesses and have steadily been proved indispensable for quicker development of an economically viable bioprocess. Classical optimization methods, e.g. One – Factor at A Time (OFAT), Sequential Removal Approach, etc., where one component is added or removed sequentially is still being practiced. OFAT is slow, requires more number of runs, utilizes an array of components repeatedly, is inefficient and results in pseudo-optimal optimization (Kanwar, 2005; Cocaign et al, 1995). These methods were frequent in optimization before statistical designs proved their worth. OFAT depends on the trial and error approach and fails to take into account the interaction between the process variable or the components involved. Response Surface Methodology (RSM) in a collection of models and designs that are increasingly being used for the bioprocess optimization (Gresham & Inamine, 1986). RSM takes into account the interaction(s) that occurs between the process components. RSM involves statistics and mathematical modeling tools that help in the optimization of a multi-factorial/multi-component system. Statistical methods

*Corresponding author: Arshad Jawed

have successfully been employed for the production of Carotenoids (Singh, *et. al.,* 2016), derivatization of Colchicine into demethylated forms (Jawed A *et. al.*, 2015), production of microbial metabolites, extraction of ovotransferrin from egg white (Al Shammari *et. al.,* 2015), etc. CCD is the most commonly used design for biological process development. RSM has been used for the optimization of a number of processes, right from upstream to downstream processes. RSM employes Fishers F-value, p-values, lack of fit, ANOVA, etc for generating an mathematical model equation that is able to predict the production levels at infinite combinations of the process variables, without actually executing them in real time. Owing to these benefits statistical methods prove to be faster, require less number of runs, produce more accurate results, navigate to complete design space and quantify interaction(s) between different components in the process investigated (Haque *et.al.,* 2016[a],).

Recombinant proteins are foreign proteins that are expressed inside the cells. In order to recover the protein, the cells need to be lysed or broken apart and the protein of interest is then extracted by various means. Selection of a cell lysis process depends on the type of cells, the sturdiness of the cell wall, type of the protein expressed, its thermo-stability, half-life, pH tolerance, expression levels and location, and the end use of the product (Middleberg, 1995). There are many cell lysis processes available at lab scale, e.g. ultrasonication where the breaking of cells is achieved by ultrasonic waves, chemical lysis where a mix of chemicals mostly urea with chaotropic salts are used for cell lysis, quick decompression where cells are pressurized and then almost instantly decompressed. Decompression lyses the cells like an explosion. All these methods suit processes where cell densities are low and volumes are smaller. When the cell densities and process volumes both increase, these labs – scale methods lose efficiency and result in partial cell lysis. This results in very low recoveries and wastage of valuable protein expresses and still remaining with the unlysed cells (Middleberg, 1995). Bead milling grinds the cells in a cylindrical chamber filled with glass beads or zirconium beads. The cylindrical chamber kept cold with the help of chilled circulating water as grinding beads generate a lot of heat that may denature the protein of interest. Bead milling is a method of choice for large scale cell lysis processes that are intended to be taken to a commercial level. Bead mills don't involve any chemicals, are milder, handle large volumes of cell slurry at much-concentrated cell slurries. Bead mills are easy to operate, control but not easy to optimize. A lot of parameters work in sync while bead milling proceeds. Major parameters include bead loading, cell loading and run times (Haque et al., 2016[b]). In this article, we investigate a model process comparing classical and statistical methods for COD recovery. This article will give an insight and provide the researchers with an easy methodology to switch from classical to statistical methods for bioprocess development.

1.1 *Materials and methods*

All the chemicals used were of analytical grade or molecular biology grade. These were obtained from Sigma India, Avantor INC, India, HiMedia (India), *E. coli* BL21 was used as the host for transformation with pRT24b(+) with recombinant COD gene. IPTG inducible Cholesterol oxidase gene (COD) was cloned in *E. coli* as reported by Haque *et.al.*, 2016[a]. The expression of COD was quantified using the protocol as reported (Volante *et.al.,* 2010). The cell culture was initially grown in Luria Bertani (LB) containing 30μg/ml Kanamycin in all the cell culture medium used. LB medium was used to develop the inoculums and Terrific Broth (TB) was used as production medium [0.8% (v/v) glycerol, 1.2% (w/v) Bacto-tryptone, 2.5% (w/v) Yeast Extract, 16 mM KH_2PO_4, 54 mM K_2HPO_4 and 0.2% $MgCl_2$. Erlenmeyer flasks (500ml) was added with 90 ml TB medium and inoculated with 8h old (OD_{600nm} = 0.6) at 10% (v/v) of starter culture. Cells were grown at 37°C for 14 h at 200 rpm in a rotary orbital shaker. The cells were induced at 0.8 OD with IPTG for the expression of COD.

1.2 *Batch fermentation*

5L jacketed glass fermenter Biostat® C, from Sartorius AG, Germany was used for production. The culture temperature was kept at 37 °C with 200 – 700 rpm agitator speed cascaded with dissolved Oxygen (DO) + Air. Aeration was controlled from 0.5 – 2 volume of air per volume of medium (vvm). The starter culture was grown overnight and diluted to 0.6 OD600nm before addition to the fermenter. The inoculum was added at a final concentration of 10% v/v. 1 mM IPTG was added after 4 h of incubation for inducing COD expression. Samples were collected intermittently for analysis of cell growth and COD expression. The COD activity was determined as demonstrated by *Haque et al.,* 2016[a]. One unit of COD activity was defined as the amount of enzyme that produces 1μmol of H_2O_2 per minute at 25°C.

1.3 *Determination of maximum extractable COD*

Cells were centrifuged and the pellet obtained was washed twice with and resuspended in Tris-HCl buffer (pH 7.5, 50 mM). 10 ml of this suspension was taken and the OD was adjusted to 10 and this suspension was subjected to sonication (15sec ON; 45 sec OFF cycle). The 6.4 mm sonication probe used was obtained from Heat systems, NY, USA, operating at 25kHz. An ice bath was used to maintain the slurry temperature at 4°C. Sonication was performed for 30 min and the amount of COD obtained was taken as maximum extractable COD and considered as a reference and 100% recovery for further experiments.

1.4 *Bead milling*

200 ml working volume grinding chamber was employed for batch processing of cell lysis. KDL type bead mill called Dynomill from W. A. Bachofen AG, Switzerland was used for studying bead milling. The grinding chamber was loaded with 0.5 – 0.75 mm diameter glass beads at 80% v/v or as per the RSM design. The cell slurry was adjusted to OD_{600} = 50 and loaded into the grinding chamber. The Bead mill was operated under batch mode. The grinding was executed at 3000 rpm for 40 min at 4°C, with intermittent sampling done every 5 min.

1.5 *Statistical design for bead milling*

With reference to previous study available (Middleberg, 1995), the crucial factors in bead milling are bead loading, cell loading, run time and feed rate. As we are performing bead milling in batch mode, feed rate becomes irrelevant. The experiment was designed with 3 parameters; namely bead loading, cell loading and run time were crucial for optimal lysis of the cells. CCD from RSM was selected for the optimization study. A 3 – factor 5 – level Central Composite Design was used for the cell lysis optimization and the study of the interaction between the parameters involved. Bead loading, cell loading and run time were coded as A, B, and C respectively. These parameters were considered independent and varied at 5 levels (– 2, – 1, 0, +1, +2). The details of the experimental space and design are shown in Table 1.

CCD generates a table with a set of trial runs with combinations of these variables at the coded levels. The experimental matrix contains axial points, Center points (in replicates) and factorial points. The CCD generates 16 runs with few replicates at center points (Table 2, replicates at center points not shown). The experiments obtained were performed thrice and the average value of COD obtained per experiment was fed against the corresponding run. Mean,

Table 1. Experimental Design.

S. no.	Factor	Units	Type	Low Level	High Level	Mean
1	Bead Loading	%-w/v	Numeric	60	80	70
2	Cell loading	%-w/v	Numeric	50	70	60
3	Run time	%-w/v	Numeric	20	30	25

standard deviation, F-values, P-values, ANOVA was calculated and a mathematical equation (Equation 1) was generated based on the amount of COD obtained per run.

$$Y = a_0 + a_1A + a_2B + a_3C + a_4AB + a_5BC + a_6AC + a_7A^2 + a_8B^2 + a_9C^2$$

Where
Y = response (obtained metabolite; COD in this case)
A, B and C = coded values of process variables.

The responses were utilized for the statistical analysis for Analysis of Variance (ANOVA). The developed statistical model was evaluated in detail per run for its significance, using Fisher F test (F – value), p – values, lack of fit value and aptness of mathematical equation in predicting the obtainable COD values. The individual variables/factors with p values ≤ 0.05 and interaction terms with p values ≤ 0.1 are considered significant and considered for analysis. The coefficient of determination R^2) is also calculated for checking the fitness of data. The developed model and the prediction surface were shown as contour plots of 3 – D surface with variables on the respective axis within the design space. The contours and the 3 – D Surface plots were generated using Design Expert™ ver 7, Statease Inc, USA. These plots are used to navigate the design space and help to visualize the trend of COD obtained at infinite combinations of the selected variables.

2 RESULTS

2.1 *Cell culture and recovery of COD*

COD production was initially checked at shake flask level and fermenter. The cells were separated by centrifugation and the supernatant was analyzed for any leaky expression of COD. No COD activity in the culture medium. The intracellular expression was measured to be 58.98 µg/ml per unit OD_{600nm}. This was calculated by directly lysing the cells and quantifying it densitometrically on an SDS-PAGE. (Data not shown) The molecular weight of cholesterol Oxidase enzyme was calculated to be 46.5 kDa.

2.2 *Quantification of cell lysis and COD obtained*

By using ultrasonication, the maximum obtainable COD was calculated. The maximum COD obtained via ultrasonication was found to be 58.8 µg/ml per unit OD_{600nm}, which corresponds to 91.4 units of enzyme activity. Figure 1 shows ultrasonication for 20 min resulted in the highest release of COD with negligible increment observed beyond 25 min. This was considered maximum obtainable COD and 100 % cell lysis for further recovery calculations.

2.3 *Bead milling*

Three sets of bead-milling runs in the batch mode were performed with 0.5 – 0.75 mm glass beads. The 200 ml grinding chamber was loaded with glass beads at 80% v/v concentration. The remaining volume (empty + void volume) was filled with cell slurry, adjusted to OD_{600nm} = 50. The gap setting was adjusted to 0.2 mm for drawing samples. Samples were drawn intermittently every 5 min for calculating the amount of COD released (Figure 2). The increase in the amount of COD released was steady in the initial part of bead milling but reached a plateau after 20 min (Figure 2). The amount of COD recovered was ~7490 Units corresponding to ~36.6 µg/ml per OD_{600nm}. The release of COD was steady initially and around 25 min it reached its saturation. As calculated by these results, the variables and their ranges for RSM were selected as Bead loading: 60 – 80% v/v, Cell loading: 50 – 70 (OD_{600nm}) and Run time: 20 – 30 min (Table 1).

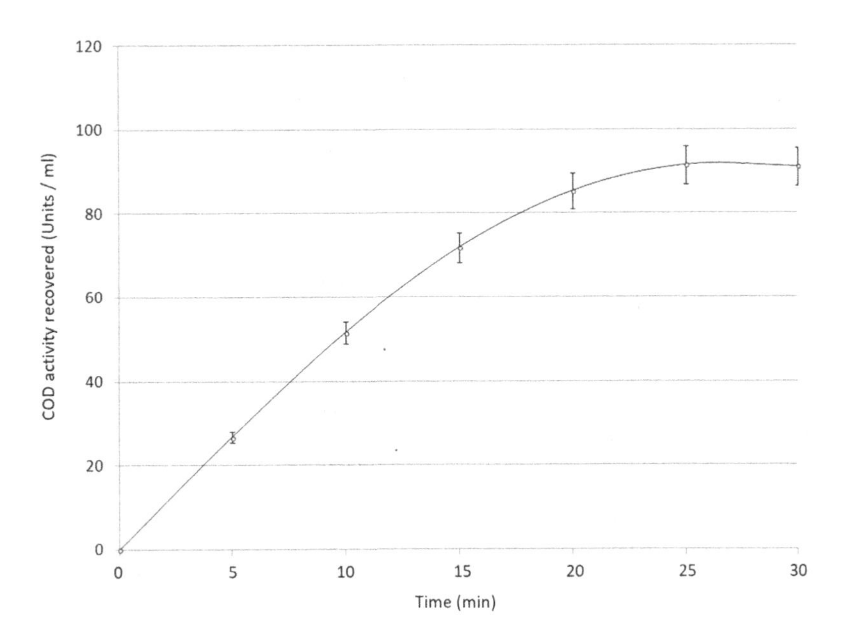

Figure 1. Ultrasonication for the quantification of COD released.

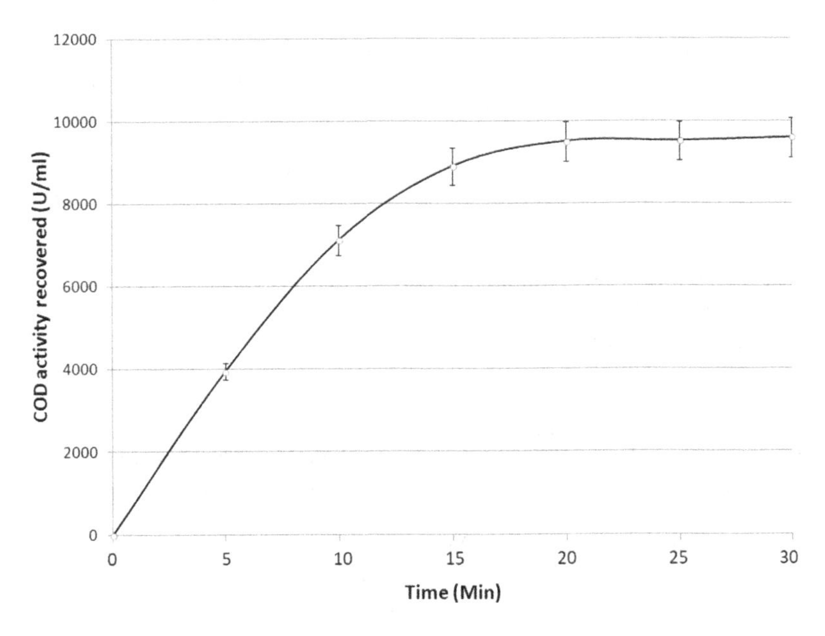

Figure 2. Bead milling for the quantification of COD released.

3 STATISTICAL DESIGN

Central Composite Design was selected under the Response Surface Methodology for the optimization study. A full factorial run with 3 parameters at 5 levels was performed, resulting in a total of 20 trial runs (Table 1 & 2). Each run was performed in triplicate and the average of the amount of COD obtained was fed into the corresponding trial runs.

Table 2. Actual Experiment Runs.

Run	A:Bead (%, v/v)	B:Cell Conc. (OD600nm)	C:Run time (min)	COD Recovery (mg/ml)
1	70.00	60.00	25.00	1.95
2	60.00	70.00	20.00	2.34
3	53.18	60.00	25.00	1.68
4	80.00	50.00	30.00	2.26
5	70.00	60.00	25.00	2.54
6	80.00	70.00	30.00	3.06
7	70.00	60.00	16.59	1.13
8	70.00	60.00	25.00	2.54
9	80.00	50.00	20.00	0.78
10	70.00	43.18	25.00	1.48
11	70.00	60.00	25.00	2.54
12	80.00	70.00	20.00	0.82
13	60.00	70.00	30.00	2.73
14	70.00	60.00	33.41	2.10
15	60.00	50.00	20.00	1.64
16	70.00	60.00	25.00	2.54
17	86.82	60.00	25.00	1.94
18	70.00	76.82	25.00	2.77
19	60.00	50.00	30.00	0.99
20	70.00	60.00	25.00	2.54

The responses spanned from 0.78 obtained in the trial no. 09; while the highest COD recovered was with trial no. 06 mounting to 3.06 mg/ml of COD obtained. The ratio of the highest to the lowest response is 3.92, which suggests that no power transformation is needed. The model selected automatically by the software was quadratic, with F-value = 20.05 and p-value < 0.001. It had 9 degrees of freedom and the lack-of-fit was found to be 'not significant'. The high F-value and not significant lack of fit showed that the model is highly significant; there is less than 0.01% chance that this value can occur due to noise and the mathematical equation generated was accurate in predicting the value amount of COD obtained at various levels of the variables included in the experiment. The individual terms A, B, C, interaction terms AB, AC and BC, and Quadratic terms A^2, B^2 and C^2 were found to be significant. The adjusted coefficient of determination (R^2) showed to be around 0.9 with predicted R^2 of 0.7959. Adequate Precision (AP) which is essentially signal to noise ratio, was 14.24 and the standard deviation was 0.9475. The degree of precision of the design is shown by CV (%). The current study CV was 10.78 that confirm the results to be accurate.

The model demonstrated a good fit between the predicted and the observed responses showing the model to be valid and usable. The response obtained is generally plotted as contours or 3 – D surface. Hence, the graph is called the 3-D response Surface Plot and the methodology draws its name as Response Surface Methodology (RSM). Out of all the variables, 2 are selected on the x- and the y-axis and the response(s) are plotted on the z-axis. As the system takes into account multiple regression and complex calculations, the yield, i.e., the amount of COD obtained is calculated at every combination of x- and y- values and shown in the form of a 3D surface. All these values affirm the validity of the model and suggest the mathematical equation generated to be accurate in predicting the amount of COD recovered.

$$Y = 2.44 - 0.024*A + 0.40*B + 0.37*C - 0.20*A*B + 0.50*A*C + 0.22*B*C - 0.22*A^2 - 0.11*B^2 - 0.29*C^2$$

Where

Y = Amount Of COD recovered. A = Bead loading (%, v/v), B = Cell loading (%, v/v), C = Run time (min)

As observed responses when plotted for predicted vs. actual COD recoveries, tend to aggregate towards the central line Figure 3.

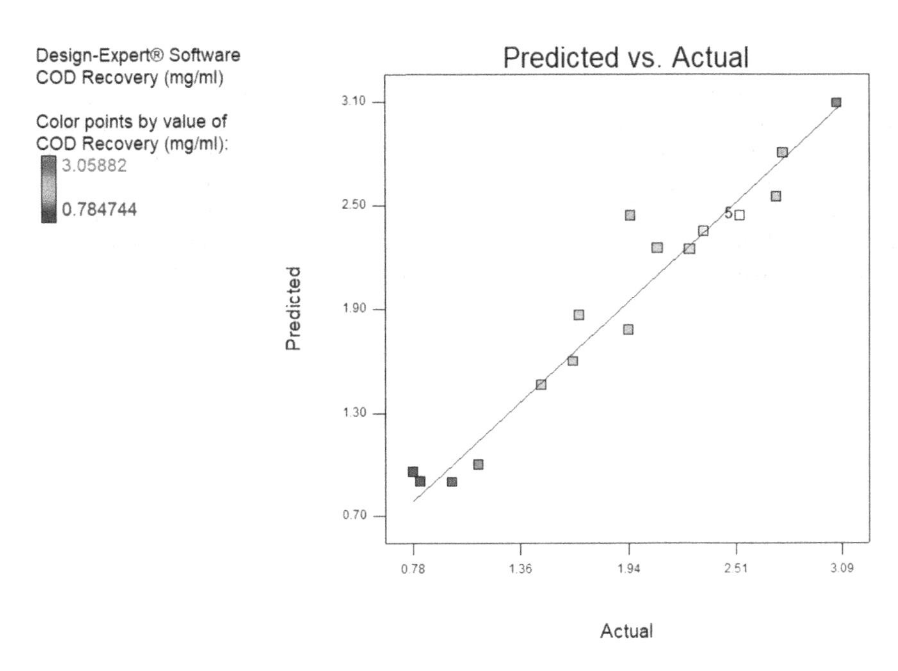

Figure 3. Graph showing COD recoveries with predicted vs. actual values for each trial run.

This indicates that the predicted values of COD and the obtained amounts are very close to each other. Run no. 8 yields maximum COD as all the parameters are being operated at their maximum (Bead loading = 80 % v/v, Cell loading = 70 OD_{600nm}, run time = 30 min) while run no. 09 and 12 are amongst the runs providing lower yields. All the other combinations vary within this range. The interaction components show high significance with p-values < 0.1. This indicates that there is considerable interaction between components and change in one affects the requirement for the other (Table 3).

According to the RSM plots, the recovery of COD increases with increase in bead loading up to 75% v/v and it decreases if the loading is increased further (Figure 4A). Alternatively,

Table 3. ANOVA table for Quadratic model analysis.

Analysis of variance table [Partial sum of squares - Type III]

Source	Sum of Squares	df	Mean Square	F Value	p-value Prob > F	
Model	9.82	9	1.09	20.05	< 0.0001	significant
A-Bead Loading	0.009	1	0.009	0.17	0.0686	
B-Cell Loading	2.49	1	2.49	45.83	< 0.0001	
C-Run time	2.17	1	2.17	39.92	< 0.0001	
AB	0.37	1	0.37	6.8	0.0261	
AC	2.27	1	2.27	41.75	< 0.0001	
BC	0.46	1	0.46	8.53	0.0153	
A^2	0.79	1	0.79	14.55	0.0034	
B^2	0.19	1	0.19	3.55	0.0888	
C^2	1.37	1	1.37	25.17	0.0005	
Residual	0.54	10	0.054			
Lack of Fit	0.21	5	0.043	0.65	0.6788	not significant
Pure Error	0.33	5	0.066			
Cor Total	10.37	19				

the recoveries increase with increase in the cell loading (OD_{600nm}) from 50 to 70. No decline was observed with increase in cell loading in the design space. Comparing bead loading with run time increase in run time as well as bead loading proves to be beneficial for higher recoveries (Figure 4B). While increase in run time beyond 25 min had little or no effect on the maximum recoveries of COD (Figure 4C) with increase in cell loading (OD_{600nm}). This suggests that cell loading becomes the limiting factor for the maximization of COD recoveries. Further increase in bead loading as well as cell loading would increase the number of cells available for lysis and may tend to result in higher cell lysis and higher COD recoveries.

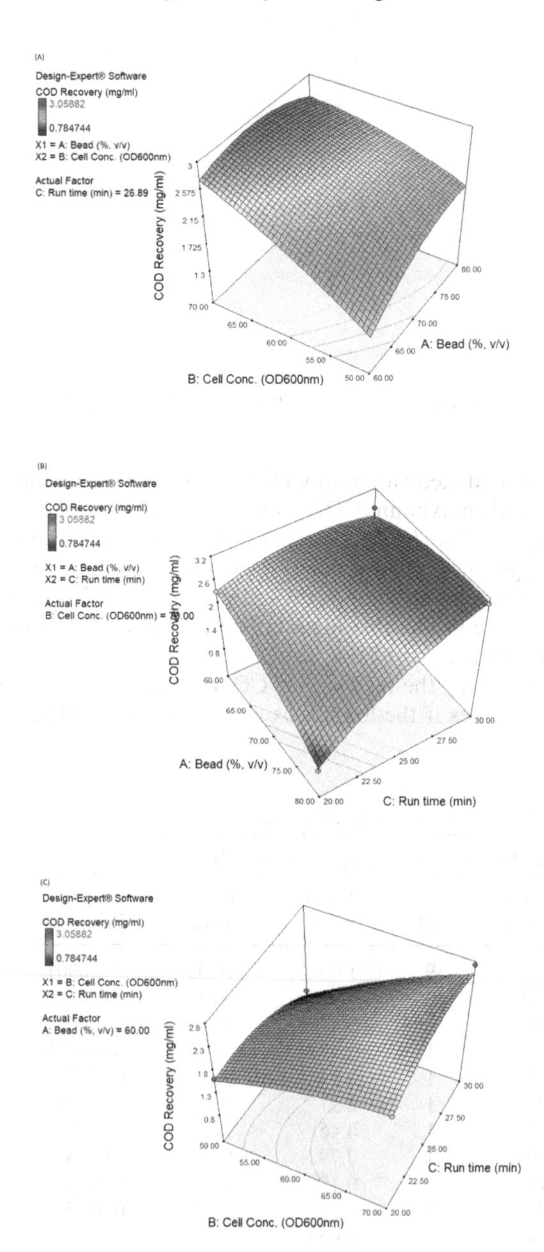

Figure 4. (A)3D RSM graph showing COD recovery with cell concentration vs. Bead loading. (B) 3D RSM graph showing COD recovery with run time vs. bead loading. (C) 3D RSM graph showing COD recovery with cell concentration vs. run time.

4 DISCUSSION

One of the major bottlenecks in the development of a commercially feasible biological process is downstream processing (Rosano & Ceccarelli, 2014). In very simple terms, what is being added needs to be removed for obtaining the protein of interest in its pure form. Upstream processing has been attempted by many groups successfully, for maximizing the production of recombinant proteins. *E. coli* as a host is a preferred microbe for the intracellular expression of foreign proteins. *E. coli* has shorter duplication time, easy to manage, has many expression systems developed and optimized, etc. Scale up process with this organism is straightforward, with bioreactor systems in place. Expression of foreign proteins in *E. coli* is essentially intracellular in nature. In order to recover the intracellular protein expressed, cell lysis becomes a critical step (Shokri et al., 2003).

Laboratory scale methods are best for processing small volumes of cell slurries with lower concentrations. Employing laboratory scale methods, at a larger scale result in inefficient cell lysis, lower recoveries, cumbersome process, etc. Ultrasonication is said to be effective but it can handle lower volumes only. The ultrasonic waves lose their intensity on traveling longer distances inside the cell slurry and result in partial cell lysis. The process generates too much heat; hence it needs to be chilled all the time to avoid protein denaturation. Ultrasonication is a slow, discontinuous process that requires ON/OFF cycles which increases the processing time (Fish & Lily, 1984). Chemical lysis process is not feasible as these require a large number of chemicals that increase the cost, add extra steps in downstream processing for their removal, and may have a detrimental effect on the protein of interest.

Cell lysis by bead milling is one of the processes that suit the needs of large scale cell lysis. It is mild, needs no addition of chemical agents, can handle higher cell concentration, higher viscosities, results in larger cell debris that is easy to separate, it is easy to operate and the mill can be used in batch as well as in continuous mode. The cells lyse when the beads are churned in the grinding chamber at high speeds with the help of an axially mounted agitator. The cells are said to be broken when they are crushed between two beads. These beads possess momentum due to their movement and inherent mass. Smaller beads tend to float as the cells lyse. Lysis of cells releases intracellular components, mainly nucleic acids that further increase the viscosity of the lysate (Middleberg, 1995). Bigger beads resist floatation by virtue of their mass but result in lower efficiencies due to the increase in void spaces. Therefore, bead loading becomes critical parameters for the optimization of the bead milling process. Higher the bead load, better is the cell lysis efficiency. Bead loading beyond a certain concentration affects the cell lysis and hence recovery. Higher bead loading leaves less space for the cell slurry to be loaded and requires higher agitator power, compromising the process economics. Therefore, bead loading was selected to be optimized in the study. Cell loading, on the other hand, is a direct parameter. Higher the cell load, more is the amount of protein recovered (Haque et al., 2016[b]). The volume of cell slurry added in the grinding chamber depends on the volumes left after loading the dry beads. The void volume between the beads is not considered in calculation; approx 40% of the bead loading volume remains available for cell slurry to be added. Hence, accurate volumes of cell slurry loaded are always a case to case calculation. It is however best to consider specific yield rather than total yield. It's clear that with an increase in cell loading, the amount of COD recovered increases. Cell loading showed the highest F-value indicating it to be the most crucial parameter for the bead milling process. Cell loading increases the amount of COD available to be extracted. On the other hand, it makes the system more viscous, thick and hard to grind. It also affects the efficiency of grinding as beads face increasingly higher resistance to movement. Therefore, higher grinding speeds are recommended for concentrated cell slurries. Run time is one of the parameters that directly affect cell lysis and process economics in a substantial way. Higher bead load, higher viscosities tend to increase the power required to drive the agitator shaft. Cell lysis increases with run time increasing the COD release. After all the cells lyse, the increase in COD release becomes zero as there are no further cells to be broken apart. Hence, running the bead mill after a certain time will increase in the process cost with negligible increase in the amount of COD release.

Therefore, run time needs to be optimized for maximum protein recoveries without compromising the yield. As evident from the study, bead load, cell load and run time are critical towards the development of large scale cell lysis and product recovery process. Maximum extraction of intra-cellularly expressed recombinant protein becomes critical when the lysis process aims to cater to large scale production. Lab scale cell lysis processes don't yield higher recoveries and suffer limitations in terms of the process volume. At the cell concentrations used in bead mills, chemical lysis and ultrasonication methods fail to provide a viable recovery.

In conclusion, the current study affirms the significance and applicability of Central Composite Design (CCD) under Response Surface Methodology (RSM) for developing the cell lysis process at process scale. It also establishes the fact that the parameters under the bead milling process interact with each other significantly and changes in one affect the other. In our study, we recovered around ~90% of the total expressed/recoverable COD. The data obtained in this study can be used as a starting point for the Artificial Neural Network (ANN) models for further optimization and refinement of the process. Our findings prove the applicability of RSM in bead milling optimization process for the recovery of the intracellularly expressed protein.

The model demonstrated a good fit between the predicted and the observed responses showing the model to be valid and usable. The response obtained is generally plotted as contours or 3 – D surface. Hence, the graph is called the 3-D response Surface Plot and the methodology draws its name as Response Surface Methodology (RSM). Out of all the variables, 2 are selected on the x- and the y-axis and the response(s) are plotted on the z-axis. As the system takes into account multiple regression and complex calculations, the yield, i.e., the amount of COD obtained is calculated at every combination of x- and y- values and shown in the form of a 3D surface.

ACKNOWLEDGMENT

Authors thank Jazan University, KSA, for the facilities provided for the study.

REFERENCES

Alshammari, E. Khan, S. Jawed, A. Adnan, M. Khan, M. Nabi, G. Lohani, M. Haque, S. 2015. Optimization of extraction parameters for enhanced production of ovotransferrin from egg white for antimicrobial applications. *BioMed Research International, 2015, 934512.*

Cocaign Bousquet, M. Garrigues, C. Novak, L. Lindley, N.D. & Loubiere, P. 1995. Rational development of simple synthetic medium for the sustained growth of lactobacillus lactis. *J. Appl. Bacteriol. 79, 108 – 116.*

Fish, N.M and Lili, M.D. 1984. The interactions between fermentation and protein recovery. *Biotechnol. 2. 623 – 627.*

Gresham, R.L. Inamine, E. 1986. Nutritional improvement of processes in *Manual for Industrial Microbiology and Biotechnology,eds.* Demain, A.L. & Soloman, N.A. Washington, D.C. *ASM. 41 – 48.*

Haque, S. Khan, S. Wahid M. Dar, S.A. Soni, N. Mandal, R.K. Singh, V. Tiwari, D. Lohani, M. Areeshi, M.Y. Govender, H.G Jawed, A. 2016[b]. Artificial intelligence vs. Statistical modelling and optimization of continuous bead milling process for bacterial cell lysis. Frontiers in Microbiology. 7. *1852.*

Haque, S. Khan, S. Wahid M. Mandal, R.K. Tiwari, D. Dar, S.A. Paul, D. Areeshi, M.Y. Jawed, A. 2016[a]. Modeling and optimization of a continuous bead milling process for bacterial cell lysis using Response Surface Methodology. *RSC Advances, 6, 16348.*

Jawed, A. Dubey, K.K. Khan, S. Wahid, M. Areeshi, M.Y. Haque, S. 2015. Efficient solvent system for maximizing 3-demethylated colchicine recovery using response surface methodology, *Process Biochemistry 50 (12), 2307 – 2313.*

Kanwar, S.S. Kaushal, R.K. Jawed, A. Gupta, R. Chimni, S.S. 2005. Methods for inhibition of residual lipase activityin colorimetric assay: A comparative study. *Ind. J. Biochem & Biophy.42. 233 – 237.*

Middleberg, A.P.J. 1995. Process scale disruption of micro-organisms. *Biotecnol. Adv. 13. 3. 491 – 551.*

Rosano, G.L. and Ceccarelli, E.A. 2014. Recombinant protein expression in *E. coli:* Advances and Challenges. *Frontiers in Microbiol. 5. 172.*
Shokri, A. Sanden, A.M. Larson, G. 2003. Cell and process design for targeting of recombinant protein into culture medium of Escherichia *coli. Appl. Microbiol. Biotechnol.60.6. 654 – 664.*
Singh, G. Jawed, A. Paul, D. Bandyppadhyay, K.K. Kumari, A. Haque, S. 2016.. Concommitant production of lipids and carotenoids in Rhodosporidium toruloids under osmotic stress using response surface methodology. *Forntiers in Microbiol. 7. 1686.*
Volante, F., Pollegioni, L., Molla, G., et al., (2010), Production of recombinant cholesterol oxidase containing covalently bound FAD in *Escherichia coli, BMC Biotechnology, 10 (33),1472 – 6750.*

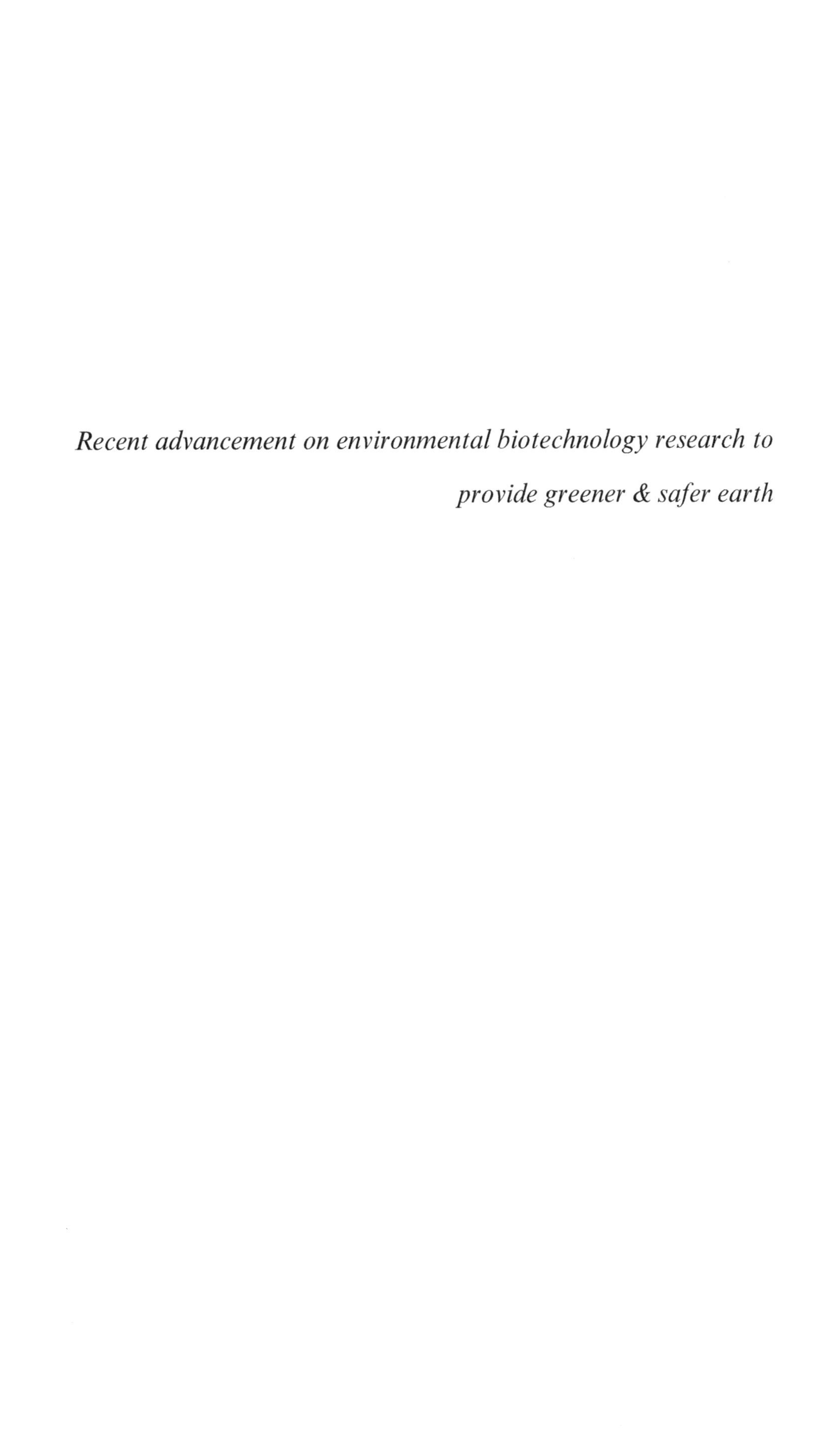

Recent advancement on environmental biotechnology research to provide greener & safer earth

Biotechnology and Biological Sciences – Sen et al. (Eds)
© 2020 Taylor & Francis Group, London, ISBN 978-0-367-43161-7

Study on antioxidant properties of bioconverted agricultural waste

Debarati Roy
Dietetics and Community Nutrition Management, Vidyasagar University, West Midnapore, West Bengal, India

Sukanya Chakraborty
NIFTEM, Sonipat, Haryana, India

ABSTRACT: The Fruits and vegetables account for the largest portion of food wastage all throughout the world. Bioconversion of waste is proving to be an efficient means of utilizing perishable matters. The bio-route for flavour consists of microbial fermentation processes with the help of microbial cells or enzymes produced by them. The present research proposes biotechnological route for development and production of flavour especially from micro-organism namely *Aspergillus oryzae* and *Penicillium species* using waste such as cabbage stalk, watermelon rind, orange peel substrate, studying their antioxidant activity and analyzing them through chromatographic technique. The antioxidant study shows rise in DPPH activity value of Aspergillus bioconverted sample and Penicillium bioconverted sample as compared to fresh samples. The FTIR results indicated the presence of new chemical component in the bioconverted sample.

1 INTRODUCTION

The increase in world population has led to a greater demand of agricultural produce in order to feed all human being. With the increase in food production the food loss and wastage is also evident. Thus there is a great need of the proper storage and transportation facility for preventing the enormous amount of food waste. Food is being wasted at the harvest point during transportation and at the processing stage that is throughout the supply chain. But the waste generated from the fruits and vegetable itself like the leaf, stalk, peel waste example Potato peel, Banana peel, Beet waste, beet pomace, orange peel, Soybean litter, soybean molasses, Apple pomace, waste apples, Pea shell are least taken care of and left unutilized. The characterization of unutilized and discarded fractions of food wastes indicate their potential of reprocessing. Modern researchers are working upon the microbial transformation of large number of organic compounds to obtain components of therapeutic and industrial interest.

Bioconversion or biotransformation is the use of microorganisms for the conversion of organic materials, like plant, food or animal waste to generate energy or some useful products. These bioconverted material not only serve for the purpose of flavour and other industry but they also have the potential of antioxidative properties which can fight several diseases like Cancer, Alzheimer etc. by preventing our body from oxidative damages.

2 MATERIALS & METHODS

The substrate like Cabbage stalk, Watermelon rind and Orange peel were inoculated with *Aspergillus oryzae* and *Penicillium* species separately and then incubated. The bioconverted samples were taken for solvent extraction with petroleum ether. The raw samples and the bioconverted one were tested for proximate analysis, antioxidant activity and FTIR.

2.1 *Proximate Analysis*

2.1.1 *Sample preparation for proximate analysis*

100g of the fresh orange sample and orange peel sample, cabbage stalk and watermelon rind samples were weighed after cleaning and macerating them in mixture grinder. The obtained samples were then considered for proximate analysis for percentage moisture, ash, crude protein, crude fat and carbohydrate content. Moisture, ash, crude fat were determined according to their respective standard methods as described in Association of Official Analytical Chemists (AOAC) 1990.

2.2 *Bioconversion*

The bioconversions of the substrate were done with fungal cultures of *Aspergillus oryzae* and *Penicillium.*

2.2.1 *Bioconversion of Orange Sample*

Raw mandarin orange peel (50g) samples were chosen as substrate and sterilized with ethanol. The orange samples were infected with the fungal strains namely *Aspergillus oryzae* and *Penicillium species,* and then incubated for seven days at 28°C. The infected samples were mashed and stirred with 10ml solvent (40-60 petroleum ether) for two hours in stopper conical. The solvent layer gets collected and the oil was allowed to separate out through rotary evaporator. The essential oil from the infected sample was thus obtained. The peel of the infected orange samples was taken out. Then the samples were macerated and stirred with 10ml solvent (40-60 petroleum ether) for two hours. The solvent layer gets collected. Then the oil was allowed to separate out through rotary evaporator from the solvent. The essential oil from the infected orange peel sample was thus obtained.

2.2.2 *Bioconversion of Cabbage Stalk*

The raw cabbage stalk samples were taken and cut into small pieces. Then they were macerated using mixture grinder. 50g of the sample was weighed. Sterile water was mixed with the sample in a ratio of 1:2. Then the solution was homogenized. Now the pH of the sample was adjusted to 5. Then the samples were equally distributed into two separate conical flasks. The samples were then autoclaved at 121psi for 15 minutes. The samples were then inoculated with strains of *Aspergillus oryzae* and *Penicillium* species separately. The inoculated samples were kept for incubation for ten days. The fermented samples are then taken for solvent extraction with pet ether.

2.2.3 *Bioconversion of Watermelon Rind*

The above procedure was repeated for watermelon rind sample.

2.3 *Antioxidant Determination*

The oil extracted from the bioconverted sample i.e. orange was taken for antioxidant study.

- Antioxidant activity analysis by DPPH assay (For cabbage stalk, watermelon rind and orange peel) [Shimamura, 2014]
- ABTS Assay (For orange peel sample)[Rajurkar, 2011]

3 RESULTS & DISCUSSION

3.1 *Proximate Compositions*

Here we analyzed the carbohydrate, protein, fat, moisture & ash contained of sample (in 100gm)

Table 1. Proximate analysis.

Sample	Carb.	Protein	Fat	Ash	Moisture
Cabbage stalk	6.5%	1.4%	0.2 %	0.85%	78%
Watermelon rind	79.67%	7.4%	1.05%	3.11%	85.12%
Orange	11.2%	0.52%	0.15%	81.80%	6.32%

3.2 *Antioxidant Assay*

Table 2. Antioxidant activity results.

DPPH Activity			
Sample	Fresh sample	A1 Sample	P1 sample
Cabbage stalk	19%	12.6%	21.3%
Watermelon rind	29.3%	35%	43.6%
Orange peel oil	16.06%	18.08%	50%
ABTS Activity			
Orange peel oil	35%	61.34%	64.28%

Aspergillus bioconverted sample: A1
Penicillium bioconverted sample: P1

3.3 *FTIR Report*

The FTIR analysis displayed formation of new bonds in the bioconverted samples in all the samples. This indicated the presence of new compounds in the test samples as compared to the raw one.

4 CONCLUSION

This project thus gives us the opportunity to explore the field of bioconversion. Work on various fruit and vegetable waste substrates. The bioconversion of these substrates can help in the improvement of the antioxidant activity or may lead to the formation of new compounds leading to the information that these waste fruits and vegetables can further be explored as a potential radical scavengers, bioactive components or bioflavours and may benefit several industrial sectors like drug and food industries.

ACKNOWLEDGMENT

The present research work was undertaken under the Alma mater IIEST Shibpur. Special thanks to our faculty guide Professor D.K.Bhattacharya and Dr. Jayati Bhowal.

REFERENCES

1. Bajpai, V.K., Kang, S.C., Heu, S., Shukla, S., Lee, S. and Baek, K.H., 2010. Microbialconversion and anticandidal effects of bioconverted product of cabbage (Brassicaoleracea) by Pectobacterium carotovorum pv. carotovorum 21. *Food and Chemical Toxicology*, *48*(10), pp.2719-2724.
2. Egbuonu, A.C.C., 2015. Comparative investigation of the proximate and functional properties of watermelon (Citrullus lanatus) rind and seed. *Research Journal of Environmental Toxicology*, *9*(3), p.160

3. Kedare, S.B. and Singh, R.P., 2011. Genesis and development of DPPH method of antioxidant assay. *Journal of food science and technology*, *48*(4), pp.412-422.
4. Perkins, C., Siddiqui, S., Puri, M. and Demain, A.L., 2016. Biotechnological applications microbial bioconversions. *Critical reviews in biotechnology*, *36*(6), pp.1050-1065.
5. Rajurkar, N.S. and Hande, S.M., 2011. Estimation of phytochemical content and antioxidant active ty of some selected traditional Indian medicinal plants. *Indian journal of pharmaceutical sciences*, *73* (2), p.146
6. Shimamura, T., Sumikura, Y., Yamazaki, T., Tada, A., Kashiwagi, T., Ishikawa, H., Matsui, T., Sugimoto, N., Akiyama, H. and Ukeda, H., 2014. Applicability of the DPPH assay for evaluating the antioxidant capacity of food additives–inter-laboratory evaluation study–. *Analytical Sciences*, *30*(7), pp.717-721.
7. Yang, C., Chen, H., Chen, H., Zhong, B., Luo, X. and Chun, J., 2017. Antioxidant and anticancer activities of essential oil from Gannan navel orange peel. *Molecules*, *22*(8),p.1391
8. Zheng, L., Zheng, P., Sun, Z., Bai, Y., Wang, J. and Guo, X., 2007. Production of vanillin from waste residue of rice bran oil by Aspergillus niger and Pycnoporus cinnabarinus. *Bioresource Technology*, *98* (5), pp.1115-1119.

Biotechnology and Biological Sciences – Sen et al. (Eds)
© 2020 Taylor & Francis Group, London, ISBN 978-0-367-43161-7

Evaluation of hypolipidemic and antioxidant potential of Ketoki joha, an aromatic rice of Assam, India

Saikat Sen*, Raja Chakraborty & Pratap Kalita
Faculty of Pharmaceutical Science, Assam down town University, Guwahati, Assam, India

ABSTRACT: Rice (*Oryza sativa*) is one of the most staple food grains with enormous nutritive value. Though, the parts of rice also contain different bioactive molecules including vitamin, minerals and antioxidants. The present work aimed to investigate hypolipidemic and antioxidant potential of ketoki joha, an aromatic rice of Assam, India. Activity of the water extract of rice seed with bran were evaluated on high fat diet & high sugar induced hyperlipidemia model using rats. Treatment with water extract of ketoki joha (200 mg and 400 mg/kg, oral) for 30 days significantly ameliorated the lipid profile compare to disease control. Apart for lipid profile the effect, reduce hepatic enzyme and lactate dehydrogenase level in serum compare to disease control. Extract increase serum ApoA1 and reduce ApoB, ApoB/ApoA1 ratio. Extract treatment significantly reduced atherogenic coefficient and cardiac risk ratio index and positively ameliorated endogenous antioxidant level. Current work, found that ketoki joha significantly reduced hyperlipidemia and exerted *in vivo* antioxidant activity

Keywords: Ketoki joha, Hyperlipidemia, rice bran, lipid profile

1 INTRODUCTION

Rice (*Oryza sativa*) is popular staple food grains that sustain more than 66% population of the world (Burlando & Cornara 2014). Parts for their nutritive value, different rice were also investigated for their medicinal value. Presence of different antioxidant molecule, vitamins, and minerals were reported in rice (Burlando & Cornara 2014). A number of studies around the world particularly from Thailand, Japan, Koear etc. have reported that different parts of rice particularly rice bran possess antioxidant, antidiabetic, anticancer, anti-inflammatory, hepatoprotective, hypolipidemic, cardioprotective and hepatoprotective activity (Burlando & Cornara 2014; Friedman 2013; Gul et al. 2015). A recent study reported that intake of whole grain reduces the risk of cardiovascular disease, and other diseases including cancer (Aune et al. 2016). Joha, a fine grain aromatic rice mainly cultivated in Assam and it have special role in several occasions of Assam (Medi et al. 2004). More than 20 different varieties of joha rice are available Assam, though very few studies are carried out focusing joha rice. Present work aimed to investigate *in vivo* hypolipidemic and antioxidant potential of ketoki joha rice of Assam, India

2 GETTING STARTED

2.1 *Rice sample*

Ketoki joha is collected from Institute of Advanced Study in Science and Technology (IASST), Guwahati.

*Corresponding author: saikat.pharm@rediffmail.com

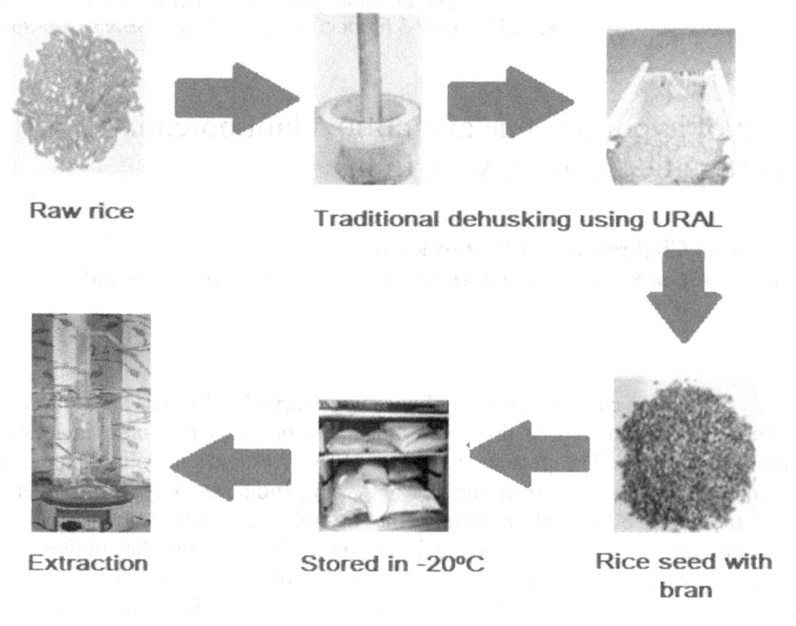

Figure 1. Flow chart - processing of rice sample.

2.2 *Processing of rice sample*

*R*ice was process for de-husking (husk was removed without removing the bran) using traditional equipment called as 'URAL' in Assamese. Rice seed thus obtain contain bran with seed. This sample was grinded using mechanical grinder to get course powder and then extracted with water using Soxhlet apparatus. Extract was dried using rotary evaporator to get water extract of ketoki joha (WEKJ).

2.3 *Experimental animals*

Wistar rats (180±20g) were used for this study. Experimental animals were housed in appropriate environmental condition for one week for acclimatization. Rats were fed with standard pellet diet and water *ad libitum*. Wister rats were used to evaluate acute toxicity and hypolipidemic activity. All animal experiments were approved by the Institutional Animal Ethical Committee, Assam down town University, Guwahati (No. AdtU/IAEC/2016/005).

2.4 *Acute toxicity study*

Acute toxicity study was carried out using the method as per OECD guideline 423 (Annexure 4d). Healthy, non-pregnant female rats were used. WEKJ (2000 mg/kg) is administered orally to three animals. Animals were observed for any mortality or side effects for 14 days (Organisation for Economic Co-operation and Development 2001).

2.5 *Hypolipidemic activity*

Hypolipidemic activity of WEKJ was studied using the high fat diet (vanaspati ghee and coconut oil in 3:1 ratio) (HFD) and high sugar (25% fructose water) (HS) induced hyper-lipidemic rat models. Animals were divided in various groups (n=6). Group 1: Normal control, Group 2: Disease control, Group 3 & 4: Water extract of ketoki joha (WEKJ) (200 and 400 mg/kg). First 15 days animals of group II-IV received HFD and HS. From day 16th to 45th, animal received respective drug treatment orally along with HFD+HS (Munshi et al. 2014). On 46th day, blood samples were collected by retro orbital route under light ether anaesthesia, and serum sample was used for determination of lipid profile, serum glutamic oxaloacetate transaminase (SGOT), serum glutamic pyruvic transaminase (SGPT), alkaline phosphatase (ALP), lactate dehydrogenase (LDH), ApoA1, ApoB with commercial biochemical diagnostic kit using Autoanalyzer (Mispaace, Agapee.). Atherogenic coefficient and cardiac risk ratio were also calculated using standard formula (Chawda et al. 2014). Serum were also analyzed for estimation of superoxide dismutase (SOD), catalase (CAT) and reduced glutathione (GSH), lipid peroxidation using standard protocol (Ellman 1952; Sinha 1972; Marklund & Marklund 1974; Ohkawa et al., 1979).

2.6 *Statistical analysis*

The data were expressed as mean ± SEM (n = 6). Statistical analysis was carried out by ANOVA followed by Tukey tests.

3 RESULT AND DISCUSSION

Traditional method of de-husking was used to retain the bran with seed. The research carried out in Japan, Korea, Thailand, China etc showed that rice bran contain numerous bioactive component including oryzanol and its derivatives, tocopherols, phenols and polyphenols, saponins, flavonoids, anthocyanin, carotenoids, saturated and unsaturated fatty acids etc. [1-3]. During milling process bran is removed from seed, and thus mostly we consume the rice without bran that could have most health promoting effect.

Extract (2000 mg/kg, b.w.) not produced any mortality or side effect when tested through acute toxicity study. Administration of HFD+HS results enhancement of lipid profile which is evident by the increase in serum triglyceride (TG), total cholesterol (TC), low density lipoprotein (LDL), very low density lipoprotein (VLDL), and reduction of high density lipoprotein (HDL) in disease control group compare to healthy group. Dyslipidemia is a key risk factor for atherosclerosis, and other cardiovascular diseases (CVDs) that is marked by the alteration of lipid profile, as we observed. Abnormal lipid profile considered as the hallmark of metabolic syndrome (Ntchapda et al. 2015). Atherosclerosis is linked with high level of TC, TG and LDL, and low level of HDL in blood. Our study showed that 30 days treatment with WEKJ alter the lipid profile significantly compare to disease control. After treatment with WEKJ (400 mg/kg) level of TG, TC, LDL, VLDL reduced significantly and found to be 110.39, 130.26, 39.00, 22.08 mg/dl respectively and HDL increased to 39.7 mg/dl (Figure 2).

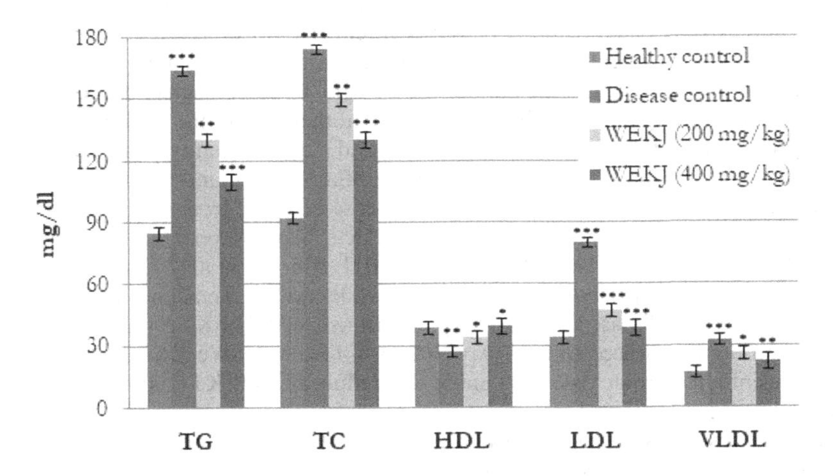

Figure 2. Effect of ketoki joha on lipid profile.

Values are given as mean ± SEM (n=6); *$p < 0.05$, **$p < 0.01$, ***$p < 0.001$ when disease control group compared with normal control while test group compared with disease control group.

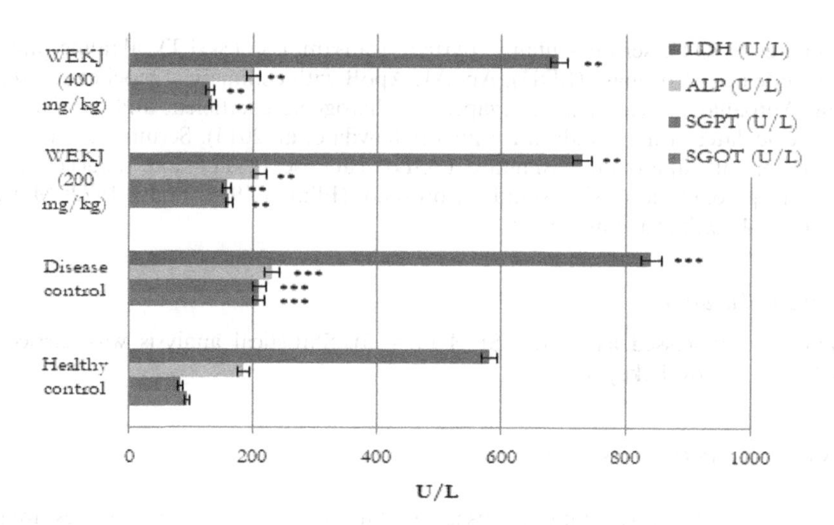

Figure 3. Effect of ketoki joha on hepatic enzymes and LDH.

Values are given as mean ± SEM (n = 6); *p< 0.05, **p<0.01, ***p<0.001 when disease control group compared with normal control while test group compared with disease control group.

HDL is inversely linked with serum cholesterol. Reduced level of HDL may enhance the development of atherosclerosis that also enhances the risk of ischaemic heart diseases, by altering clearance of cholesterol from the arterial wall (Subramani et al. 2017). Ketoki joha administration reduced the level of LDL, VLDL, TG and TC, while level of HDL increases. Results are suggesting hypotriglyceridemic and hypocholesterolemic effects of ketoki joha.

Hypercholesterolemia may induce harmful effect on vital organs like heart and lever. Retention of lipid in liver increase by the HFD that result reduced hepatic function and liver steatosis (Chawda et al., 2014). Damage of hepatic cell results increase level of SGOT, SGPT in serum. Increased level of SGPT can induce higher risk cardiac disease (Devi & Singh 2017). In present study also, was observed that HS+HFD cause reduce hepatic and cardiac function that is evident by the increase level of SGPT, SGOP, ALP and LDH in disease control group. Treatment with WEKJ averts such amelioration significantly and brings the enzyme level near to normal (Figure 3). Results showed that WEKJ also exert beneficial effect on heart and liver.

Apolipoproteins are key parts of lipoprotein particles. Current investigations concluded estimation of different forms of apolipoproteins may useful in predicting the risk of CVDs and thus useful to take step in time. Apolipoprotein B (apoB), a single molecule in VLDL, IDL, and LDL (they are considered as atherogenic lipoprotein). Thus, estimation of ApoB and its level in plasma indicates the amount of TC and, to some degree, particles that contain triglyceride (Lu et al. 2011; Florvall et al. 2006). Another important apolipoprotein is Apolipoprotein A1 (apoA1) that associated with HDL, and a chief initiator of the process of reverse cholesterol transport. ApoA1 can also exert anti-inflammatory and anti-oxidant effects, and can induce production of endothelial NO and release of PGI2 from endothelium. Therefore, level of ApoA1 related with anti-atherogenic effects. ApoB/ApoA1 ratio is considered as a powerful indicator of risk for CVDs (Lu et al. 2011; Florvall et al. 2006). Our result showed in disease control group level of ApoA1 reduced while level of ApoB increased which indicate highr risk of CVDs in disease control group. Treatment with WEKJ reverses the trend, which is also evident by the reduced ratio of ApoB/ApoA1 (Table 1). We also observed that atherogenic coefficient and cardiac risk ratio index also reduced in WEKJ (200 and 400 mg/kg) treated group (Table 2).

These results clearly showed that water extract of ketoki joha have the potential to reduce hyperlipidemia as well as in reducing the risk of CVDs.

Result of present investigation also showed that extract exhibited strong *in vivo* antioxidant activity. Level of SOD, CAT and GSH reduced in disease control group. After treatment with

Table 1. Effect of ketoki joha on ApoA, ApoB and their ratio.

Para-meters	Normal Control	Disease Control	WEKJ (200 mg/kg)	WEKJ (400 mg/kg)
Apo A1 (mg/dL)	9.32 ± 1.01	6.42 ± 0.91*	7.42 ± 0.87*	8.41 ± 1.02*
Apo B (mg/dL)	8.53 ± 1.12	13.33 ± 1.50*	12.70 ± 1.03*	10.82 ± 1.75*
ApoB/ApoA1	0.915	2.076	1.712	1.28

Values are given as mean±SEM (n=6); *p< 0.05 when disease control group compared with normal control while test group compared with disease control group

Table 2. Effect of ketoki joha on atherogenic coefficient and cardiac risk ratio index.

Parameters	Normal Control	Disease Control	WEKJ (200 mg/kg)	WEKJ (400 mg/kg)
Atherogenic Coefficient	1.205	5.038	2.831	1.788
Cardiac Risk Ratio	2.205	6.038	3.831	2.788

WEKJ the level of SOD, CAT and GSH increased. Level of GSH in disease control, WEKJ (200 mg/kg) and WEKJ (400 mg/kg) was 0.21 ± 0.02, 0.35 ± 0.09 and 0.47 ± 0.10 μg of GSH/mg of protein respectively. Level of CAT in disease control and WEKJ (400 mg/kg) group was 0.025 ± 0.01, 0.23 ± 0.05 unit/protein. SOD, an important antioxidant enzyme (endogenous) and considered as first line defense system against free radical and non radical reactive species. SOD scavenges superoxide radical to H_2O_2 and oxygen (Sen & Chakraborty 2011). CAT another antioxidant enzyme that decompose H_2O_2 to water and oxygen (Sen & Chakraborty 2011). Hyperlipidemia can cause impaired antioxidant defence system and that cause deletion of CAT and SOD (Durkar et al. 2014).

GSH is a non enzymatic antioxidant playing key role in maintaining -SH groups and detoxify foreign compounds, free radicals like hydroxyl radical and singlet oxygen, hydrogen peroxide etc. (Sen & Chakraborty 2011). Oxidative stress is considered as key mediator for number of diseases including CVDs. Results showed that extract playing important role by enhancing the level of endogenous antioxidants which may be correlated with the hypolipidemic activity of ketoki joha.

Hypercholesterolemia, hypertriglyceremia are responsible for week antioxidant defense system, resulting from oxidative stress that causes tissue damage. Increased free radical production also induces the development and progress of atherogenesis and other CVDs. MDA is an end product of lipid peroxidations and used to estimate the level of free radical generation. Hyperlipidemia can increase lipid peroxidation and increase level of MDA in high fat diet induced hyperlipidemic animals is a indicator of enhance lipid peroxidation (Prasanna & Purnima 2011). HFD+HS administration increase lipid peroxidation and treatment with WEKJ (200 and 400 mg/kg) reduced the

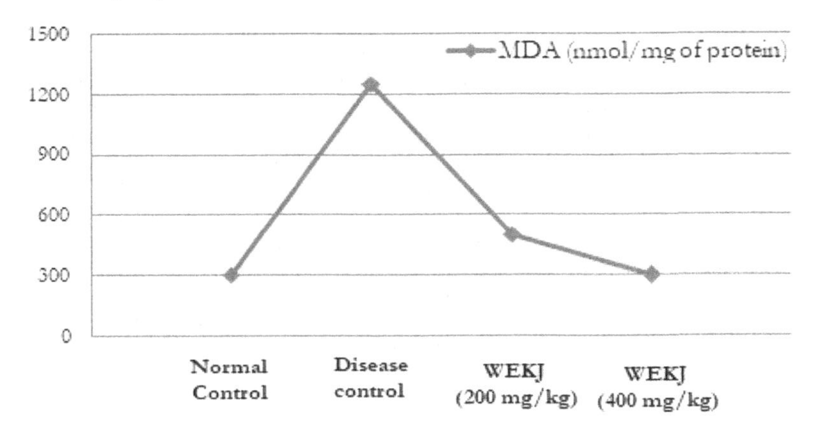

Figure 4. Effect of ketoki joha on lipid peroxidation.

level of MDA significantly (Figure 4). Results suggested that extract have potent effect to reduce lipid peroxidation and endogenous enhancing antioxidant level.

Pharmacological and phytochemical investigations on rice bran have received much interest because of the presence of numerous bioactive components. Current investigation was undertaken to find whether ketoki joha produced any beneficial effect on HFD+HS induced hyperlipidemia. The study indicated that antioxidant activity exerted by extract may responsible for its hypolipidemic activity. Enhance removal of lipoprotein, triglyceride from circulation or excretory phases of lipids may responsible for hypolipidemic activity of ketoki joha. In general, bran contain high proportion of antioxidant compounds like phenolic compounds, anthocyanidins etc., which may responsible for hypolipidemic activity of rice extract.

This result indicated that consumption of rice (ketoki joha) without removing bran could be helpful in preventing hyperlipidemia. Further studies are going on in our lab to characterize antioxidant molecule of rice and to testify its effect on human volunteer.

4 CONCLUSION

This study indicated that water extract of ketoki joha (seed with bran) helpful to control hyperlipidemia in high fat diet and high sugar induced hyperlipidemia. Further researches are going on to find bioactive molecule and possible effect of rice on human.

ACKNOWLEDGEMENT

The present research work is supported Department of Biotechnology, Govt. of India under the networking project on 'Scented Rice of NE India' (DBT-NER/AGRI/20/2015).

REFERENCES

Burlando, B. & Cornara, L. 2014. Therapeutic properties of rice constituents and derivatives (*Oryza sativa* L.): A review update. *Trends in Food Science & Technology* 40: 82-98.

Friedman, M. 2013. Rice brans, rice bran oils, and rice hulls: composition, food and industrial uses, and bioactivities in humans, animals, and cells. *Journal of Agricultural and Food Chemistry* 61: 10626−10641.

Gul, K., Yousuf, B., Singh, A. K., Singh, P. & Wani, A. A. 2015. Rice bran: Nutritional values and its emerging potential for development of functional food–a review. *Bioactive Carbohydrates and Dietary Fibre* 6: 24-30.

Aune, D., Keum, N., Giovannucci, E., Fadnes, L.T., Boffetta, P., Greenwood, D.C., Tonstad, S., Vatten, L. J., Riboli, E. & Norat, T. 2016. Whole grain consumption and risk of cardiovascular disease, cancer, and all cause and cause specific mortality: systematic review and dose-response meta-analysis of prospective studies. *British Medical Journal* 353. doi: 10.1136/bmj.i2716.

Medhi, K., Talukdar, P. & Baruah, I. 2004. Extent of genetic variation in indigenous scented rice varieties of Assam. *Indian Journal of Plant Genetics Resources* 17(2): 27-29.

Organisation for Economic Co-operation and Development (OECD). OECD guideline for testing of chemicals (test no. 423: acute toxicity—acute toxic class method), 2001. Available in: https://www.oecd-library.org/environment/test-no-423-acute-oral-toxicity-acute-toxic-class-method_9789264071001-en.

Munshi, R.P., Joshi, S.G. & Rane, B.N. 2014. Development of an experimental deit model in rats to study hyperlipidemia and insulin resistance, markers for coronary heart diseases. *Indian Journal of Pharmacology* 46: 270-276.

Chawda, H.M., Mandavia, D.R., Parmar, P.H., Baxi, S.N., Tripathi, C.R. 2014. Hypolipidemic activity of a hydroalcoholic extract of *Cyperus scariosus* Linn. root in guinea pigs fed with a high cholesterol diet. *Chinese Journal of Natural Medicines* 12: 819-826.

Sinha, A.K. 1972. Colorimetric assay of catalase. *Analytical Biochemistry* 47: 389-394.

Ellman, G.L. 1959. Tissue sulfhydryl groups. *Archives of Biochemistry and Biophysics* 82: 70-77.

Marklund, S. & Marklund, G. 1974. Involvement of the superoxide anion radical in the autoxidation of pyrogallol and a convenient assay for superoxide dismutase. *Europe Journal of Biochemistry* 47: 469-474.

Ohkawa, H., Ohishi, N. & Yagi, K. 1979. Assay for lipid peroxides in animal tissues by thiobarbituric acid reaction. *Analytical Biochemistry* 95: 35-l-358.
Ntchapda, F., Maguirgue, K., Adjia, H., Etet, P.F.S. & Dimo, T. 2015. Hypolipidemic, antioxidant and anti-atherosclerogenic effects of aqueous extract of *Zanthoxylum heitzii* stem bark in diet-induced hypercholesterolemic rats. *Asian Pacific Journal of Tropical Medicine*, 359-365. 10.1016/S1995-7645(14)60344-8.
Subramani, C., Rajakkannu, A., Rathinam, A., Gaidhani, S., Raju, I. & Singh, D.V.K. 2017. Anti-atherosclerotic activity of root bark of *Premna integrifolia* Linn. In high fat diet induced atherosclerosis model rats. *Journal of Pharmaceutical Analysis* 7: 123–128.
Devi, S. & Singh, R. 2017. Evaluation of antioxidant and anti-hypercholesterolemic potential of *Vitis vinifera* leaves. *Food Science and Human Wellness* 6: 131-136.
Lu, M., Lu, Q., Zhang, Y. & Tian, G. 2011. ApoB/apoA1 is an effective predictor of coronary heart disease risk in overweight and obesity. *Journal of Biomedical Research* 25: 266–273.
Florvall, G., Basu, S. & Larsson, A. 2006. Apolipoprotein A1 is a stronger prognostic marker than are HDL and LDL cholesterol for cardiovascular disease and mortality in elderly men. Journals of Gerontology. Series A, Biological Sciences and Medical Sciences 61: 1262–1266.
Sen, S. & Chakraborty, R. 2011. The role of antioxidants in human health. In: A. Silvana & M. Hepel (eds), *Oxidative Stress: Diagnostics, Prevention, and Therapy*: 1-37. American Chemical Society: Washington DC.
Durkar, A. M., Patil, R. R. & Naik, S. R. 2014. Hypolipidemic and antioxidant activity of ethanolic extract of *Symplocos racemosa* Roxb. In hyperlipidemic rats: A evidence of participation of oxidative stress in hyperlipidemia. *Indian Journal of Experimental Biology* 52: 36-45.
Prasanna, G. S. & Purnima, A. 2011. Protective Effect of Leaf Extract of *Trichilia connaroides* on hypercholesterolemia induced oxidative stress. *International Journal of Pharmacology* 7: 106-112.

Biotechnology and Biological Sciences – Sen et al. (Eds)
© 2020 Taylor & Francis Group, London, ISBN 978-0-367-43161-7

Characterization of soil - extracted arsenic tolerant bacteria from arsenic belt of West Bengal

U. Bhattacharyya, S.K. Sarkar & B. Roychoudhury
Department of Biotechnology, Bengal Institute of Technology, Kolkata, India

ABSTRACT: Our research is focused on understanding bacterial arsenic resistance on the molecular level. This is done by a combination of molecular genetic, biochemical and biophysical techniques. Of special interest are changes in protein structure for example upon binding of arsenite or during extrusion of arsenite from the cell. Our model organism is *Bacillus subtilis* which is a common soil bacterium with a well understood physiology and sequenced genome.

1 INTRODUCTION

Arsenic is a well known toxic chemical that the Environmental Protection Agency (EPA) and the World Health Organization (WHO) list as a known carcinogen. The deleterious effects of arsenic to human health resulting from environmental contamination have been reported worldwide (Oremland 2003 & Stolz 2005). Arsenic is found in a wide variety of chemical forms throughout the environment and can be readily transformed by microbes, changes in geochemical conditions and other environmental processes. While arsenic occurs naturally, it also may be found as a result of a variety of industrial applications leather and wood treatments and pesticides. Man-made arsenic contamination results mainly from manufacturing metals and alloys refining petroleum and burning fossil fuels and wastes (Nealson, Belz & Mckee 2002). These industrial activities have created a strong legacy of arsenic pollution throughout the United States. The World Health Organization estimates that in West Bengal (India) and Bangladesh alone, more the 112 million people are drinking water contamination with arsenic which exceed the WHO maximum permissible level of 50mg/lit. An estimated 2,00,000 – 3,00,000 people in India have arsenic included skin lesions and cancer and an estimated 2,00,000 – 2,70,000 cancer caused deaths in Bangladesh will be high levels of arsenic in drinking water (Nealson, Tsapin, & Storrie-Lombardii 2002).

Unlike organic pollution, arsenic cannot be transformed into a non-toxic material. It can be transformed into a form that is less toxic when exposed living organisms in the environment, because arsenic is a permanent part of the environment, there is long term need of regular monitoring at sites where arsenic containing waste has been disposed of and at sites where it occurs naturally at elevated levels. A range of analytical field assay methods for pollutants such as arsenic provide valuable tools to support improved site characterization initiatives, such as- triad method from EPA, Adaptive Sampling and Analysis Programs (ASAP) from Department of Energy (DOE) (Oremland & Stolz 2003, 2005). Nonetheless, certain prokaryotes use arsenic oxy-anions for energy generation either by oxidizing arsenate or by respiring arsenate. These microbes are phylogenetically diverse and occur in wide range of habitats. Arsenic cycling may take place in absence of O_2 and can contribute to organic matter oxidation (Afkar, Lisak, Saltikov, Basu, Oremland & Stolz 2003). In microbial bio films growing on the rock surface of anoxic brine pools fed up hot springs containing arsenate and sulfide at high concentration, we discovered light dependent oxidation of arsenite As(III) to arsenate As(V) occurring under anoxic condition. The communities were composed primarily of *Ectothiorhodospira* – like purple bacteria or *Oscillatoria* like cyanobacteria. (Ehrlich, 2002). The strain contained genes encoding a putative As(V) reductase but no detectable homolog of the As(III) oxidase genes of aerobic chemolithotrophs, suggesting a reverse functionality for the reductase (Ehrlich, 2002). The enzyme responsible for the respiratory oxidation of Arsenite (III) to Arsenate (V) has been found widely in various groups of Bacteria and Archaea and has been studied in details (Santini, Sly, Wen, Comrie, Wulf-Durand & Macy 2002).

2 MATERIALS & METHODS

2.1 *Material collection*

The materials i.e., the water and the soil is collected from various locations of Bantala, Bamanghata and Bhangarh, West Bengal. Microbes have cultured from these soil and water.

2.2 *Preparation and standardization*

Preparation and standardization of nutrient agar Media was done with pH 7.2 for isolation of soil microbes. Dilution plating technique is used for the separation of a dilute mixed/single population of microorganism so that individual colonies can be isolated. In this technique cell suspension of microorganisms are serially diluted and from each dilution the cell suspension is spread over the solidified agar medium with a sterile L-shaped glass rod. Some of these cells will be separated from each other by a distance sufficient to allow the colonies that develop to be free from each other. Gram staining was done for morphological identification. Isolation of DNA from selected microorganism was done. Lysozyme is a bactericidal enzyme that catalyzes the cleavage of the peptidoglycan linkages in bacterial cell walls, thereby lysing bacteria and releasing genomic DNA. Mechanical lysis mostly utilizes breaking of the cell wall by glass beads, mild sonication etc. Chemical Lysis utilizes use of various combinations of salts. To the lyophilized bacterial cells, 500ml of genomic DNA suspension buffer and vortex the cells for a minute are added. 5ml of the RNase A solution provided is added and mixed well. The tube is incubated at 65°C for 10 minutes. 650ml of genomic DNA lysis buffer is added and mixed well and is incubated for 15minutes at 65°C. The supernatant is divided into two vials containing 600ml each. 600ml of precipitation solution is added to both the vials. After mixing by slowly inverting the tubes till you can see white strands of genomic DNA separating out at the interphase of the two liquids. It is then spun at 10 at 10,000 rpm for 15minutes at room temperature and the supernatant is discarded. The pellet is washed with 70% alcohol. Add 1ml of 70% ethanol to both the vials and mixed well. It is spun at 10,000 rpm and the supernatant is discarded. The solution is incubated at 50-55°C for 10-15 minutes or at 4°C overnight. Now the genomic DNA is ready to load.

2.3 *Restriction digestion of selected microbes*

Minimum two experiments are to be carried out simultaneously. 1% agarose gel is prepared before setting up the restriction digestion reactions. The agarose is allowed to set for 1 hour at room temperature.

2.4 *Restriction digestion reaction*

Modified pUC18 plasmid DNA is digested with three restriction enzymes. Three separate reactions have to be set up. The plasmid DNA is taken out; assay buffer is kept in 20°C and put them on crushed ice. The components are allowed to get thawed on ice. All the components for the mixture are added and tapped 2-3 times for mixing. A short spin if given at 10,000 rpm for 20 sec. The mixture is incubated at 37°C dry bath for 1 hour. 1% Agarose gel is prepared before setting up the restriction digestion reaction.

3 RESULTS & DISCUSSION

The majority of inorganic arsenic is distributed in the biosphere by water in the Arsenate (V) or Arsenite (III) form. Arsenite enters the cell via the uptake system for phosphate (Pit) and blocks the oxidative phosphorylation since it cannot stable high–energy compounds. The uptake of arsenate is thought to occur via transporters of glycerol (GLPF). Bacterial resistance systems are often of the extrusion type, i.e. the metal ions are actively pumped out of the cell. The genes encoding the resistance machineries are generally arranged in operons and come in several variants. The minimum set of genes needed for arsenic resistance is ars RBC.

Best development of bacterial culture was observed in 2X Dilution. The slides are examined under microscope. All the slides showed clusters of violet colored rod-shaped and spherical bacteria (Table 1 and Table 2). From the observation it can be concluded that the bacteria in the slides were all Gram Positive bacillus or cocci. Liquid Broth medium was prepared, in which bacterial strain from the parent culture are inoculated and maintained for sub culturing, as for molecular characterization fresh culture if necessary. The genome of arsenic tolerant microbes have been isolated and quantified by running on 1% agarose gel. From the banding pattern, one can note that each restriction enzyme has specific recognition site and cuts only at specific position .The recognition site is unique for a particular enzyme. (Figure 1). After the restriction digestion four lane containing DNA bands have been noticed. The third lane contains the genomic DNA of Isolated bacteria and Fourth lane contains the marker DNA. The first and Second lanes are digested with Eco RI and Hind III respectively. The first lane contains seven fragments among which the first five is of moderately higher molecular weights and rest two is medium in molecular weight. In the second lane five bands were found among which all of them are high molecular weight. No fragment of lower molecular weight is found. (Figure 2). The enzyme responsible for the respiratory oxidation of Arsenite (III) to Arsenate (V) has been found widely in various groups of Bacteria and Archaea and has been studied. Our research is focused on understanding bacterial arsenic resistance on the molecular level. This is done by a combination of molecular genetic, biochemical and biophysical techniques.

Table 1. Bacterial growth in different media.

Component	M1% Conc.	M2% Conc.	M3% Conc.*	M4% Conc.
Beef extract	0.2	0.2	0.5	0.7
Peptone	0.5	0.7	1	0.2
NaCl	0.5	0.5	0.5	0.5
D-Glucose	1	1	1	1
pH	7.2	7.2	7.2	7.2
Agar	2	2	2	2

*Bacterial growth was found to occur in the M3 medium

Table 2. Bacterial growth for different dilution of soil samples.

Dilution	Growth
2x	+++
5x	+
10x	±

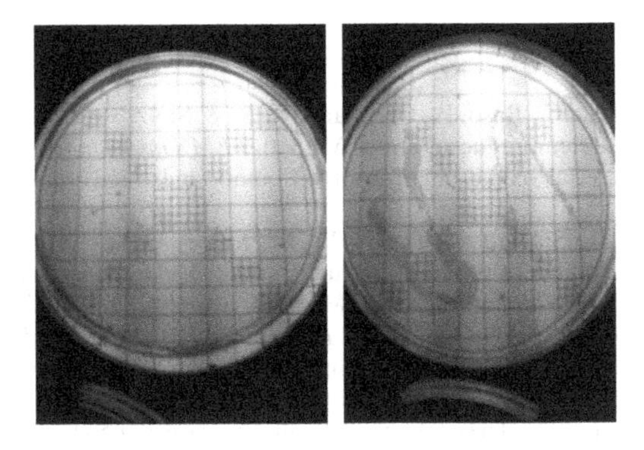

Figure 1. Comparison of control bacteria with Arsenic Tolerant Bacteria in Arsenic medium (15ml/lit arsenic concentration).

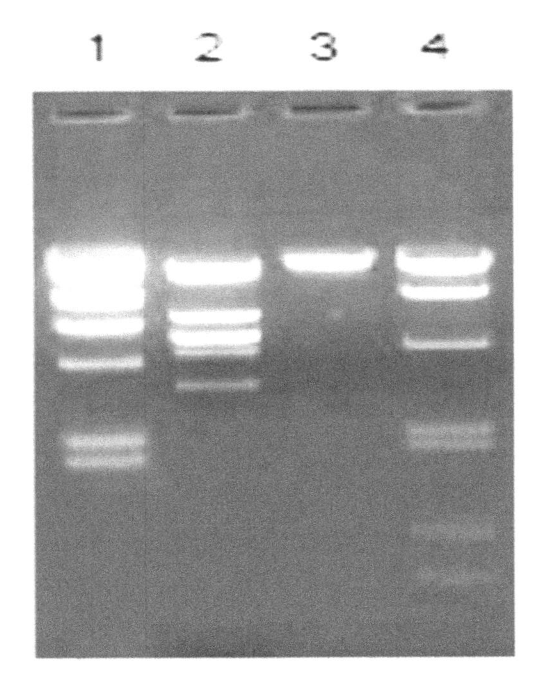

Figure 2. Agarose gel electrophoresis of isolated bacteria.

4 CONCLUSION

The spore of the bacteria mixing with the organic manure can be directly applied into the field as a part of inter cultural operation of the cropping system, where these microbes germinate and can remove Arsenic from the field.

In the second line of thought, if the gene responsible for Arsenic degradation, can be identified, a lot of beneficial activities can be carried out like manufacture of Genetically Modified Crops, which are cultivated commercially after testing the bioseptical measures.

REFERENCES

Afkar, E., Lisak, J., Saltikov, C., Basu, P., Oremland, R. S., & Stolz, J. F. 2003. The respiratory arsenate reductase from Bacillus selenitireducens strain MLS10. *FEMS microbiology letters*, 226(1),107-112.

Ehrlich, H. L. 2002. Bacterial oxidation of As (III) compounds. *Environmental chemistry of arsenic*, 313-328.

Muller, D., Lievremont, D., Simeonova, D.D., Hubert, J.C. & Lett. M.C. 2003. Arsenite oxidase aox genes from a metal-resistant beta-proteobacterium, *Journal of Bacteriology* 185(1): 135-141.

Nealson, K.H., Tsapin, A. & Storrie-Lombardi. M. 2002. Searching for life in the Universe: unconventional methods for an unconventional problem, *International Microbiology* 5(4): 223-230.

Nealson, K.H., Belz, A. & McKee, B. 2002. Breathing metals as a way of life: geobiology in action, *Antonie Van Leeuwenhoek* 81(1-4): 215-222.

Oremland, R.S. & Stolz, J.F 2003. The ecology of arsenic, *Science* 300(5621): 939-944.

Oremland, R.S. & Stolz, J.F 2005. Arsenic, microbes and contaminated aquifers, *Trends in Microbiology* 13(2): 45-49.

Santini, J. M., Sly, L. I., Wen, A., Comrie, D., Wulf-Durand, P. D., & Macy, J. M. 2002. New Arsenite-Oxidizing Bacteria Isolated from Australian Gold Mining Environments–Phylogenetic Relationships. *Geomicrobiology Journal*, 19(1),67-76.

Biotechnology and Biological Sciences – Sen et al. (Eds)

Monsoonal variability of phytoplankton in a tide dominated river of Sundarban

Saranya Chakraborti, Tanaya Das & Goutam Kumar Sen
School of Oceanographic Studies, Jadavpur University, Kolkata

Joydeep Mukherjee
School of Environmental Studies, Jadavpur University, Kolkata

ABSTRACT: A study has been carried out to understand the dynamics in monsoonal variation pattern of phytoplankton cell count with different physicochemical parameters with the help of a simple empirical model that has been developed over a tide dominated river, Jagaddal in the Sundarbans Estuarine System or ***SES*** (West Bengal, India). Real time data were also collected from the river during monsoon (July, 2015) which has been used in the model.

Keywords: Phytoplankton cell count, Simple empirical model, tide dominated river

1 BACKGROUND

Sundarbans estuarine system (SES) is one of the most sensitive ecosystems and also represents a unique ecosystem at the land-ocean boundary between the Bay of Bengal and the major river system in India and Bangladesh. It is the largest macrotidal estuary of India and jointly shared between India and Bangladesh. The Indian Sundarbans falls within the geographical limits of the Dampier and Hodges line (imaginary line drawn in 1831) in the northwest, the river Hooghly in the west, the river Icchamati - Kalindi - Raimangal in the east, and the Bay of Bengal in the south (Gurmeet Singh 2009). The ecosystem of this area is complex and dynamic due to strong gradients of nutrients dissolved in water as a result of the variation in tidal dynamics, salinity and other complex hydrodynamic processes. Furthermore, seasonal forcing is important in modifying the local ecosystem of SES.

In recent times, attempts have been made to understand the seasonal variability of tide, salinity, temperature, nutrients and other ecological parameters in SES. However, no study on the factors controlling the nutrient and phytoplankton distribution has yet been carried out in the SES.

Thus, the objective of this study is to understand the effect of temperature, salinity and nutrients on the distribution of surface phytoplanktons (diurnal and seasonal) in the SES. The second objective was to develop a simple empirical model to predict the phytoplankton density in the estuary.

In this approach we first tried to understand how nutrient distribution are exhibited at various locatios with salinity. Then we tried to link up the phytoplankton distribution pattern with respect to salinity. Finally a composite diagram has been formed where linkages betwwen phytoplankton distribution and nutrients becomes clearly understandable.

2 MATERIALS & METHODS

2.1 *Study area*

Our study area was confined to the region of Jagaddal river (Figure 1), a small estuary of the river Saptamukhi (East) which is one of the principal estuaries of SES. The study area is extended from

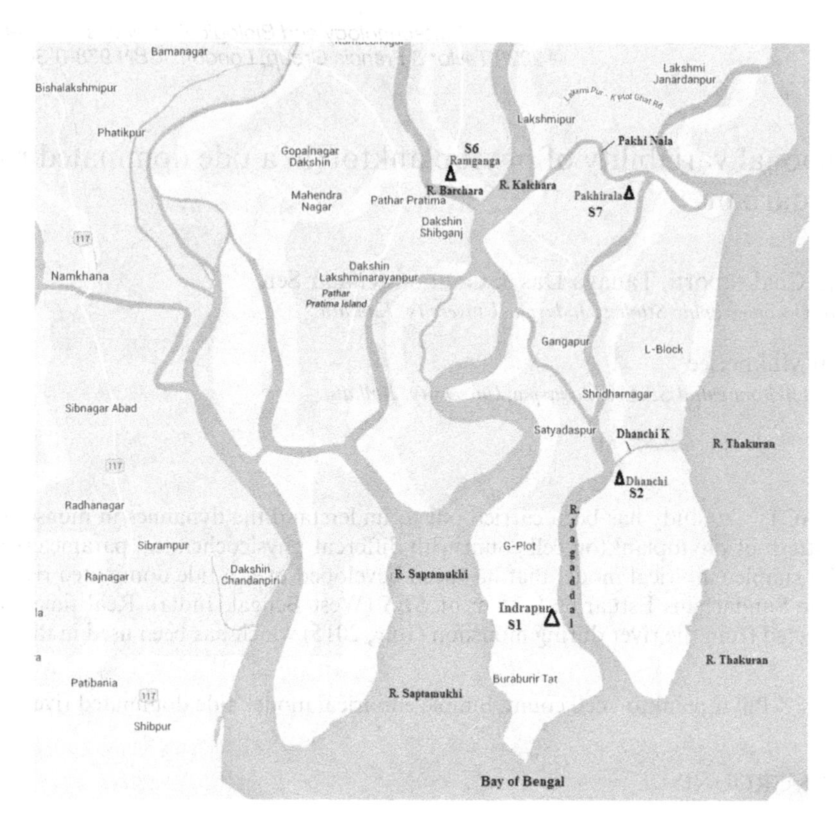

Figure 1. Study Area: Jagaddal River Estuary. Triangles show the observation stations along the river; Abbreviation R. is used to refer "River" ***(Goutam et.al., 2015)***.

Ramganga (21.79944° N, 88.36658° E) to a location **Indrapur**, 15 km away from the Bay of Bengal (21.66872°N, 88.41263°E).

3 SAMPLING AND ANALYTICAL PROCEDURE

Water samples were collected during monsoon (July, 2015 and 2016), at eleven stations which were subdivided into six different zones (S1-S6, where S6-Ramganga, S1-Indrapur) between 21°41.077' to 21°47.9666' N latitude and 88°21.995' to 88°24.758' E longitude (Figure 1) on Jagaddal river (S6:head, north; S1: mouth, south) in Sundarbans estuarine system. Sampling was performed along 15 km stretch to record various physico-chemical (like water temperature, salinity, transparency, nitrate concentration, phosphate concentration etc.) and biological characteristics (phytoplankton cell counts) of water.

4 DATA ANALYSIS

We have used a simple empirical model to predict phytoplankton density. With the objective to understand major contributing factors to phytoplankton density, a correlation matrix was prepared and the corresponding eigen values were computed. Then the loading factors (l.f values) corresponding to the higher eigen values were evaluated. The parameters having l.f values greater than 0.35 were considered to contribute significantly on the phytoplankton density. These loading factors (corresponding to the higher eigen values) were thought to be the most important contributors to the phytoplankton density. To understand the combined effect of these loading factors an index (b3) was calculated. These b3 values show the cumulative effect of the controlling parameters as obtained from the computation.

5 RESULT AND DISCUSSION

Figure 2 compare the index values (b3) with the observed phytoplankton density (for 6 locations, head to mouth) during monsoon and winter. During monsoon, b3 showed the combined effects of **surface temperature, surface nitrate and surface phosphate** which, according to our analysis, have highest influence on the determination of surface phytoplankton density. Now,

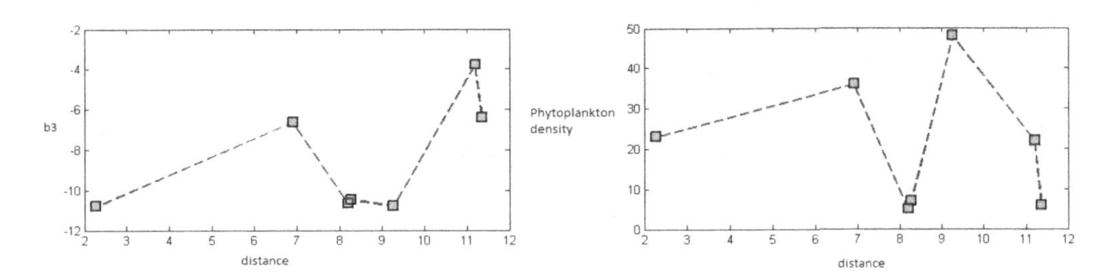

Figure 2. Monsoon observation; b3 curve shows the combined effect of surface temperature, surface nitrate and surface phosphate.

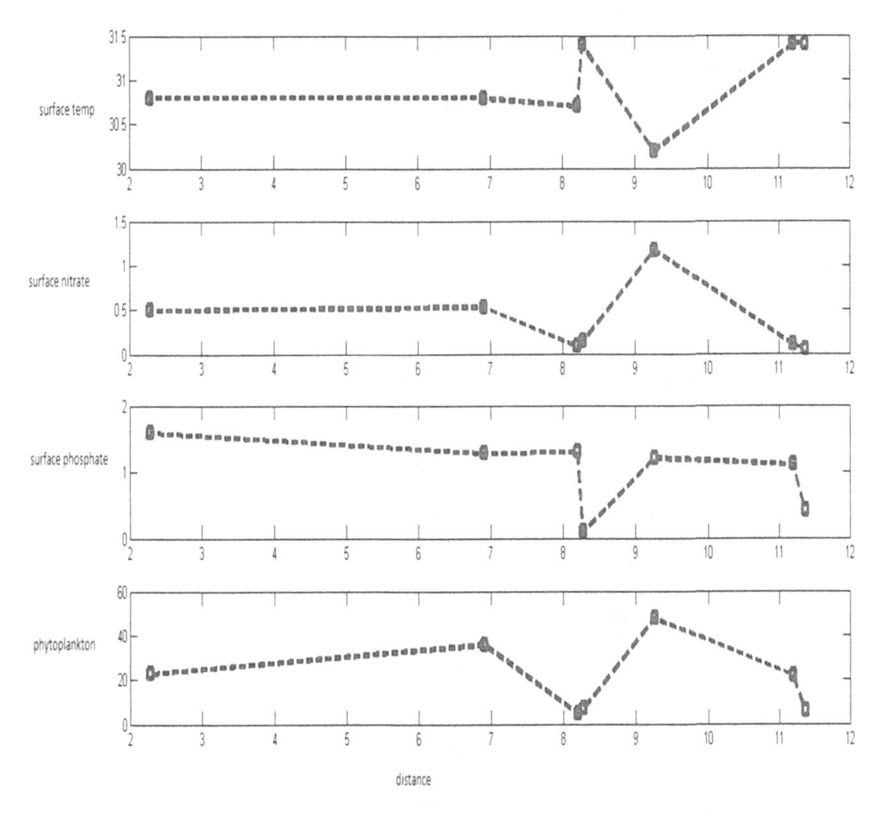

Figure 3. Monsoonal variation of surface temperature, surface nitrate, surface phosphate and phytoplankton.

if we compare both the b3 and phytoplankton density curves, they follow almost similar pattern indicating that surface nutrient concentrations are proportional to the surface phytoplankton density. This fact is in agreement with previous studies (Rahaman et al., 2013; Bhattacharjee et al., 2013) which reported the presence of high nutrients (mainly from mangrove litter which is the principal source of nutrients in mangrove areas) and corresponded to high phytoplankton density.

We further considered the pattern of individual dominant parameters having maximum contribution on b3 curves. We have also plotted the phytoplankton density curve to understand the effect of the individual dominant parameters on phytoplankton density.

During monsoon, surface temperature, surface nitrate and surface phosphate were the dominant controllers on the surface phytoplankton density. It is observed from the Figure 3 that both the surface nitrate as well as phytoplankton density curves followed the same trend which may be indicative of the fact that surface nitrate concentration was the primary contributor to the surface phytoplankton density.

It should be noted that during data collection period in monsoon, there was heavy precipitation. This fact combined with the strong southern wind that blew from Bay of Bengal resulting an unstable situation which was reflected in the fluctuating concentration of the nutrients and phytoplankton at different locations over the study area in monsoon data. The average phytoplankton abundance was lower during monsoon probably due to this unstable condition that restricted the phytoplankton survival to a great extent.

6 CONCLUSION

It is evident from our study that, stuarine environmental factors like surface salinity, surface temperature and nutrients have considerable effect on the variation of phytoplankton biomass or cell count. In this attempt, we have studied the spatial variability of the salinity, nutrients, phytoplankton and water temperature during monsoon. From the raw data sets it can be concluded that a strong salinity and phytoplankton fluctuation was present during the study period. Differential contributions of the influencing parameters has lead us to formulate an index (b3) for predicting monsoonal phytoplankton variability. This index can also be used as a predictor of phytoplankton density in future studies.

REFERENCES

Journal Articles

1. Banerjee K., Mondal K., Mitra A. Spatial band tidal variations of physic-chemical parameters in the lower Gangetic Delta Region, West Bengal, India. Journal of Spatial Hydrology. Vol.11, No.1 Spring 2011.
2. Bhadury P., Bhattacharya D., Samanta B., Danda A., Temporal succession of phytoplankton assemblage in a tidal creek system of the Sudarbans Mangroves: An integrated approach; International Journal of Biodiversity. doi.org/10.1155/2013/824543.
3. Bhattacharya M., Manna S., Chowdhury K., Bhattacharya S. Dynamics of Sundarban estuarine ecosystem: eutrophication induced threat to mangroves. Saline Systems 2010, 6:8.
4. Bhunia A.B., Choudhury A. Observations on the hydrology and the quantitative studies on benthic microfauna in a tidal creek of Sagar island, Sundarbans, West Bengal, India. Proc. Indian natn. Sci. Acad. B47 No. 3 pp. 398–407 (1981).
5. Chatterjee M, Shankar D, Sen GK, Sanyal P, Sundar D (2013) Tidal variation in the Sundarbans Estuarine System, India. J Earth Syst Sci 122(4): 899–933.
6. Rahaman S.Md.B., Sarder L., Rahaman Md. S., Ghosh A.K., Biswas S.K., Siraj S.M.S., Huq K.A., Hasanuzaman A.F.Md., Islam S.S. Nutrient dynamics in the Sundarbans mangrove estuarine system of Bangladesh under different weather and tidal cycles. Ecological Processes 2013,2:29.
7. Rahaman SMB., Golder J., Rahaman MS., Hasanuzzaman AFM., Huq KA., Begum S., Islam SS., Bir J. Spatial and temporal Variations in Phytoplankton abundance and species diversity in the Sundarbans mangrove forest of Bangladesh; Marine Science Research & Development 2013, 3:2. Doi. 10.4172/2155-9910.10000126.

8. Rahaman SMB., Golder J., Rahaman MS., Hasanuzzaman AFM., Huq KA., Begum S., Islam SS., Bir J. Spatial and temporal Variations in Phytoplankton abundance and species diversity in the Sundarbans mangrove forest of Bangladesh; Marine Science Research & Development 2013, 3:2. Doi. 10.4172/2155-9910.10000126.
9. Ray S., Mandal S., Ghosh P.B., 2012. Modelling the impact of mangroves on fish population dynamics of Hooghly–Matla estuarine system, West Bengal, India. Procedia Environmental Sciences 13 (2012) 414–444.
10. Trivedi S., Mitra A., Zaman S., Pramanick P., Chakraborty S., Pal N., Fazli P., Banerjee K. Decadal variation of nutrient level in two major estuaries in Indian Sundarbans. Jordan Journal of Biological Sciences; Volume 8, Number 3, September. 2015; ISSN 1995–6673; Pages 231–236.

Books and Documents

1. Danda A.A, Sriskanthan G, Ghosh A, Bandopadhyay J, Hazra S. Indian Sundarbans Delta: A Vision. New Delhi, World Wide Fund for Nature, India. March, 2011.
2. Agricultural Research Data Book, Indian Council of Agricultural Research, New Delhi, 2002.
3. Qasim S.Z., Indian Estuaries. Allied Publishers Pvt. Limited, 2003.
4. UNEP WCMC 1987 (updated May 2011) Sundarbans National Park, West Bengal, India; World Conservation Monitoring Centre, 11p.

Online documents

1. Bandopadhyay B., Planktonic ecology in mangrove forest like Sundarbans estuary West Bengal, India. 0689-B2; http://www.fao.org/docrep/ARTICLE/WFC/XII/0689B2..HTM; 06–04–2015.
2. Bhattacharya M., Manna S., Chaudhuri K., SenSarma K., Naskar P., Bhattacharya S. Interplay of physical, chemical and biological components in estuarine ecosystem with special reference to Sundarbans, India. www.intechopen.com.

Biotechnology and Biological Sciences – Sen et al. (Eds)
© 2020 Taylor & Francis Group, London, ISBN 978-0-367-43161-7

Evaluation of biomolecular characteristics and phytotoxic effects of organic fish co-composts developed from fish industrial processing wastes

L. Aranganathan
Centre for Ocean Research, Sathyabama Institute of Science and Technology, Chennai

Radhika Rajasree. S.R.
Centre for Ocean Research, Sathyabama Institute of Science and Technology, Chennai
Department of fish processing Technology, Kerala University of Fisheries and Ocean Studies (KUFOS)

R.R. Remya
Centre for Ocean Research, Sathyabama Institute of Science and Technology, Chennai

T.Y. Suman
Centre for Ocean Research, Sathyabama Institute of Science and Technology, Chennai
College of Life Science, Henan Normal University, Xinxiang, China

S. Gayathri
Centre for Ocean Research, Sathyabama Institute of Science and Technology, Chennai

ABSTRACT: The present study demonstrated the bio-conversion of fish industrial processing wastes into organic fertilizer through co-composting technology using Sugarcane bagasse and Cowdung substrates. The matured fish co-composts were evaluated for physico-chemical and spectroscopic properties and seed germination assays to confirm non-toxicity for application as organic soil input. Chemical analysis revealed high levels of macronutrients (NPK) in fish and sugarcane bagasse co-compost. In contrast, high level of micronutrients (Mn, Cu, Fe) was estimated in fish and cowdung co-compost. The absorption ratio (E_4/E_6) values confirmed high humification degree and good maturity levels. FTIR analysis detected presence of different functional moieties such as hydroxyl group of alcohols (3280 cm^{-1}); amide I band (1636 cm^{-1}) etc. A sharp intense peak located at 1031 cm^{-1} indicated better degradation of polysaccharides into simple carbohydrates confirming the maturity of the organic matter. Phytotoxicity assays performed with different concentration of aqueous extracts of co-composts demonstrated good seed germination effect on *Trigonella foenum*-graceum confirming non-toxicity and growth promotion effects. Based on the findings of the study, sugarcane bagasse and cowdung could act as better substrates for effective transformation of fish wastes into organic fertilizer for application as organic amendment to enrich soil fertility.

Keywords: Fish waste, Co-compost, Macronutrients, FTIR, soil fertility

1 INTRODUCTION

Sea food processing industries dispose fishery discards and by-products that have composition of variety of species, trash fishes from by-catch and damaged commercial species. The disposal

*Corresponding Author: E-mail: radhiin@gmail.com

of these residual organic matter ranges between 50-70% of the processed fish (Rustad et al., 2011) and of the waste generated, approximately half is equivalent to organic materials (Bugallo et al., 2012). Generally, the processed solid wastes which are rich source of nitrogenous compounds such as protein, amino acids along with oil (Ghaly et al., 2013) are dumped in landfill sites or improperly disposed on lands (Murado et al. 1994). Such activity of disposing raw wastes adversely affect environment by the production of leachate and contamination of soil. Besides, these industries also release effluents that contains large amount of organic matter, small particles of flesh, suspended solids, body fluids etc (Islam et al., 2004) that creates environmental problem due to release of bad odour and emission of noxious gases such as ammonia. Conventionally, fish wastes are used for the production of high-protein animal feeds (Faid et al.,1997) and fish meal. Numerous studies have reported biological method of converting these wastes to ensure effective organic waste management such as biological fermentation of the fish waste using bacterial and fungi species for conversion into liquid fertilizer (Yamamoto et al., 2004) and composting process with different bulking agents such as sawdust and wood shavings (Laos et al.,1998) and seaweed (Illera-Vives et al.,2015) for the successful transformation of the wastes into valuable organic inputs (Han et al.,2014). Also, composting fish wastes was recognized as valid method to convert the waste into valuable soil amendment for utilization in agriculture (Frederick, 1989). Hence the present study involves biological conversion of fish processing wastes using substrates: sugarcane bagasse and cowdung and to analyze the physico-chemical, spectroscopic characteristics, seed germination effects of the mature co-composts which may offer better management of biowastes to produce organic inputs to improve soil nutrition.

2 MATERIALS AND METHODS

2.1 *Co-composting fish wastes*

Marine fish wastes (head, gills, viscera etc.,) were collected from urban fish processing centres of Chennai. The wastes were mixed with substrates:Cowdung and Sugarcane bagassse in 1:2 ratio in well aerated plastic containers and allowed for co-composting process. The mature fish co-composts was collected separately and used for experimental analysis.

2.2 *Physico-chemical analysis*

The matured fish co-composts were analyzed for physic-chemical parameters (The Fertilizer (Control) Order -1985).

2.3 *Spectroscopic analysis*

2.3.1 *Determination of humification*

1g of the fish co-composts were dissolved in 10 ml of 0.5 M NaOH and left for 24 hours in orbital shaker and filtered with Whatmann filter paper (No.2). The filtrate was measured for absorption at 472 nm and 664 nm using photometer and the absorbance ratio E_4/E_6 was calculated for determining the degree of humification (Albrech et al., 2011).

2.3.2 *FTIR*

FTIR spectra was recorded with KBr pellets for each fish co-compost in the range of 4000 cm^{-1}-400 cm^{-1} wavelength using FTIR spectrometer (Perkin Elmer).

2.4 *Seed germination assay*

Seeds of *Trigonella foenum-graceum* were treated with aqueous extract of fish co-composts at different concentrations (0.25 to 1%) in Whatman filter paper placed inside the petridishes and incubated for 72 hours in dark condition (Tiquia et al., 1996). The relative seed germination (RSG) was calculated using the formula;

$$\text{Relative seed germination (RSG\%)} = \frac{\text{No.of seeds germinated in extract}}{\text{No.of seeds germinated in control}} \times 100$$

3 RESULTS AND DISCUSSION

3.1 *Characteristics of the mature co-composts*

The marine fish co-composts developed with sugarcane bagasse and cowdung substrate attained complete maturity after 45 and 30 days of composting process. Both co-composts appeared blackish brown color with fine texture and less odour intensity confirming the organic matter maturity.

3.2 *Physico-chemical analysis*

The fish co-composts developed with sugarcane bagasse and cowdung susbstrates showed variations in the chemical constituents. The level of macronutrients (N,P,K) was found to be slightly high in fish and sugarcane bagasse co-compost but in contrast level of micronutrients (Fe, Mn, Cu) were high in fish and cowdung co-compost. Organic Carbon was also found to be high in fish and sugarcane bagasse co-compost (Table 1).Variations in the concentration of the macro and micro nutrients might be due to the influence of the substrates used in the co-composting process. However, the level of the organic nutrients was found to be satisfactory for application as economically viable organic inputs to agricultural soil for the enrichment of organic nutrients.

3.3 *Spectroscopic characterization*

3.3.1 *Humification index*

The absorbance ratio (A_{472}/A_{664}) represents the degree of organic matter maturity. The ratio value being under 5 is the characteristic of humified material. The HI value was calculated to be 1 and 0.81 for both the co-composts confirming the high level maturity. The low ratio value obtained in the co-composts indicated high degree of aromatic condensation and a higher level of organic material humification (Zbytniewski and Buszewki 2005).

Table 1. Physiochemical analysis of fish co-composts developed with Sugarcane bagasse and Cowdung substrates.

Parameters	Fish waste and Sugarcane bagasse co-compost	Fish waste and Cowdung co-compost
pH	7.40	7.75
EC ms/cm	11.45	6.1
Moisture %	40.71	43.69
Total N %	2	1.48
Phosphorus%	1.95	1.53
Potassium%	0.78	0.35
Calcium%	2.47	1.59
Magnesium%	0.17	0.15
Sulfur%	0.21	0.13
Zinc mg/kg	29.44	26.61
Manganese mg/kg	30.84	109.51
Iron mg/kg	7	624.03
Copper mg/kg	7	11.87
Organic C %	47	39.93
Lead mg/kg	8.48	11.1
Nickel mg/kg	0.81	1.7

3.3.2 *FTIR*

FTIR spectra of the marine fish waste co-compost developed with sugarcane bagasse and cowdung substrates were represented in Figure 1a and b. A broad peak located at 3280 cm^{-} region that corresponds to H-bonded OH group of alcohol or organic acids (Huang et al.,2006). A weak shoulder at 2921 cm^{-1} observed in fish waste and cow dung co-compost was ascribed to aliphatic C-H stretching (Provenzano et al.,2015). A short peak at 1636 cm^{-1} was attributed to amide I band . A weak shoulder at 1542 cm $^{-1}$ recorded in fish and cowdung co-compost was assigned to amide II bond of peptidic material (Ouatmane et al.,2000). A weak signal recorded at 1416 cm^{-1} in sugarcane bagasse trash based compost indicate the C-O stretching of polysaccharide such as cellulose and hemicelluloses. Another minor peak located at 1391 cm^{-1} in cowdung based co-compost indicated the deformation vibration signals of CH groups in cellulose and hemicellulose. A sharp intense peak located at 1031 cm^{-1} represents C-O bonds in polysaccharide (Grube et al.,2006).

3.4 *Seed germination assay*

Phytotoxic studies showed that *Trigonella foenum-graceum* seeds treated with Cowdung based fish co-compost at dose of 0.25% and 1% showed 88 % relative seed germination (RSG). Beside, seeds treated with the same co-compost extract at 0.5 and 0.75% showed 100% RSG. Sugarcane bagasse based fish co-compost at 0.25% showed 100 % RSG but the germination rate decreased with increase in concentration (Figure 2). However, seeds maintained as Control (treatment with distilled water) in both the treatments showed similar RSG (90%).

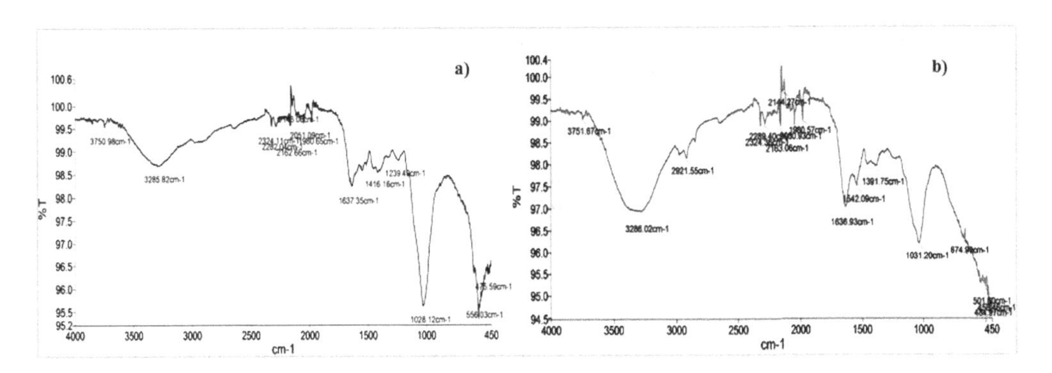

Figure.1. FTIR spectra of marine fish co-compost developed with a) sugarcane bagasse and b) cowdungsubstrate.

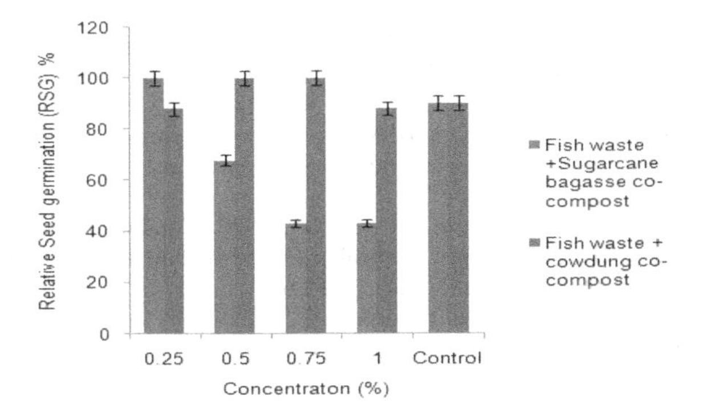

Figure 2. Germination index of *Trigonella foenum-graceum* seeds treated with different concentration of Sugarcane bagasse and Cowdung based fish co-composts.

4 CONCLUSION

Biological transformation of fish processing wastes into valuable organic co-composts was effectively developed using sugarcane bagasse and cow dung substrates. Physico-chemical analysis of the mature co-composts revealed satisfactory level of nutrients especially organic carbon (OC) suitable for soil application. Photometric ratio ($Q_{4/6}$ =1) revealed high humification levels of the organic matter. FTIR spectra of the fish co-composts depicted intense peak at 1031 cm^{-1} (C-O stretch) indicating better degradation of polysaccharides into simple carbohydrates that confirmed greater degree of organic matter maturity. Phytotoxic assays also confirmed no toxicity which further ensured growth stimulation activity of the fish co-composts. Hence it is concluded that co-composting technology could be adapted as a cost effective method for biological transformation of fish processing wastes into valuable organic fertilizer to enrich soil nutrition.

ACKNOWLEDGEMENTS

The authors gratefully acknowledge Department of Science & Technology (No. SSD/SS/033/2011/G) for the financial support and the Management of Sathyabama Insitutute of Science and Technology, Chennai for providing necessary facilities to carry out this work.

REFERENCES

Albrecht, R., Le Petit, J., Terrom, G. and Périssol, C., 2011. Comparison between UV spectroscopy and nirs to assess humification process during sewage sludge and green wastes co-composting. *Bioresource Technology*, *102*(6), pp.4495-4500.

Bugallo, P.M.B., Stupak, A., Andrade, L.C. and López, R.T., 2012. Material Flow Analysis in a cooked mussel processing industry. *Journal of food engineering*, *113*(1), pp.100-117.

Faid, M., Zouiten, A., Elmarrakchi, A. and Achkari-Begdouri, A., 1997. Biotransformation of fish waste into a stable feed ingredient. *Food chemistry*, *60*(1), pp.13-18.

Frederick, L.L., 1989. Compost solution to dockside fish wastes. University of Wisconsin Sea Grant Advisory Services, Madison, WI, pp. 1–12.

Ghaly, A.E., Ramakrishnan, V.V., Brooks, M.S., Budge, S.M. and Dave, D., 2013. Fish Processing Wastes as a Potential Source of Proteins. *Amino Acids and Oils: A Critical Review. J Microb Biochem Technol*, *5*(4), pp.107-129.

Grube, M., Lin, J.G., Lee, P.H. and Kokorevicha, S., 2006. Evaluation of sewage sludge-based compost by FT-IR spectroscopy. *Geoderma*, *130*(3-4), pp.324-333.

Han, W., Clarke, W. and Pratt, S., 2014. Composting of waste algae: a review. *Waste management*, *34*(7), pp.1148-1155.

Huang, G.F., Wu, Q.T., Wong, J.W.C. and Nagar, B.B., 2006. Transformation of organic matter during co-composting of pig manure with sawdust. *Bioresource Technology*, *97*(15), pp.1834-1842.

Illera-Vives, M., Labandeira, S.S., Brito, L.M., López-Fabal, A. and López-Mosquera, M.E., 2015. Evaluation of compost from seaweed and fish waste as a fertilizer for horticultural use. *Scientia Horticulturae*, *186*, pp.101-107.

Islam, M.S., Khan, S. and Tanaka, M., 2004. Waste loading in shrimp and fish processing effluents: potential source of hazards to the coastal and nearshore environments. *Marine pollution bulletin*, *49* (1-2), pp.103-110.

Laos, F., Mazzarino, M.J., Walter, I. and Roselli, L., 1998. Composting of fish waste with wood by-products and testing compost quality as a soil amendment: experiences in the Patagonia region of Argentina. *Compost Science & Utilization*, *6*(1), pp.59-66.

Murado, M.A., Siso, I.G., Gonzalez, P. and Montemayor, I., 1994. A simple form of immobilisation and its effects on morphologic trends and metabolic activity of pellet-forming microfungi. *Bioresource technology*, *48*(3), pp.237-243.

Ouatmane, A., Provenzano, M.R., Hafidi, M. and Senesi, N., 2000. Compost Maturity Assessment Using Calorimetry, Spectroscopy and Chemical Analysis. *Compost Science & Utilization*, *8*(2).pp 124-134.

Provenzano, M.R., Carella, V. and Malerba, A.D., 2015. Composting Posidonia oceanica and sewage sludge: chemical and spectroscopic investigation. *Compost Science & Utilization*, *23*(3), pp.154-163.
Rustad, T., Storrø, I. and Slizyte, R., 2011. Possibilities for the utilisation of marine by-products. *International Journal of Food Science & Technology*, *46*(10), pp.2001-2014.
The Fertilizer (Control) Order (1985), "Ministry of agriculture and rural development", (Department of Agriculture and Cooperation), Government of India, Document No.11-3/83-STU, New Delhi, pp. 1-91.
Tiquia, S.M., Tam, N.F.Y. and Hodgkiss, I.J., 1996. Effects of composting on phytotoxicity of spent pig-manure sawdust litter. *Environmental Pollution*, *93*(3), pp.249-256.
Yamamoto, M., Saleh, F., Ohtsuka, A. and Hayashi, K., 2005. New fermentation technique to process fish waste. *Animal Science Journal*, *76*(3), pp.245-248.
Zbytniewski, R., Buszewski, B., 2005. Characterization of natural organic matter (NOM) derived from sewage sludge compost. Part 1: chemical and spectroscopic properties. *Bioresource Technology* 96,pp. 471-478.

Biotechnology and Biological Sciences – Sen et al. (Eds)
 ISBN 978-0-367-43161-7

Effect of soil arsenic on stress markers of plant – The analysis between a hyper and non-hyper accumulator plant of West Bengal, India

Sonali Paul, Tirtha Tarafdar, Ankita Dutta, Monidipa Roy & Susmita Mukherjee*
Department of Biotechnology, University of Engineering & Management, New Town, India

ABSTRACT: Arsenic (As) is a long drawn problem of the country and state of Bengal. Arsenic contamination can happen by natural as well as anthropogenic sources. Contamination through food chain happens primarily due to the soil As. Reclamation of As-contaminated soil by the engineered device is expensive hence the general practice is phytoextraction by hyperaccumulator plants as these do not show much symptoms of stress. Present study involves a hyper and a non-hyperaccumulator plant for phytoextraction study to analyze the stress responses. The results show that the non-hyperaccumulator plant has increased accumulation of Proline, MDA, phenol the stress biomarkers and lower value of the concentration of protein and chlorophyll compared to the hyperaccumulator Brassica used in the study. Total biomass was more in the non-hyperaccumulator plant in lower As dose. Hence as an alternative to the conventional approach, selected varieties of non-hyperaccumulator plants may be used for phytoextraction of As.

Keywords: Hyper & Non-hyper accumulator, Proline, Malonaldehyde, Oxidative stress, Acclimatization

1 INTRODUCTION

Arsenic (As) has long back been identified as a carcinogen and the elevated levels of As in the ecosystem is a matter of great concern both for environment and public health point of view (Hingston et al, 2001). As is a non-essential element for plants, plants take up As from soil majorly in inorganic forms with the help of various transporter proteins depending on the concentration gradient between the source and the sink. The mechanism of As uptake by plants varies with the As species, it has been reported that As (V) uses Pi channels and replaces Pi (Ghosh et al, 2015).

As stress can affect growth and productivity of the plants by a plethora of physiological and biochemical alterations and the most damaging one is the production of reactive oxygen species (ROS) (Singh et al, 2017). In order to cope with such stress plants sometimes develop various tolerance and adaptive mechanisms that also involve a series of physiological and biochemical changes that can effect As uptake by plants (Khalid et al, 2017).

In most of the work involving phytoextraction of As from contaminated site hyperaccumulator plants are used, but sometimes hyperaccumulator plants are not biomass generating, hence exploring the use of non-hyperaccumulator as an alternative approach might be helpful (Lucas et al, 2013). In a study by Selva raj et al (2015) between *Brassica juncea* and *Abelmuscus esculanta,* a hyper and hypo accumulator plant, it was found that, when co-cultivated *Abelmuscus* shows less toxicity as *Brassica* being hyper accumulator saves *Abelmuscus esculentus.*

*Corresponding author: E-mail:susmita.mukherjee@uem.edu.in

Under such circumstances the present work was planned to compare the extent of change on the biochemical stress markers like, total protein, pigment content and so on between *Brassica juncea* (Indian black mustard) and *Arachis hypogaea* (Indian peanut) both treated with two equal doses of As, in order to understand the difference in stress responses and also to understand the stress adaptations between a hyper and non-hyper accumulator plant to get an idea whether non-hyperaccumulator can be used for phytoextraction.

2 MATERIALS & METHOD

Choice of plants: *Brassica juncea* belongs to family brassicaceae is a hyperaccumulator plant and is a very important oil crop. Peanut (*Arachis hypogaea*) is another oilseed plant used as a non-hyperaccumulator of the family fabaceae.

The seeds of the sample plants were sown in the earthen pots in triplicates. Both the types of plants were exposed to the two different concentrations of Arsenic, 20 mg/kg as tolerant and 40 mg/kg as toxic dose. The plants were exposed under natural photoperiod of 12 to 14 hours and temperature of 32 ± 2°C. The toxic dose exposure was given for 7 days and tolerant dose exposure was for 15 days. After treatment the plants were sacrificed for biochemical assay.

2.1 *Assay of biochemical parameters*

2.1.1 *Estimation of proline content*

The sample extract was prepared in 3% aqueous sulfosalicylic acid then the required reagents were added followed by 1 hour treatment in ice bath. The absorbance was recorded at 520 nm. and concentrations were calculated by plotting against a standard curve (Sadasivam and Manickam 2008).

2.1.2 *Estimation of malondialdehyde content*

The sample extract was prepared and 20% of TCA and 0.5% TBA were added to it. The mixture was boiled and then quickly cooled on ice and centrifuged. The supernatant was collected and the absorbance was recorded at 532 nm. The concentration was calculated against standard (Zhang and Huang 2013).

2.1.3 *Estimation of chlorophyll content*

The leave extract was prepared by repeated centrifugation till the residue becomes colorless. The absorbance was measured at 645 nm and 663 nm to calculate the total chlorophyll concentration (Sadasivam and Manickam, 2008)

2.1.4 *Estimation of total phenol*

The sample extract was prepared in 80% ethanol. Folin-Ciocalteu reagent and sodium carbonate were added to the aliquots, it was treated with boiling water bath and cooled. The absorbance was recorded in 650 nm. And the concentrations were calculated (Sadasivam and Manickam 2008).

2.1.5 *Estimation of total protein content*

The sample extract was collected to it reagents were added and finally, Folin – ciocalteau reagent was added and incubated at room temperature in the dark for 30 min. The intensity of the blue colour was estimated by measuring absorbance at 660 nm. and the protein concentrations were calculated (Sadasivam and Manickam 2008).

3 RESULTS & DISCUSSION

All experiments were conducted in triplicate, the standard deviation was determined, and Student's t test at $p < 0.05$ was applied to find out the statistical significance.

The vegetative growth of both the herbs were compared and it was found that in 20 mg total biomass of Arachis is more than that of Brassica, but in 40 mg the germination rate of Arachis is less compared to Brassica.

Abiotic stress leads to reduced water uptake by root system which causes disturbances in metabolic pathways resulting into oxidative stress and generation of more Reactive Oxygen Species (ROS) (Kosová et al., 2011). Arsenic stress induces cascade of lipid peroxidation reaction resulting into higher concentration of MDA (Singh et al, 2006). In Figure 2 shows high accumulation of MDA in Arachis compared to Brassica, which suggests that Brassica was able to maintain homeostatic control under As stress.

Proline is well known to get accumulated on exposure to abiotic stress in wide variety of organisms ranging from bacteria to higher plants (Saradhi et al., 1993). Proline is a well-known osmoprotectant, cell wall plasticizer, protects cell from ROS mediated damages (Chandrakar et al, 2016). In Figure 2 with increase in As dose proline content increases in both the plants but the increase in proline content in Arachis is 2 times more than Brassica in 20 mg dose.

Arsenic is widely reported to inhibit the rate of photosynthesis in plants (Gusman et al, 2013). Figure 3 shows the comparison between the two plants in chlorophyll content and the symptoms of stress is much on Arachis compared to Brassica.

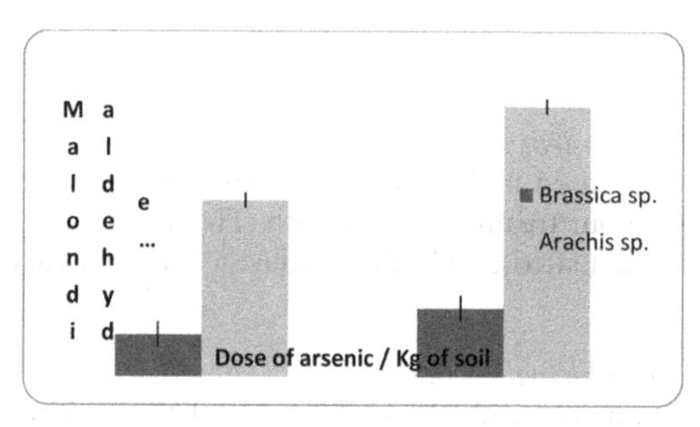

Figure 1. Change in MDA content with different doses of As.

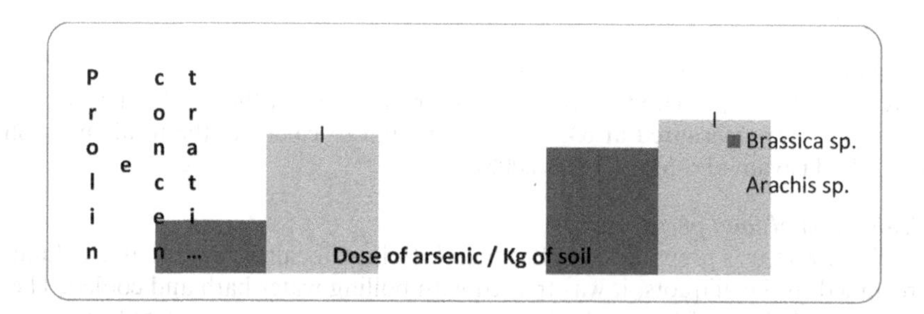

Figure 2. Change in Proline content with different doses of As.

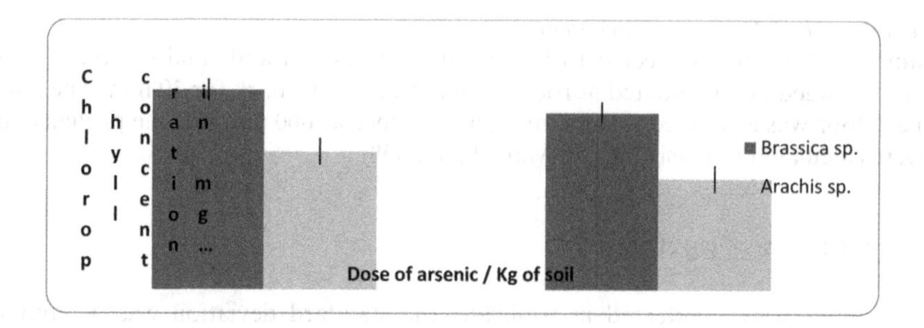

Figure 3. Change in Chlorophyll content with different doses of As.

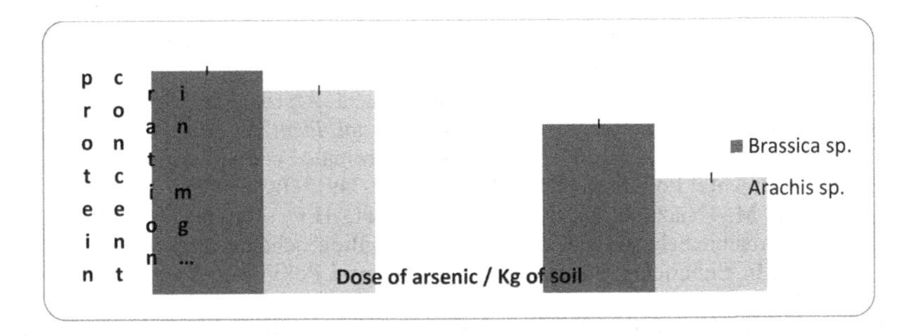

Figure 4. Change in Protein content with different doses of As.

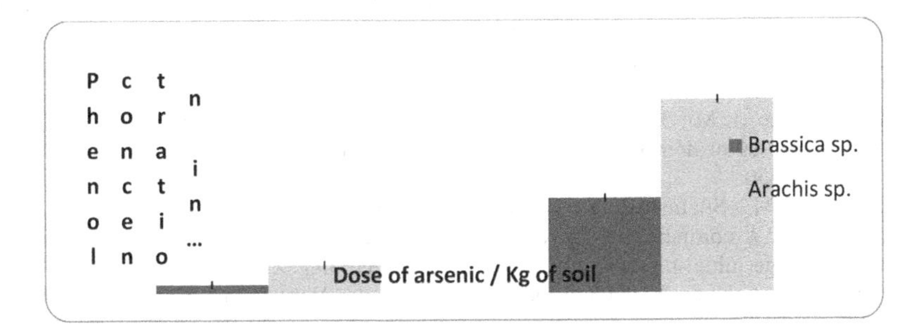

Figure 5. Change in Phenol content with different doses of As.

Kosova et al (2011) showed that even a small change in the ambient environment can lead to profound change in the total protein content to acclimatize stress response. In the experimental plants (Figure 4) decrease in protein content in higher dose may be due to increased protease activity which happens in increased lipid peroxidation an effect of oxidative stress as an initial coping mechanism (Palma et al., 2002).

Under As exposure, plants can induce oxidative stress, which causes an imbalance between ROS generation and ROS scavenging (Rafiq et al, 2017). Phenols are associated with radical quenching activity hence it has got a direct relation with stress in plants. In Figure 5 the phenol content is increased in Arachis compared to Brassica in both the doses of As. It shows Arachis has more stress but it also developed stress quenching mechanism in terms of excess phenol generation.

4 CONCLUSION

From the above analysis if we compare both the plants in two major categories, stress symptoms (Proline, MDA, Chlorophyll and total protein content in this case) and stress quenching response in terms of phenol content. It is clear that stress symptoms are more pronounced in Arachis but the interesting observation is that antioxidant activity is more pronounced in Arachis. Hence comparing the vegetative growth between the two and the biochemical parameters it can be said that Arachis may be used as a new approach for phytoextraction by non-hyperaccumulator in certain cases.

REFERENCE.

Chandrakar, V., Naithani, S.C., Keshavkant, S. 2016. Arsenic-induced metabolic disturbances and their mitigation mechanisms in crop plants: A review. *Biologia*. 71, 367–377.

Ghosh, P., Rathinasabapathi, B., Ma, L.Q. 2015. Phosphorus solubilization and plant growth enhancement by arsenic-resistant bacteria. *Chemosphere* 134, 1–6.

Gusman, G.S., Oliveira, J.A., Farnese, F.S. 2013. Cambraia, J. Arsenate and arsenite: The toxic effects on photosynthesis and growth of lettuce plants. *Acta Physiol. Plant*. 35, 1201–1209.

Hingston J. Collins C.D. Lester J.N. 2001. Leaching of Chromated copper arsenate wood prteservative: A Review. Environmental Pollution. DOI: 10.1016/S0269-7491(00)00030-0

Khalid, S., Shahid, M., Niazi, N.K., Rafiq, M., Bakhat, H.F., Imran, M., Abbas, T., Bibi, I., Dumat, C. 2017. Arsenic behaviour in soil-plant system: Biogeochemical reactions and chemical speciation influences In Enhancing Cleanup of Environmental Pollutants. Springer: Berlin, Germany, 2017; pp. 97–140.

Kosová, K., Vítámvás, P., Prášil, I. T., and Renaut, J. 2011. Plant proteome changes under abiotic stress - Contribution of proteomics studies to understanding plant stress response. *J. Proteomics* 74, 1301–1322. doi: 10.1016/j.jprot.2011.02.006

Lucas A. S, Fernando A. P., Roberta C. N., Ricardo A. A. 2013. Use of non-hyperaccumulator plant species for the phytoextraction of heavy metals using chelating agents, *Sci. Agric*. v.70, n.4, p.290-295.

P. Lee, K. Sreekanth, T. 2010. Heavy metals, occurrence and toxicity for plants: A review. *Environ. Chem. Lett*. 8, 199–216.

Palma, J.M., Sandalio, L.M., Javier Corpas, F., Romero-Puertas, M.C., McCarthy, I., del Río, L.A. 2002. Plant proteases protein degradation and oxidative stress: role of peroxisomes. *Plant Physiol. Biochem*. 40, 521–530.

Rafiq, M., Shahid, M., Shamshad, S., Khalid, S., Niazi, N.K., Abbas, G., Saeed, M.F., Ali, M., Murtaza, B. 2017. A comparative study to evaluate efficiency of EDTA and calcium in alleviating arsenic toxicity to germinating and young Vicia faba L. seedlings. *J. Soils Sedim*. 2017, 1–11.

Sadasivam, S., & Manickam, A. 2008. Biochemical methods. New Delhi: New Age International (P) Ltd., Publishers.

Saradhi, P.P., Alia., Vani, B., 1993. Inhibition of mitochondrial electron transport is the prime cause behind proline accumulation during mineral deficiency in *Oryza sativa. Plant Soil*. 155/156, 465–468.

Selvaraj K, Sevugaperumal R, Ramasubramanian V. 2015. Phytoextraction: Using Brassica as a Hyper Accumulator. *International Journal of Biochemistry & Physiology*, OMICS International, open access, ISSN: 2168-9652.

Singh, N., Ma, L.Q., Srivastava, M., Rathinasabapathi, B. 2006. Metabolic adaptations to arsenic-induced oxidative stress in Pterisvittata L. and Pterisensiformis L. *Plant Sci*. 2006, 170, 274–282.

Singh, A.P., Dixit, G., Kumar, A., Mishra, S., Kumar, N., Dixit, S., Singh, P.K., Dwivedi, S., Trivedi, P.K., Pandey, V. 2017. A protective role for nitric oxide and salicylic acid for arsenite phytotoxicity in rice (*Oryza sativa* L.). *Plant Physiol. Biochem*. 115, 163–173.

Zhang, Z., & Huang, R. 2013. Analysis of malondialdehyde, chlorophyllproline, soluble sugar, and glutathione contentin Arabidopsis seedling. *Bioprotocol*., 3,1–8.

Biotechnology and Biological Sciences – Sen et al. (Eds)
© 2020 Taylor & Francis Group, London, ISBN 978-0-367-43161-7

A comparative study on the effect of metal induced stress in two major vegetable crops of West Bengal, India – Brinjal (*Solanum melongena*) and Chili (*Capsicum annuum*)

Susmita Mukherjee, Surbhi Agarwal, Ankit Chakraborty, Sahil Mondal, Adil Haque, Debangana Bala, Arunima Ghosh, Sweta Singh & Pratik Talukder*
Department of Biotechnology, University of Engineering & Management, Kolkata (Newtown), West Bengal, India

ABSTRACT: Anthropogenic activities introduce different types of heavy metals into the soil. Lead (Pb) is one of the most hazardous heavy metals in terms of environmental load and toxicity. Plants are exposed to lead from soil and aerosol sources. It is a stable heavy metal and any positive role of it in plant growth and metabolism is still not quite known. It can be taken up by plants through cation channels present in root and transported to stem and leaves. Pb toxicity in plants causes induction of oxidative stress and either directly or indirectly induces DNA damage and protein degradation. Taken together, all these detrimental effects result in reduction of crop yield and quality. Copper (Cu) is an essential micronutrient for plant. Plants maintain low copper concentration in the cells because at higher concentrations it behaves as a cytotoxic stress factor by generating Reactive Oxygen Species (ROS) and other free radicals. These free radicals are not only capable of introducing DNA damage but also the other macromolecules including protein; thus renders the cell both structurally and functionally vulnerable. This present study depicts the very initial stages of investigation of cellular macromolecular damage due to metal stress.In recent years, great efforts are made to increase Pb and Cu stress tolerance capacity of plants. This present study depicts the initial stages of investigation of cellular macromolecular damage due to metal stress. This will enable to understand the basic mechanism of metal induced toxicity on plant cell physiology.

Keywords: Heavy metals, Copper, Lead, Chilli, Brinjal, *Solanum melongena*, *Capsicum annuum*

1 INTRODUCTION

Plants and the metals in the soil have had a long and intimate evolutionary association. Due to a sudden surge in unrestricted and unregulated industrialisation and excessive use of agricultural fertilizers, the amount of heavy metal contamination is increasing at an alarming rate. Plants, due to their immobility, are vulnerable to this heavy metal exposure. Heavy metals are defined as those metals and metalloids which have atomic density 5 g cm^{-3} and above and atomic number greater than 20 and should possess properties of metals. Their concentration of less than 0.1% in soil becomes toxic to the plants. Essential Heavy metals are Fe, Mn, Cu, Zn, Co, Mo and non- essential heavy metals are Cd, Pb, Ur, Tl, Cr, Ag, Hg and metalloids such as As, Se. Copper is an essential micronutrient. Being a redox active metal excess

*Corresponding author: pratiktalukder@gmail.com

accumulation of copper generates Reactive Oxygen Species (ROS) [1,2]. At higher concentration it causes cellular toxicity by generating high amount of ROS [3]. Copper toxicity is accompanied with high rate of oxidative stress and lipid peroxidation [4]. Several studies have documented high amount of ROS production with [5]. Lead (Pb) is a potential pollutant and most common among the heavy metals which gets readily accumulated in soil. Lead (Pb), despite not being an essential element for plants, it gets readily absorbed and accumulated in different parts of the plant. Mining, smelting activites, Pb containing paints and gasoline, and municipal wastes enriched in lead has considerably increased the amount of lead accumulation in the environment [6]. In Industrial areas, significant increase in Lead content is observed, and it accumulates in the upper layers of soil and its concentration decreases with the increase in depth of soil [7]. In this study two important vegetable plant which is consumed by us in our daily diet namely, Chilli (*Capsicum annum*) and Brinjal (*Solanum melongena*) were exposed to different concentrations of sub lethal doses of Copper and Lead, to understand the impact of metal stress on plant health.

2 MATERIALS AND METHOD

2.1 *Plant materials and treatments*

Chilli *(Capsicum annum)* (Cultivar BCKV–C1) and Brinjal *(Solanum melongena)* (Cultivar BCKV–B2) [Bidhan Chandra Krishi Viswavidyalaya, India] seeds were used for this study. The seedlings were grown aseptically in the agar-sucrose medium at standard laboratory conditions (temperature 22°C – 25°C, relative humidity 55 – 60% and illumination at 1500 Lux for 16/8 h duration of light/dark photoperiods). 10 day old seedlings were transferred to the soil.

Copper Sulphate Pentahydrate - $CuSO_4$, $5H_2O$ and Lead Nitrate - $Pb(NO_3)_2$ were used as stress inducing elements. Both of these compounds were used in 2 different concentrations – 20 μM, 4 & 60 μM to treat both the brinjal and chilli plants as the LD 50 value were 70 μM and 80 μM respectively for Cu and Pb.

2.2 *DNA extraction*

Genomic DNA was extracted following the method of Edward *et al.*,(1991) with minor modifications [8]. 1 g of tissue of was weighed of the sample and 4 ml freshly prepared DNA extraction buffer was added. The homogenate was then distributed in micro centrifuge tubes and centrifuged at 10000 rpm for 10 minutes at room temperature then supernatant was taken and phenol-chloroform was added 1:1 ratio, then it was again centrifuged. The upper aqueous phase was without disturbing the inter phase of unwanted debris of protein. Then again equal volume (1:1) of phenol-chloroform was added to it. After 10 to 15 min of mixing it was again centrifuged at 10000 rpm for 10 minutes. This step was done twice. Finally the clear upper aqueous layer was collected. To it, 1/10th volume of 3M ammonium acetate was added and then equal volume isopropanol was mixed and shaken gently. The mixture was centrifuged at 10000 rpm for 10 minutes and supernatant was discarded. Any traces of supernatant were completely drained out using tissue paper. 200 μl of 70% chilled ethyl alcohol was added to the pellet and the tubes were tapped gently for 4-5 times to dissolve the pellet completely. The tubes were centrifuged at 10000 rpm for 10 minutes and the supernatant was drained out completely and the pellet was air dried. The air dried pellet was then dissolved in 100 μl of sterile triple distilled water. The DNA was stored at - 20°C.

2.3 *Quantification of genomic DNA*

For quantization of genomic DNA, the absorbance of each of the DNA samples at 260 nm was recorded. 1 OD at 260 nm corresponds to a concentration of 50 μg/ml of DNA. Hence, the DNA content of the solution was calculated by the following formula:

DNA content (µg/ml) = 50 × A_{260} x Dilution factor

Where, dilution factor is (× + n/n), if n µl of DNA solution is mixed with × µl of distilled water for spectrophotometric analysis.

2.4 *Agarose gel electrophoresis of genomic DNA*

The genomic DNA samples were electrophoresed in 1% Agarose gel at 72V for 1 hour and stained with ethidium bromide to observe the quality of DNA, whether clear bands could be visualized or not.

2.5 *Preparation of plant extracts*

The extraction of plant materials was done according to the method of Brolis *et al.*, (1998) with little modifications [9]. For preparation of plant extracts, about 100 mg of tissue from each of the samples was finely crushed and homogenized and dissolved in 1 ml of 50% aqueous ethanol. Then it was taken in graduated tube and ultrasonicated for 20 minutes followed by centrifugation at 10000 g for 5 minutes at room temperature. The supernatant was collected in fresh autoclaved tube and volume was made up to 1 ml with 50% aqueous ethanol.

2.6 *Determination of total protein*

Protein estimation was done according to the method given by Okutucu B. (2017) [10]. 1 ml of Plant extract was mixed with 1ml distilled water and 2 ml of alkaline copper sulphate solution in test tubes. The test tubes were incubated at room temperature for 30 minutes. Then in each test tube, 500 µl of Folin's reagent was added. Again the test tubes were incubated at room temperature for 30 minutes. Absorbance was recorded using spectrophotometer at 660 nm. The concentration of proteins was estimated as BSA equivalent by using BSA Standard curve.

2.7 *Preparation of BSA (Bovine Serum Albumin) standard curve*

BSA was used as a standard to express total protein content in the tissues. For this purpose, a BSA calibration curve was constructed beforehand by taking a range of standard BSA concentrations from 0-1.0 mg/ml. Each of the standard BSA solutions was reacted with alkaline copper sulphate solution and Folin's reagent using the same procedure discussed above and absorbance were recorded at 660 nm.

3 RESULTS AND OBSERVATIONS

3.1 *Determination of concentration of DNA in control and lead treated chilli plants*

SAMPLE	OD VALUE	CONCENTRATION
Chilli Control	0.355	3.57 mg/ml
Chilli Lead	0.026	0.26 mg/ml

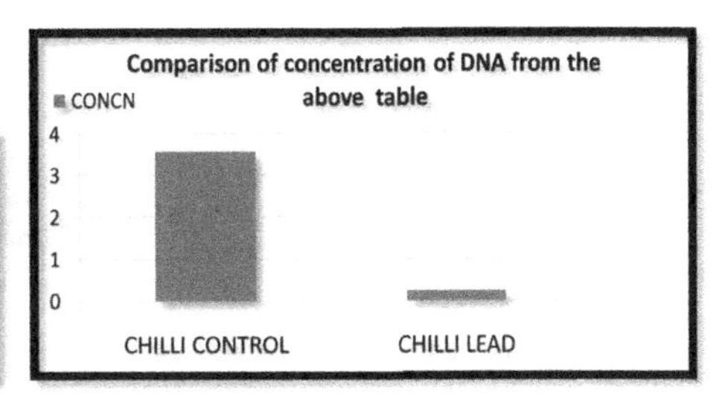

3.2 *Determination of concentration of DNA in control and lead treated Brinjal plants*

SAMPLE	OD VALUE	CONCENTRATION
Brinjal Control	0.36	3.62 mg/ml
Brinjal Copper	0.064	0.67 mg/ml

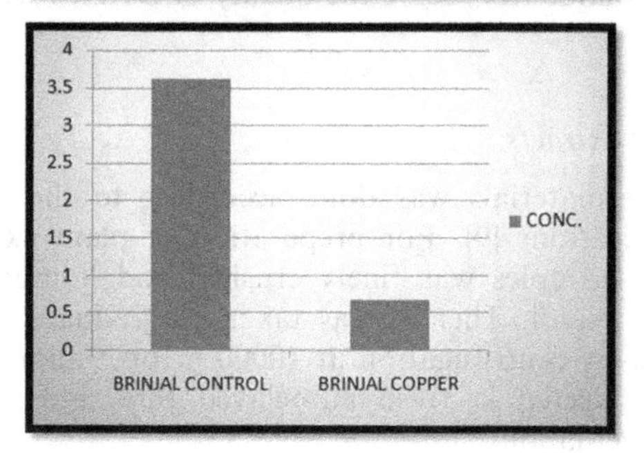

3.3 *Gel electrophoresis*

Agarose Gel electrophoresis of genomic DNA extracted from Brinjal plant

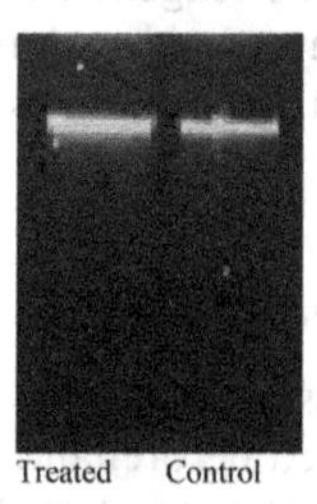

In the control sample the DNA is somewhat intact but it formed a smear in the treated sample.

3.4 *Agarose Gel electrophoresis of genomic DNA extracted from chilli plant*

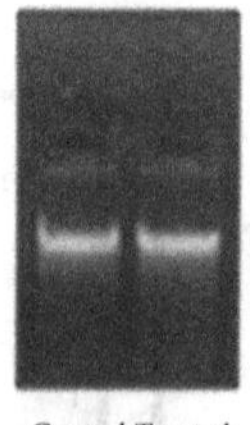

In the control sample the DNA is somewhat intact but it formed a smear in the treated sample therefore higher DNA damage has occurred.

3.5 *Standard curve of BSA*

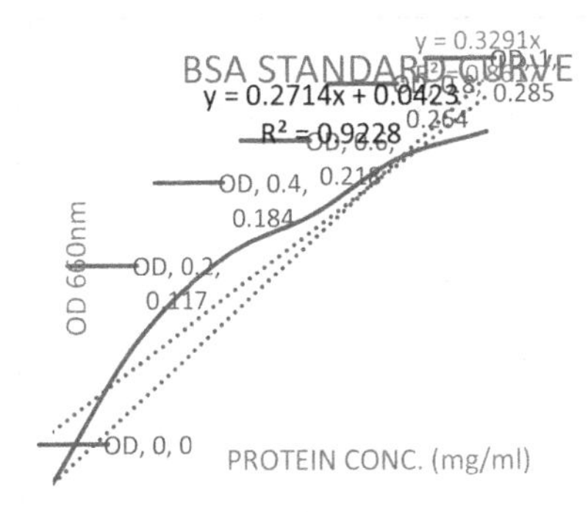

3.6 *Determination of total protein content*

Total protein content in chilli plant

SAMPLE	OD	CONCN.
Chilli control	0.71	1.6 mg/ml
Chilli Lead	0.22	0.41mg/ml

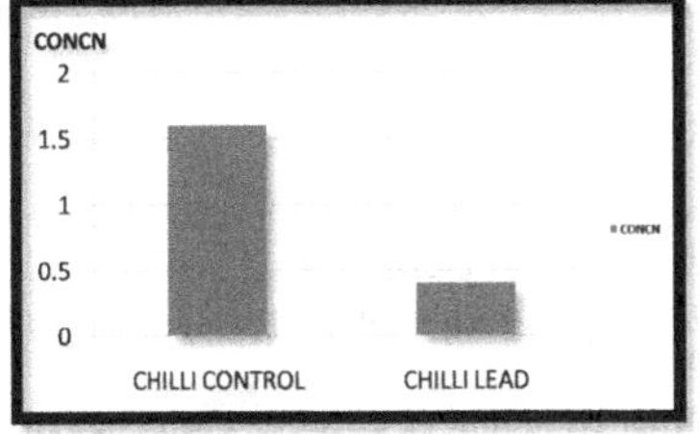

3.7 *Total protein content in Brinjal plant*

SAMPLE	OD	CONC.
Brinjal Control	0.96	2.21 mg/ml
Brinjal Copper	0.78	1.77 mg/ml

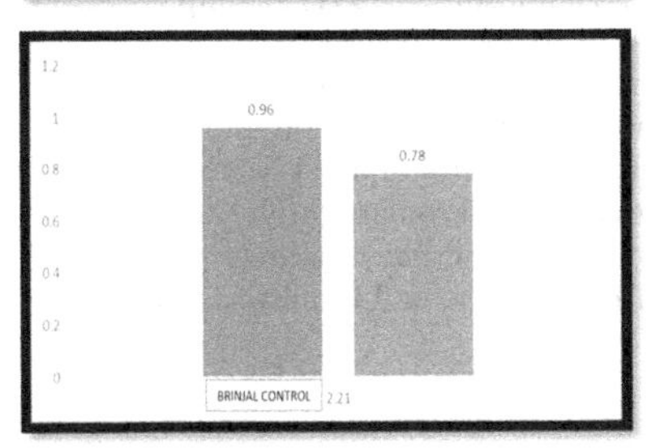

4 DISCUSSION

Heavy metals mainly affect the plants by the following ways- Bind to essential –SH groups of enzymes and inactivate their function, Replace functional elements in prosthetic groups in enzymes and interrupts their catalysis, Generates Reactive Oxygen Species (ROS) like O2.-, OH., H2O2 either by Fenton's reaction or Haber- Weiss reaction. Plants, due to their immobility, are vulnerable to this heavy metal exposure. Heavy metals such as Lead (Pb) generate free radicals and Reactive Oxygen Species (ROS) via Fenton-Haber-Weiss pathway. Metals like Copper (Cu), that are micronutrients required for plant growth can also bring upon the same fate at a higher concentration. The data shows that at higher metal stress (60 μM Cu treatment, 60 μM Pb treatment) both higher DNA damage and lesser amount of protein production were observed.

The results suggest that higher dose of copper and lead has negative effect on plant health. A positive relation between DNA damage and lesser protein production was observed. This study would provide key insight in to the regulatory aspects of cellular macromolecules under transition metal stress and under abiotic stress.

5 CONCLUSION

This summarises the adaptive response of plant under sub lethal dosages of metal stress. It clearly shows that higher concentration of heavy metals causes enhancement in reactive oxygen species (ROS) production which negatively effects the DNA and protein production in the plants. This present study would enable us to understand how the plants respond to metal stress and try to adapt with the increasing concentration of metals in the soil and adjacent environment.

REFERENCES

[1] Cho UH, Seo NH (2005) Oxidative stress in *Arabidopsis thaliana* exposed to cadmium is due to hydrogen peroxide accumulation. Plant Science 168:113–120.

[2] Stohs SJ, Bagchi D (1995) Oxidative mechanisms in the toxicity of metal ions. Free Radical Biology & Medicine 18.2:321–336.

[3] Yruela I (2005) Copper in plants. Brazilian Journal of Plant Physiology17:145–156.

[4] Díaz J, Bernal A, Pomar F, Merino F (2001) Induction of shikimate dehydrogenase and peroxidase in pepper (*Capsicum annuum* L.) seedling response to copper stress and its relation to lignification. Plant Science 161:179–188.

[5] Drazkiewicz M, Skórzyñska- Polit E, Krupa Z (2004) Copper-induced oxidative stress and antioxidant defence in *Arabidopsis thialana*. BioMetals 17:379–387.

[6] Chaney RL, Ryan JA (1994) Risk based standards for arsenic lead and cadmium in urban soils. Dechema, Frankfurt, Germany.

[7] de Abreu CA, de Abreu MF and de Andrade JC (1998) Distribution of lead in the soil profile evaluated by DTPA and Mehlich-3 solutions. Bragantia 57:185–192.

[8] Edwards K, Johnstone C, Thompson C (1991) A simple and rapid method for the preparation of plant genomic DNA for PCR analysis. Nucleic Acids Research 19(6):1349.

[9] Brolis M, Gabetta B, Fuzzati N, Pace R., Panzeri F, Peterlongo F (1998) Identification by high-performance liquid chromatography–diode array detection–mass spectrometry and quantification by high performance liquid chromatography–UV absorbance detection of active constituents of *Hypericum perforatum*. Journal of Chromatography 825:9–16.

[10] Okutucu B. (2017) Comparison of five methods for determination of total plasma protein concentration. Journal of Biochemical and Biophysical Methods. 70(5): 709–711.

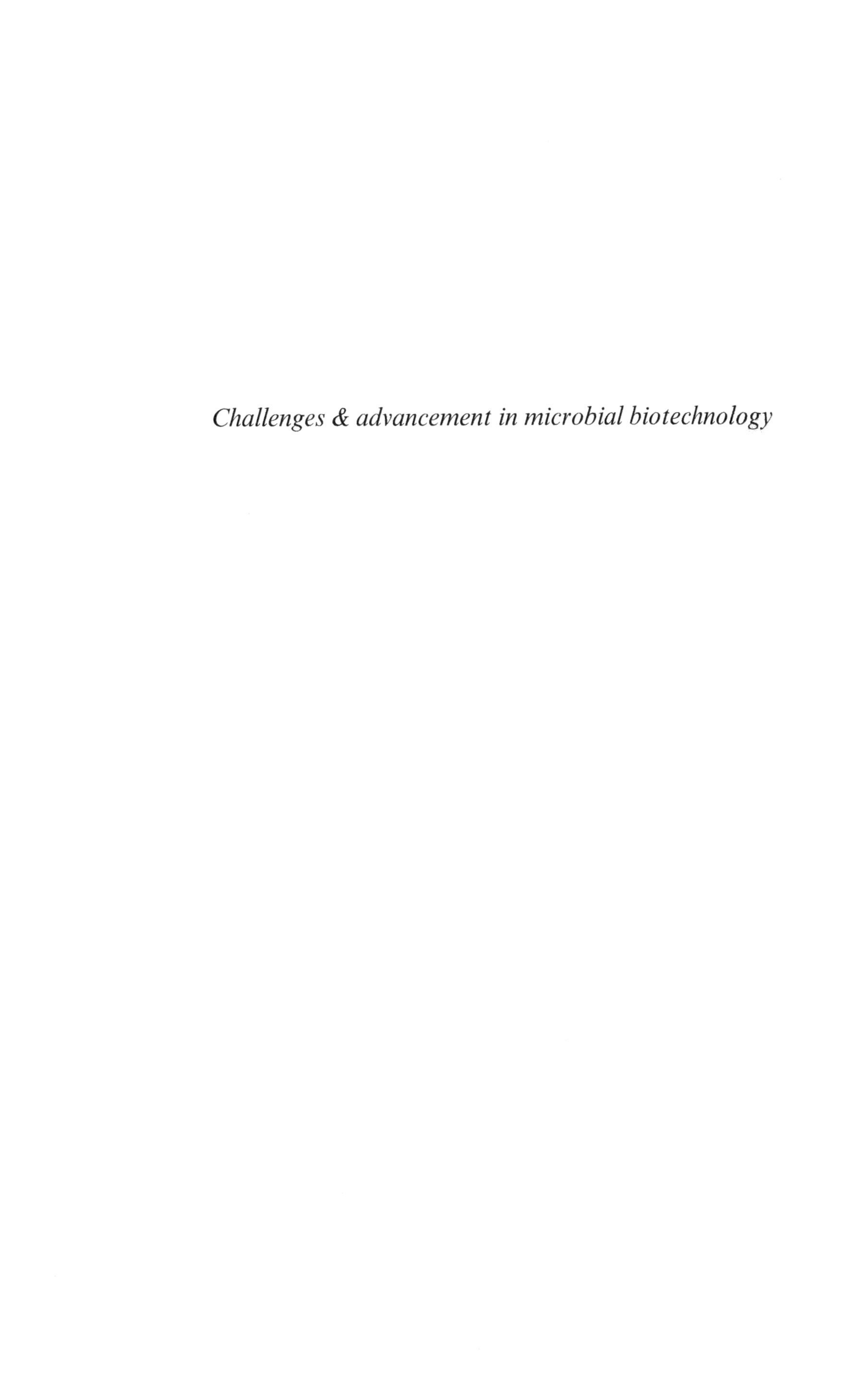

Challenges & advancement in microbial biotechnology

Biotechnology and Biological Sciences – Sen et al. (Eds)
© 2020 Taylor & Francis Group, London, ISBN 978-0-367-43161-7

Study on efficacy of nano-antimicrobials using automated susceptibility testing device

Archi Ghosh
Department of Microbiology,Institute of Post-Graduate Medical Education & Research, India

Mahaua Ghosh Chaudhuri
School of Materials Science & Nanotechnology, Jadavpur University, India

Prasanta Kumar Maiti
Department of Microbiology,Institute of Post-Graduate Medical Education & Research, India

ABSTRACT: Altered cut-off MICs after addition of test drug at a sub-MIC effective concentration with bacterial emulsion fluid has been noted for combination effect study by VITEK for a new drug with all commonly used drugs. A shifting from R to S range can be considered as synergism indicator. Thus tests with several resistant strains bacteria may give interaction results for all drugs. Wide range nonspecific synergism of silver nanoparticles (AgNPs) with different antibiotics were demonstrated by conventional microbroth dilution method. For all practical purpose, an indicative break point synergism with sub-MIC AgNPs and batteries of antibiotics against different multi drug -resistant strain bacteria or yeast can be performed more reliably, using automated susceptibility testing device than manually performed checkerboard method.

1 INTRODUCTION

Silver was known to use for making water portable as early as 1000.B.C. In 1881, German Obstetrician Carl S.F. Crede first used aqueous solution of 1% silver nitrate (AgNO3) for the treatment of ophthalmia-neonatorum. The 1–2% solution of AgNO3 was also used as topical germicides on mouth ulcers, root canals, and for dressing burn wounds[1]. Like many other heavy metals, it attacks bacteria mainly by protein coagulation and blocking sulphydryl group enzymes. Of all other antimicrobial heavy metals, silver is least toxic with potentiality to exert stronger antimicrobial action after conversion into colloidal nanoparticles in reduced stabilized state [2,3]. Silver nano-particles (AgNPs) have greater affinity and permeation on microbial cell membrane due to their greater surface–volume ratio and high positive charge on the capping substance. This results in some structural and functional alterations on surface and subcellular structures of the microbe leading to osmotic dysbalance and altered cell physiology. Oxygen transport enzymes are altered by unstable zero valent silver (Ago) and produce excess reactive oxygen species (ROS) which in turn cause essential, macromolecule inactivation, enzymatic and nuclear dysfunction, apoptosis and membrane disruption of cells [4,5,6,7,8]. But by de-capping and re-oxidation of reduced silver, weak antimicrobial action is likely to happen predominantly in mammalian cells with a tolerable toxicity at low concentrations. This difference may be due to more organized mitochondria and nucleus in eukaryotes. The special importance of AgNPs is intervention of biofilm colonization by greater permeability across matrix and damaging dormant bacteria within biofilms [9]. The nonspecific synergism of multi-targeting AgNPs with most of the antibiotics [10] with selective target of action has increased importance of nano-

*Corresponding Author: E-mail: archighoshkmc@gmail.com

antimicrobials for their use in combinations at low subtoxic dose. Automated antimicrobial susceptibility testing device gives accurate results for batteries of drugs with interpretative MIC values in a rapid effortless manner. Results of an indicative synergism of nano-antimicrobial particles with all pre-programmed drugs can be obtained from same machine by lowering MIC values in pair tests against multi-drug resistant microbes. Modified tests are planned to perform with ultra-diluted nano-preparation at sub-MIC effective concentration in emulsion fluid or sub-MIC nano treated washed microbial cells..

Most of the earlier used nanoparticles prepared with bio-incompatible ingredients, have been restrictedly used topically with risk of much side-effects following systemic absorption and crossing of blood–brain barrier. But for the management of chronic kerato-conjunctivitis, infections of burn wound or skin ulcer, use of biocompatible colloidal AgNPs coupled with antibiotics may be rewarding [11,12], even if used empirically.

The chance of resistance development may also be low due to less drug application and least chance of molecular adaptation by microbes against unusual reaction by reduced state of silver (Ago). The use of an eye tolerable drug, carboxymethyl cellulose (CMC) for nano preparation as capping agent may have advantage of least toxicity following absorption into blood and may be suitable for application on eye or skin. PVP capped silver nanoparticle are suitable for dermatological purposes. Citrate capped silver nano particle may be suitable for systemic use [12].

2 MATERIALS & METHODS

2.1 *Preparation of silver nanoparticles*

To prepare CMC capped silver nanoparticles by chemical reduction method, sodium carboxy methyl cellulose is used as primary capping and reducing agent and dextrose as additional reducing agent. In the preparation of poly vinyl pyrrolidone (PVP) capped silver nanoparticle, PVP is used as stabilization agent & glucose is used as as reducing agent. In the preparation of human serum capped AgNPs, trisodium citrate is also used as capping and reducing agent.

2.2 *Physical Characterization of AgNPs*

Physical characterization of prepared silver nanoparticles for hydrodynamic size were determined by UV–Vis absorption spectrophotometer (Jasco V 650 UV VIS Spectrophotometer, Japan). A narrow size distribution of the AgNPs was documented the lognormal size distribution curve obtained from DLS (Malvern Zen 3600 Zetasizer, USA) . It was further confirmed by transmission electron microscopy (JEOL JEM 2100 HR with EELS, USA) Zeta potential (Malvern Zen 3600 Zetasizer, USA) ensured high aggregation stability of such mixture of AgNPs in aqueous dispersion, while actual size and shape were determined by scanning electron microscope, transmission electron microscope images.

2.3 *Study on anti-microbial properties*

MIC values of nano-silver preparations were performed by micro broth dilution method against ATCC strains & multidrug resistant clinical isolates *of bacteria and yeast*. Methicillin resistant (MRSA) reference strain *S. aureus ATCC-43300* (Microbiologics-Inc, USA) and multidrug resistant clinical isolates,ATCC strain of *S. aureus, Escherichia coli, Pseudomonas. aeruginosa, Candida albicans* were included as test organisms.

Minimum inhibitory concentrations (MIC) for native silver nitrate solution and equivalent silver-containing nano-particles were determined by twofold serial broth micro-dilution method. For each microorganism, 100 μl aliquots of serially diluted AgNPs in de-ionized water were added in one row of 96 wells plates to 100 μl of 0.5McFurland standard (Mf) bacterial/1 Mf yeast suspension indouble strength Mueller–Hinton broth. Thus, about 105–106CFU/mL bacteria or yeast were added into each well. Afterincubation at 37 °C for 24 h/(48 h for yeast), plates were observed for turbidity to note end points as MICs. The endpoints were also cross-checked by subculture.

Their cut-off MIC values for "susceptible" or "resistant" status with batteries of antimicrobial agents were obtained from VITEK-2 automated system. Same were repeated after adding PVP, CMC capped AgNP solution to bacterial emulsion fluid up to ½ MIC so that effective concentration for combination study attained ¼ MIC. Due to rejection of serum capped AgNPs by VITEK -2, microbes emulsified with that enriched solution were incubated at 37° C for 3 hours, washed thrice with emulsion solution and subjected for testing of nano-primed cells in VITEK system after turbidity adjustment.

Synergism study of silver nanoparticles with antibiotic or antifungal agents are performed by checkerboard method (manually). Synergism study of silver nanoparticle with antibiotic or antifungal agents by automated VITEK -2 & manually checkerboard method are done to compare the experimental results.

TEM study are also performed to evaluate the synergistic effect of silver nano particle with antimicrobial agents on the microbes.

3 RESULTS & DISCUSSION

PVP, CMC and serum & citrate capped AgNPs showed average size of 20,9.5,10 nm with triangular, triangular, triangular shape respectively. Their UV-absorption spectrum were 410 nm, 409nm, 420 nm respectively. Zeta potential values of prepared silver nano particle were -28 mV,-25 mV, -30 mV respectively.

Nano transformation was confirmed by physical parameters of CMC, PVP, Citrate stabilized silver nano-antimicrobial. By biological characterization, it showed 128–512-fold higher

Table 1. Shifting of MICs after addition of ¼ MIC AgNP (Break-point synergism) with different resistant antimicrobials in VITEK study.

	Enhanced antimicrobial activities in combination with ¼ MIC AgNPs, compared with untreated [value within () indicates cutoff MIC for the drug alone in ≥ mg/L for susceptible & ≤ for resistant drugs]						
Organism	No indication of change for sensitive drugs with cutoff ≥	≥2 folds dil.	≥4 folds dil	≥8 folds dil.	≥16 folds dil.	≥32 folds dil.	≥64 folds dil.
MDR *S. aureus*	T(1),TS(10),GEN (0.5), CIP(0.5),E (0.25)	TP(1),NFT(32), LF(0.25),BP (0.25)	LZ (8)	OXA(4)	DM(8),CD(4)		VAN (32),RIF (4), TG (32)
MDR *E. coli*	TG(0.5),CL(0.5)			NA(32),NFT (128), CPS(64)	AP(32), AM-CL(32), CIP (4), TS(320), GEN(16)	PP-TZ (128), AK (64)	CFM (64), CFM-AX (64), CTR (64), CPM (64), MPM (16),
MDR P. aeruginosa	CL(0.5)		DOR (8), LF (8)	CPS(64),CFZ (64), AZ(64), IPM(16), MIN (16),AZ(64)	CIP (4), TCCL (128), GEN (16), TG (16), TS(320)	AK (64), TG (16)	CPM (64), MPM (16)
MDR *C. albicans*		VCZ(0.25), CPF(0.25)		FC(8),MF(0.5)	AMB (8)		FCZ (64)

MDR = Multi drug resistant; AK = Amikacin, AM-CL = Amoxicillin Clavulanic acid, AMB = amphotericin B, AP = Ampicillin, AZ = Aztreonam, BP = Benzyle Penicillin, CD= Clindamycin, CFM = Cefuroxime, CFMAX = Cefuroximeaxetil, CFZ= Ceftazidime, CIP = Ciprofloxacin, CL = Colistin, CPM = Cefepime, CPF = capsafungin, CPS = Cefoperazone-Salbactum, CTR = Ceftriaxone, DM = Daptomycin, DOR=Doripenem, E = Erythromycin, ERT = Ertapenem, FC = flucytosine, FCZ = Fluconazole, GEN = Gentamicin, IPM = Imipenem, LF= Levofloxacin, LZ = Linezolid, MF = micafungin., MIN= Minocycline, MPM = Meropenem, NA= Nalidixic acid, NFT = Nitrofurantoin, OXA= Oxacillin, PP-TZ = Piperacillin- Tazobactam, RIF = Refampicin T = Tetracycline, TCCL= Ticarcillin- clavulanic acid, TG = Tigecycline, TP = Teicoplanin, TS= Trimethoprim-Sulfamethoxazole, VAN = Vancomycin, VCZ = voriconazole,

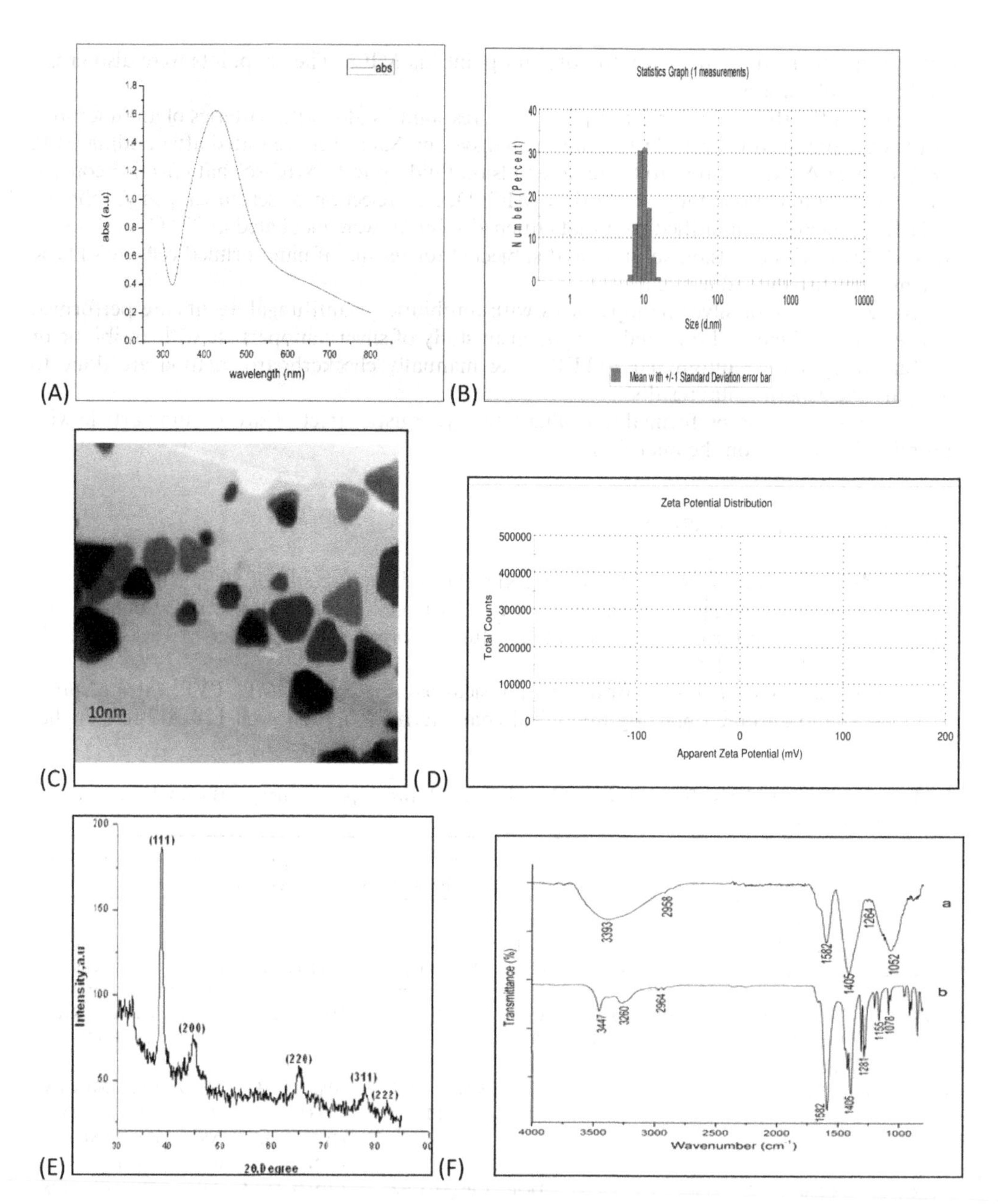

Figure 1. (A) UV-Vis absorption spectra of citrate capped silver nanoparticles at 410 nm and narrow distribution (B) TEM image of triangular shaped citrate capped silver nanoparticles, (C) Size distribution obtained from DLS measurements of Citrate capped silver nanoparticles. (D) Zeta potential of citrate capped silver nanoparticles . (E) XRD of Citrate capped silver nanoparticles. F) (a) FTIR spectrum of citrate capped silver nanoparticles. (b)FTIR spectrum of pure trisodium citrate.

antimicrobial action than that of equivalent ionic silver solution, what we call as “bonus effect” and taken as the microbiological marker of nano-antimicrobials. For individual micro-organism “bonus effect” was almost constant for test repetitions, as AgNPs exerted multi-targeted action. Slight difference for different organisms might be due to their difference of cell wall integrities. The MICs of the AgNPs for different test bacteria were about 0.02_mg/L and 0.01_mg/L for candida strain.

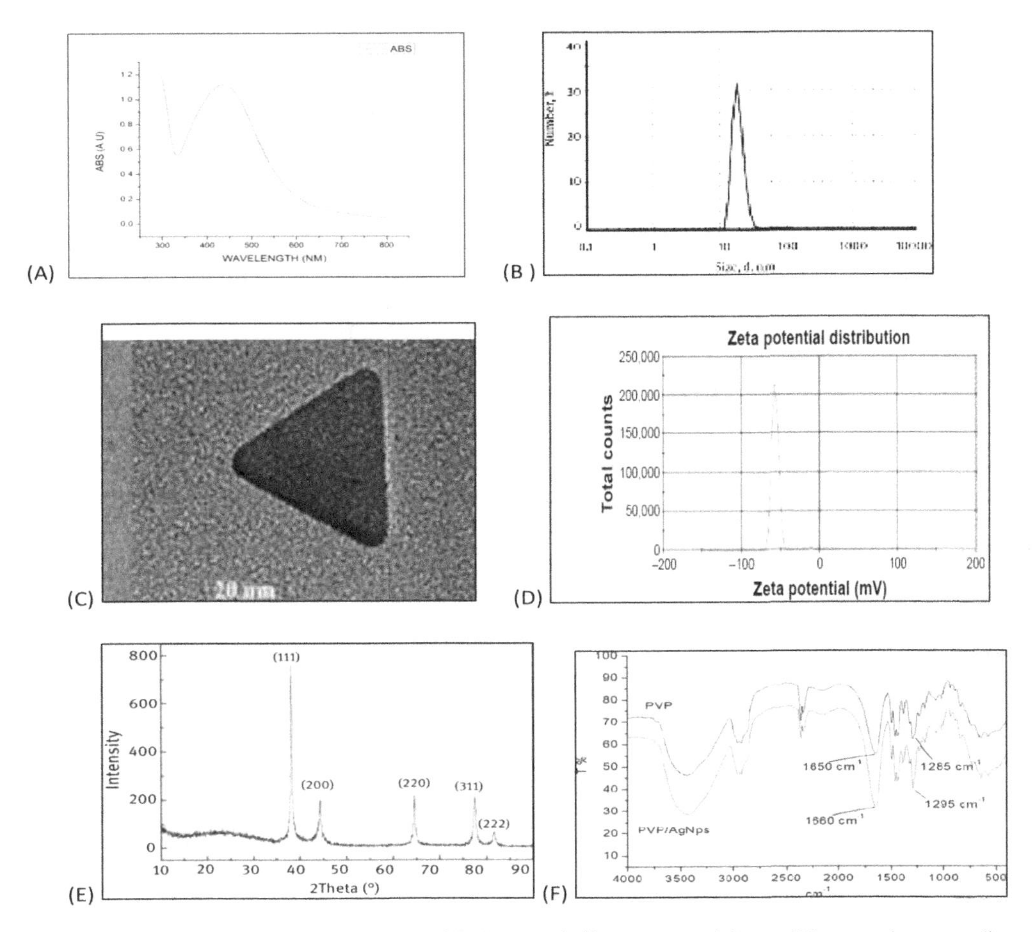

Figure 2.(A). UV-Vis absorption spectra of PVP capped silver nanoparticles at 420 nm and narrow distribution (B) TEM image of triangular shaped PVP capped silver nanoparticles, (C) Size distribution obtained from DLS measurements of PVP capped silver nanoparticles. (D) Zeta potential of PVP capped silver nanoparticles. (E) XRD of PVP capped silver nanoparticles. (F) FTIR spectrum of pure PVP and PVP capped silver nanoparticles.

Instead of checkerboard method, the serial dilutions of two drugs to observe combined effect of another antibiotic, when we carried out simultaneous study for good number of antibiotics individually coupled with a fixed sub-MIC level nano-antimicrobials, a synergism indicative results were obtained in automated susceptibility testing device for all resistant drugs, indicating several folds lowering of MIC values and all reached to cutoff- sensitive MIC level. In combination with ¼ MIC AgNPs with emulsion mixture or nano-primed cells, all organisms shifted from resistance cut-off to sensitive level for all resistant drugs indicating mutual synergisms. However cut-off MIC values for sensitive drugs in automated system remained unchanged.

The synergistic MIC value of antibiotic penicillin and silver nano particle against Pseudomonas aeruginosa, Klebsiella pneumoniae is (A/4 + N/8) and (A/8 + N/4) respectively . The synergistic MIC value of antibiotic penicillin and silver nano particle against Staphylococcus aureus, Enterococcus faecium is (A/4 + N/16) & (A/16 + N/8) respectively in checkerboard analysis.

The change of MIC levels or shifting of "resistant" cutoff to "sensitive" range is qualitative evidence of strong synergism. This simpler method is more reproducible than conventional checkerboard method for two drugs efficacy study by broth dilutions. As wide range of antibiotics with different modes of action and different chemical structures are included, results

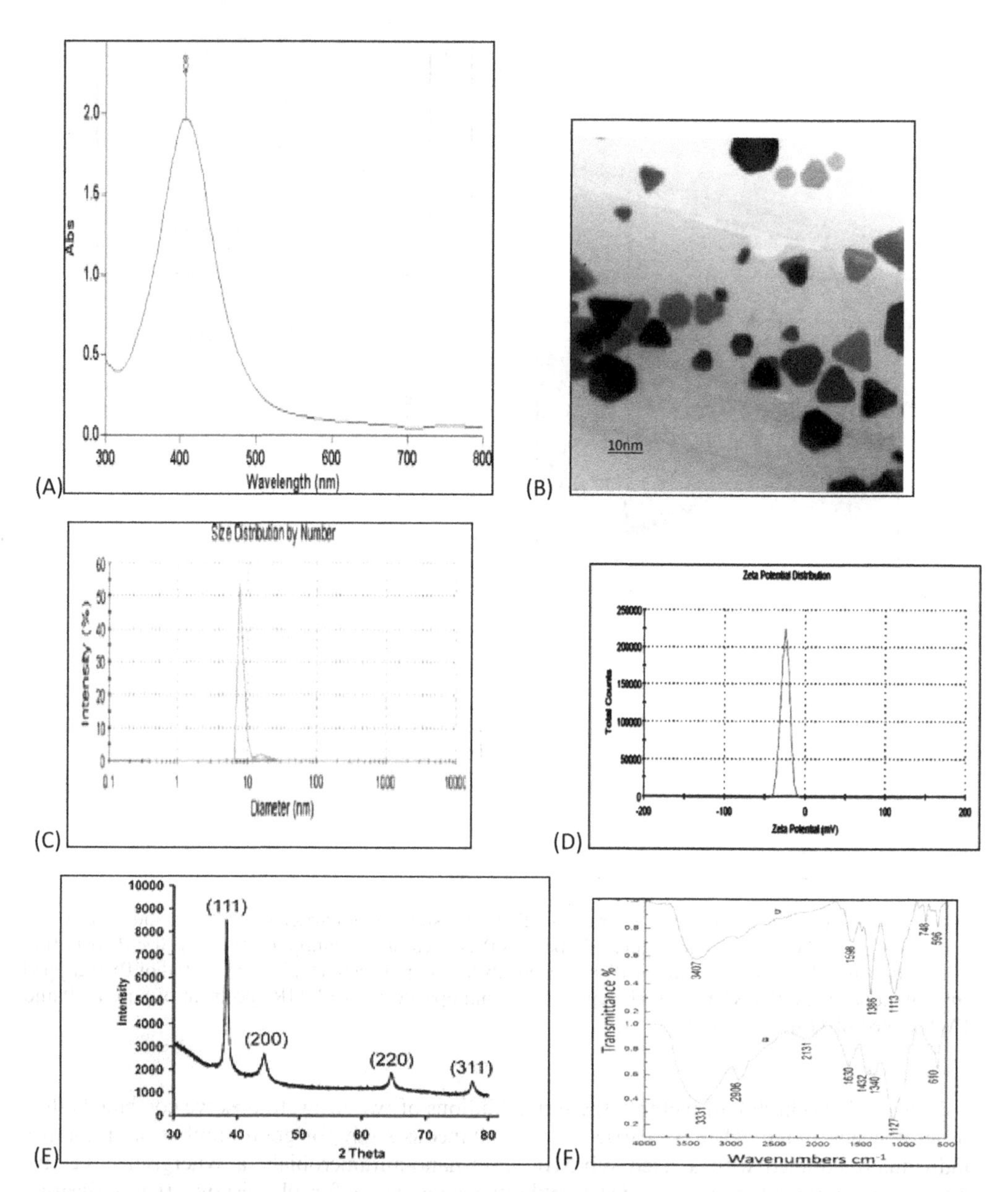

Figure 3. (A) UV-Vis absorption spectra of carboxy methyl cellulose capped silver nanoparticles at 409 nm and narrow distribution [Reference 5](B) TEM image of triangular shaped carboxy methyl cellulose capped silver nanoparticles [Reference 5], (C) Size distribution obtained from DLS measurements of carboxy methyl cellulose capped silver nanoparticles. [Reference 5] (D) Zeta potential of carboxy methyl cellulose capped silver nanoparticles[Reference 5] .(E) XRD of carboxy methyl cellulose capped silver nanoparticles.(F)(a)FTIR spectrum of pure Carboxy methyl cellulose;(b) FTIR spectrum of carboxy methyl cellulose capped silver nanoparticles.

indicate strong non-specific synergism of AgNPs with all resistant antibiotics. Therefore, in two drugs combination study, low or no additive effect is reflected for sensitive drug while for resistant drug greater lowering of MIC points towards degree of synergism, though exact combined effect may be even higher.

4 CONCLUSION

For wide range synergism of antibiotics with silver nano-particles, recovery may be enhanced. Microbes may not develop drug resistance against antibiotic if used in combinations. Exploiting the nonspecific synergism of nanoparticles with wide range anti-microbials, a therapeutic policy can be developed for combination drugs one being sub-toxic concentration of nano-antimicrobials.

REFERENCES

[1] Wright JB, Lam K, Hansen D, Burrell RE (1999) Efficacy of topical silver against fungal burn wound pathogens. Am J Inf Cont 27:344–350.

[2] Matsumura Y, Yoshikata K, Kunisaki S, Tsuchido T (2003) Mode of bactericidal action of silver zeolite and its comparison with that of silver nitrate. Appl Environ Microbiol 69(7):4278

[3] Franci G, Falanga A, Galdiero S, Palomba L, Rai M, Morelli G, and Galdiero M (2015). Silver Nanoparticles as Potential Antibacterial Agents. Molecules; 20, 8856-8874; doi:10.3390/molecules20058856.

[4] Shrivastava S, Bera T, Roy A, Singh G, Ramachandrarao P, Dash D (2007) Characterization of enhanced antibacterial effects of novel silver nanoparticles. Nanotechnology 18(22):225103. https://doi.org/10.1088/0957-4484/18/22/22510 3.

[5] You C, Han C, Wang X, Zheng Y, Li Q, Hu X, Sun H (2012) The progress of silver nanoparticles in the antibacterial mechanism,clinical application and cytotoxicity. Mol Biol Rep 39:9193–9201. https://doi.org/10.1007/s1103 3-012-1792-8PMID:22722 996.

[6] Song K, Lee S, Park T, Lee B (2009) Preparation of colloidal silver nanoparticles by chemical reduction method. Korean J Chem Eng 26:153–155.

[7] Yan X, He B, Liu L, Qu G, Shi J, Hu L, Jiang G (2018) Antibacterial mechanism of silver nanoparticles in pseudomonas aeruginosa: proteomicsapproach. Metallomics 10:557–564.

[8] Akter M, Sikder MT, Rahman MM, Ullah AKMA, Hossain KFB, Banik S et al (2018) A systematic review on silver nanoparticlesinduced cytotoxicity: physicochemical properties and perspectives. J Adv Res 9:1–16.

[9] Ansari MA, Khan AA, Cameotra SS, Alzohairy MA (2015) Anti-biofilm efficacy of silver nanoparticles against MRSA and MRSE isolated from wounds in a tertiary care hospital. Indian J Med Microbiol 33:101–109.

[10] Hwang IS, Choi H, Kim KJ, Lee DG (2012) Synergistic effects between silver nanoparticles and antibiotics and the mechanisms involved.J Med Microbiol 61:1719–1726.

[11] Seil JT, Webster TJ (2012) Antimicrobial applications of nanotechnology: methods and literature. Int J Nanomed 7:2767–2781. https://doi.org/10.2147/IJN.S2480 5.

[12] Maiti PK, Ghosh A, Parveen R, Saha A, Choudhury MG (2019). Preparation of carboxy-methyl cellulose capped nanosilver particles and their antimicrobial evaluation by automated device. Applied Nanoscience. 2019, 9:105-111 doi.org/10.1007/s13204-018-0914-6.

Biotechnology and Biological Sciences – Sen et al. (Eds)
© 2020 Taylor & Francis Group, London, ISBN 978-0-367-43161-7

Microbial cellular machinery for process intensification in production of nutraceuticals

S. Sengupta
School of Community Science and Technology, Indian Institute of Engineering Science and Technology, Shibpur, Howrah

S. Chatterjee
University School of Biotechnology, Guru Gobind Singh Indraprastha University, Sector - 16C, Dwarka, New Delhi, India

ABSTRACT: Microorganisms act as a potential source for the production of different value-added products like enzymes, nutraceuticals, etc. An emerging as well as an important area of research is the food development, food bioprocessing, nutraceuticals, biocolors and bioflavors using microbes. Large scale use and production of nutraceuticals has been restricted due to supply limitations and extraction difficulties. The study provides a comprehensive overview of the role of microorganisms in cellular machinery used for the process intensification. For the production of different nutraceuticals, it may also prove to be an environment friendly alternative approach. Metabolic genes form an integral part in the development of genetic engineering of food-grade microorganisms. Microbial cell factories are designed for the production of nutraceuticals through the application of metabolic engineering. Nutraceuticals form an essential ingredient during the designing of different value-added functional foods. These innovative technologies may serve as a pillar for the process intensification of such compounds.

Keywords: metabolic engineering, nutraceuticals, cellular machinery, value-added products

1 INTRODUCTION

Production of different food constituents, enzymes and nutraceuticals by using microorganisms is quite an interesting area for microbiologists, food technologists and researchers (McNeil, 2013). The name 'nutraceuticals' as coined by Stephen De Felici in the late 20th century, exemplifies an extensive scope of food and different constituents of food possessing various health perks (Pszczola, 1992). The increasing needs and passion in sustaining human wellbeing through the medium of food intake have remarkably boosted the success of the nutraceutical field. As per a latest report issued by the Global Information Inc.', the world nutraceutical share in the market was predicted to outpace $171.8 billion in 2014 and grasp $241.1 billion by 2019, whereas the US market share was approximately $75.9 billion (Jain, 2013). Metabolic construction of nutraceuticals caters an alluring substitute for chemical amalgamation and extraction that facilitates enantiomerically natural compounds to be formed at favorable settings without the need for application of extreme heat or high pressure (Lopes, 2017). A number of important nutraceuticals are complicated molecules that can be produced in an improved manner by means of microbial co-cultures or synthetic leagues (Liu, 2018). The practice of developing genetic and regulatory processes in cells professes remodeling of fundamental metabolism to produce metabolites at a surplus rate. In dispersion through these biomolecules, because of the health caring or disease inhibiting characteristics of nutraceuticals, they have gained appreciable interest. Similarly, the objective of microbial engineering is to produce nutraceuticals which will remain unaffected by the conventional restraints of small production from extractions and complicated chemical reactions (Yuan, 2019).

2 MATERIALS AND METHODS

2.1 *Phytochemicals*

They are a colossal paramount reserve for nutraceuticals consisting of polyphenolic compounds (curcuminoids, isoflavonoids, stilbenoids, flavonoids, etc), terpenoids (monoterpenes, tetraterpenes, carotenoids, saponins, lycopenes, etc), alkaloids and their derivatives (Jain, 2013). The manufacture of nutraceuticals from simple carbon sources are being executed by employing microorganisms as in Figure 1 (Yuan, 2019).

2.2 *Microorganisms as cell factories*

One of the widely studied lactic acid bacteria is the *L.lactis*. Classic as well as dynamic genetic instruments or machineries have been matured to make use of their application in this bacterium. These tools are in high demand because of their application in metabolic engineering blue prints that intents in the inactivation of blackballed genes and/or overexpression of existing genes (Hugenholtz, 2002). Previously, metabolic engineering of *L. lactis* was centered well on the rerouting of pyruvate metabolism. Through the combinative approach, deactivation of ilvB gene which was encrypting for α-acetolactate decarboxylase was accomplished along with the production of diacetyl from glucose and lactose (Hols, P. 1992). Production of nutraceuticals biologically through metabolic engineering strategies has been depicted in Figure 2. Metabolic engineering (ME) can be explained as the guided development of any product design or cellular characteristics via alteration of certain biochemical reactions or the influx of new ones with the application of recombinant DNA technology (RDT) (Gonzalez, 2013).

2.3 *Needs and requirements for process intensification*

In spite of a large number of advancements that has been made in producing and enforcing bioprocesses in industrial markets, still a rising number of conceivably intriguing bioprocesses from an industrial point of view undergoes through a trivial problem. The enzymes (and cells enclosing them) have been previously improved by the transformation in nature over many years. Ideally, utilizing newer and improved (bio)catalysts would seem just perfect until one becomes aware of the fact that the unbiased role in nature is rather posing a contrasting picture from that in a process plant (Burton, 2002).

There are several cases of bioprocess intensification stated in various research studies, although they occasionally unfold the scope of advancements or developments in respect of 'intensification'. Few recorded intensification theories and assimilation of unit operations are largely engaged, even though for other reasons to those in traditional chemical process technology. In contrast, unlike traditional chemical processes, other approaches for intensification have become feasible applying biological catalysts via modification of the enzyme (protein

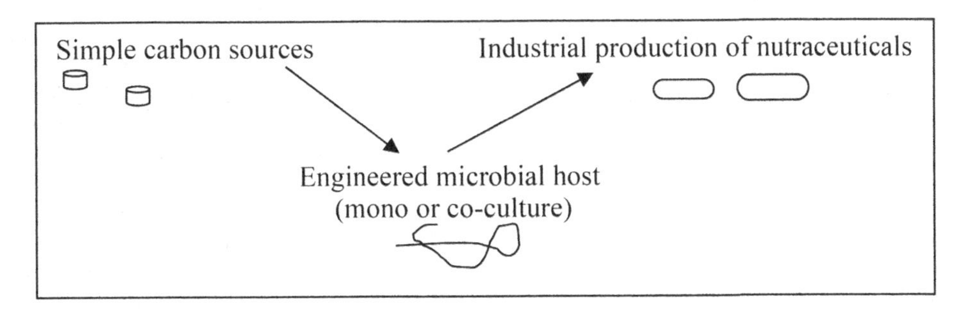

Figure 1. Microbial production of nutraceuticals.

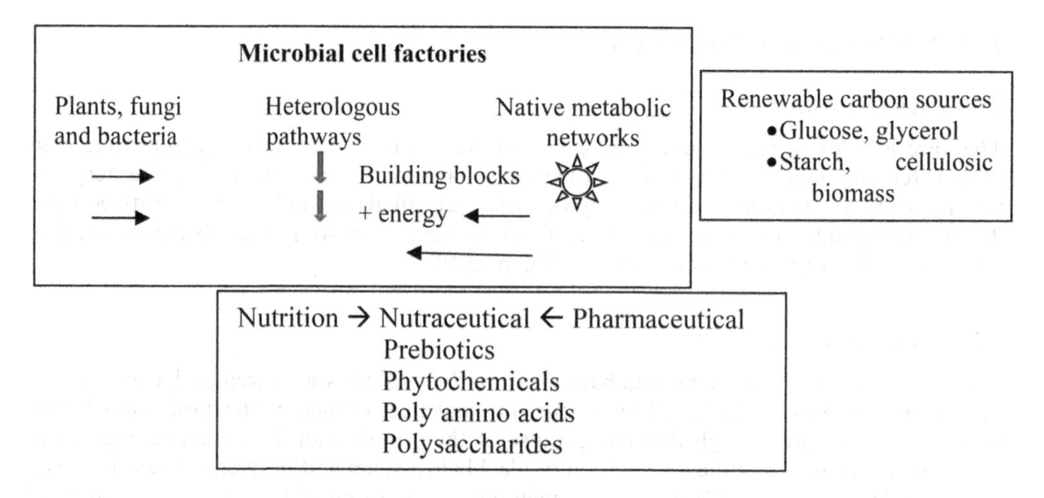

Figure 2. Microbial production of value-added nutraceuticals.

engineering) or in circumstances where cells are the favored biocatalysts by modification of the cell machinery (metabolic engineering) (Woodley, 2017).

2.4 *Production of nutraceuticals through metabolic engineering of microbial cellular factories*

The production of different nutraceutical products employing the concept of metabolic engineering gives an alluring substitute to synthesis of compounds which were otherwise produced through chemical means (Lopes, 2017).

2.4.1 *EPA and DHA*

The biosynthetic pathways of EPA and DHA are usually carried out through the aerobic desaturase/elongase route even though the synthesis is possible via an anaerobic polyketide synthase (PKS) pathway (Xue, 2013). *Y. lipolytica* was used by the DuPont researchers as they exploited this aerobic pathway to produce a strain suited for yielding EPA at 56 % of TFA and about 15 % of the dry cell weight. Considering all known sources of EPA, this was recorded as the highest percentage (Xue, 2013). These researchers subsequentially was able to mature a modified commercial strain (*Y. lipolytica* Z5567) which was able to enhance carbon flux towards the biosynthetic pathway of EPA (Xie, 2016). An EPA enriched oil was produced by this strain when it was cultured utilizing a two-stage fed-batch fermentation process (Zhu, 2015).

2.4.2 β-*Alanine*

From the biochemical aspect, β-alanine is composed by the processes of l-aspartate-α-decarboxylase which had the power to catalyse the decarboxylation phase of l-aspartate. Albeit, a latest research stated that a favorable enzymatic modification was made possible through ADC with a good efficiency (97.2%). However, the process demanded a big ticket as huge amount of enzyme and the precursor l-aspartate were necessary to carry out the work (Shen, 2014). Therefore, this cutting-edge microbial production of β-alanine utilizing inexpensive carbon sources holds a promising future for the commercial enterprises. Future endeavors for developing β-alanine production are now intensified on bacterial hosts as fungal organism (viz. *S. cerevisiae*) and not having orthologs of panD or high fluxes towards precursors (Borodina, 2015). A de novo biosynthetic pathway was lately fashioned for the highest recorded production of β-alanine (32.3 g/L) from glucose utilizing *Y. lipolytica* (Song, 2015).

2.4.3 *Tagatose*

Tagatose is basically a carbohydrate which is claimed to be a very promising sucrose replacer. Till date, there is only one such biochemical pathway; breaking of lactose by lactic acid bacteria or some other microorganisms. Tagatose 6-phosphate pathway (Van Rooijen, 1991) is liable for the degeneration of galactose moiety utilizing L. lactis during lactose metabolism. It was a preferred promising cell factory for the generation of tagatose due to its comparatively simple carbon metabolism, potent gene assimilation structure and lastly, a very good productive system for the surplus production of enzymes. L. lactis was able to revamp it into a tagatose (phosphate) producing cell factory in the most favorable way. Nevertheless, smart approaches will have to be adopted to make its use in pragmatic (food) situations feasible by dephosphorylation of tagatose diphosphate (Hugenboltz, 2002).

3 RESULTS AND DISCUSSION

Functional foods like nutraceuticals, modified foods, fortified foods, etc. shows a promising future trend for developing overall health or well-being by giving certain perks beyond that of the traditional nutrients it contains. This study was able to highlight the assorted technologies used for the manufacture of bioactive compounds which are required in the formulation of nutraceuticals and functional foods. Metabolic Engineering is bestowed here as a biotechnological gizmo to enhance the supply of nutraceuticals. A paradigm shift has been noticed in the production of nutraceuticals via microbial cellular machinery. This has been achieved to meet the huge needs for nutraceuticals to protect humans to fight against life-threatening diseases. Metabolic engineering employing microbes was successfully shaped in the laboratory scale arrangement as well as in industrial scale manufacture of complex produce such as production of carotenoids from simple carbon sources. Specially by the application of biosynthetic avenues and evolution of synthetic biology approaches, unique or more convoluted value- added constituents may be generated using microorganisms (Wang, 2016).

Due to intelligent handling of genes and endurance of heterologous enzymes, microbes have an innate capacity to produce varied nutraceuticals by virtue of exploiting their fundamental metabolic grids. Keeping in mind the disadvantages also, it is believed that with the continual advancement in metabolic engineering as well as synthetic biology, the generation of nutraceuticals will indeed reach a remarkable height through microbial based platforms. Modified foods are gaining interest because of the transformations in eating traits and more alertness about health. Further exploration is required in the upcoming days to establish the potency with more in vivo investigations in order to make rapid progress in the formulation of new modified foods and nutraceuticals.

4 CONCLUSION

A remarkable progress in the production of nutraceuticals was observed by dint of metabolic engineering of microbial planted platforms. Different novel and value-added products can be produced at the hand of biosynthetic pathways along with the slow evolution of various synthetic biology strategies. The tractability of the different modification and alteration involving genes and vigor of heterologous enzymes facilitated microorganisms to harvest varied nutraceuticals by virtue of exploiting their fundamental metabolic grids. The concept of microbial cellular machinery is like a rising star with serious benefits for the blossoming of nutraceuticals, exclusively those that frame from more convoluted precursors. Continuous and steady progress are noticed in both spheres of basic genetic engineering and synthetic biology which are as a result expanding the form and amount of feasible host organisms to fit in to the interest for present and prospective nutraceutical products. The production of nutraceuticals and other functional foods can be achieved through continuous development and improvement in ongoing strategies in synthetic biology and metabolic engineering.

ACKNOWLEDGEMENT

The authors would like to thank Indian Institute of Engineering Science and Technology, Shibpur, India.

REFERENCES.

Borodina, I., Kildegaard, K.R., Jensen, N.B., Blicher, T.H., Maury, J., Sherstyk, S., Schneider, K., Lamosa, P., Herrgård, M.J., Rosenstand, I. 2015. Establishing a synthetic pathway for high-level production of 3-hydroxypropionic acid in Saccharomyces cerevisiae via β-alanine. *Metab Eng.* 27:57–64.

Burton, S.G., Cowan, D.A., Woodley, J.M. 2002. The search for the ideal biocatalyst. *Nature Biotechnology*. 20: 37-45.

González, R. 2013. Metabolic engineering: Use of system-level approaches and application to fuel production in Escherichia coli. *Electronic Journal of Biotechnology*. 16(3).

Hols, P., Kleerebezem, M., Schank, A.N., Ferain, T., Hugenholtz, J., Delcour, J., de Vos, W.M. 1999. Conversion of Lactococcus lactis from homolactic to homoalanine fermentation through metabolic engineering. *Nat Biotechnol.* 17:588-592.

Hugenholtz, J and Smid, E.J. 2002. Nutraceutical production with food-grade microorganisms. *Current Opinion in Biotechnology*. 13:497–507.

Jain, N., Ramawat K.G. 2013. Nutraceuticals and antioxidants in prevention of diseases. *Natural Products, Springer*. 2559-2580.

Liu, X., Li X, B., Jiang, J., Liu, Z.N., Qiao, B., Li, F.F., Cheng, J.S., Sun, X., Yuan, Y.J., Qiao, J., Zhao, G.R. 2018. Convergent engineering of syntrophic Escherichia coli coculture for efficient production of glycosides. *Metab Eng.* 47: 243–253.

Lopes, D.B., Madeira, Jr. J.V., de Castro Reis, L.V., Macena Leão, K.M., Alves Macedo, G. 2017. Chapter 1—Microbial production of added-value ingredients: state of the art. Microbial production of food ingre- dients and additives. *Grumezescu AM: Academic Press*: 1–32.

McNeil, B., Archer, D., Giavasis, I., Harvey, L. 2013. Microbial Production of Food Ingredients, Enzymes and Nutraceuticals. *Woodhead Publishing Series in Food Science, Technology and Nutrition, 1st Edition*: 1-656.

Pszczola, D.E. 1992. The nutraceutical initiative: a proposal for economic and regulatory reform. *Food Biotechnol.* 46:77-79.

Shen, Y., Zhao, L., Li, Y., Zhang, L., Shi, G. 2014. Synthesis of β-alanine from l-aspartate using l-aspartate-α-decarboxylase from Corynebacterium glutamicum. *Biotech Lett.* 36:1681–1686.

Song, C.W., Lee, J., Ko, Y.S., Lee, S.Y. 2015. Metabolic engineering of Escherichia coli for the production of 3-aminopropionic acid. *Metab Eng.* 30:121–129.

Van Rooijen, R.J., Van Schalkwijk, S., de Vos, W.M. 1991. Molecular cloning, characterization and nucleotide sequence of the tagatose-6phosphate pathway gene cluster of the lactose operon of Lactococcus lactis. *J Biol Chem.* 266:7176-7181.

Wang, J., Wang, S.G., Mattheos AG, Koffas., Yajun, Y. 2016. Microbial production of value-added nutraceuticals. *Current Opinion in Biotechnology*. 37: 97–104.

Woodley, J.M. 2017. Bio Process Intensification for the Effective Production of Chemical Products. *Computers and Chemical Engineering*. 1-39.

Xie, D., Miller, E., Tyreus, B., Jackson, E.N., Zhu, Q. 2016. Sustainable production of omega-3 eicosapentaenoic acid by fermentation of metabolically engineered *Yarrowia lipolytica. Quality Living Through Chemurgy and Green Chemistry*. 17-33.

Xue, Z., Sharpe, P.L., Hong, S.P., Yadav, N.S., Xie, D., Short, D.R., Damude, H.G., Rupert, R.A., Seip, J.E., Wang, J., et al. 2013. Production of omega-3 eicosapentaenoic acid by metabolic engineering of *Yarrowia lipolytica. Nat Biotechnol.* 31: 734.

Yuan, S-F., Alper, H.S. 2019. Metabolic engineering of microbial cell factories for production of nutraceuticals. *Microb Cell Fact.* 18:46. pp 1-11.

Zhu, Q., Jackson, E.N. 2015. Metabolic engineering of *Yarrowia lipolytica* for industrial applications. *Curr Opin Biotechnol.* 36: 65–72.

Biotechnology and Biological Sciences – Sen et al. (Eds)
© 2020 Taylor & Francis Group, London, ISBN 978-0-367-43161-7

Inhibitory characteristics of environmentally preformed biofilm to colonization of other biofilm producers by an in vitro study

Ananda Sanchayeeta Mandal*, Kumkum Bhattacharryya & Prasanta Kumar Maiti
Department of Microbiology, IPGMER, Kolkata, India

ABSTRACT: The problem in eliminating a chronic infection associated with biofilm production lies in fact that biofilm bacteria by virtue of reduced permeability and dormant persistence are able to survive greater antibiotic concentration. With the objective of identifying naturally formed biofilm producers, inhibiting further colonization of other biofilm producers on them, a strong biofilm producing *Aeromonas sobria* was isolated from slimy layer of Indian cat fish, confirmed by vitek-2 automated system. Biofilm formation of *Aeromonas* over cut silicon catheter chips and on 96 well microtitre plates was quantified by OD values followed by subsequent growth of a known biofilm positive *Staphylococcus aureus* (methicillin resistant) over the *Aeromonas* biofilm catheter chips was assessed. The results showed inhibition of growth of *Staphylococcus* in the background of *Aeromonas* both under microscopy and significant difference in OD values from 0.347892 to 0.03521

Keywords: Strong biofilm, *Aeromonas sobria*

A biofilm is an assemblage of surface-associated microbial cells that is enclosed in an extracellular polymeric substance matrix. Van Leeuwenhoek, using his simple microscopes, first observed microorganisms on tooth surfaces and can be credited with the discovery of microbial biofilms.

A biofilm is a thick layer of prokaryotic organisms that combine to form a colony. The colony attaches to a surface with a slime layer which helps in protecting the microorganisms from harsh conditions. This idea led to the thought that the slimy layer of catfish is an environmentally preformed biofilm.

The basic structural unit of the biofilm is the microcolony. Proximity of cells within the microcolony (or between microcolonies) provides an ideal environment for creation of nutrient gradients, exchange of genes, and quorum sensing.

The discovery that microbial biofilms are ever present in nature has resulted in the study of a number of chronic infectious diseases from a biofilm perspective. Cystic fibrosis, native valve endocarditis, otitis media, periodontitis, and chronic prostatitis following TURP all seem to be caused by biofilm-forming microorganisms. A range of indwelling medical devices used in hospital set up have shown to form biofilms, resulting in increasing incidences of medical device-associated infections.

Characteristics of biofilms that are significant in infectious diseases a) detachment of cells or microcolonies from biofilm can cause bloodstream or urinary tract infections or formation of emboli, b) transfer of plasmid resistance genes within biofilms, c) cells in biofilms have decreased susceptibility to antimicrobials, d) biofilm forming gram-negative organisms can synthesize endotoxins and e) biofilms are resistant to clearance by host immune system

*Corresponding author: Dr.Sanchayeeta@gmail.com

Table 1. Variables important in cell attachment and biofilm formation.

Properties of the substratum	Properties of the bulk fluid	Properties of the cell
Texture or roughness	Flow velocity	Cell surface hydrophobicity
Hydrophobicity	pH	Fimbriae
Conditioning film	Temperature	Flagella
	Cations	Extracellular polymeric substances
	Presence of antimicrobial agents	

Table 2.

Infections associated with *Staph.aureus*	
Staphylococcus aureus	Artificial hip prosthesis
	Central venous catheter
	Intrauterine device
	Prosthetic heart valve

The different control mechanisms (such as medical devices treated with antimicrobial agents and antimicrobial locks for inhibiting biofilm colonization of medical devices, the success of these treatments, the function of biofilms in antimicrobial resistance, biofilms as a source for persistor planktonic organisms, and biofilms in causing chronic diseases were assessed.

The most productive way to prevent these infections is removal of the biofilm contaminated medical device but it may not be possible many times.

1.1 *The objectives of the study are*

1. to develop a reasonable method for prevention of medical device related biofilm colonization.
2. to determine extent of biofilm inhibition by preformed biofilm coat on surface of device materials.

2 METHODOLOGY

2.1 *Materials*

(a) for bacterial culture-Mac conkeys media, blood agar media
(b) for growth of biofilm: Brain Heart Infusion Broth
(c) biofilm formation surface: silicon rubber catheter cut into pieces-1cm * 1cm * 1mm
(d) to study biofilm formation:

1. 96 well microtitre plates
2. Micropipettes for inoculation of microorganisms
3. Phosphate buffer saline
4. Cold methanol for fixation
5. Crystal violet for staining
6. Decolorising solution-70%ehanol
7. Colorimeter to measure optical density

2.2 *To isolate and identify bacteria*

Aeromonas sobria is isolated from the surface of Indian catfish by gentle scraping of the slime layer on its skin

Plated on mac conkeys agar and blood agar
On mac conkey agar: typicaly non lactose fermenting colonies
On blood agar: large round opaque, beta hemolytic colonies
Motility:motile.
Biochemical properties: Catalase and oxidase=positive,
Indole: positive
TSI: akali/acid
Citrate: negative
Urea: negative
Mannitol fermentation: positive
Hugh liefson test: oxidation: positive fermentation: positive
Vitek 2 automated system: *Aeromonas sobria* with (96% propability); multidrug sensitive.
The isolate is stocked.
Staphylococcus aureus ATCC strain is collected and the isolate is stocked.

3 PROCEDURE

Inoculum preparation:(stepanovic method) stock culture are grown on mac conkey s agar and incubated overnight at 37°c, 3-4 well isolated colonies are suspended in 5ml of brain heart infusion broth in two test tubes containing cut silicon rubber catheter pieces 1cm*1cm*1mm.1st with AS and 2nd with SA, incubated with shaking for 24 hrs at 37°c

The 1st test tube with AS is washed thrice with PBS, to that a fresh bacterial suspension of SA is added on the preformed AS biofilm, incubated for 24 hrs at 37°c, washed thrice with PBS, fixed with methanol, stained with crystal violet, dye extraction done by ethanol, OD is measured by spectrophotometer at 570 nm with microtitre plates. The OD of AS and SA on catheter chips alone are also measured in the same way as before. Negative control is taken as BHI broth in a test tube and OD measured. The catheter chips from all the test tubes taken out with sterile forceps are plated on agar and colony characteristics, Grams stain smear are examined.

4 RESULTS AND DISCUSSION

The mean OD of SA is 0.3478 reduced to 0.03521 for mean OD of SA on AS. Results showed that the difference in OD values of SA biofilm on AS precoat biofilm is significant.

Microscopy also showed inhibition of colonies of SA in the background of AS colonies.

The AST report of *Aeromonas sobria* showed sensitivity to almost all the antibiotics tested therefore it may imply that it is a strong biofilm naturally present in nature which prevents further colonisation of any other bacteria present in its vicinity and preventing resistance transfer gene from other multidrug resistance bacteria, also treatable by common antibiotics if required.

5 CONCLUSION

The natural biofilms are so perfectly designed so as to inhibit further biofilm on their surface that no technology can match with that in terms of nano level surface modification and surface charge alteration. This may help in developing strategies in preventing biofilm related human infections and future application in anti-biofilm property. Further studies with different microorganisms and on different medical device can be carried out in future

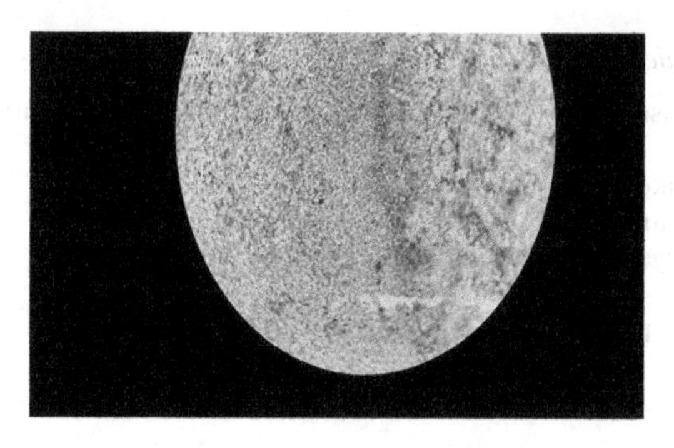

Figure 1. *S.aureus*29213 on *A.sobria* (colony Grams stain smear).

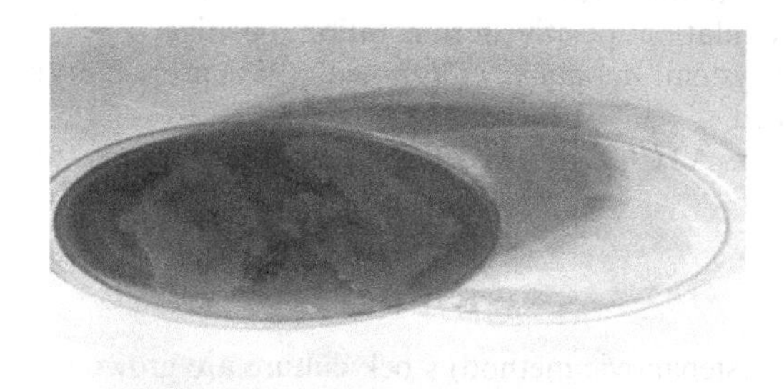

Figure 2. *A.sobria* growth on Mac conkeys agar from squeezed catheter chips.

ACKNOWLEDGEMENT

The present research work is not funded by any organization

REFERENCES

Andersson DI, Hughes D. Antibiotic resistance and its cost: Is it possible to reverse resistance? *Nat. Rev. Microbiol.* 2010, 8, 260–271.

Biofilm - Definition, Function and Structure | Biology Dictionary https://biologydictionary.net Ecology.

Costerton JW, Stewart PS, Greenberg EP. Bacterial biofilms: a common cause of persistent infections. *Science.* 1999, 284, 1318–1322.

Dawson CC, Intapa C, Jabra-Rizk MA. 'Persisters': Survival at the cellular level *PLoSPathog*. 2011, 7, e1002121.

Donlan RM. Biofilms and device-associated infections. Emerg. Infect. Dis. 2001, 7, 277–281.

Flemming H.C., Wingender J. The biofilm matrix. *Nat. Rev. Microbiol.* 2010, 8, 623–633.

Rodney M. Donlan Biofilms: Microbial Life on Surfaces Emerg Infect Dis. 2002 Sep; 8(9): 881–890.

S. Stepanović, D. Vuković, I. Dakić, B. Savić - Journal of, 2000 – Elsevier. A modified microtiter-plate test for quantification of staphylococcal biofilm.

Stoodley P, Sauer K, Davies DG, Costerton JW. Biofilms as complex differentiated communities. *Annu. Rev. Microbiol.* 2002, 56, 187–209.

Tollefson DF, Bandyk DF, Kaebnick HW, Seabrook GR, Towne JB. Surface biofilm disruption. Enhanced recovery of microorganisms from vascular prostheses. *Arch. Surg. Chic. Ill 1960*. 1987, 122, 38–43.

Biotechnology and Biological Sciences – Sen et al. (Eds)
 ISBN 978-0-367-43161-7

Assessing mycoparasytic activity exhibited by phyto-friendly-fungi (PFF) in combating phytopathogenic fungi by producing various glucanases

Dhavalkumar Patel
Department of Biotechnology and Microbiology, Parul Institute of Applied Science and Research, Parul University, Ahmedabad, India
Department of Biochemistry and Biotechnology, St. Xavier's College (Autonomous), Ahmedabad, India

Shreya Patel, Sony Tekwani, Mahima Patel, Sudeshna Menon & Sebastian Vadakan
Department of Biochemistry and Biotechnology, St. Xavier's College (Autonomous), Ahmedabad, India

Dhaval Acharya
Department of Biotechnology and Microbiology, Parul Institute of Applied Science and Research, Parul University, Ahmedabad, India

Dweipayan Goswami*
Department of Biochemistry and Biotechnology, St. Xavier's College (Autonomous), Ahmedabad, India
Department of Microbiology and Biotechnology, School of Sciences (SoS), Gujarat University, Ahmedabad, India

ABSTRACT: Soil-borne pathogenic fungi are extensively spread and liable to cause serious damage and are accountable for great losses in agricultural area. Excessive and irrational use of pesticides brought infrequent toxicity and adverse effects to soil and ground water. To overcome this, biocontrol agent of great potential in contradiction of soil-borne phyto-pathogens is a substitute way to artificial fungicides. Rhizospheric soil of agricultural farm was collected from the farm at Anand district, Gujarat. Seclusion of fungi was carried out from the soil and were further sub cultured. Qualitative and quantitative analysis of glucanase enzyme was performed using Carboxy Methyl Cellulose (CMC) media. Qualitative and quantitative analysis of chitinase enzyme was performed using colloidal chitin media. Assessment of mycoparasitic activity against pathogen like *Aspergillus niger, Rhizoctonia solani* and *Fusarium oxysporum* was carried out of potent fungi based on the activity of glucanases enzymes. Few PFF had shown decent outcome against various phyto-pathogen by constraining their proliferation.

Keywords: *Phyto-friendly-fungi, pathogens, glucanases*

1 INTRODUCTION

Parasitism is a communal form of symbiosis among many groups of heterotrophic organisms, fundamentally involves a nutritional relationship that is favourable to existence of the parasite. Mycoparasitism is one of kind in which fungi testify to the constant struggle among organisms occupying the identical habitat [1]. Basically, it indicates the interrelationship of a fungus parasite and a fungus host. Soil endured pathogenic fungi are extensively disseminated besides are accountable for solemn impairment to many horticultural as well as agricultural crops

*Corresponding author: dweipayan79@gmail.com

worldwide [2]. Biological deterrence of soil-borne phyto-pathogens is a latent substitute to the practice of chemical pesticides, which have recognized detrimental to the environment [3].

Chitin and β-glucan are the foremost structural component of fungal cell-wall. Biological control is a process of controlling pests that inculde insects, weeds, mites and plant ailments that are caused by pathogens [4]. Mycoparasitic activity of a fungi is one of the chief tackles of biocontrol agent. Quite a few mycorrhizal fungi have been shown to produce chitinase and glucanase enzymes and they play a noteworthy part in biological control [5].

Fungi belonging to the genus *Trichoderma* (mycoparasitic) have stood rummage-sale for bio-controller of pathogens of plants; likewise, soil-borne (e.g. *R. solani* and *Sclerotium rolfsii*) and foliar pathogens together with *Botrytis cinerea*, that taints cucumber [6]. *Talaromyces flavus* could constrain and decline rotting disease of beans (due to their metabolites) triggered by *Sclerotium rolfsii* [7]. *Talaromyces flavus* can also control wilt ailment triggered by *Verticillium dahlia* besides tomato wilt that is instigated by *Verticilium albo-atrum* [8].

2 MATERIALS & METHODS

Fungal strains were isolated and were purified from rhizospheric soil that was collected from the farm Anand District, Gujarat (22.457308, 72.905985). Isolation of fungi was carried out using Rose Bengal chloramphenicol agar (RBCA) and sub-culturing was done on Potato Dextrose Broth (PDB). Identification of fungal isolates was done on gene level. Fungal isolates were sent to GSBTM for 18S rRNA identification.

Qualitative analysis of β-glucanase enzyme was performed using carboxy methyl cellulose (CMC) agar plate technique. Media was prepared by 1/4th PDB with 1% CMC and Zone of clearance was observed after Congo red staining method [9]. Quantitative estimation of β-glucanase activity was evaluated by the liberated glucose using 3,5-dinitrosalicylic acid (DNSA) method [10]. One unit of enzyme activity was expressed as μmoles of glucose ml^{-1} hr^{-1}.

Qualitative analysis of chitinase enzyme was performed using chitin agar plates. Transparent zone of clearance around colony was observed [11]. Quantitative estimation of chitinase activity was evaluated by the liberated N-acetyl glucosamine (GlcNAc) using DNSA method [12]. One unit of enzyme activity was expressed as μmoles of N-acetyl glucosamine ml^{-1} hr^{-1}.

Assessment of mycoparasitic activity was carried out using dual culture plate technique [13]. Phytopathogens *Rhizoctonia solani, Fusarium oxysporum* and *Aspergillus niger* were used for the study of Antagonism.

3 RESULTS & DISCUSSION

On isolating fungi from the soil total 32 isolates were obtained. On identifying them on bases of their morphological character and microscopic structure, isolates belong to *Aspergillus, Trichoderma, Penicillium, Fusarium* and *Talaromyces.* All the isolates were preserved on PDB slants and glycerol stocks were prepared

Qualitative estimation was performed of all 32 fungal isolates for β-glucanase enzyme as well as for Chitinase enzyme. Out of 32 isolates 11 isolates showed good zone of hydrolysis and were further screened for quantitative assay. Those 11 isolates were sent to GSBTM for 18S rRNA identification and the strains used in current research are displayed in Figure 1

Before carrying the quantitative assay for β-glucanase enzyme and Chitinase enzyme, standard curve of Glucose as well as N-Acetylglucosamine was performed as the final enzyme-product in formed respectively. Linear Regression (R^2) value for Glucose obtained is 0.9977 and for GlcNAc is 0.9982.

On performing assay, the enzyme activity (μmoles ml^{-1} h^{-1}) of each fungal isolate is summarised in Table 1. *Talaromyces pinophilus* showed the maximum enzymatic activity 27.92 ± 0.19 and 29.29 ± 0.17 for β-glucanase enzyme and chitinase enzyme correspondingly, which was latter followed by *Trichoderma* sp. M20, *Trichoderma longibrachitum, Penicillium citrinum* and *Penicillium chrysogenum.* Other strain showed normal enzymatic activity.

After quantitative assay, antagonism was checked against plant pathogen. All fungi isolates showed decent compatibility. Optimum result was exhibited by *Talaromyces pinophilus*, *Penicillium citrinum* and *Trichoderma* sp. M20 which is displayed in Figure 3.

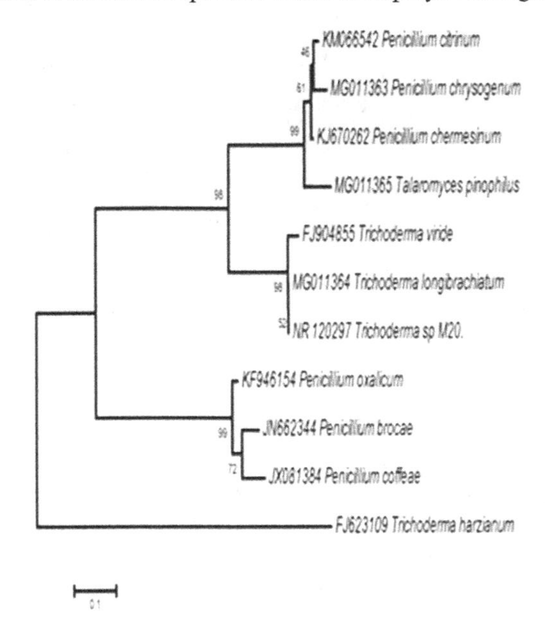

Figure 1. Display of phylogeny of screened fungal isolates used in current study.

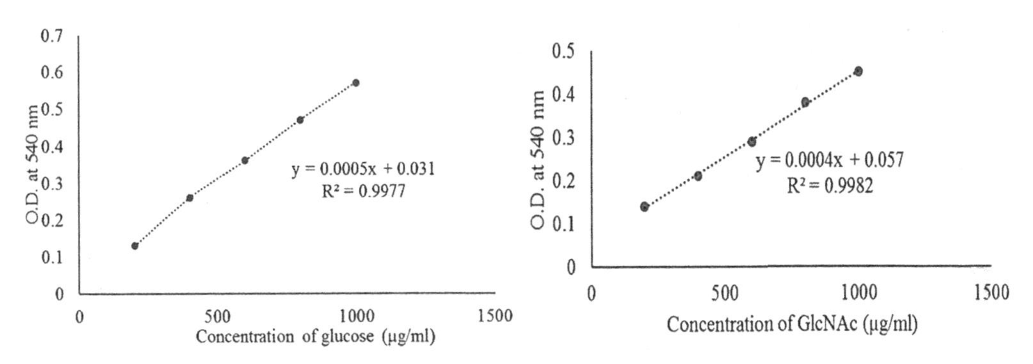

Figure 2. Graphical representation of glucose standard curve and GlcNAc curve.

Table 1. Fungal strains used under current study.

Strains	Cellulase (µmoles/ml/h)	Chitinase (µmoles/ml/h)
Penicillium oxalicum	5.92 ± 0.05	4.23 ± 0.03
Penicillium citrinum	12.32 ± 0.07	14.36 ± 0.12
Penicillium chrysogenum	10.32 ± 0.15	9.54 ± 0.08
Penicillium coffeae	5.52 ± 0.02	6.37 ± 0.07
Penicillium chermesinum	8.72 ± 0.11	7.62 ± 0.06
Penicillium brocae	8.02 ± 0.08	6.21 ± 0.06
Trichoderma viride	7.72 ± 0.05	4.99 ± 0.05
Trichoderma longibrachitum	15.12 ± 0.13	14.23 ± 0.15
Trichoderma harzianum	9.32 ± 0.09	7.23 ± 0.07
Trichoderma sp. M20	19.92 ± 0.12	17.23 ± 0.15
Talaromyces pinophilus	27.92 ± 0.19	29.29 ± 0.17

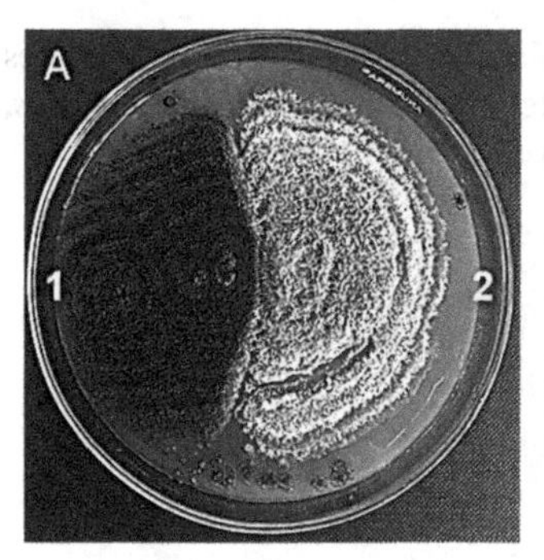

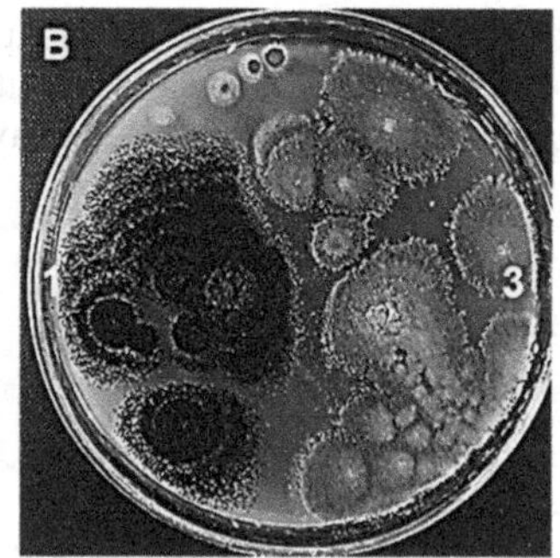

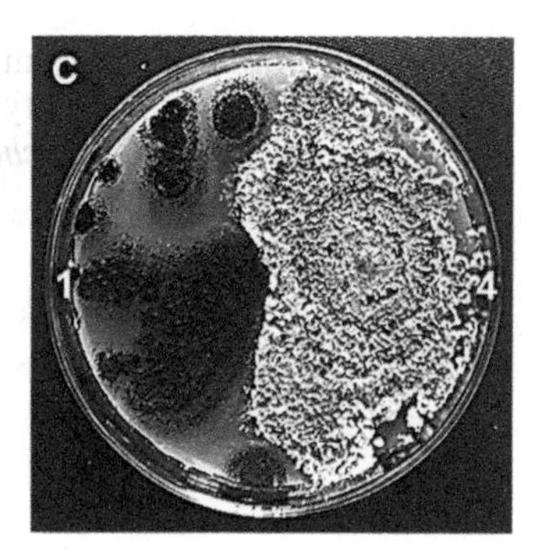

Figure 3. Display of mycoparasitic activity of (A) *Aspergillus* sp. (1) against *Talaromyces pinophilus* (2), (B) *Aspergillus* sp. (1) against *Penicillium citrinum* (3) and *Aspergillus* sp. (1) against *Trichoderma* sp. M20 (4).

Vivas et al. (2018) explored the antagonistic activity of the *Hansfordia pulvinata* (H-611 fungi), *Acremonium* sp (strain A-617, A-602 and A-598), *Simplicillium lanosoniveum* (strain S-599), *Lecanicillium lecanii* (strain L-622), and *Sarocladium implicatum* (strain I-609) on the plant pathogenic fungus *A. caricae* which causes papaya black spot. They checked protease enzyme activity while here we opted for β-glucanase enzyme and chitinase enzyme as fungi's cell wall is composed of chitin and the cell wall of bacteria is lipo-polysaccharides [14].

Qualhato et al. (2013) deliberated the interface between four *Trichoderma* species and phytopathogenic fungi: *Sclerotinia sclerotiorum, Rhizoctonia solani* and *Fusarium solani. Trichoderma* spp. were the utmost operative antagonists in contradiction of the pathogens as they were able to produce and secrete β-glucanase, chitinase, NAGAse, acid proteases, acid phosphatase and alginate lyase [15].

Larena et al. (2002) experimented with conidia of *Penicillium oxalicum* that acted as biocontrol agent against *Fusarium oxysporum.* They used fresh as well as 180 days old conidia which were stored at - 20°C and noted that there was reeducation in *Fusarium* wilt of tomato by 49 and 61%, correspondingly. They concluded that *Penicillium* spp. are an auspicious biocontrol agent for *Fusarium* wilt and other tomato illnesses [16].

Naraghi et al. (2018) evaluated their cotton and potato growth promotion by means of *Talaromyces flavus.* They got amplified root and crown length, plant height, and plant dry weight by 1.71, 1.09, 1.45 and 3.75-fold, individually and concluded that the overall results of their research is may be by consuming the antagonistic fungus *Talaromyces flavus* [8].

4 CONCLUSION

Overall in the current research *Talaromyces pinophilus, Trichoderma* sp. M20, *Trichoderma longibrachitum, Penicillium citrinum* and *Penicillium chrysogenum* exhibited β-glucanase and chitinase enzyme that occupies the significant role in mycoparasitic activity of fungi. Mycoparasite inhibits the growth of pathogen by producing these enzymes and can be used as biocontrol agent. Thus, *Talaromyces pinophilus* is novel fungal isolate that can be used as biocontrol agent against various phytopathogens

ACKNOWLEDGMENT

Authors are thankful to the Gujarat State Biotechnology Mission (GSBTM) for providing the funding under FAP 2017-18 GSBTM/MD/PROJECTS/SSA/5041/2017-18 project and St. Xavier's College Ahmedabad for providing necessary facilities.

REFERENCES

[1] Batygina, T. B. (Ed.). (2019). Embryology of Flowering Plants: Terminology and Concepts, Vol. 3: Reproductive Systems. CRC Press.

[2] Panabieres, F., Ali, G. S., Allagui, M. B., Dalio, R. J., Gudmestad, N. C., Marie-Line, K. U. H. N., & Zampounis, A. (2016). *Phytophthora nicotianae* diseases worldwide: new knowledge of a long-recognised pathogen. *Phytopathologia Mediterranea*, *55*(1), 20–40.

[3] De Buck, A. J., Van Rijn, I., Roling, N. G., & Wossink, G. A. A. (2001). Farmers' reasons for changing or not changing to more sustainable practices: an exploratory study of arable farming in the Netherlands. *The Journal of Agricultural Education and Extension*, *7*(3), 153–166.

[4] Strand, J. F. (2000). Some agrometeorological aspects of pest and disease management for the 21st century. *Agricultural and Forest Meteorology*, *103*(1-2), 73–82.

[5] Jung, S. C., Martinez-Medina, A., Lopez-Raez, J. A., & Pozo, M. J. (2012). Mycorrhiza-induced resistance and priming of plant defenses. *Journal of chemical ecology*, *38*(6), 651–664.

[6] Verma, M., Brar, S. K., Tyagi, R. D., Surampalli, R. Y., & Valero, J. R. (2007). Antagonistic fungi, *Trichoderma* spp.: panoply of biological control. *Biochemical Engineering Journal*, *37*(1), 1–20.

[7] Punja, Z. K. (1997). Comparative efficacy of bacteria, fungi, and yeasts as biological control agents for diseases of vegetable crops. *Canadian Journal of Plant Pathology*, *19*(3), 315–323.

[8] Naraghi, L., Negahban, M., Heydari, A., Razavi, M., & Afshari-Azad, H. (2018). Growth Inhibition of *Fusarium oxysporum* f. sp. *lycopercisi*, the Causal Agent of Tomato *Fusarium* Wilt Disease by Nano formulations Containing *Talaromyces Flavus. Ekoloji*, *27*(106), 103–112.

[9] Gohel, H. R., Contractor, C. N., Ghosh, S. K., & Braganza, V. J. (2014). A comparative study of various staining techniques for determination of extra cellular cellulase activity on Carboxy Methyl Cellulose (CMC) agar plates. *Int J Curr Microbiol App Sci*, *3*, 261–266.

[10] Teixeira, R. S. S., da Silva, A. S. A., Ferreira-Leitão, V. S., & da Silva Bon, E. P. (2012). Amino acids interference on the quantification of reducing sugars by the 3, 5-dinitrosalicylic acid assay mislead carbohydrase activity measurements. *Carbohydrate research*, *363*, 33–37.

[11] Chernin, L. S., Winson, M. K., Thompson, J. M., Haran, S., Bycroft, B. W., Chet, I., ... & Stewart, G. S. (1998). Chitinolytic activity in *Chromobacterium violaceum*: substrate analysis and regulation by quorum sensing. *Journal of bacteriology*, *180*(17), 4435–4441.

[12] Yang, S., Fu, X., Yan, Q., Guo, Y., Liu, Z., & Jiang, Z. (2016). Cloning, expression, purification and application of a novel chitinase from a thermophilic marine bacterium *Paenibacillus barengoltzii. Food chemistry*, *192*, 1041–1048.

[13] Abo-Elyousr, K. A., Abdel-Hafez, S. I., & Abdel-Rahim, I. R. (2014). Isolation of *Trichoderma* and evaluation of their antagonistic potential against *Alternaria porri. Journal of Phytopathology*, *162* (9), 567–574.

[14] Vivas, J. M. S., da Silveira, S. F., dos Santos, P. H. D., Carvalho, B. M., de Sousa Poltronieri, T. P., Jorge, T. S., & de Moraes, R. (2018). Antagonism of fungi with biocontrol potential of papaya black spot caused by *Asperisporium caricae'. Australian Journal of Crop Science*, *12* (5),827.

[15] Qualhato, T. F., Lopes, F. A. C., Steindorff, A. S., Brandao, R. S., Jesuino, R. S. A., & Ulhoa, C. J. (2013). Mycoparasitism studies of *Trichoderma* species against three phytopathogenic fungi: evaluation of antagonism and hydrolytic enzyme production. *Biotechnology letters*, *35*(9), 1461–1468.

[16] Larena, I., Melgarejo, P., & De Cal, A. (2002). Production, survival, and evaluation of solid-substrate inocula of *Penicillium oxalicum*, a biocontrol agent against *Fusarium* wilt of tomato. *Phytopathology*, *92*(8), 863–869.

Biotechnology and Biological Sciences – Sen et al. (Eds)
© 2020 Taylor & Francis Group, London, ISBN 978-0-367-43161-7

Study of the effectiveness of antibiotics in management of bacterial wilt in plants

Shamayeeta Sarkar*
Department of Botany, Rishi Bankim Chandra College, Naihati

ABSTRACT: This study aims to determine the effectiveness of some of the commonly used antibiotics in management of bacterial wilt in plants caused by enterobacterial strains. The enterobacterial strains were isolated from wilt infected vegetable crops and weeds. Agar diffusion method was used to determine the growth inhibition caused by the drugs streptomycin, tetracycline and terramycin following the MIC parameters. It was observed that about 85% of the pathogenic Enterobacterial strains were resistant to one of the antibiotics (streptomycin) and about 27% showed resistance to more than two drugs. Statistical interpretation of the data using F Test showed that out of the three drugs, terramycin and tetracycline performed significantly better than streptomycin but the effectiveness of tetracycline and terramycin were comparable. Kendell's Test was done to see if there was a correlation between virulence of the bacterial strains and their resistance to antibiotics. The result showed that bacteria that were found to be more virulent on tomato and brinjal were also more resistant to the three antibiotics tested on them.

1 INTRODUCTION

Antibiotics, after their discovery, were mainly used as human medicine and in animal agriculture. Study of effectiveness of antibiotics in control of plant diseases was initiated in the 1950s. Though initially it seemed to be the best possible way of controlling bacterial infestations in susceptible host plants, several drawbacks were identified later. Antibiotic resistance is one of the serious issues in today's world. Use of antibiotics in controlling bacterial diseases of plants is far less than that used for animal diseases. A study in year 2009 shows that in United states of America, antibiotics used on orchards is only 0.12% of the total antibiotics used in animal agriculture. However, in spite of this, antibiotic use has been in use for more than 50 years in plants and it has been found to be active on plants for about a week, thus showing no adverse effect on consumption of crops after harvest. [Stockwell 2012]

2 MATERIAL AND METHOD

Isolation and Identification of bacterial strains: An exhaustive screening of plants infested with bacterial wilt was done. The samples were tested for bacterial oozing, were surface sterilized (using NaOCl solution) and bacteria were isolated from vascular bundles of the infected stem. The isolated bacteria were repeatedly transferred to differential media and single colonies were cultured to obtain pure strains. The bacterial strains were identified using biochemical tests and molecular identification tests. They were also tested for their pathogenicity on a set of five solanaceous plants viz. tomato, potato, eggplant, chilli and tobacco under greenhouse condition [Sarkar, 2015].

Antibiotic Assay: Molten medium was cooled and seeded with the test inoculums (~10^6 CFU/mL of bacterial suspensions) and poured into plates. On solidifying, the agar medium was bored

* Corresponding author: E-mail: shamayeeta.s gmail.com

(10mm diameter) and drug solution was poured into the wells, at the following concentrations: 50μg/ml, 100μg/ml, 200μg/ml, 250μg/ml, 300μg/ml. Each well could house 500μl of the antibiotic solutions. Three different drugs tested were Streptomycine (aminoglycosides), Tetracycline (tetracyclines) and Terramycin (tetracyclines). The plates were incubated at 28°C for 3-4 days and the inhibition zone diameter was measured. Each measurement was compared to the zone-size interpretive chart derived from the M100 chart of antimicrobial susceptibility for Enterobacteriaceae and grouped as susceptible, intermediate or resistant. According to the NCCLS (The National Committee for Clinical Laboratory Standards) standards of antimicrobial activity and susceptibility testing, M100 contains tables for interpreting the results of disk diffusion and MIC tests [CLSI M100 S26:2016]. Statistical interpretations was done to compare the performance of the three antibiotics on the bacterial strains. Kendall's Test for correlation was performed between the AUDPC values of Greenhouse Pathogenicity Test shown by the thirty bacterial strains and the Inhibition diameter for the three antibiotics for the same bacterial strains.

3 RESULT

The collection of samples was done based on the known visible symptoms. Most of the strains were isolated from symptomatic plants while a few were collected from asymptomatic weeds growing in the vicinity of infected symptomatic crop plants and known from literature. The sensitivity of bacterial strains to antibiotics was tested for three antibiotics whose combinations are commonly used against phytopathogens viz. terramycin, streptomycin and tetracycline. From the data presented in Table 1, it could be concluded that tetracycline followed by

Table 1. Inhibition zone diameter (mm) at drug concentration of 300 ppm.

Isolates	Terramycin	Streptomycin	Tetracycline
Amavir1	17	4	17
SomeUD	3	1	3
CapBurd	12	12	14
Some2	22	8	25
JatR	15	4	16
LiesM	7	3	12
Rum1	26	9	17
Amsp2	21	2	20
Lies1	20	4	13
Lies5	3	9	6
Cap3	35	11	4
Pha1	21	5	19
CU4	31	11	22
Castor1	22	18	20
Euod1Pot	10	9	19
Cap4	22	14	22
CrdiUD	0	5.5	10
CapUD	12	8.5	15
Euod3	22	8	14
Cap11	3	13	19
Parth1	15	11	20
Gna1	19	14.5	30
Brol2	9	3	12
Cain2	11	5	16
Some Mur	3	3	4
Un4	14	15	25
Amsp1	18	10	17
SN1	11	6	14
LiesDD	17	13	18
Curlong1	1	2	3

terramycin were more effective and significant inhibition of bacterial growth of most of the strains tested was observed compared to streptomycin.

From the data it was observed that at the highest dose of drug used, 26.7% of the 30 strains used in the experiment showed multiple drug resistance to all the three drugs used. 36.7% of the bacterial trains were resistant to highest dose of terramycin used and almost 83.3% of the strains were resistant to streptomycin. The F Test showed that out of the three drugs, terramycin and tetracycline performed significantly better than streptomycin but the effectiveness of tetracycline and terramycin were comparable (Table 2). Thus it could be predicted that these phytopathogens showed maximum resistance towards streptomycin while tetracycline fared well in management of the bacterial growth among the three drugs used.

Table 2. Analysis of variance between the effectiveness of the three drugs, comparing two at a time (maximum dosage has been taken into consideration). Drug 1- Terramycin, Drug 2- Streptomycin, Drug 3- Tetracycline.

F-Test Two-Sample for Variances
Drug 1 vs drug3 (dosage5)

	Variable 1	Variable 2
Mean	14.733333	15.533333
Variance	78.409195	45.567816
Observations	30	30
df	29	29
F	1.7207144	
P(F≤f) one-tail	0.0749112	
F Critical one-tail	1.8608144	

Greater than 0.05 so not sig different

F-Test Two-Sample for Variances
drug 1 vs drug2

	Variable 1	Variable 2
Mean	14.733333	8.0833333
Variance	78.409195	20.29454
Observations	30	30
df	29	29
F	3.8635611	
P(F≤f) one-tail	0.0002467	
F Critical one-tail	1.8608114	

Sig different

F-Test Two-Sample for Variances
drug 2 vs drug3

	Variable 1	Variable 2
Mean	8.0833333	15.533333
Variance	20.29454	45.567816
Observations	30	30
df	29	29
F	0.44537	
P(F≤f) one-tail	0.0165422	
F Critical one-tail	0.5374	

Sig different

4 CONCLUSION

Antibiotics have been used for controlling bacterial diseases in plants since 1950s. Use of such antibiotics has been increasing over the last decade, especially in south asian countries. Not only the massive exposure leads to its resistance among pathogens, many of them are not easily degraded and continue to stay in the environment for longer periods e.g tetracycline [Larsson, 2014]. It has been observed that often an overdose of antibiotic is used, especially in animal farming and agriculture (in the form of sprays and animal feed mix) than that used as human medicines. These get accumulated in the crops, poultry and meat, in waterbodies (surface water and rivers) and even in soil.

The present work aims to analyse the effectiveness of antibiotics in management of a common plant disease- 'bacterial wilt', caused by a diverse froup of bacteria belonging to the Enterobacteriaceae. The antibiotics chosen for the test are known to be some of the most commonly used antibiotics in combatting plant diseases caused by eneterobacterial members like *Erwinia* sp. However, it has already been reported that emergence of streptomycin-resistant strains of *Erwinia amylovora, Pseudomonas* spp., and *Xanthomonas campestris* has impeded the control of several important diseases [McManus etal, 2002]. As seen from Table 1 and Table 2, it is observed that most of the wilt causing eneterobacterial strains show resistance to streptomycin (83.3%) and about 37% of the strains show resistance to the highest dose of terramycin, which is the best performing drug used in the experiment. Many of the strains (~27%) show multiple drug resistance i.e more than two drugs. From this result, it can be implied that drugs like streptomycin need to be re-assessed before using them for managing diseases that were previously reported to be controlled through antibiotic use. Some of the drugs e.g terramycin or tetracycline should also be used in moderation i.e at proper cases where effective and in proper doses to prevent overexposure of these drugs in nature and thereby development of drug resistance. In fact, the discovery of homologous resistance determinants in bacteria associated with plants, soil, animals, and humans has led to intriguing hypotheses regarding the origins of resistance [Sundin, 1996] and to maintain the efficacy of antibiotics as human medicine, random use of all antibiotics on plants and food animals has been scrutinized [McManus etal, 2002]. This study reinforces on re- examination of practiced norms related to antibiotic uses in agriculture.

The bacterial group used in the experiment has a wide host range consisting of mostly vegetable crops and allied weeds and the strains are highly virulent on tomato, moderately virulent on brinjal and potato and less virulent on chilli, as observed from their greenhouse pathogenicity data [Sarkar, 2015] . This available data has been utilized to find out any correlation between pathogenicity of a bacteria to its resistance towards antibiotics on the basis of the multiple literatures reported on the cotransmission of the pathogenecity genes and drug resistance genes as packets of Pathogenicity islands [Beceiro et al., 2013]. Kendall's Test for Correlation (Table 3) has shown that the inhibition diameter is negatively correlated to the pathogenicity marker i.e AUDPC values obtained on hosts tomato and brinjal. Thus, it can be concluded that bacterial strains with high virulence are more resistant to antibiotics (smaller inhibition diameter). However, it is difficult to predict such correlation as the test gave inconsistent values (both negative and positive) for the correlation of the drug sensitivity with pathogen virulence on potato and chilli.

Table 3. Kendall's Test for Correlation of Pathogenicity and resistance towards drugs.

	AUDPC brinjal	AUDPC chilli	AUDPC potato	AUDPC tomato
Terramycin	-0.15	-0.05	0.11	-0.22
Streptomycin	-0.24	0.03	0.08	-0.17
Tetracyclin	-0.11	-0.13	-0.05	-0.2

REFERENCES

1. Stockwell V.O., Duffy B. 2012. Use of antibiotics in plant agriculture. Rev. sci. tech. Off. int. Epiz, 31 (1), 199-210.
2. McManus P.S., Stockwell V.O., Sundin G.W. & Jones A.L. 2002. Antibiotic use in plant agriculture. Annu. Rev. Phytopathol., 40, 443–465.
3. Shamayeeta Sarkar & Sujata Chaudhuri. 2015. New report of additional enterobacterial species causing wilt in West Bengal, India.Canadian Journal of Microbiology, 61(7): 477-486, https://doi.org/10.1139/cjm-2015-0017.
4. CLSI M100 S26:2016 — Performance Standards for Antimicrobial Susceptibility Testing; 26th Edition.
5. Larsson D.G. 2014. Antibiotics in the environment. Ups J Med Sci.;119(2):108–112. doi:10.3109/03009734.2014.896438.
6. Sundin G.W., Bender C.L. 1996. Molecular analysis of closely related copper- and streptomycin resistance plasmids in Pseudomonas syringae pv. Syringae. Plasmid.;35(2):98-107.
7. Beceiro Alejandro., Tomás María ., Bou Germán. 2013. Antimicrobial Resistance and Virulence: a Successful or Deleterious Association in the Bacterial World? Clinical Microbiology Reviews Apr, 26 (2) 185–230; DOI: 10.1128/CMR.00059-12.

Biotechnology and Biological Sciences – Sen et al. (Eds)
© 2020 Taylor & Francis Group, London, ISBN 978-0-367-43161-7

In-vitro antibiofilm activity of ethyl acetate fraction of *Monochoria hastata* (L.) Solms

Debabrata Misra & Manab Mandal
Department of Botany, University of Gour Banga, India
Sukhendu Mandal
Department of Microbiology, University of Calcutta, Kolkata, India
Vivekananda Mandal
Department of Botany, University of Gour Banga, India

ABSTRACT: An aquatic ethno-medicinal herb, *Monochoria hastata* (L.) Solms belonging to the family Pontederiaceae is used in Indian sub-continent as remedy of several ailments like wounds and boils, gastropathy, hepatopathy etc The aim of the present work was to assess the antibiofilm activity of ethyl acetate extract of the aerial part of this plant.. TLC and LC-MS was performed to visualize the presence of active principles with their mass profile and Antibiofilm activities of the purified compound were assessed following the standard biofilm production assay protocol (Djordjevic *et al.*, 2002) against the Gram-positive strain, *Staphylococcus epidermidis* MTCC 3086 (*S. epidermidis*) and Gram negative strain, *Vibrio cholerae* MTCC 3906 (*V. cholerae*). The extracted compounds were centralized in a single spot (S_1) at retardation factor (Rf) value of 0.92 in TLC and peaks with their characteristic features and relative abundances were detected at different retention times in LC-MS and exhibited selective antibiofilm activities without any cytotoxic effect in human cell line.

Keywords: Antibiofilm Activity, Cytotoxicity, Gastro-enteritis, LC-MS, *Monochoria hastata* (L.) Solms

1 INTRODUCTION

Bacteria including vector-borne and food-borne agents are the potent pathogens to induce human disease of varying severity and epidemiology with varying means of transmission (Christou, 2011). Multidrug-resistant *Bacillus cereus, Clostridium perfringens, Escherichia coli, Staphylococcus aureus, Vibrio cholerae*, and *Vibrio parahemolyticus* bacteria in the food chain can adversely affect the food safety of animal and human health (Friedman, 2015). Bio-film producing microfloras have a 'feast or famine' lifestyle (Carlsson, 1983) which is another severe problem. Continuous searching of novel antimicrobial compounds is an important line of research due to antibiotic resistance acquired by several microorganisms. Plants are the potent source of antimicrobial compounds. Due to the occurrence of many side effects by use of synthetic drugs for various diseases, medicinal plants are considered as the main source of new drugs as they have less or no side effects and a number of important modern or allopathic drugs have been derived from ethnomedicinal plant sources which are used by indigenous people (Balick and Cox, 1996; Fabricant and Farnsworth, 2001). Various plant secondary metabolites like alkaloids including phenylalkylamines, pyrrolidines, pyrrolizidines, tropanes and purine alkaloids, several flavonoids and tannins, alcoholic, aldehydal, phenolic, ketonic and esterified derivatives of terpenoids, quinines and resins have antimicrobial properties against many fungi and bacteria

*Corresponding Author: vivekugb@gmail.com

(Compean and Ynalvez, 2014). The general research methods includes proper selection of medicinal plants, preparation of crude extracts, biological screening, detailed chemo pharmacological investigations, toxicological and clinical studies, standardization and use of active moiety as the lead molecule for drug design (Wink *et al.*, 2005). Mass-spectroscopy based metabolomic data can be analyzed by various bioinformatic tools and databases (Sugimoto *et al.*, 2012). An emergent hydrophyte *Monochoria hastata* (L.) Solms belonging to the family Pontederiaceae which grows in wide geographical range and leaf of this plant has antibacterial efficacy (Misra *et al.*, 2018). The aerial parts of *Monochoria hastata* (L.) Solms were collected and shade dried to prepare the herbarium sheet as voucher specimen (Tag No. UGB/DM/01) for authentication by the Central National Herbarium of Botanical Survey of India, Howrah-711 103, West Bengal, India for identification. Dried powder sample of aerial parts of the plant was taken for experiments.

2 MATERIALS AND METHODS

2.1 *Powder sample preparation and extraction*

For powder preparation, the collected plant materials were washed thoroughly under running tap water and rinsed with distilled water. The aerial parts including the leaves with long petioles were segregated from the underground part and were dried in hot air chamber at 50°C for one week. The dried samples were then grinded into powder using a mixer- grinder machine and the powder sample was stored at 4°C in an air tight container. The dried powder sample was extracted in ethyl acetate (MhEa) using a Soxhlet apparatus in 40°C for 48 Hrs and filtered through filter paper and charcoal column.

2.2 *Chromatographic analysis*

LC-MS analysis of the MhEa extract was performed on a ZORBAX EXT LC-MS instrument having a column (4.6mm×50mm, 5μ) and using NH_4OAc solvent (10mmol L^{-1}) to evaluate the molecular weights of the molecules present. TLC was performed using MERCK TLC Silica gel 60 F_{254} aluminium sheets and eluted by a standardized mixture of n-hexane and ethyl acetate in a ratio of 3:2 as mobile phase and visualized under 254nm UV light. R_f value was determined by calculating the ratio of the distance run by stationary phase to the distance run by mobile phase.

2.3 *Evaluation of in-vitro antibiofilm activity*

Antibiofilm activity of MhEa was assessed following the standard biofilm production assay protocol (Djordjevic *et al.*, 2002) against the Gram-positive strain, *Staphylococcus epidermidis* MTCC 3086 (*S. epidermidis*) and Gram negative strain, *Vibrio cholerae* MTCC 3906 (*V. cholerae*) procured from the Microbial Type Culture Collection and Gene Bank (MTCC), Chandigarh, India and the Microbial Culture Collection (MCC), National Centre for Cell Science, Pune, India and maintained on nutrient agar (1.3% nutrient broth in 2% agar) plates at 4°C. A loop full of culture was grown in 1.3% Nutrient broth (NB) and incubated at 32°C for 18 hrs to use as seed culture. Overnight cultures were transferred (1%) to 10 ml of NB and 100 μl of this culture was transferred into 96-well microtiter plate and treated with the test sample and grown in 37°C for overnight with agitation at 180 rpm. Medium was carefully removed from wells without disturbing the bio films. Wells were washed five times with sterile distilled water to remove loosely associated bacteria. Plates were air dried for 45 min. Each well was stained with 150 μl of 1% crystal violet solution in water for 45 min. Plates were de-stained with sterile distilled water five times. At this point, bio films were visible as purple rings formed on the side of each well. The quantitative analysis of bio film production was performed by adding 200 μL of 95% ethanol to de-stain the wells and 100 μL from each well was transferred to a new microtiter plate. The purified compound and ethyl acetate solvent was taken as test sample and positive control, respectively. Only microorganisms without solvent and test sample were taken as negative control. Absorbance was taken at 595 nm using a microtiter plate reader (Biotek). The average optical density (OD) from the control wells were subtracted from the OD of all test wells.

2.4 *Assessment of in-vitro cytotoxicity*

Cytotoxicity of MhEa was evaluated by cell proliferation assay based on the reduction of 3-[4, 5-dimethylthiazol-2-yl]-2, 5-diphenyl tetrazolium bromide (MTT), a yellow coloured tetrazole dye in living cells which can be measured colorimetrically following standard protocol (Alley *et al.*, 1986). On the first day of experiment, trypsinized viable hypo-triploid human epithelial lung carcinoma cells (A 549 ATCC CCL-185™) which was procured from ATCC, USA, were added to 5 ml media and centrifuged in a sterile 15 ml falcon tube at 500 rpm in the swinging bucked rotor (~400 x g) for 5 min. Media were removed and cells were re-suspended to 1.0 ml with complete media and cells were counted and recorded as count/ml. Then it was diluted to 75,000 cells per ml and 100 µl i.e., 7500 nos. of total cells were added into each well of a 96 well micro-plate and incubated overnight. On the second day, cells were treated with active compound and final volume made up to 100 µl per well. On the last day of experiment, 20 µL of 5 mg/ml MTT added to each well. One set of wells were included with MTT but no cells, i.e., treated as control set. All the sets were incubated for 3.5 hours at 37° C in culture hood. Media were carefully removed without disturbing cells or rinsing with PBS. 150 µl of MTT solvent was added and covered with tinfoil cells agitated on orbital shaker for 15 min. Absorbance was read at 590 nm with a reference filter of 620 nm.

2.5 *Validation of data and statistical analysis*

All the experiments had been performed in triplicate. Arithmetic Mean (AM) and standard error of mean (SEM) of all the results were calculated. The data were statistically validated as arithmetic mean ± standard error of mean (AM ± SEM).

Figure 1. TLC chromatogram of MhEa: S_1 is the eluted compounds with Rf value of 0.92.

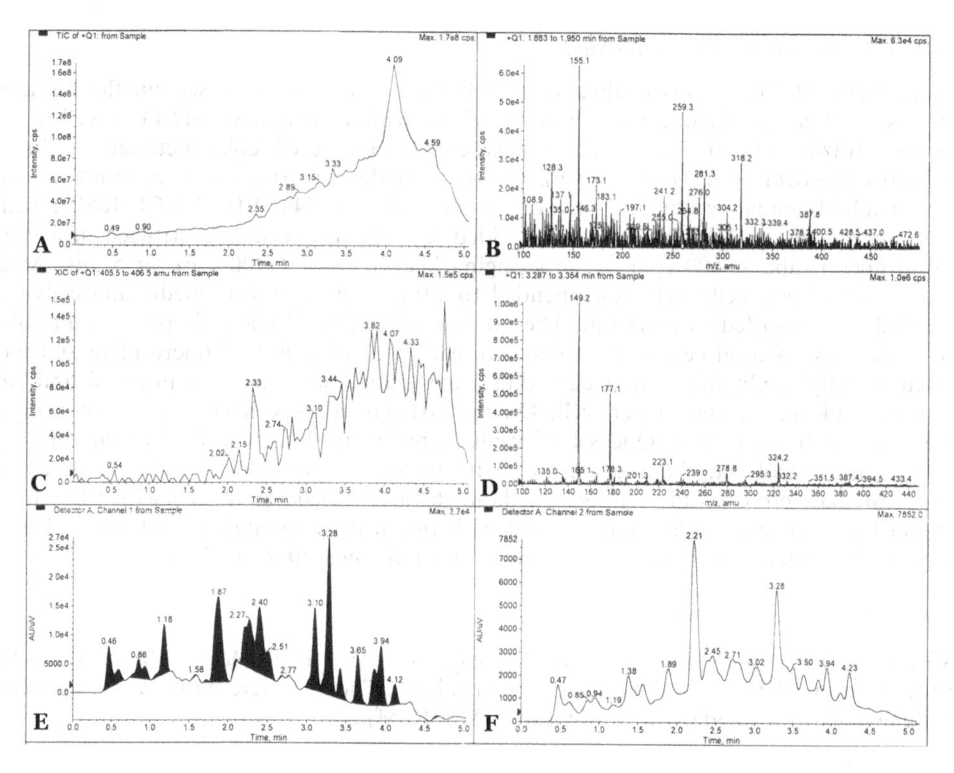

Figure 2. LC-MS chromatograms of the ethyl acetate extract.
(A) Total spectrum from sample; (B) Mass spectra from 1.883 min to 1.950 min; (C) Spectra from 405.5 amu to 406.5 amu from sample; (D) Mass spectra from 3.287 min to 3.354 min; (E) Compound peaks at 220 nm; and (F) Compound peaks at 260 nm.

3 RESULTS AND DISCUSSION

3.1 *Chromatographic analysis*

The TLC of MhEa eluted in a solution of n-hexane and ethyl acetate (3:2 v/v) with few drops of formic acid was visibleunder 254nm short UV light with single spot (S_1) at retardation factor (Rf) value of 0.92 as shown in the Figure 1. The LC-MS analysis of MhEa confirmed the presence of a number of compounds with different % Area and % Height as shown in the Figure 2. The peaks with their characteristic features and relative abundances which were detected at different retention times as mentioned in Table-1.

3.2 *In-vitro anti-biofilm activity*

In the antibiofilm activity assay the OD value at the concentration level 60μg/ml was 1.40 and 2.40 for *V. cholerae* and *S. epidermidis* respectively and at 480μg/ml concentration, it was 0.75 for both the microorganisms while the absorbance of cells without solvent and that with solvent had the same absorbance, *i.e.*, 1.6 and 2.7 for *V. cholera* and *S. epidermidis* respectively as shown in the Figure 3.

3.3 *In-vitro cytotoxicity activity*

The results of the MTT cytotoxicity assay with the human epithelial lung carcinoma cells were between 0.10 to 0.16 OD values along ascending concentration gradient ranging between 37.5μg/ml and 300μg/ml as shown in the Figure 4.

Table 1. LC-MS peak profile of the ethyl acetate extract.

S.No.	Time (min)	Area (counts)	% Area	Height	% Height	Width (min)	Type
1	0.4625	3.5656e4	4.9675	7353.7320	4.9891	0.1717	Valley
2	0.5933	1.2186e4	1.6978	2319.9669	1.5740	0.1267	Valley
3	0.8571	1.5344e4	2.1377	3303.9013	2.2415	0.1650	Valley
4	0.9355	9438.6174	1.3150	2214.5917	1.5025	0.1167	Valley
5	1.1823	3.9388e4	5.4875	8571.1971	5.8150	0.2383	Base to Base
6	1.5825	773.3578	0.1077	248.3737	0.1685	0.0700	Base to Base
7	1.7208	1474.5870	0.2054	468.5403	0.3179	0.0783	Valley
8	1.8725	1.0681e5	14.8807	1.4765e4	10.0171	0.2650	Valley
9	2.1036	2944.4931	0.4102	587.0031	0.3982	0.0750	Base to Base
10	2.2105	1.9538e4	2.7220	6791.131	4.6074	0.0833	Valley
11	2.2698	4.8675e4	6.7812	8401.3192	5.6998	0.1150	Valley
12	2.3987	6.8599e4	9.5570	1.1446e4	7.7655	0.1433	Valley
13	2.5088	2.0370e4	2.8379	4513.5536	3.0622	0.1267	Valley
14	2.6850	681.8419	0.0950	241.3933	0.1638	0.0683	Base to Base
15	2.7696	324.4959	0.0452	202.8974	0.1377	0.0450	Base to Base
16	3.1006	6.6279e4	9.2338	1.4355e4	9.7388	0.2117	Valley
17	3.2818	1.2209e5	17.009	2.7191e4	18.4471	0.1967	Valley

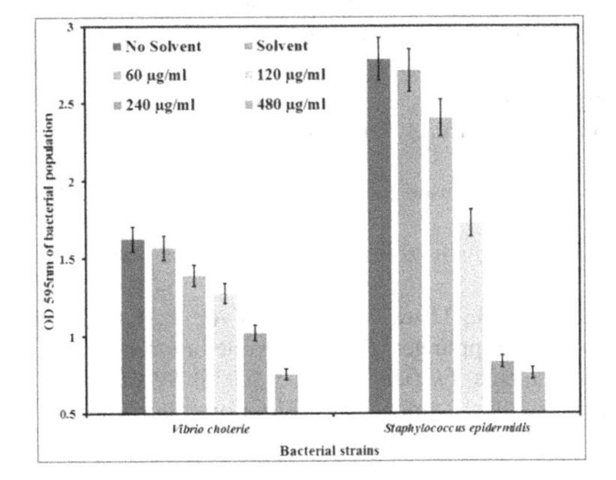

Figure 3. Antibiofilm activity of the purified compound: Concentration of the test sample ranged from 60µg/ml to 480µg/ml.

Figure 4. Is the effect of the compound on hypotriploid human epithelial lung carcinoma cells A 549 ATCC CCL-185™.

4 CONCLUSION

The MhEa fraction was yellow coloured, semi solid and having UV sensitivity at 254.1 nm with a *Rf* value of 0.92 and contained an array of active molecules. The LC-MS chromatogram of the sample showed that the peak at the retention time 3.28 min was the base peak with peak area of 17% and another peak was detected at the retention time 1.87 min with peak area of 14%. The mass to charge ratio (m/z, amu) at these two points were 149.2 and 155.1 respectively. The anti-biofilm activity assay in different concentrations demonstrated drastic change in opacity at each higher concentration level. The absorbance of cells without solvent and that with solvent had the same absorbance, *i.e.*, 1.6 and 2.7 for *V. cholera* and *S. epidermidis,* respectively. Hence, there was no any interference of the solvent in the experiment. The OD value at the concentration level 60µg/ml was 1.40 and 2.40 for *V. cholera* and *S. epidermidis,* respectively and at 480 µg/ml concentration; it was 0.75 for both the microorganisms. So, the experiment confirmed that the bioactive fraction had potential antibiofilm activity against both the organisms and the biofilm production was much hampered in Gram-positive strain, *S. epidermidis* than the Gram-negative strain, *V. cholera* by the purified compound. The toxicological study applying MTT cytotoxicity assay with A 549 ATCC CCL-185™ cells did not show any drastic change regarding proliferation of the hypo-triploid human epithelial lung carcinoma cells. All the results were between OD values 0.10 to 0.16 along ascending concentration gradient. Hence, it can be stated that the purified compound had no any cytotoxicity effect even in 300 µg/ml concentration up to which the experiment commenced. So, it is pharmacologically safe for use (Edziri *et al.*, 2012). Hence, MhEa was more susceptible to inhibit the bio-film produced by the Gram-positive strains than that of the Gram-negative strains without any cytotoxicity effect to human.

ACKNOWLEDGEMENT

We are thankful to Mr. Raihanuddin, Research Scholar of Dept. of Microbiology, University of Calcutta (W.B.), India, for his cooperation and CHEMBIOTEC, Kolkata, for LC-MS study.

REFERENCES

[1] Christou, L., The global burden of bacterial and viral zoonotic infections. *Cl. Micro. Infect.*, 17(3), (2011), 326-330.

[2] Friedman, M., Antibiotic-Resistant Bacteria, Prevalence in Food and Inactivation by Food-Compatible Compounds and Plant Extracts. *J. Agric. Food Chem.*, 63 (15), (2015), 3805–3822.

[3] Carlsson, J., Regulation of sugar metabolism in relation to feast-and-famine existence of plaque. *In Cariology Today*, Ed: Guggenheim B., Basel: Karger, (1983).

[4] Balick, M., Cox, P., Plants, People and Culture: The Science of Ethnobotany. Scientific American Library, New York, (1996).

[5] Fabricant, D.S., Farnsworth, N.R., The value of plants used in traditional medicine for drug discovery. *Environ. Heal. Pers.*, 109, (2001), 69–75.

[6] Compean, K.L., Ynalvez, R.A., Antimicrobial activity of plant secondary metabolites, a review. *Res. J. Med. Pl.*, 8(5), (2014), 204–213.

[7] Wink, M., Alfermann, A.W., Franke, R., Wetterauer, B., Distl, M., Windho¨vel, J., Krohn, O., Fuss, E., Garden, H., Mohagheghzadeh, A., Wildi, E., Ripplinger, P., Sustainable bioproduction of phytochemicals by plant in vitro cultures: anticancer agents. *Pl. Genet. Res.*, 3, (2005), 90–100.

[8] Sugimoto, M., Kawakami, M., Robert, M., Soga, T., Tomita, M., Bioinformatics tools for mass-spectroscopy-based metabolomic data processing and analysis. *Curr. Bioinform.* 7, (2012), 96–108.

[9] Misra, D., Mandal, M., Ghosh, N.N., Mandal, V., Pharmacognostic standardization of an ethnomedicinal aquatic herb, *Monochoria hastata* (L.) Solms for its antibacterial potentiality. *Pharm. J.* 10 (3), (2018), 533–540.

[10] Djordjevic, D., Wiedmann, M., McLandsborough, L.A., Microtiter plate assay for assessment of *Listeria monocytogenes* biofilm formation. *Appl. Environ. Microb.* 68, (2002), 2950–2958.

[11] Alley, M.C., Scudiere, D.A., Monks, A., Czerwinski, M., Shoemaker, R., Boyd, M.R., Validation of an automated microculture tetrazolium assay (MTA) to assess growth and drug sensitivity of human tumor cell lines. *Proc Am Assoc Cancer Res.* 27, (1986), 389–391.

[12] Edziri, H., Mastouri, Maha., Mahjoub, M.A., Mighri, Z. Mahjoub, A., Verschaeve, L., Antibacterial, antifungal and cytotoxic activities of two flavonoids from *Retama raetam* Flowers. *Molecules*, 17, (2012), 7284–7293.

Biotechnology and Biological Sciences – Sen et al. (Eds)
 ISBN 978-0-367-43161-7

A comparative assessment of silver nanoparticles synthesised using aqueous and hydromethanolic Licorice extracts for its antimicrobial and cytotoxicity potential

Arwa Gheewala & Sudeshna Menon
Department of Biochemistry and Biotechnology St. Xavier's College (Autonomous), Ahmedabad

ABSTRACT: Nanotechnology has found application in various fields including biology. Chemically synthesized nanoparticles have been studied for possible biological activities. Due to increasing resistance against antibiotics, silver nanoparticles (AgNPs) are being used as an alternative for antibiotics. Moreover, green synthesis of nanoparticles has also found application in cancer biology as potential cytotoxic agents. In this study, green synthesis of AgNPs using powdered Licorice roots and two extraction methods have yielded nanoparticles. The bioreduction of silver nitrate by both the plant extracts (aqueous and hydromethanolic extracts) was studied by UV-Vis spectroscopy and characterized by Dynamic Light Scattering (DLS). The AgNPs synthesized using aqueous extract were of an average size of 81.89 nm while AgNPs synthesized using hydromethanolic (HM) extract were of an average size of 84.81 nm. Assessment of potential antimicrobial and antiproliferative activities of these nanoparticles showed that the AgNPs synthesised using aqueous extract showed higher antimicrobial potential against bacteria and the cytotoxicity potential of AgNPs, synthesized from aqueous extract, against *S. cerevisiae* increased with increasing dosage.

1 INTRODUCTION

One of the concerns in the present century is the diversity of diseases and their causative pathogens. Several drugs have been developed over the years and yet today there is an exploration for inexplicable number of means to combat such diseases. Hence, there is a concerted effort to identify efficient remedies. Silver is the most preferred metal over other noble metals because of its unique properties such as chemical stability, good conductivity, catalytic activities and most importantly, antibacterial, antiviral, antifungal in addition to anti-inflammatory activities. Silver has been used in various forms such as metallic silver, silver nitrate and silver sulfadiazine for the treatment of wounds, burns and several bacterial infections (Tawfeeq et al. 2015).

1.1 *Synthesis of nanoparticles*

Nanoparticles are synthesized from a variety of chemical and physical methods that are quite expensive and hazardous to the environment. There are two approaches for the synthesis of silver nanoparticles: "top to bottom" approach and "bottom to top" approach. In "bottom to top" approach, the atoms self-assemble and form new nuclei and then the particles are formed into nanoscale (Murray et al. 2001). In "top to bottom" approach, bulk material is broken down into fine particles that are in nano range (Murray et al. 2001). The reduction of silver ions in an aqueous solution generates a colloidal solution with Ag particles and the particle diameter is of several nanometers. When the particle size is reduced agglomeration occurs and oligomeric clusters are formed (Kapoor et al. 1994). These clusters make a colloidal solution of silver with Ag^{0}. When the colloidal particles are much smaller than the wavelength of visible light, the colour of the solution changes from transparent-clear to yellow and gives a sharp distinct band in the 380 nm-400 nm range in the absorption spectrum (Tessier et al. 2000; Cao et al. 2002; Rosi and Mirkin 2005).

In the current study, the AgNPs are synthesised from the aqueous and HM extract of Licorice roots and were assessed for its antimicrobial and cytotoxic potential.

2 MATERIALS AND METHODS

2.1 *Collection and preparation of the plant extracts*

The roots of *Glycyrrhiza* sp. were collected from a local shop based in Banswara, Rajasthan in September. The collected roots were washed thrice in running tap water and then with distilled water and allowed to shade dry. After the roots were completely dry, they were crushed and then blended to obtain the powder. The powder obtained was sieved through the mesh and two textures of the powder was found: coarse powder and fine powder.

This method was adapted from Mohammed et al (2018). The aqueous extract was prepared from both the powders by decoction method. An amount of 0.05 grams of both the powders were taken and mixed with 5mL of distilled water. One set was kept at 80°C while the other set was kept at 90°C for 1 hour in boiling water bath. The extract obtained was filtered through Whatmann No. 1 filter paper (pore size 25 μm) and stored at 4°C.

The hydromethanolic (HM) extract was prepared by Soxhlet extraction method using 10 grams of fine powder and 5 grams of coarse powder. The extraction was done using 250 ml Methanol at 60°C. The fine powder took 6 cycles while the coarse took 4 cycles until a light-yellow colored solution was obtained. The solution was then poured into petri plates and the methanol was allowed to evaporate. After the evaporation powder was left behind which was scrapped out and stored in the vials. Whenever needed the powder was dissolved in distilled water and then it was used as HM extract.

2.2 *Synthesis of silver nanoparticles (AgNPs)*

The components present in the plant extract have redox potential (Ahmad and Sharma 2012). So, plant extract was used to reduce the silver into its elemental form Ag° or Ag^{+}. This method was adapted from Girón-Vázquez et al. (2019). 50mL of 1mM silver nitrate was mixed with 1mL of both the extracts and kept on magnetic stirrer for at least 24 hours.

2.3 *Characterization of the AgNPs*

The synthesised AgNPs were characterized by UV Spectroscopy and Dynamic Light Scattering techniques.

2.4 *Assessment of antimicrobial potential*

The antimicrobial potential was determined by agar cup diffusion method. This method was adapted from Mohammed et al. (2018). This was determined against *Pseudomonas aeruginosa, Staphylococcus aureus, Serratia marcescens* and *Bacillus megaterium.* In this method, the media for the bacteria were prepared and autoclaved. When the temperature of the media was bearable the bacterial culture was added to it. The media was then poured into the petri plates and allowed it to solidify. After the solidification of the agar, wells were bored in the plate and then the growth inhibitors such as the AgNPs synthesised from both the extracts and 1mM silver nitrate were added. The solution of the AgNPs would diffuse from the well and the bacterial growth would be inhibited in the area in which the solution has been diffused. The plates were kept in the incubator at 37°C for 24 hours. After 24 hours of the incubation the diameter of the zone of inhibition was observed.

2.5 *Cytotoxicity of AgNPs*

The cytotoxicity of AgNPs was studied by 3-(4,5-dimethylthiazo l-2-yl)-2,5- diphenyltetrazolium bromide (MTT) assay against *S. cerevisiae*. This method was adapted from Pandian and Chidambaram (2017). For MTT assay, the culture was prepared by inoculating a loopful of *S. cerevisiae* in the YPD media and grown to a log phase at an optical density of 0.4 at 600 nm. 100 μL of the culture was then transferred to 96-well plate containing 1 mM silver nitrate, AgNPs synthesised from the two plant extracts with their respective plant extracts. The volume of the growth inhibitors varied from 10 μL-50 μL followed by 70 μL-30 μL phosphate-buffered saline (PBS) and incubated at 37°C for 24 hours. Blank and control was prepared by adding 100 μL of the media and culture respectively followed by autoclaved distilled instead of growth inhibitors and PBS in the same volume as per the sample wells. After the incubation, 20 μL MTT dye was added to all the wells and then incubated for another 4 hours. After 4 hours of incubation, dimethyl sulfoxide (DMSO) was added to solubilize the formazan crystals formed by the reduction of MTT by the reducing factors and enzymes such as NADH or NADPH, succinate dehydrogenase or Cytochrome *c* present in the cytoplasm, mitochondria or endoplasmic reticulum of the cell. The following formula was used to calculate % viable cells:

$$\% \text{ Viable Cells} = \frac{(Abs\ sample) - (Abs\ blank)}{(Abs\ control) - (Abs\ blank)} X\ 100$$

3 RESULTS

3.1 *Synthesis of AgNPs and its characterization*

The aqueous extract and the hydromethanolic extract reduced 1mM silver nitrate to its nano range within 48 hours as the colour of the solution changed from yellow to dark brown. The bioreduction of silver was confirmed by UV-Vis spectroscopy at 350-500 nm. The AgNPs synthesised from aqueous extract gave the peak at 450 nm in 27 hours (Figure 1) while the AgNPs from HM extract gave the peak at 440 nm in 41 hours (Figure 2). Ahmad and Sharma showed that the AgNPs gave sharp distinct peak at 430nm due to the Surface Plasmon Resonance. They also showed that as the reaction time increased the peak became sharper and sharper. This could be because of the synthesis of more nanoparticles due to the reduction process going on. The slight variation in the absorption peak in our study could be because of the plant extract.

The particle size analysis was done by DLS technique, which gave the average size of the AgNPs. The average size of the AgNPs synthesised from aqueous extract was 81.89 nm with PDI value 0.377 (Figure 3) while the average size of the AgNPs synthesised from the HM extract was 84.81 nm with PDI value 0.395 (Figure 4). This suggests that the particles obtained

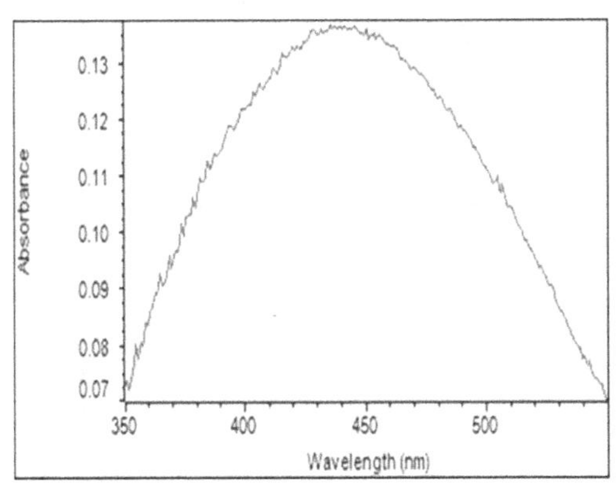

Figure 1. Peak of AgNPs from aqueous extract at 27th hr.

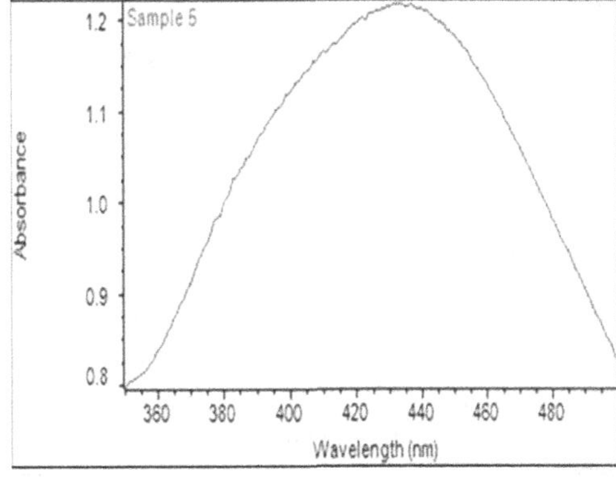

Figure 2. Peak of AgNPs from HM extract at 41st hr.

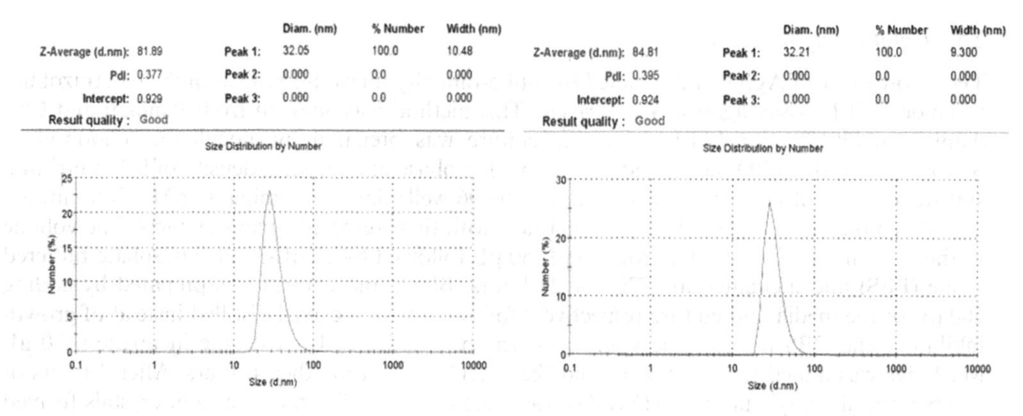

Figure 3. Particle size analysis of AgNPs from aq. extract.

Figure 4. Particle size analysis of AgNPs from HM extract.

are in nano range with mid-range size distribution, viz. the particles present in the colloidal solution has both larger as well as smaller particles which is shown by the PDI value.

3.2 *Assessment of antimicrobial potential*

The diameter of the zone of inhibition was observed and reported in mm. The AgNPs synthesised from aqueous extract showed higher antimicrobial potential against Gram positive as well as Gram negative bacteria as compared to the AgNPs synthesised from HM extract and silver nitrate (Table 1). As observed in Figure 5, the antimicrobial potential of the green synthesised silver nanoparticles has enhanced antimicrobial potential. MubarakAli et al. showed that the AgNPs synthesised from *M. piperita* showed higher antibacterial activity against Gram negative *E. coli* as compared to the Gram positive *S. aureus*.

3.3 *Cytotoxicity of AgNPs*

The growth inhibitors such as the AgNPs from both the extracts and silver nitrate decreased the cell viability as the concentration increased. As the concentration of growth inhibitors increased from 10 µL to 50 µL the cell viability decreased from 100% to 40% in case of AgNPs synthesized from aqueous extract of fine powder of Liquorice (Table 2). The other preparations too decreased viability when the volume of nanoparticles was increased five times. Pandian and Chidambaram in their study too stated that the AgNPs synthesised from *Glycyrrhiza glabra* showed potent cytotoxic activity against HeLa cells. The concentration of synthesized AgNPs at 7.8 µg/ml, 15.6 µg/ml, 31.2 µg/ml, 62.5 µg/ml, 125 µg/ml showed cytotoxic activity of 60.8%, 57.2%, 54.13%, 51.13% respectively.

Table 1. Zone of inhibition of AgNPs.

Micro organism	Silver nitrate (mm)	AgNPs from FE-S (mm)	AgNPs from FE 90 (mm)	AgNPs from FE 80 (mm)	Plant extract (mm)
P. aeruginosa	11.5 ± 0.45	11.5 ± 0.42	13 ± 1.93	13 ± 2.46	8.6 ± 0.40
S. aureus	13.5 ± 0.59	15.4 ± 1.78	16.5 ± 1.62	15 ± 0.35	9 ± 0.45
S. marcescens	13.4 ± 0.31	14.5 ± 0.2	16.2 ± 1.54	15.2 ± 1.08	12 ± 1.3
B. megaterium	11.4 ± 0.79	13.6 ± 2.49	15.2 ± 1.29	14.1 ± 1.5	8 ± 1

FE-S: HM extract of fine powder, FE 90: aqueous extract of fine powder treated at 90°C, FE 80: aqueous extract of fine powder treated at 80°C

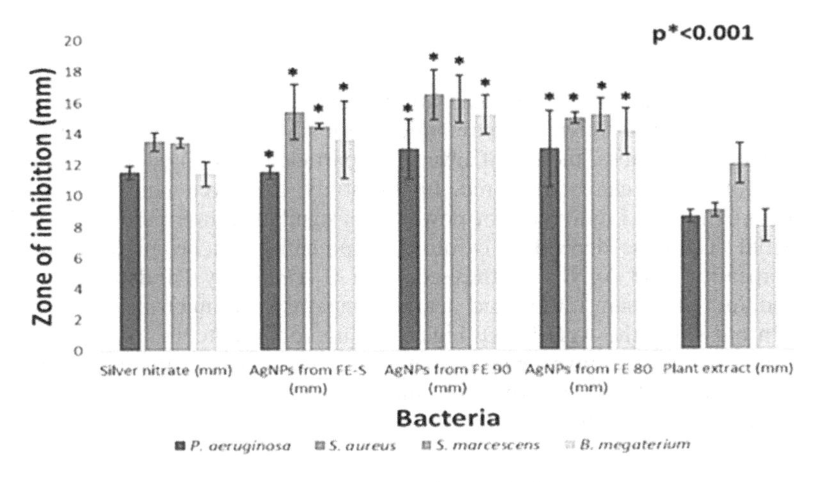

Figure 5. Graphical representation of agar cup diffusion method.

FE-S: HM extract of fine powder, FE 90: aqueous extract of fine powder treated at 90°C, FE 80: aqueous extract of fine powder treated at 80°C, *: denotes the p-value

Table 2. Cell viability of *S. cerevisiae* with increasing volume of AgNPs by MTT assay.

	10 μL	20 μL	30 μL	40 μL	50 μL
$AgNO_3$	165% ± 10%	45% ± 10%	23% ± 10%	156% ± 10%	42% ± 10%
CE-S	171% ± 13%	86% ± 13%	75% ± 13%	93% ± 13%	39% ± 13%
CE-S PE	186% ± 12%	93% ± 12%	105% ± 12%	134% ± 12%	100% ± 12%
FE 90	104% ± 15%	64% ± 15%	40% ± 15%	39% ± 15%	36% ± 15%
FE 90 PE	173% ± 17%	156% ± 17%	157% ± 17%	167% ± 17%	137% ± 17%

CE-S: AgNPs from HM extract of coarse powder, CE-S PE: HM extract of coarse extract FE 90: AgNPs from aqueous extract of fine powder treated at 90°C, FE 90 PE: aqueous extract of fine powder treated at 90°C.

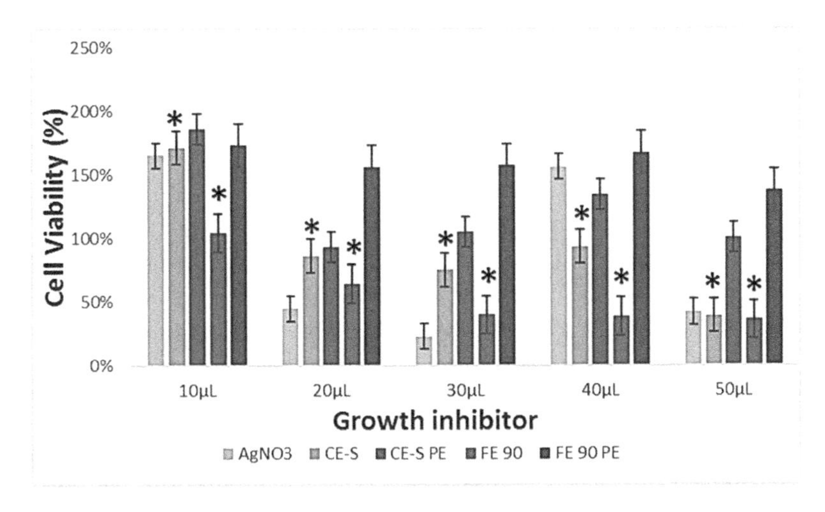

Figure 6. Graphical representation of cytotoxicity of AgNPs against *S. cerevisiae*.

CE-S: AgNPs from HM extract of coarse powder, CE-S PE: HM extract of coarse powder plant extract, FE 90: AgNPs from aqueous extract of fine powder treated at 90°C, FE 90 PE: aqueous extract of fine powder treated at 90°C, *: denotes the p-value.

4 CONCLUSION

This study is an affirmation of enhanced antimicrobial and anti-proliferative (cytotoxicity) potential of plant extracts when in combination with silver. In therapeutics, with increasing resistance against antibiotics, the use of silver nanoparticles synthesised using plant extracts is thus an alternative that can be explored. In this study, the silver nanoparticles synthesised using aqueous extract of Liquorice root powder has significant antimicrobial as well as cytotoxic potential. The dose dependent cytotoxicity potential can be investigated further for cancer therapeutics. The AgNPs synthesised from *G. glabra* by Pandian et al. (2017) and Anburaj and Jothiprakasam (2018) had nearly comparable antimicrobial potential with the antibiotic. In our study the AgNPs had higher antimicrobial potential as compared to silver nitrate. Further studies can investigate the molecular machinery of the liquorice synthesized nanoparticles, which in turn can give an insight into its use as a drug. There is also a need to ascertain minimum side effects.

REFERENCES

Ahmad, N., & Sharma, S. (2012). Green synthesis of silver nanoparticles using extracts of *Ananas comosus*. Green and Sustainable Chemistry, 2(04), 141.

Anburaj R., & Jothiprakasam V. (2018). *Glycyrrhiza glabra* as a potential synthesizer of silver nanoparticles and their microbicidal action. Int. J. Pharm. Sci. Rev. Res., 51(1),137–142.

Kaur R, Kaur H, Dhindsa A S, G *Glycyrrhiza glabra:* A phytopharmacological review. International Journal of Pharmaceutical Science and Research 2013; 4: 2470–22477.

Mohamad N A N, Arham N A, Jai J, Hadi A. Plant extract as reducing agent in synthesis of metallic nanoparticles: A Review. Advanced Materials Research. Vol 832 (2014);350–3355.

Mohammed, A., Al-Qahtani, A., Al-Mutairi, A., Al-Shamri, B., & Aabed, K. (2018). Antibacterial and cytotoxic potential of biosynthesized silver nanoparticles by some plant extracts. Nanomaterials, 8(6), 382.

Mohammed, A., Al-Qahtani, A., Al-Mutairi, A., Al-Shamri, B., & Aabed, K. (2018). Antibacterial and cytotoxic potential of biosynthesized silver nanoparticles by some plant extracts. Nanomaterials, 8(6), 382.

MubarakAli, D., Thajuddin, N., Jeganathan, K., & Gunasekaran, M. (2011). Plant extract mediated synthesis of silver and gold nanoparticles and its antibacterial activity against clinically isolated pathogens. *Colloids and Surfaces B: Biointerfaces*, *85*(2), 360–365.

Pandian, N., & Chidambaram, S. (2017). Antimicrobial, cytotoxicty and anti cancer activity of silver nanoparticles from Glycyrrhiza glabra. *Int J Pharm Sci Res*, *8*(4), 1633–1641.

Tawfeeq, A. T. (2018). Genotoxicity of Silver Nanoparticles synthesized by Laser Ablation Method in Vivo. Iraqi Journal of Cancer and Medical Genetics, 8(1).

Biotechnology and Biological Sciences – Sen et al. (Eds)
© 2020 Taylor & Francis Group, London, ISBN 978-0-367-43161-7

Crude polysaccharides from two Russuloid myco-food potentiates murine macrophage by tuning TLR/NF-κB pathway

Somanjana Khatua & Krishnendu Acharya
Molecular and Applied Mycology and Plant Pathology Laboratory, Department of Botany, University of Calcutta, Kolkata, West Bengal, India

ABSTRACT: During recent field survey, two morphologically unique myco-food were collected that have ethically been praised for health promoting effects. Such customary practice tempted us to investigate on them where one taxon emerged as novel namely *Russula alatoreticula* K. Acharya, S. Khatua, A.K. Dutta & S. Paloi, sp. nov.; while the other specimen i.e. *Russula senecis* S. Imai was recorded as tribal delicacy for the first time. Alongside, water soluble crude polysaccharides were isolated from each macrofungus consisting β-glucan as the principal component. Subsequently, the fractions demonstrated strong immune-boosting property evident by augmentation of macrophage viability, phagocytosis, NO production, ROS generation and pseudopod formation. Thereafter, through RT-PCR analysis, significant increase in expression of TLR-2, TLR-4, NF-κB, COX-2, TNF-α, Iκ-Bα, IFN-γ and iNOS were observed explaining mode of action through TLR/NF-κB pathway. Thus, the results scientifically validate ethnic use of the studied mushrooms and suggest further study for development of novel nutraceuticals.

1 INTRODUCTION

Since time immemorial, mushrooms have captured increasing attention of humankind across the globe being included in gourmet cuisine due to their organoleptic merits. Later, folks also recognized the significant healing properties and considered fungi as powerful drug to fight illness (Khatua & Acharya 2016). However, many countries including India did not have any written document and such extensive knowledge has been verbally transmitted to the next generation. Consequently, mushrooms still play an important role in rural alimentary strategies and economic activities. Today researchers have explored that mushrooms possess about 100 different medicinal activities of which immune enhancing effect is a key property. In this context, fungal polysaccharide particularly β-glucan is known as the most powerful immune stimulant as it can activate macrophages (Ayeka 2018). Macrophages are the first cell in human body to fight against invading pathogens and act as bridge between innate as well as adaptive immune responses (Tabarsa, Karnjanapratum, Cho, Kim & You 2013).

In this context, India harbors a treasure house of basidiomycetes being blessed with diverse agro-climatic zones. Amongst 29 states, West Bengal is the only territory being extended from Himalaya in northern extreme to Bay of Bengal down in south, with plateau and Ganges delta intervening in between. Besides, West Bengal is the home of various ancestral mycophagy groups like Santal, Munda, Lodha, Bhutia etc. who are till now living in forest areas isolated from mainstream communities. Many such jungle is dominated with Sal, *Shorea robusta*, that facilitates fruiting of several ectomycorrhizal mushrooms in rainy season among which members of *Russula* are very common (Pradhan, Dutta, Roy, Basu & Acharya 2012).

The genus is considered as one of the most widely distributed group, characterized by fairly large and colorful fruit bodies. Several members have traditionally been cherished as food and medicine from time immemorial (Das, Dowie, Li & Miller 2014). However despite the wide distribution and ethnic value, members of *Russula* are not prized by elite societies. It may be explained by the myth that common people recognize intensely colorful macrofungi as poisonous and this misconception may play a significant role behind their disapproval. In addition, lack of knowledge also play a key role in this situation. Thus, scientific exploration on these macromycetes is essential not only to reduce gap between our traditional practice and modern life-style but also to convey the precise information regarding edibility of these bio-resources to civilized publics.

2 MATERIALS AND METHODS

2.1 *Collection and authentication of basidiocarps*

Two morphologically unique *Russula* sp possessing red and yellow colored pileus were collected under Sal tree in lateritic areas of West Bengal during monsoon season. Edibility of the taxa was registered in consultancy with inhabited tribal people and local market survey. Identification of the gathered specimens was accomplished after thorough characterization based on taxonomy and phylogenetic position. Subsequently, representative voucher specimens were deposited in Calcutta University Herbarium.

2.2 *Extraction of crude polysaccharide*

10 gm of the powdered fruit bodies were steeped in absolute ethanol to remove fat and the dehydrated filtrate was then refluxed with 400 ml of distilled water under boiling condition. After 7 h, the extract was cooled and filtered through nylon cloth to isolate the remnants. Further, four volume of absolute ethanol was added to the mixture and left at 4°C overnight to precipitate macromolecules. Next, the extract was centrifuged and isolated pellet was repeatedly washed with ethanol as well as acetone to acquire hot water extracted crude polysaccharide.

2.3 *Determination of architecture of isolated polysaccharides*

Total sugar content was measured by phenol sulphuric acid method using glucose as standard. Quantity of total glucan and its types were estimated using Mushroom and Yeast β-Glucan Assay kit as per the manual. Protein concentration was determined using Bradford reagent. Further, the polymers were subjected to molecular composition analysis by high performance thin layer chromatography (HPTLC) and gas chromatography mass spectroscopy (GC-MS). Fourier transform infrared (FT-IR) spectra were recorded on PerkinElmer Precisely Spectrum 100 Model (USA) in frequency range 400-4000 cm^{-1}. Finally, helical structure of carbohydrate backbone was analyzed by characterizing Congo red-polysaccharide reaction (Khatua & Acharya 2016).

2.4 *Determination of immune-stimulatory potential*

RAW 264.7 murine macrophages were purchased from National Centre for Cell Science, India. The cells were maintained in Dulbecco's Modified Eagle Medium (DMEM) supplemented with 10% fetal bovine serum, 0.5% PenStrep (5,000 IU/ml penicillin and 5 mg/ml streptomycin) and 0.25% amphotericin B (250 μg/ml). Effect of polysaccharides on cellular viability and phagocytic uptake was determined by using water soluble tetrazolium (WST) and neutral red reagent respectively. Influence on nitric oxide (NO) and reactive oxygen species (ROS) production were estimated by Griess and 2′,7′-dichlorofluorescin diacetate (DCFDA) reagents respectively. Cellular morphology was viewed and photographed with the help of fluorescent microscope. The total RNA of macrophage cells was extracted and reverse transcribed into cDNA. The cDNA was further amplified using primers (Table 1) specific to Toll like receptor (TLR)-4, TLR-2, nuclear factor (NF)-κB, IκB-α, cyclooxygenase (COX)-2, inducible nitric oxide synthase (iNOS), tumor necrosis factor (TNF)-α and interferon (IFN)-γ

Table 1. Primer sequences used to study immune-stimulation activity.

Gene	Primer sequence	Tm (°C)
TLR-4	F: 5′CAGCTTCAATGGTGCCATCA3′ R: 5′CTGCAATCAAGAGTGCTGAG3′	54
TLR-2	F: 5′CACCACTGCCCGTAGATGAAG3′ R: 5′AGGGTACAGTCGTCGAACTCT3′	57
NF-κB	F: 5′AGAAGGCTGGGGTCAATCTT3′ R: 5′CTCAGGCTTTGTAGCCAAGG3′	51
IFN-γ	F: 5′CCTCAAACTTGGCAATACTCA3′ R: 5′CTCAAGTGGCATAGATGTGGA3′	54
Iκ-Bα	F: 5′CTTGGTGACTTTGGGTGCTGAT3′ R: 5′GCGAAACCAGGTCAGGATTC3′	57
iNOS	F: 5′GAGCGAGTTGTGGATTGTC3′ R: 5′GGGAGGAGCTGATGGAGT3′	55
COX-2	F: 5′CCCCCACAGTCAAAGACACT3′ R: 5′GAGTCCATGTTCCAGGAGGA3′	57
TNF-α	F: 5′ATGAGCACAGAAAGCATGATC3′ R: 5′TACAGGCTTGTCACTCGAATT3′	56
β-Actin	F: 5′GCTGTCCCTGTATGCCTCT3′ R: 5′TTGATGTCACGCACGATTT3′	55

genes where β-actin was used as control (Tabarsa, Karnjanapratum, Cho, Kim & You 2013; Hou, Ding, Hou, Song, Wang, Wang & Zhong 2013). For each gel, ImageJ software was applied for quantitative estimation of band intensity.

2.5 *Statistical analysis*

Results presented herein are expressed from three experiments (n = 3). Differences in mean values between the groups were analyzed by a one-way analysis of variance (ANOVA) with *post-hoc* Tukey HSD test using IBM SPSS statistics (IBM Corp., Armonk, NY, United States).

3 RESULTS AND DISCUSSION

In general, to identify a macrofungal specimen, at least 67 macroscopic characters are noted in field and minimum 40 anatomical features are recorded under microscope. According to the protocol, thorough characterization was performed where the taxon with red colored cap appeared as a novel mushroom which was then entitled as *Russula alatoreticula* K. Acharya, S. Khatua, A.K. Dutta & S. Paloi, sp. nov. On the other hand, the specimen with yellow colored upper surface emerged as *Russula senecis* S Imai evident by morpho-anatomical features and phylogenetic position. Interestingly, consultancy with mycophagy elders revealed that they consider these macrofungi as health promoting food that purportedly enhances overall immunity against seasonal hazards. Such ethnic importance of these mushrooms under consideration was so far hidden from science until our recent expedition.

Alongside the research was further extended, considering the traditional customary of *R. alatoreticula* and *R. senecis* to explore their myco-chemical composition and therapeutic efficacy. For that, a conventional hot water process was followed for each specimen to isolate crude polysaccharides. In the view of molecular composition, both the fractions were consisted of similar range of total carbohydrate suggesting it as the major constituent along with small amount of protein. Experiment was further carried out to quantify glucan content being a major component in mushroom polysaccharides. Total glucan content in both the fractions was enumerated in similar array where β-glucan was detected as the predominant component.

The same output was reflected in FT-IR spectra that presented specific bands for polysaccharides, protein and β-glucan. Further HPTLC and GC-MS were implemented for preliminary characterization of the monomer types present in the fractions. Compared to the controls it could be deducted that the crude polysaccharides extracted from *R. alatoreticula* and *R. senecis* were mainly composed of 3 and 5 monomers respectively where glucose was the main structural unit. Besides, Congo red assay was performed to determine presence of any helical conformation in carbohydrate backbone. Results indicated that the polysaccharides could be ascribed to arrangement into triple strand helical chains in water and such conformation is considered important for biological activities.

To determine immune modulatory potential, RAW 264.7 cell line was used as a model system and in each assay lipopolysaccharide (LPS) was considered as a standard. At first we wanted to know whether the fractions have immune boosting effects or not. For that, cells were challenged with both types of polysaccharides at a range of concentrations. WST results showed that treatment with macromolecules increased cell viability within 24 h in comparison to negative control. Interestingly, extended incubation for another 24 h revealed cell number amplification by more than 3 fold compared to blank set indicating time course activity of fractions (Figure 1a). Thus, data suggested that the polymers may possess immune stimulation activity demanding in-depth studies.

In immune response, the first and imperative defense function of macrophages is represented by phagocytosis that causes ingestion as well as elimination of pathogens. Thus, increase in engulfment power can definitely help to boost immunity and also signals for macrophage stimulation. As shown in the graph 1b, treatment of the studied fractions increased phagocytosis indices, detected by neutral red assay in a time dependent manner.

NO is membrane-permeable an inorganic gas as well as a free radical. It is synthesized from L-arginine by nitric oxide synthase and released by macrophages in response to pathogens. Therefore, increase in NO production is considered as an indication of macrophage activation. As shown in the Figure 1c, negative control cells secreted minimal amount of NO, while treatment of crude polysaccharides at concentrations between 50 and 200 μg/ml upregulated NO

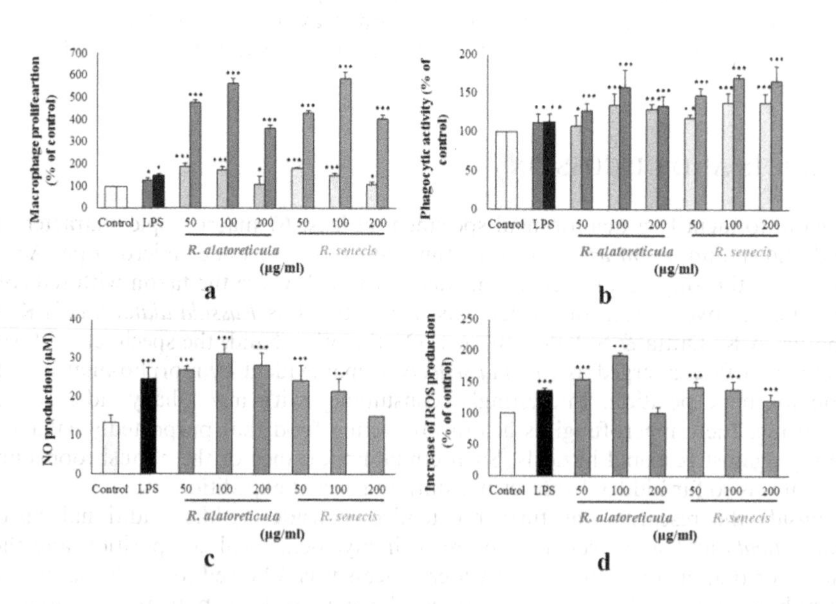

Figure 1. Effect of water extracted crude polysaccharides isolated from studied Russuloid fungi on activity of macrophages such as (a) proliferation (b) phagocytic uptake (c) nitric oxide production (d) reactive oxygen species generation. In all assays, LPS at the concentration of 5 μg/ml was used as a positive control. Values are presented herein as mean ± standard deviation of at least three independent experiments. (*p< 0.05, ***p< 0.001, unpaired t-test).

level significantly. Of note, extract from *R. alatoreticula* presented the best effect maximizing at 100 μg/ml dosage.

Besides, superoxide possess ability to react with NO leading to formation of peroxynitrite that increases antibacterial and cytotoxic effects of macrophages. Thus synthesis of ROS by activated monocytes is regarded as an important cytostatic/cytotoxic mechanism of non-specific immunity that has been assessed herein by measuring DCF fluorescence. The fluorescence intensity in cells treated with studied fractions was significantly increased as peaks shifted to the right. When compared with the control group, fluorescence intensity was noticeably enhanced and the highest activity was exhibited by *R. alatoreticula* fraction at 100 μg/ml concentration (Figure 1d).

After facing stimuli, macrophage cell morphology is changed due to development of lamellipodia and filopodia from exterior boundaries. Thus to determine whether the fractions possess any effect on morphology, cells were incubated with the extracts at a range of concentrations for 24 h. Negative control set exhibited typical macrophage cells whereas the treated cells were larger in size, irregular in shape and more adherent to 6 well dishes. Most importantly, the challenged monocytes displayed dendritic outline with microvilli-like structures from the entire surface presenting definite sign for immune-stimulation. Scientists have reported that NF-κB induces psseudopod formation and morphological changes in macrophage cells.

It could be inferred from above results that isolated preparations definitely possess murine macrophage cell stimulatory property, though mechanism of action was still not clear. Thus further research was devoted to confirm whether the effect is managed by cellular transcription level and to identify molecular targets. In this context, it was plausible to assume that the polysaccharides bind to certain receptor and trigger signaling cascade to regulate monocyte function. Among various receptors present on macrophage surface, TLRs are the best characterized that play important roles in identifying pathogens and activating innate immune system. Recent studies have showed that binding of ligands to TLR2 and TLR4 consequently trigger cellular transduction, led to activation of a variety of transcription factors, for example, NF-κB. Activated NF-κB then controls expression of pro-inflammatory mediators such as iNOS, COX-2 and TNF-α. For instance, TNF–α can bind to upstream promoter response element in iNOS gene and trigger transcription resulting NO production. These functional cytokines are related to secondary immune responses such as proliferation of T and B cells, as well as activation of macrophages. In addition to these inflammatory mediators, NF-κB also boosts transcription of its own inhibitor, Iκ-B, indicating auto-regulation. Accordingly to determine mode of action of each extract, cells were exposed to the polymers for 24 h and subsequently RNA was isolated as well as RT-PCR was performed. Visual observation of gel pictures clearly showed that the treatment resulted increase in transcriptional level of all investigating genes namely TLR-2, TLR-4, NF-κB, COX-2, iNOS, TNF-α, Iκ-Bα and IFN-γ. Thus, our study conveyed that both the studied fractions exhibited immune enhancing property by triggering TLR/NF-κB pathway.

4 CONCLUSION

In summary, the present work could be regarded as milestone expanding ethnic myco-knowledge reinforced by scientific evidences. In that note, the research contributes detailed systematic study on two wild edible colorful Russuloid fungi from West Bengal viz. *R. alatoreticula* and *R. senecis*, of which one is new to science. The effort also highlights that the investigated taxa are blessed with several bioactive compounds like carbohydrate (β-glucan in particular), protein and so forth. As a result, the mushrooms endow outstanding immune enhancing properties mediated by TLR/NF-κB pathway. The endeavor undertaken would be regarded successful if it can enhance societal use and scientific concern on these neglected macrofungi as functional food or dietary supplements to serve humankind.

REFERENCES

Ayeka, P.A. 2018. Potential of mushroom compounds as immunomodulators in cancer immunotherapy: A review. *Evidence-Based Complementary and Alternative Medicine* Article ID 7271509, 9 pages.

Das, K., Dowie, N.J., Li, G.J. & Miller, S.L. 2014. Two new species of *Russula* (Russulales) from India. *Mycosphere* 5(5): 612-622.

Hou, Y., Ding, X., Hou, W., Song, B., Wang, T., Wang, F. & Zhong, J. 2013. Immunostimulant activity of a novel polysaccharide isolated from *Lactarius deliciosus* (L. ex Fr.) Gray. *Indian Journal of Pharmaceutical Sciences* 75: 393-399.

Khatua, S. & Acharya, K. 2016. Influence of extraction parameters on physico-chemical characters and antioxidant activity of water soluble polysaccharides from *Macrocybe gigantea* (Massee) Pegler & Lodge. *Journal of Food Science and Technology* 53(4): 1878-1888.

Pradhan, P., Dutta, A.K., Roy, A., Basu, S.K. & Acharya, K. 2012. Inventory and spatial ecology of macrofungi in the *Shorea robusta* forest ecosystem of lateritic region of West Bengal. *Biodiversity* 13 (2): 88-99.

Tabarsa, M., Karnjanapratum, S., Cho, M., Kim, J.K. & You, S. 2013. Molecular characteristics and biological activities of anionic macromolecules from *Codium* fragile. *International Journal of Biological Macromolecules* 59: 1-12.

Biotechnology and Biological Sciences – Sen et al. (Eds)
© 2020 Taylor & Francis Group, London, ISBN 978-0-367-43161-7

Establishment of microbial electrochemical systems as microbial peroxide producing cells for oxidative depolymerization and dye decolorization

Dhruva Mukhopadhyay & Pratima Gupta*
Department of Biotechnology, NIT, Raipur

ABSTRACT: Hydrogen peroxide is one of the most versatile and eco-friendly molecules as it is reduced to water and oxygen. A dual chamber microbial electrochemical system with the bacteria *Shwanella peutrifascians* was used as a microbial peroxide producing cell for the oxidative depolymerization and dye decolorization purposes. They consisted of electrodes made of carbon cloth in the anode whereas, the cathode electrode was made of stainless steel mesh. The net volume of the system was approximately 70 mL, combining both the anode and the cathode volume. The anode consisted of synthetic media, whereas, the catholyte consisted of 50 mM sodium sulphate at pH 3. The maximum OCV was around 190 mV on the 3rd day of operation and the maximum H_2O_2 concentration was around 110 mM. The system was run in a batch cycle for a period of 5 days. Once the system was established for H_2O_2 production, the dyes were added in the same set of systems, for testing of dye decolonization. The dyes used were, crystal violet, methylene blue and azo-dye Eriochrome Black-T. The present set up of MPPC resulted in successful decolorization of these dyes. It can also be useful method for *in-situ* depolymerization of different classes of aromatic polymer by oxidative means in the cathode.

1 INTRODUCTION

Microbial electrochemical systems are a type of electrochemical systems which are sustainable and renewable technologies for the production of electricity for various applications like energy production, hydrogen production, fermentation, dye decolorization and production of hydrogen peroxide.(Young *et al.*, 2017) Principally, MECS is based on types of reaction called Oxygen Reduction Reaction (ORR) in which the oxygen is reduced in the cathode.(Kodali *et al.*, 2018; Santoro *et al.*, 2018)

Hydrogen peroxide is an eco-friendly and versatile compound with applications which include wastewater treatment(Rosenbaum *et al.*, 2011), sewage water treatment (Gajda *et al.*, 2018), dye decolorization(Fu, S. J. You, *et al.*, 2010), oxidative depolymerization(Sharma, Mukhopadhyay and Gupta, 2018). Its end product is water and oxygen. However, it is synthesized by Anthraquinone Oxidation method(Campos-Martin, Blanco-Brieva and Fierro, 2006), which is toxic and carcinogenic in nature. An alternative to this can be the MPPC system(Fu, S. J. You, *et al.*, 2010; Dier *et al.*, 2017; Young *et al.*, 2017; Sharma, Mukhopadhyay and Gupta, 2018). However, the main disadvantage of these systems is the short hydraulic retention time and lower concentration of H_2O_2.

In this paper, we will discuss the applications of double chamber MPPC systems for the dye decolorization and lignin depolymerization.

*Corresponding author

2 MATERIALS & METHODS

MPPC Setup and H_2O_2 production analysis A dual chamber MPPC was constructed according to our previous work(Sharma, Mukhopadhyay and Gupta, 2018). It consisted of an anode chamber of 40 mL and a cathode chamber of 30 mL working volume, connected by a salt bridge of 3% agar and a super saturated salt concentration. The anode consisted of carbon electrode and the media was glucose-yeast extract media. The cathode electrode was made of stainless steel mesh and consisted of 50 mM Na_2SO_4 at pH 3. The set up was operated for five days. Hydrogen peroxide production was analyzed by starch-iodine colorimentric test (Hollo and Szejtli, 1957). A standard H_2O_2 curve was prepared by using 30% H_2O_2 (w/v) H_2O_2.

2.1 *Dye Decolorization*

100 ppm of Azo-dye Eriochrome Black-T was added in catholyte for its decolorization experiment. 0.0025% $FeCl_3$ was added to induce Fenton Reaction. Two sets, control without culture and other with microbial culture i.e., the working setup was run.

2.2 *Lignin Depolymerization*

0.01% (w/v) kraft lignin was added to the catholyte for lignin depolymerization Two set ups first control without culture and 0.01% lignin and second with microbial culture i.e., the working setup with 0.01% lignin was run. FTIR spectroscopy conducted in the range of 1800-400 cm^{-1} of samples withdrawn from both setups at 24 hr interval and compared with 0.01% lignin spectra.

3 RESULTS

3.1 *Voltametric Output of the Electrochemical Cell*

The Open Circuit Voltage (OCV) on the first day of operation was 170 mV. However, it reached a peak OCV value of 478 mV on the 2nd day of operation. The voltage of the system then gradually started to decrease over the course of time till it reached a value of 300 mV till the 5th day of operation.

3.2 *Concentration of H_2O_2*

In the 5 days of operation, the concentration of H_2O_2 was recorded highest on the 2nd day of operation, i.e., 86 mM after which, it experienced a decline. The concentration of H_2O_2 was in correlation with the OCV of the system. As the OCV experienced a rise from day one to day two, the concentration of H_2O_2 also experienced a rise. However, it experienced a decline in its concentration by the 5th day of operation. As the OCV increased from 170 mV to 478 mV, the concentration of hydrogen peroxide increased from 40 mM to 86 mM, however, after the 2nd day, the OCV experienced a fall, and the concentration of H_2O_2 started to decrease till the 5th day of operation to almost 30 mM.

3.3 *Dye Decolorization*

The azo-dye Eriochrome Black-T of concentration 100 mgL^{-1} was used for the purpose of dye decolorization. At the first day of operation, no decolorization was witnessed. However, it started to happen after the second day of operation. By the end of 5th day of operation, the color (dark purple) of the dye disappeared as shown in Figure 1. whereas in the control set no decolorisation was seen this indicated the dye decolorisation catalyzed by H_2O_2.

3.4 *Lignin Depolymerization*

The lignin depolymerization was analyzed using FTIR spectroscopic systems as shown in Figure 2. The fingerprint region was from 1500-400 cm^{-1} region. Over the course of 5 days,

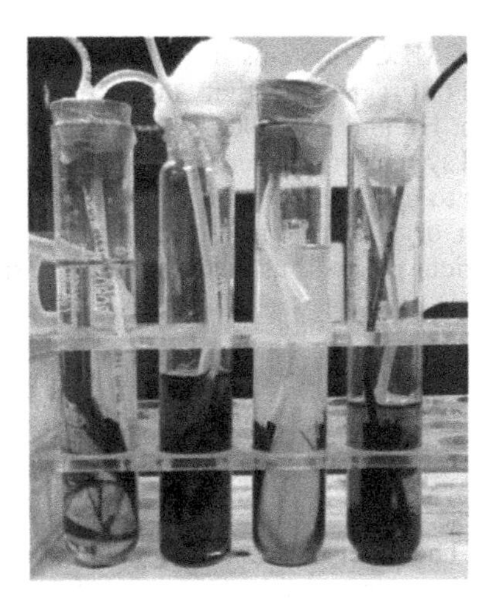

Figure 1. The MPPC system for dye decolorization. The setup on the left is the control and on the right is working system.

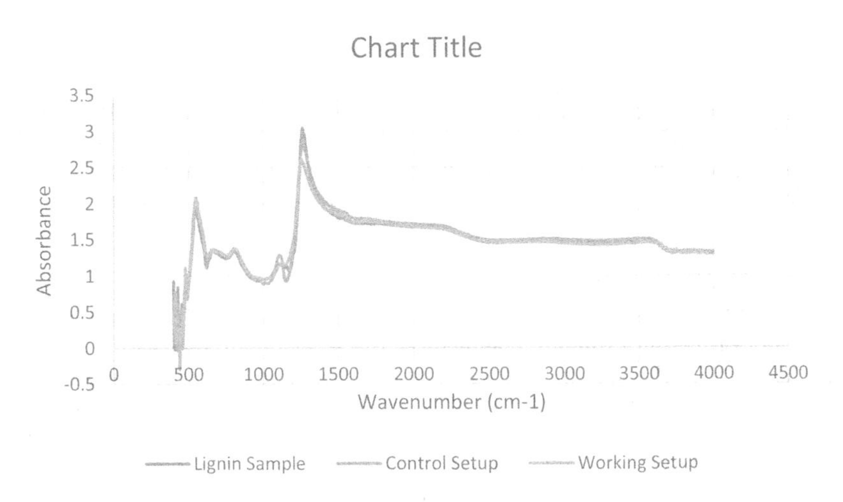

Figure 2. Depicting the Kraft Lignin Depolymerization using 0.01% (w/v) Kraft Lignin.

the characteristic peak reduced, however, the degree of depolymerization was not to a great extent. The peak observed around 1270 cm^{-1} was that of the guaicyl group. The peak intensity at this wavenumber compared to control reduced hence, indicating a dissociation of the respective bonds. Aromatic groups and non-aromatic alkenes came in the wave number region of 805 cm^{-1} and 667 cm^{-1} respectively. The peak intensity exhibited a slight increase in these bonds. However, the aromatic alkenes in the region of 544 cm^{-1} showed decrease in the peak intensity.

4 DISCUSSIONS

Voltametric Output of the Electrochemical Cell: The MPPC system was run for a period of 5 days on a batch cycle. The system exhibited an increase in the OCV during the second day of operation however, started to gradually decline after that. Previous research has shown similar effect on the performance of MECT systems in which the electrochemical activity reduced after a certain period of time but regained activity after a new batch cycle was started or was run in a fed-batch mode(Fan, Hu and Liu, 2007; Zhu *et al.*, 2011). However, on a pilot scale the microbial electrochemical systems have been reported and the life span of the systems were reported from a few months to a year on a pure batch cycle(Kim *et al.*, 2009). This may be because of the phenomenon of the prolonged death phase in which the microorganisms increase their life span by various means thus increasing the life span of the MECT system.

4.1 *Concentration of H_2O_2*

The concentration of H_2O_2 was in direct correlation to the OCV of the system. When the OCV was around 486 mV, the hydrogen peroxide concentration peaked around 86 mM, however, as the voltage started to drop, the concentration of Hydrogen Peroxide also started to drop. This is primarily because, hydrogen peroxide is a thermodynamically unstable molecule (Fu, S.-J. You, *et al.*, 2010). The other reason for the reduction in the H_2O_2 concentration is the reduction in the OCV of the system. As the OCV reduces, the number of available protons and electrons for the synthesis of H_2O_2, its concentration reduces(Tartakovsky and Guiot, 2006). Often in the case of the 2 electron step ORR, there is an additional 2 electron step process which leads to transfer of excess of electrons to the cathode which leads of the decay of H_2O_2 and subsequently leading to the formation of water(O Z, no date).

4.2 *Dye Decolorization*

Azo dye Eriochrome Black-T was used for the decolorization experiments. Principally it operates by the breaking of the azo bond (-N=N-) group in the molecule, thus making it less environmentally toxic(de Luna *et al.*, 2013). The color of the dye changed from dark purple to almost colorless.

4.3 *Lignin Depolymerization*

Lignin is one of the most versatile and complex organic molecules. It is the most abundant naturally occurring aromatic chemical in the world. It is the second most abundant organic chemical in the world. It is a very complex and versatile molecule which can yield many organic materials of commercial importance(Kumar and Sharma, 2017). The fingerprint region as stated previously, is in the region of 1500-500 cm^{-1}. One of the major bonds which exhibited a characteristic peak were in the region of 1266 cm^{-1}, which indicate the presence of guaiacyl group(Diop *et al.*, 2015). Primary, secondary and tertiary alcohols along with phenols ethers and esters are found in the region of 1104 cm^{-1}(Watkins *et al.*, 2015). The other bonds that were detected were in the region of 805, 667 and 544 cm^{-1}, respectively(Diop *et al.*, 2015) which were in the region of aromatic C-H, non-aromatic alkenes and aromatic alkenes. These smaller bonds were shown to have an increase in the peak, i.e., increase in the concentration. This maybe because of the dissociation of bonds in lignin led to an increase in the concentration of the smaller molecules. The guaiacyl group C-O stretch was that of the ester groups (Dialkyl Ether Linkage) and in the 1104 cm^{-1} was ether and ester linkage. These are the major linkages found in the lignin molecules (65-70% of the bonds), hence, according to various literatures, the breakage of these bonds indicate the depolymerization of these bonds (Diop *et al.*, 2015).

5 CONCLUSION

The present experiment has demonstrated the ability of the system to apply *in-situ* hydrogen peroxide produced in MPPC for with dye decolorization and lignin depolymerization. The reported concentration of the produced H_2O_2 is approximately 86 mM. This amount may be less compared to the standard method of production i.e., Anthroqinone Oxidation method, however, it is cheaper and eco-friendly in nature.

REFERENCE

Campos-Martin, J.M., Blanco-Brieva, G. and Fierro, J.L. G. (2006) 'Hydrogen peroxide synthesis: An outlook beyond the anthraquinone process', *Angewandte Chemie - International Edition*, 45(42), pp. 6962–6984. doi: 10.1002/anie.200503779.

Dier, T.K.F. *et al.* (2017) 'Sustainable Electrochemical Depolymerization of Lignin in Reusable Ionic Liquids', *Scientific Reports*. Springer US, 7(1), pp. 1–12. doi: 10.1038/s41598-017-05316-x.

Diop, A. *et al.* (2015) 'Kraft lignin depolymerization in an ionic liquid without a catalyst', *BioResources*, 10(3), pp. 4933–4946. doi: 10.15376/biores.10.3.4933-4946.

Fan, Y., Hu, H. and Liu, H. (2007) 'Enhanced Coulombic efficiency and power density of air-cathode microbial fuel cells with an improved cell configuration', *Journal of Power Sources*. doi: 10.1016/j. jpowsour. 2007.06.220.

Fu, L., You, S.J., *et al.* (2010) 'Degradation of azo dyes using in-situ Fenton reaction incorporated into H2O2-producing microbial fuel cell', *Chemical Engineering Journal*. Elsevier B.V., 160(1), pp. 164-169. doi: 10.1016/j.cej. 2010.03.032.

Fu, L., You, S.-J., *et al.* (2010) 'Synthesis of hydrogen peroxide in microbial fuel cell', *Journal of Chemical Technology & Biotechnology*. doi: 10.1002/jctb.2367.

Gajda, I. *et al.* (2018) 'Improved power and long term performance of microbial fuel cell with Fe-N-C catalyst in air-breathing cathode', *Energy*. Elsevier Ltd, 144, pp. 1073-1079. doi: 10.1016/j.energy. 2017.11.135.

Hollo, J. and Szejtli, J. (1957) 'THE MECHANISM OF STARCH-IODINE REACTION I. Critical investigation of actual viewpoints', *Periodica Polytechnica Chemical Engineering*, 1(2), pp. 141–145. Available at: https://pp.bme.hu/ch/article/viewFile/3724/2829.

Kim, J. R. *et al.* (2009) 'Development of a tubular microbial fuel cell (MFC) employing a membrane electrode assembly cathode', *Journal of Power Sources*. doi: 10.1016/j.jpowsour. 2008.11.020.

Kodali, M. *et al.* (2018) 'Enhancement of microbial fuel cell performance by introducing anano-composite cathode catalyst', *Electrochimica Acta*. Elsevier Ltd, 265, pp. 56–64. doi: 10.1016/j.electacta. 2018.01.118.

Kumar, A.K. and Sharma, S. (2017) 'Recent updates on different methods of pretreatment of lignocellulosic feedstocks: a review', *Bioresources and Bioprocessing*. Springer Berlin Heidelberg, 4(1). doi: 10.1186/s40643-017-0137-9.

de Luna, M.D.G. *et al.* (2013) 'Adsorption of Eriochrome Black T (EBT) dye using activated carbon prepared from waste rice hulls-Optimization, isotherm and kinetic studies', *Journal of the Taiwan Institute of Chemical Engineers*. Taiwan Institute of Chemical Engineers, 44(4), pp. 646-653. doi: 10.1016/j.jtice.2013.01.010.

O Z, É. A. R. (no date) 'Hydrogen Production with a Microbial Biocathode'. doi: 10.1021/es071720+.

Rosenbaum, M. *et al.* (2011) 'Cathodes as electron donors for microbial metabolism: Which extracellular electron transfer mechanisms are involved?', *Bioresource Technology*. Elsevier Ltd, 102(1), pp. 324-333. doi: 10.1016/j.biortech. 2010.07.008.

Santoro, C. *et al.* (2018) 'Influence of platinum group metal-free catalyst synthesis on microbial fuel cell performance', *Journal of Power Sources*. Elsevier, 375 (August 2017), pp. 11-20. doi: 10.1016/j.jpowsour. 2017.11.039.

Sharma, R.K., Mukhopadhyay, D. and Gupta, P. (2018) 'Microbial fuel cell-mediated lignin depolymerization: a sustainable approach', *Journal of Chemical Technology & Biotechnology*, (October). doi: 10.1002/jctb.5841.

Tartakovsky, B. and Guiot, S.R. (2006) 'A comparison of air and hydrogen peroxide oxygenated microbial fuel cell reactors', in *Biotechnology Progress*. doi: 10.1021/bp050225j.

Watkins, D. *et al.* (2015) 'Extraction and characterization of lignin from different biomass resources', *Journal of Materials Research and Technology*. Korea Institute of Oriental Medicine, 4(1), pp. 26-32. doi: 10.1016/j.jmrt.2014.10.009.

Young, M.N. *et al.* (2017) 'Understanding the impact of operational conditions on performance of microbial peroxide producing cells', *Journal of Power Sources*, 356, pp. 448-458. doi: 10.1016/j.jpowsour. 2017.03.107.

Zhu, N. *et al.* (2011) 'Improved performance of membrane free single-chamber air-cathode microbial fuel cells with nitric acid and ethylenediamine surface modified activated carbon fiber felt anodes', *Bioresource Technology*. doi: 10.1016/j.biortech. 2010.06.046.

Biotechnology and Biological Sciences – Sen et al. (Eds)
 ISBN 978-0-367-43161-7

Isolation and characterization of lipase producing microbial strain from coastal banks of Bhavnagar

Rajvi Panchal & Jignesh Prajapati
Department of Biochemistry and Biotechnology, St. Xavier's College (Autonomous), Ahmedabad, India

Dhavalkumar Patel
Department of Biochemistry and Biotechnology, St. Xavier's College (Autonomous), Ahmedabad, India
Department of Biotechnology and Microbiology, Parul Institute of Applied Science and Research, Parul University, Ahmedabad, India

Dweipayan Goswami*
Department of Biochemistry and Biotechnology, St. Xavier's College (Autonomous), Ahmedabad, India
Department of Microbiology and Biotechnology, School of Sciences (SoS), Gujarat University, Ahmedabad, India

ABSTRACT: Lipases are predominantly significant enzyme owing to the fact that they precisely hydrolyze acyl glycerol, greases and oils, which is one of the great interests for diverse industrial applications. Lipase producing bacterial strains were isolated from saline soil of costal banks of Bhavnagar, Gujarat, India by employing enrichment culture techniques. Tributyrin oil and minimal salt containing media was employed for isolation and sub-culturing bacteria for primary screening. All isolated bacterial strains were screened by the ratio of the lipolytic halo radius and the colonies radius. The strain having the maximum ratio was recognized morphologically, microscopically and on gene level by sequencing of 16S rRNA gene. The most productive strain produced lipase in olive oil containing complex media and the crude lipase preparation was capable of hydrolyzing *p*-nitrophenyl palmitate over a range of wide pH. The crude lipase presented maximal activity (6.16 ± 0.17 U mL^{-1} min^{-1}) at pH of 9.0. Based on the outcomes of current study, lipase of isolated bacteria is a possible alkaline lipase and an applicant for industrial claims such as laundry, leather, detergent and fine chemical industries.

Keywords: *lipases, saline soil, p-nitrophenyl*

1 INTRODUCTION

Lipases (Triacylglycerol lipases, EC 3.1.1.3) are hydrophilic enzymes which have the capacity to hydrolyse triacylglycerol to discharge free fatty acids and glycerol. Microbes and animal are very good source form which lipases and isolated as well as purified [1]. Of all these, bacterial lipases are most economical and stable [2]. Lipases constitute a significant cluster of biotechnologically valuable enzymes, essentially owing to the adaptability of their applied properties [3]. Lipases has wide range of application in the fields ranging from dairy and food industry to synthesis of biodegradable material [4,5]. Bacterial lipases are used in milk fat hydrolysis, lipolysis of butter fat and cream as well in cheese ripening and flavour enhancement [5]. It's lipolysis characteristic makes stain removal easier and so it is used as additive in detergent industry and in removal of grease from leather in leather industry [6,7]. The transesterification property of lipases can be

*Corresponding author: Email: dweipayan79@gmail.com

used as enhancer of fabric absorbency in textile industry and in synthesis of biodiesel [8–10]. Other uses of lipases include, pharmaceutical industry, enhancement of product quality in cosmetic industry and in paper & pulp industry [11–13]. The persistence of the current study is to discover an innovative bacterial strain from alkaline salt containing soil that is accomplished of making lipase. Lipases from halotolerant have the unique properties making them robust biocatalysts choice for enzymatic processes performed at high salt concentration, extreme pH and temperatures, where most enzyme display a severe reduction in their activity [14].

2 MATERIALS & METHODS

Bacterial strains were isolated from alkaline soil of Bhavnagar, Gujarat. The isolation procedure was completed by sequential dilution of samples on tributyrin agar plates conferring standard techniques 15. Tributyrin agar medium is mainly composed of 5.0 g peptone, 3.0 g yeast extract, 10 ml tributyrin, along with 15 g agar (per liter). Plates were incubated at 37±1°C and periodically examined after 120 hrs. Colonies showing zone of clearance around them were screened out and were sub-cultured on tributyrin agar plates and slants. Isolates that showed high lipase production were recognized by its' colony morphology, cell morphology, and gram staining.

Qualitative analysis of lipase was done by Zone of hydrolysis index. For detection of lipase activity tributyrin agar was used. Bacterial culture from pure slants were streaked on plates in a vertical manner and were incubated at 37°C for 48 h. Zone of hydrolysis index was calculated by following formula and bacterial strain which was giving largest clear zone was taken for further studies.

$$\text{Hydrolysis index} = \frac{\textit{Diameter of clear zone of hydrolysis } (mm)}{\textit{Diameter of bacterial streak } (mm)}$$

Quantitative analysis of Lipase enzyme was measured spectrophotometrically described by Palacios et al. (2014) with some modifications. The reaction assay system contained 3.3 mL of 50 mmol L^{-1} various buffer as shown in Table 1 (pH 7.0 – pH 12.0) containing 1 gL^{-1} gum arabic and 0.2 mL of 30 mmol L^{-1} *p*-nitrophenyl palmitate (pNPP) dissolved in 2-propanol. The mixture was prewarmed at 37°C, and then 1 mL of crude enzyme was added. After 30 min of incubation at 37°C the reaction was stopped by adding 1.5 mL of Marmur solution (chloroform:isoamyl alcohol, 24:1). The sample was centrifuged at 7000 rpm for 5 min at 4°C and the clear supernatant (aqueous phase), was taken off. The optical density of supernatant was then measured at 410 nm. Controls in which the enzyme solution was substituted by 50 mmol L^{-1} various buffer containing 1 gL^{-1} gum arabic were assayed in all cases to deduct any nonenzymatic activity.

One unit of activity (U) was defined as the amount substrate liberates i.e 1 mmol of pNP per minute of the enzyme 16.

Table 1 . Buffers for various pH.

pH	Buffer
7	50 mmol Phosphate Buffer
8	50 mmol Tris-cl Buffer
9	50 mmol Glycine-NaOH Buffer
10	50 mmol Glycine-NaOH Buffer
11	50 mmol KCl-NaOH Buffer
12	50 mmol KCl-NaOH Buffer

3 RESULTS & DISCUSSION

On isolating bacteria from the alkaline soil total four isolates were obtained. After isolation, five sector streaking was done to get pure culture and to check colony morphology which is displayed in Figure 2. Morphology of each isolate is summarised in Table 2.

Qualitative estimation was performed of all 4 bacterial isolates for Lipase enzyme and displayed in Figure 3. Out of 4 isolates, isolate 1 showed highest hydrolysis index and was further screened for quantitative assay. Isolate 1 was sent to GBRC for 16S rRNA identification, accession number yet to be come.

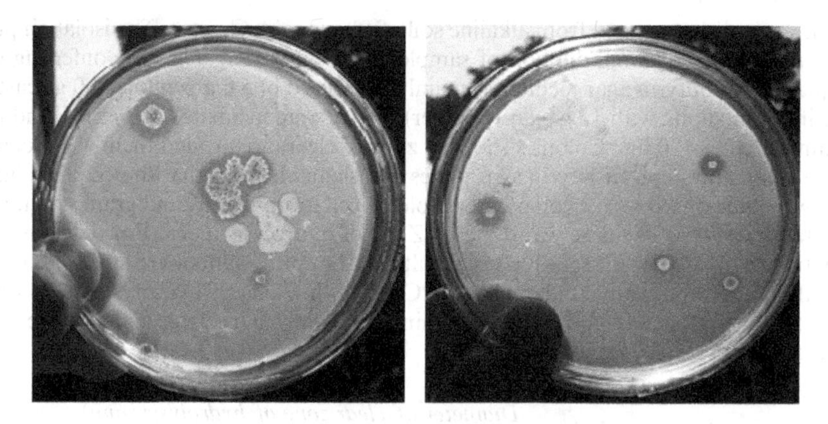

Figure 1. Display of isolation on tributyrin agar plate.

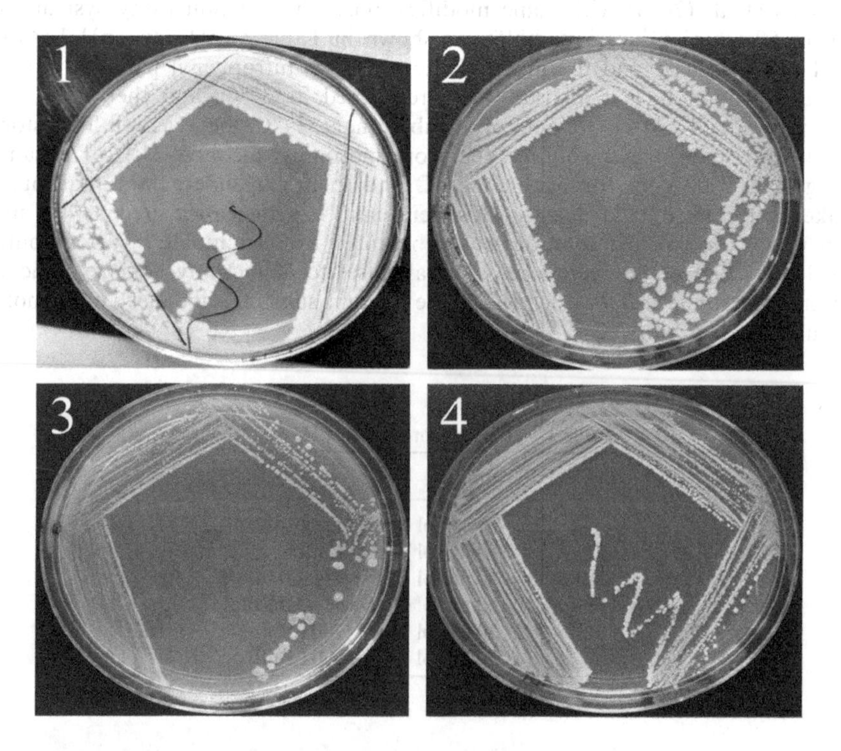

Figure 2. Display of morphological examination of four isolates.

Table 2 . Morphological examination of each isolates.

Bacterial strain	Colony Morphology	Negative staining	Gram staining
Isolate 1	Small, Circular, Entire, Flat, Opaque, Dry, Off-white	Bacilli (Short rods)	Gram Positive
Isolate 2	Small, Circular, Entire, Umbonate, Opaque, Dry, Off-white	Bacilli (Short rods)	Gram Positive
Isolate 3	Small, Irregular Curled, Raised, Translucent, Dry, Light-orange	Bacilli (Short rods)	Gram negative
Isolate 4	Punctiform, Circular, Entire, Flat, Opaque, Dry, Off-white	Bacilli (Short rods)	Gram Positive

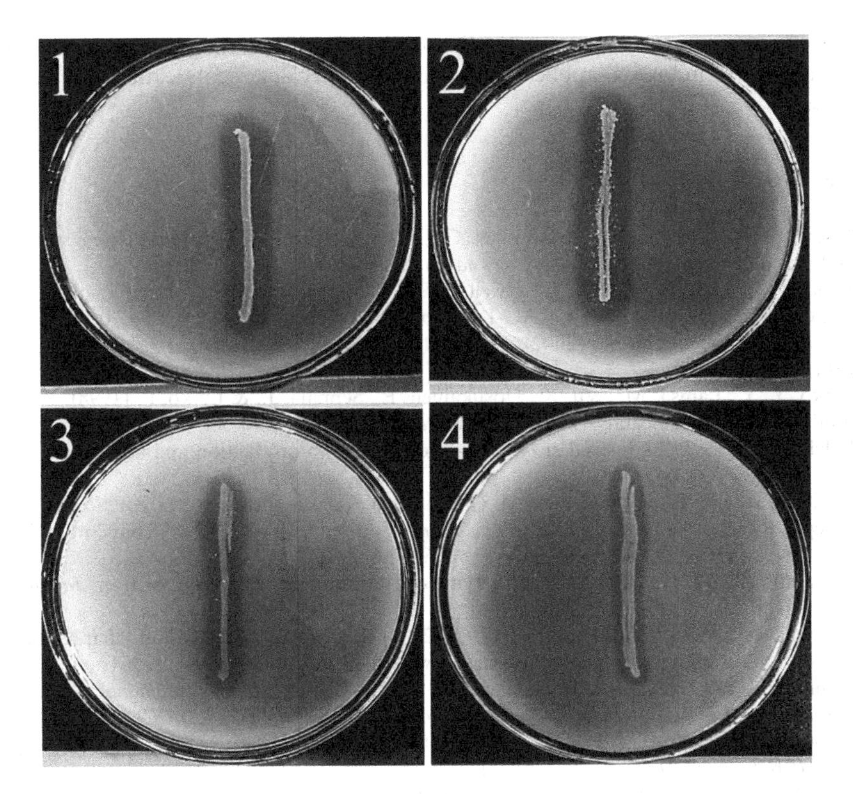

Figure 3. Display of zone of hydrolysis of four isolates.

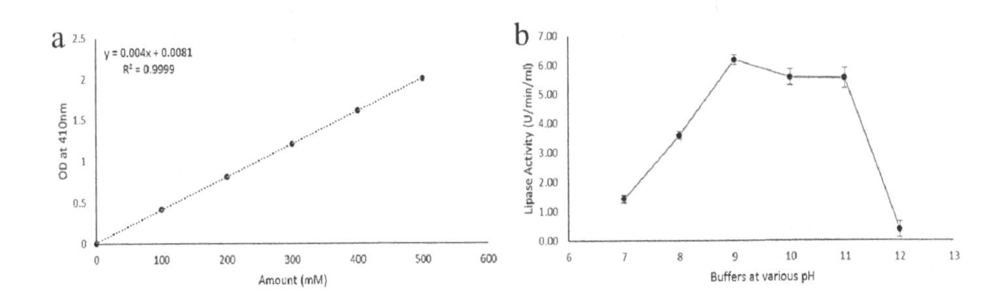

Figure 4. (a) Graphical representation of p-nitrophenol standard curve, (b) Graphical representation of lipase activity at various pH.

Before carrying the quantitative assay for Lipase enzyme, standard curve of p-nitrophenol was performed. Linear Regression (R^2) value for *p*-nitrophenol obtained is 0.999. On performing assay, the enzyme activity (U mL^{-1} min^{-1}) at various pH is shown in Figure 4b. The crude lipase presented maximal activity (6.16 ± 0.17 U mL^{-1} min^{-1}) at pH of 9.0.

Over-all bacterial lipases are constant in an inclusive range of pH i.e 4 to 11. Gupta et al. (2004) did a complete review of entirely bacterial lipase and stated that supreme activity of lipase enzyme at alkaline pH i.e higher than 7 has been detected in many cases. In present study the enzyme was showing its optimum activity at pH 9.0 (Figure 4b), rising activity from pH 7-9 can be categorized as an alkalophilic enzyme 17.

4 CONCLUSION

Based on the outcomes of current study, lipase of isolated bacteria is a possible alkaline lipase and a claimant for industrial applications such as laundry detergent, leather and fine chemical industries.

REFERENCES

[1] Joseph, B., Ramteke, P. W., & Thomas, G. (2008). Cold active microbial lipases: some hot issues and recent developments. *Biotechnology advances, 26*(5), 457–470.

[2] Snellman, E. A., Sullivan, E. R., & Colwell, R. R. (2002). Purification and properties of the extracellular lipase, LipA, of *Acinetobacter sp.* RAG-1. *European Journal of Biochemistry, 269*(23), 5771–5779.

[3] Jaeger, K. E., & Eggert, T. (2002). Lipases for biotechnology. *Current opinion in biotechnology, 13*(4), 390–397.

[4] Linko, Y. Y., Lämsä, M., Wu, X., Uosukainen, E., Seppälä, J., & Linko, P. (1998). Biodegradable products by lipase biocatalysis. *Journal of biotechnology, 66*(1), 41–50.

[5] Falch, E. A. (1991). Industrial enzymes—developments in production and application. *Biotechnology advances, 9*(4), 643–658.

[6] Fujii, T., Tatara, T., & Minagawa, M. (1986). Studies on applications of lipolytic enzyme in detergency I. Effect of lipase from *Candida cylindracea* on removal of olive oil from cotton fabric. *Journal of the American Oil Chemists' Society, 63*(6), 796–799.

[7] Nakamura, K., & Nasu, T. (1990). Enzyme containing bleaching composition. *Japanese Patent, 2*(208), 400.

[8] Sharma, R., Chisti, Y., & Banerjee, U. C. (2001). Production, purification, characterization, and applications of lipases. *Biotechnology advances, 19*(8), 627–662.

[9] Noureddini, H., Gao, X., & Philkana, R. S. (2005). Immobilized *Pseudomonas cepacia lipase* for biodiesel fuel production from soybean oil. *Bioresource technology, 96*(7), 769–777.

[10] Hasan, F., Shah, A. A., & Hameed, A. (2006). Industrial applications of microbial lipases. *Enzyme and Microbial technology, 39*(2), 235–251.

[11] Seitz, E. W. (1974). Industrial application of microbial lipases: a review. *Journal of the American oil chemists' society, 51*(2), 12–16.

[12] Bajpai, P. (1999). Application of enzymes in the pulp and paper industry. *Biotechnology progress, 15*(2), 147–157.

[13] Higaki, S., & Morohashi, M. (2003). *Propionibacterium acnes* lipase in seborrheic dermatitis and other skin diseases and Unseiin. *Drugs under experimental and clinical research, 29*(4), 157–159.

[14] Samaei-Nouroozi, A., Rezaei, S., Khoshnevis, N., Doosti, M., Hajihoseini, R., Khoshayand, M. R., & Faramarzi, M. A. (2015). Medium-based optimization of an organic solvent-tolerant extracellular lipase from the isolated halophilic *Alkalibacillus salilacus. Extremophiles, 19*(5), 933–947.s

[15] N.R. Krieg, Enrichment and isolation, in: P. Gerhardt (Ed.), Manual of Methods for General Bacteriology, American Society for Microbiology, Washington, DC, 1981, pp. 112–142.

[16] Palacios, D., Busto, M. D., & Ortega, N. (2014). Study of a new spectrophotometric end-point assay for lipase activity determination in aqueous media. *LWT-Food Science and Technology, 55*(2), 536–542.

[17] Gupta, R., Gupta, N., & Rathi, P. (2004). Bacterial lipases: an overview of production, purification and biochemical properties. *Applied microbiology and biotechnology, 64*(6), 763–781.

Biotechnology and Biological Sciences – Sen et al. (Eds)
© 2020 Taylor & Francis Group, London, ISBN 978-0-367-43161-7

Far ranging antimicrobial and free radical scavenging activity of Himalayan soft gold mushroom; *Cordyceps* sp.

Loknath Deshmukh
Fungal Biotechnology and Invertebrate Pathology Laboratory, Department of Biological Science, R.D. University, Jabalpur, India

Rajendra Singh
Bio-Design Innovation Centre, Department of Biological Science, R.D. University, Jabalpur, India

Sardul Singh Sandhu
Fungal Biotechnology and Invertebrate Pathology Laboratory, Department of Biological Science, R.D. University, Jabalpur, India

ABSTRACT: The miraculous medicinal effect of secondary metabolite produced from *Cordyceps* Sp., known for traditional use to cure deadly diseases. *Ophiocordyceps sinensis* and their best alternate *Cordyceps militaris* is widely using for the medicinal purposes in pharmaceutical industries. In the present research both the entomopathogenic fungal metabolites with various solvent fractions were tested against six bacterial strains and three *Candida* Sps. using agar well diffusion method to examine the antibacterial and anticandidal activity. Extracted metabolites were also tested for free radical scavenging potentiality by using DPPH scavenging activity. Metabolites extracted from *Ophiocordyceps sinensis* and *Cordyceps militaris* showed maximum zone of inhibition against all six bacterial strains ranges from 09 mm to 10 mm and 12 mm to 14 mm respectively. The result of anticandidal activity of *Cordyceps militaris* were also leading than *Ophiocordyceps sinensis* by showing 18mm and 15mm clear zone of inhibition respectively. 5 µl metabolite of *Cordyceps militaris* and *Ophiocordyceps sinensis* were vigorously scavenged the methanolic solution of DPPH up to 95% to 97% respectively. Positive controls were also maintained. Therefore the extracted metabolites from *Ophiocordyceps sinensis* and *Cordyceps militaris* can be industrially produced and bioactive compounds can be purified to use as far ranging antimicrobial and anti-aging bioagents for health and pharmaceutical industries.

Keywords: *Ophiocordyceps sinensis*, *Cordyceps militaris*, Antibacterial activity, Anticandidal activity, Free radical scavenging potentiality

1 INTRODUCTION

As the number of inhabitant on the earth are continuously expanding and experiencing different medical issues brought about by certain medication opposition like bacteria, fungi and several parasites. Hence, a needful step required for the development of novel drugs to solve these issues naturally (Aharwal, 2018). Natural medications play an extensive role and are the premise of customary frameworks for fix and treatment of ailments. Entomopathogenic organisms *Cordyceps* Spp. are one of the exceptional and significant wellsprings of bioactive metabolites which help in treatment of different infectious diseases (Deshmukh, 2019). The

*Corresponding author: loknath.deshmukh3108@gmail.com

Cordyceps is insect born fungi belonging to the family *Clavicipitaceae* and Order Hypocreales. *Cordyceps* spp. infects and parasitize on larvae of Lepidoptera (Chen, 2005). The mystery that a worm in winter is transformed into one herb in summer has been recorded in Chinese traditional medicines for more than one thousand years. This miraculous living thing, prominently named as winter-worm-summer-grass (Dong Chong Xia Cao in Chinese), is a growth hatchling harmonious, fruiting body of *Cordyceps* (Sung, 2007). It has far ranging microbial activity and multitude of pharmacological activity due to the presence of highly potential active compound Cordycepin (Tuli, 2013). The medicinal mushroom *Cordyceps* contains multiple bioactive compounds, like as cordycepin, adenosine, sterols and polysaccharides; because of its various pharmacological activities, it is presently used for manifold medicinal intentions (Chan, 2015). Several recent studies reveals that multiple species in this *Cordyceps* genus possess far-ranging pharmacological properties, like antimicrobial, antioxidant, antitumor, hypoglycaemic, hepatoprotective, immunomodulatory, hypocholesterolemic and nephroprotective activities, and effects on apoptotic homeostasis (Yue, 2013). As a nutraceutical and utilitarian sustenance, *Cordyceps* has likewise pulled in research enthusiasm for late years because of its cell reinforcement action. Looked at defensive impacts against oxidative harm demonstrated that *C. militaris* extract and *C. sinensis* extract might be intense hydrogen donators and the defensive impacts against oxidative harm are a consequence of their free radical scavenging capacities. Both fermented metabolites demonstrated a more grounded impact on 1, 1-diphenyl-2-picrylhydrazyl (DPPH) contrasted with positive control. Due to illegal harvesting of *Ophiocordyceps sinensis* (OCS), this supernatural power is going to be endangered from their natural habitat. Hence, another species *C. militaris* (CM) came with same or more pharmacological activities as their best alternate member. Both the fruiting bodies and their submerged culture mycelia of *C. militaris* and *Ophiocordyceps sinensis* are as of now accessibly available in market. Regardless of whether these species have comparative pharmacological impacts is an inquiry that worries consumers (Dong, 2014; Deshmukh, 2019).

In the present study, a comparative analysis of the antibacterial, anticandidal and antioxidant activity of the solvent extracted fractions from fermented mycelial free cell filtrate (MFCF) of *C. militaris* and *O. sinensis* were performed and authenticated their far ranging activity by using statistical analysis.

2 MATERIALS AND METHODS

2.1 *Sample Collection and Maintenance*

Ophiocordyceps sinensis was collected from Trisul region (Mountain) Uttarakhand, India and active vegetative culture of *Cordyceps militaris* (MTCC No. 3936) was procured from microbial type culture collection, Chandigarh, India. Both collected samples were grown on optimised media and incubated at 20°C for 07 days under fungal incubator. Potato dextrose agar with peptone was used for *Ophiocordyceps sinensis* and Sabouraud maltose agar supplemented with yeast extract was used for *C.* militaris. Grown pure cultures were stored at 4°C for further experimental uses (Figure 1).

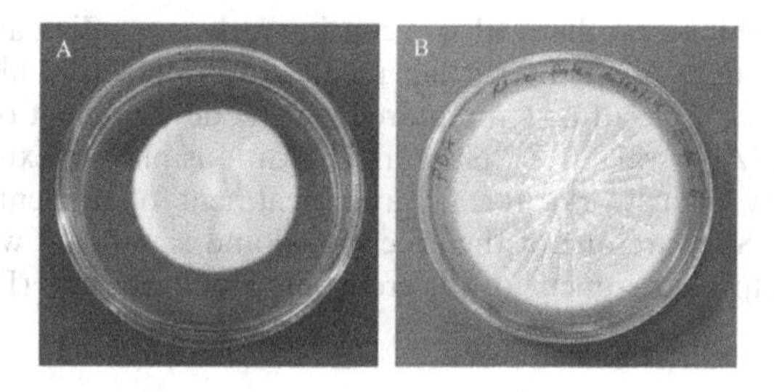

Figure 1. Pure culture plate (A) *Cordyceps militaris*, (B) *Ophiocordyceps sinensis.*

2.2 *Production of secondary metabolites and extraction*

07 days old mycelial discs of OCS & CM were inoculated in 250ml flask containing PPD (Potato Dextrose with Peptone) and SMYE (Sabouraud, Maltose, Yeast extract with Peptone) broth media and incubated at 20 ± 2°C for designed incubation period to produce required bioactive metabolites. The mycelial beds were harvested and liquid portion were filtered by using Whatman filter No. 01 and mycelia free cell free (MFCF) filtrates were processed by solvent extraction. The MFCF were used to check their antibacterial, anticandidal and antioxidant potentiality.

2.3 *Antimicrobial and antioxidant activity*

Agar well diffusion method was used for antimicrobial activity and DPPH method was used for free radical scavenging activity. Standard protocols from literatures were used with minor modifications. Antibacterial potentialities of processed extracts were tested against six pathogenic bacteria *i.e. Vibrio cholerae, Salmonella typhimurium, Escherichia coli, Bacillus subtilis, Klebsiella pneumoniae* and *Staphylococcus aureus*. Thin layer bacterial film was prepared for by taking 25 µl bacterial spore suspension seeded on nutrient agar media. 08 mm wells were prepared by sterilised cork borer and filled by 80 µl processed MFCF produced by both fungal cultures. Likewise three frequent disease causing *Candida i.e. Candida albicans, Candida auris* and *Candida krusei* were used to check anticandidal activity by using agar well diffusion method. Where, ethanolic dilutions of MFCF were used to determine free radical scavenging activity posses in secondary metabolites produced from OCS & CM. The decreasing absorbance was recorded at 517 nm and scavenging activity percentage was calculated by using standard formula. The results were positively compared with market available positive controls and statically analysed by using standard deviation and t-test approaches.

3 RESULTS

Antibacterial screening: Both the processed MFCF (After 21 days of static incubation) showed greater zone of inhibition against all 06 pathogenic bacteria. CM showed maximum 14 mm clear zone of inhibition against *Bacillus* subtilis and 12 mm zone of inhibition against *Vibrio cholerae*. It showed least 6 mm activity against *Klebsiella pneumoniae*. OCS showed 10 mm maximum zone of inhibition against two pathogenic bacteria *i.e. Bacillus subtilis* and *Vibrio cholerae*. It showed moderate results up to 07 mm to 09 mm diameter clear zone of inhibition against rest bacteria. Positive control (Chloramphenicol) was less effective than fungal extracts (Figure 2 & Table 1).

Anticandidal screening: Solvent extracted filtrate of CM showed 18mm zone of inhibition against *Candida auris* and 10 mm, 09 mm clear zone of inhibition showed against *Candida albicans* & *Candida krusei* respectively. Likewise OCS extract also showed maximum 15mm clear zone of inhibition against *Candida krusei*. Where, 14 mm and 10 mm clear zone recorded against *Candida albicans* & *Candida auris*. These results were more competitive than fluconazole used as positive control (Figure 3 & Table 2).

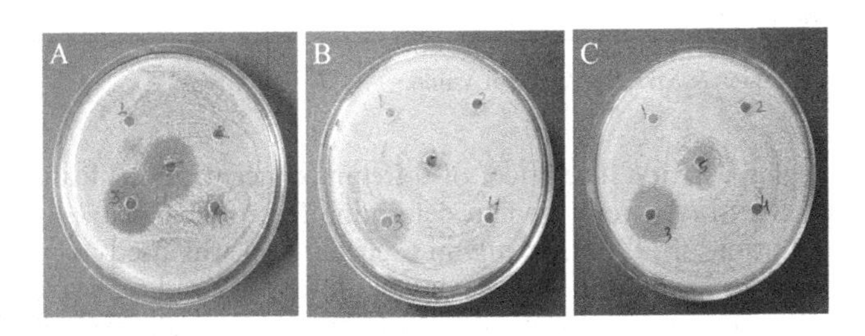

Figure 2. Antibacterial activity; Clear zone of inhibition against (A) *B. subtilis*, (B) *V. Cholerae* and (C) *S. aureus*.

Table 1. Antibacterial activity with statistical analysis.

Fungal Strain	Candidal Sps.	Clear zone of inhibition (mm)	Control (mm)	Statistical Analysis: Average Value	Statistical Analysis: Standard Deviation	t-testValue
Cordyceps militaris (CM)	*V. cholerae*	12	11	SET A	SET A	
	S. typhimurium	08	07	10.16667	2.857738	
	E. coli	11	10			
	B. subtilis	14	12			0.521203809
	K. pneumoniae	06	06	SET B	SET B	
	S. aureus	10	09	09.1666	2.316606	
Ophiocordyceps sinensis (OCS)	*V. cholerae*	10	09	SET A	SET A	
	S. typhimurium	08	07	8.833333	1.169045	
	E. coli	09	10			
	B. subtilis	10	10			0.535850119
	K. pneumoniae	07	06	SET B	SET B	
	S. aureus	09	09	8.333333	1.505545	
Incubation periods: 14 days for CM = 21 days for OCS						

Note: (A) = Sample Value & (B) = Control Value..

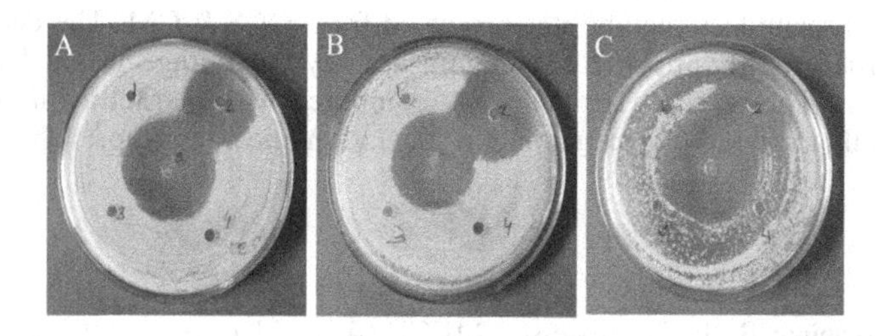

Figure 3. Anticandidal activity; Clear zone of inhibition against (A) *C. krusei*, (B) *C. albicans* and (C) *C. auris.*

Table 2. Anticandidal activity with statistical analysis.

Fungal Strain	Candidal Sps.	MFCF Filtrate Incubation Days	Clear zone of inhibition (mm)	Control (mm)	Statistical Analysis: Average Value	Statistical Analysis: Standard Deviation	t-test Value
Cordyceps militaris (CM)	*C. albicans*	14	10	07	12.3333 (A)	4.932883 (A)	0.566421666
	C. auris	14	18	16	09.6666 (B)	5.507570 (B)	
	C. krusei	14	09	06			
Ophiocordyceps sinensis (OCS)	*C. albicans*	14	14	12	13 (A)	2.645751 (A)	0.501471249
	C. auris	14	10	10	11.6666 (B)	1.527525 (B)	
	C. krusei	14	15	13			

Note: SET A = Sample Value & SET B = Control Value.

Free radical scavenging activity: The effect of different concentration (Ranges from 5 µl to 45 µl) of CM and OCS extracts were used to determine free radical scavenging activity (%) to reduce ethanolic solution of DPPH. 10 mg/ml ascorbic acid was used as positive control and blank DPPH solution was used as negative control. Potent free radical scavenging activity percentage found by four different concentrations of extracts *i.e.* 5 µl, 20 µl, 30 µl and 40 µl. CM showed maximum 93.333% scavenging activity with the 5µl concentration of MFCF extract. Positive control showed 73.333% scavenging activity against 5µl. Similarly OCS also showed

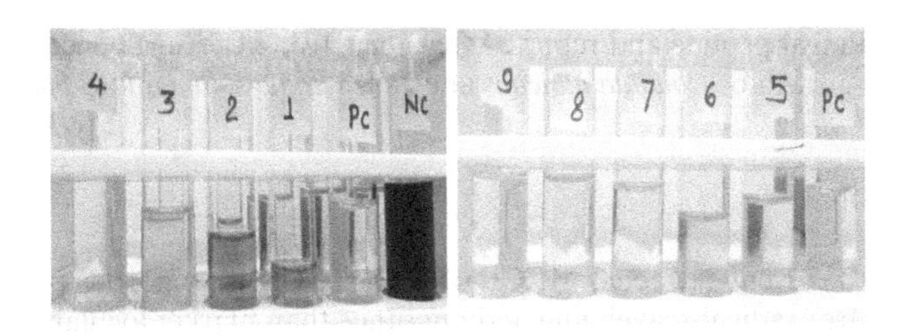

Figure 4. Free radical scavenging activity; Different dilutions and comparative scavenging percentage with PC & NC.

Table 3. Free radical scavenging activity with statistical analysis.

Fungal Strain	Incubation Period	Extract Const.	Free Radical Scavenging capacity (%)	Control (%)	Statistical Analysis		
			A	B	Average Value	Standard Deviation	t-testValue
Cordyceps militaris (CM)	07 Days	05 µl	93.333	73.333	66.654 (A)	23.19838 (A)	0.593501881
	14 Days	40 µl	51.229	45.217	57.368 (B)	14.44082 (B)	
	21 Days	20 µl	55.401	53.554			
Ophiocordyceps sinensis (OCS)	07 Days	05 µl	90.654	81.504	87.017 (A)	12.95037 (A)	0.539141649
	14 Days	05 µl	97.760	95.614	77.504 (B)	20.40564 (B)	
	21 Days	20 µl	72.637	55.395			

Note: (A) = Sample Value & (B) = Control Value.

maximum 97.760% scavenging activity with 5 µl concentration of filtrate. 95.614% scavenging activity recorded from positive control (Figure 4 & Table 3).

Statistical analysis: Experimental results were statistically analysed for their acceptance and rejection. The broad results of antibacterial, anticandidal and antioxidant activity were expressed by using average, standard deviation and student t-test. All data were compared with control results and t-test outputs showed ≥ 0.5 that is acceptable according to the hypothesis. T-test of antibacterial activity of CM and OCS with positive control recorded as 0.521203809 and 0.535850119 respectively. Likewise 0.566421666 and 0.501471249 were t-test value of anticandidal activity of CM and OCS respectively. Three higher free radical scavenging data were analysed for t-test and accepted by 0.593501881 and 0.539141649 value for CM and OCS respectively. Calculated values were greater or equal than acceptance value. All values are depicted in table 01, 02 & 03.

4 DISCUSSION

Cordycepin is the main compound of *Cordyceps* that is nucleoside adenosine analogue responsible for the broad spectrum activity. Cordycepin firstly isolated from *Cordyceps* militaris and explored for their pharmacological properties (Paterson and Russell, 2008). Different approaches were studied to check free radical scavenging activity of CM-hs-CPS2 and those were performed by *in vitro* condition. Wu *et al.* (2011), reported the 89% scavenging activity by using DPPH method at 515 nm. Zhang *et al.* (2017) worked on antibacterial potentiality of polysaccharides extracted from *Cordyceps*. 05 pathogenic bacteria were treated with water soluble polysaccharides and recorded broad spectrum activity against all used bacterial strains. 0.10 mg/mL concentration was reported as minimum inhibitory concentration. Wong *et al.*, (2011) purified cordymin an antifungal peptide and reported that their IC_{50} of 55 μM concentration was effectively inhibited the growth of *Candida albicans* and *Rhizoctonia solani* and *Bipolaris maydis.*

5 CONCLUSION

The fermented and solvent extracted MFCF of CM & OCS has vigorous potent antibacterial, anticandidal and free radical scavenging activity other than market available chemical agents. Bioactive constituent cordycepin is prominently active compound present in *Cordyceps* and that is responsible for all the activity explored in this research work. The statistical data indicate the authenticated result outputs, hence further purification of responsible novel active compound and polysaccharides will be subjected to investigate.

ACKNOWLEDGEMENT

The authors owe huge thankfulness to the Vice Chancellor, R.D. University Jabalpur, (M.P.). An endless acknowledge to the Head, Department of Biological Science, R.D. University, Jabalpur, (M.P.), for empowering us to complete this experimentation. We offer concede toward Bio-Design Innovation Centre, Department of Biological Science, R.D. University, Jabalpur, (M.P.), India, for giving workplaces and sponsoring to complete this investigation and conveying it to the world.

REFERENCES

Aharwal, R.P., Kumar, S., Thakur, Y., Deshmukh L., Sandhu, S.S. (2018). Evaluation of antibacterial activity of endophytic fungi aspergillus japonicus isolated from tridax procumbens. Asian j pharm clin res, 11(9), 212–221.

Chan, J.S.L., Barseghyan, G.S., Asatiani, M.K., Wasser, S.P. (2015). Chemical composition and medicinal value of fruiting bodies and submerged cultured mycelia of caterpillar medicinal fungus Cordyceps militaris CBS-132098 (Ascomycetes). Int J Med Mushrooms, 17, 649–659.

Chen, Y.C., Huang, Y.L., Huang, B.M. (2005). Cordyceps sinensis mycelium activates PKA and PKC signal pathways to stimulate steroidogenesis in MA-10 mouse Leydig tumour cells. Int. J. Biochem. Cell Biol, 37 (2005), 214–223.

Deshmukh, L., Gupta, D., Sandhu, S.S. (2019). Development of Marker in the Soft Gold Mushroom-Cordyceps spp. for Strain Improvement. In: Kundu R.,Narula R. (eds) Advances in Plant & Microbial Biotechnology. Springer, Singapore.

Dong, C., Yang, T., Lian, T. (2014). A Comparative Study of the Antimicrobial, Antioxidant, and Cytotoxic Activities of Methanol Extracts from Fruit Bodies and Fermented Mycelia of Caterpillar Medicinal Mushroom Cordyceps militaris (Ascomycetes). Int J Med Mushrooms, 16(5), 485–495.

Paterson, M., Russel, M. (2008). Cordyceps- A traditional Chinese medicine and another fungal therapeutic biofactory? Review. Phytochemistry, 69, 1469–1495.

Sung, G.H. (2007). Phylogenetic classification of Cordyceps and the clavicipitaceous fungi. Studies Mycol, 57, 5–59.

Tuli, H.S., Sharma, A.K., Sandhu, S.S., Kashyap, D. (2013). Cordycepin: a bioactive metabolite with therapeutic potential. Life Sci, 93, 863–869.

Wong, J.H., Ng, T.B., Sze, S.C., Zhang, K.Y., Li, Q., Lu, X. (2011). Cordymin, an antifungal peptide from the medicinal fungus Cordyceps militaris. Phytomedicine, 18(5), 387–392.

Wu, H.Y., Xiaoning, M., Junqing, Jia., Guozheng, Zhang., Xijie, G., Zhong, Z.G. (2011). Structural characterization and antioxidant activity of purified polysaccharide from cultured Cordyceps militaris Fengyao. African J Microbiol Res, 5(18), 2743–2751.

Yue, K., Ye, M., Zhou, Z., Lin, X. (2013). The genus Cordyceps: a chemical and pharmacological review. J Pharm Pharmacol, 65, 474–493.

Zhang, Y., Wu, Y.T., Zheng, W., Han, X.X., Jiang, Y.H., Hu, P.L., Tang, Z.X., Shi, L.E. (2017). The antibacterial activity and antibacterial mechanism of a polysaccharide from Cordyceps cicadae. J Func Foods, 38 (2017), 273–279.

Biotechnology and Biological Sciences – Sen et al. (Eds)
© 2020 Taylor & Francis Group, London, ISBN 978-0-367-43161-7

Extraction and characterization of siderophores from *Pseudomonas* sp. and assessing the PGPR activity of *Pseudomonas* sp.

Dhavalkumar Patel
Department of Biotechnology and Microbiology, Parul Institute of Applied Science and Research, Parul University, Ahmedabad, India
Department of Biochemistry and Biotechnology, St. Xavier's College (Autonomous), Ahmedabad, India

Mahima Patel, Shailee Patel & Bhavya Kansara
Department of Biotechnology and Microbiology, Parul Institute of Applied Science and Research, Parul University, Ahmedabad, India

Dweipayan Goswami*
Department of Biotechnology and Microbiology, Parul Institute of Applied Science and Research, Parul University, Ahmedabad, India
Department of Microbiology and Biotechnology, School of Sciences (SoS), Gujarat University, Ahmedabad, India

ABSTRACT: Siderophore are small, high affinity iron chelating compounds. Siderophore facilitates iron availability in soil, it promotes the plant growth by sequestering the free iron molecules and help with transportation of the same. *Pseudomonas* sp. are well known siderophore producer. The current research covers the siderophore production by *Pseudomonas* sp. in different media. The production of siderophore is in presence of an inducer (CAS) as well as suppressor ($FeCl_3$). Siderophore production from *Pseudomonas* sp. was confirmed by carrying out CAS assay. The extraction of siderophore from the *Pseudomonas* sp. is carried out through liquid-liquid extraction using chloroform. Further detailed analysis of the extract is carried using TLC and HPTLC. The antibacterial property of siderophore was also checked through Agar cup assay along with antagonistic effect on various pathogenic strains as well its biocompatibility with other soil dwelling microbes was checked. The plant growth promoting activity of *Pseudomonas* sp. on seeds of *Cicer aritenum* for 16 consecutive days was also studied. The cumulative property of siderophore (as anti-microbial agent) and *Pseudomonas* sp. (as a bio-control agent) can be utilized for enhancing the efficiency of biofertilizers.

Keywords: *Pseudomonas* sp, *Siderophore, TLC, HPTLC, PGPR*

1 INTRODUCTION

Pseudomonas belongs to the primogenital bacterial genera. *Pseudomonas* is a genus of gram negative, that are prevalent in the environment. *Pseudomonas* spp. are known to produces siderophores[1]. Siderophores are low molecular weight ferric ion chelating molecules formed under iron-limiting environments. An extensive variety of siderophores are formed by several types of microorganisms[2]. Siderophore upsurge the fertility of soil and are also act as

* Corresponding authors: dweipayan79@gmail.com

biocontrol agents which assistance to avert infection in contradiction of pathogenic fungi. It helps in confiscating the metal contamination as of contaminated soil and water. They have an influence on soil mineral weathering and plant growth promotion (PGP)[3].

Siderophore formation be contingent upon the environment in which the microbes are rising. Soils that comprise low amount of Fe will augment the siderophore production and vice versa. Siderophore formation is delimited at gene level too and the amount in which they are formed depends on an inducer (CAS) or suppresses ($FeCl_3$)[4]. Siderophore are generally perceived by using CAS agar plate and they can also be detected by TLC as it is a chromatography technique rummage-sale to separate non-volatile mixtures[5].

Rhizobacteria *Pseudomonas* spp. has a positive effect by inhabiting the plant root tissue surface besides provides compounds that are advantageous to plants. Some of these bacteria go in more to the network and into endophyte deprived of triggering damage or morphological deviations in plants instead increase the overall growth of the plant[6]. In current study *Pseudomonas* sp. has been isolated. Siderophore production was assessed using different media and its' properties were checked. Even the PGP activity of *Pseudomonas* sp. was studied.

2 MATERIALS & METHODS

Pseudomonas sp. used in the current study was isolated and was purified from rhizospheric farm soil from Anand District, Gujarat (22.45 N, 72.90 E). Serial dilutions were prepared from the collected soil sample (up to 10^{-7}) and were spread on nutrient agar and Luria broth (LB). Once pure colony of *Pseudomonas* was isolated, sub-culturing of the *Pseudomonas* sp. was carried out by four-flame method and Luria broth (LB).

For production and extraction of siderophore *Pseudomonas* was inoculated in LB media and then were put on the rotary shaker at 150 rpm for 48 hours for proper aeration and optimum growth. Liquid-liquid extraction was carried out using chloroform as an organic solvent for the separation of siderophore from the broth containing *Pseudomonas*. Chloroform the ratio of 1:1 was used for extraction and later centrifuged. The chloroform layer was extracted[7].

Latter different media were prepared to enhance the siderophore production. Five different media were used in total. (1) Luria Broth (2) Luria Broth with Glycerol (3) Peptone Broth (4) Luria Broth with Chrome Azurol S (CAS) and (5) Luria Broth with $FeCl_3$. Same procedure was repeated for extraction.

For assessing the siderophore produced, CAS agar plate assay was carried out. LB media was prepared, sterilised and CAS dye was added to the media. Loopful of *Pseudomonas* was inoculated and incubated at 28°C for 48 hours[8].

Thin layer Chromatography (TLC) was performed of the extract obtained. Solvent system was optimised by various trial and error method. Final solvent system used was Isopropyl alcohol: Butanol: Ammonia: Water (10:6:3:1)[9]. Amalgamation of TLC and CAS assay was carried out for confirmation of siderophore production. Once the TLC plate was run and air-dried and then was dipped in CAS dye (diluted 1:1).

Anti-microbial assay was carried out of the extract. The extract that was obtained was allowed to dry out and then it was resuspended in sterile distilled water and anti-microbial assay was carried out using agar cup method. Anti-fungal assay was carried out against *Aspergillus niger* and biocompatibility *Pseudomonas* sp. and *Staphylococcus aureus* was assessed.

The plant growth promoting (PGP) activity of *Pseudomonas* sp. was studied on *Cicer arietinum* (Chick-pea). The broth of *Pseudomonas* was centrifuged, and the pellet was re-dissolved in distilled water that contain spores. Seeds of *C. arietinum* were soaked in the spore for 4 hours along with control as distilled water. Seed were kept on cotton in two different petri-plates. Both petri-plates was sprayed with distilled water at regular time interval to maintain optimum moisture. Seed germination was studied for 4 days and latter same seeds were transferred in soil. Morphological charters were studied regularly.

Figure 1. Displays growth of *Pseudomonas* in (A) LB (B) LB with Glycerol (C) Peptone Broth (D) LB with CAS (E) LB with $FeCl_3$ and (F) Liquid-liquid extraction.

Table 1. Different media used for siderophore production.

Media	Siderophore production
Luria Broth	+++
LB + Glycerol	++++
Peptone	++
LB + CAS dye	+++++
LB + $FeCl_3$	+

3 RESULTS & DISCUSSION

Pseudomonas sp. was isolated from the soil and was identified on bases of typical morphological charters and microscopic shape and arrangement. Gene level identification is also carried out as it is sent to GSBTM for 16S rRNA identification

After 48 hours of incubation of *Pseudomonas* sp. in five different media the pigmentation and turbidity was observed as displayed in **Figure 1 (A, B, C, D and E)** and liquid-liquid extraction was carried out using chloroform that is also displayed in **Figure 1 (F).** Light bluish colour extract was obtained.

As displayed in the **Table 1** maximum production of siderophore was observed in media containing (LB + CAS dye) and least production was seen in the media containing (LB + $FeCl_3$). Plus sign (+) indicates the relative amount of siderophore production.

Once extraction was completed the extract was allowed to air dry to evaporate chloroform and then it was resuspended in distilled water. The extract was loaded on TLC (F_{254}) plate and the solvent system was allowed to run 80% and the plate was air dried and observed. The four bands obtained on the TLC plate

Under UV light showed the separation of various compounds present in the extract as displayed in **Figure 2 (A).** Latter when the plate was developed using CAS dye, the colour changed from blue to orange indicated the siderophore production in **Figure 3 (B)** [10].

The extract was latter checked for its' anti-microbial activity. The siderophore has its own anti-microbial property so, it was checked against *Bacillus* sp. and along with it anti-fungal

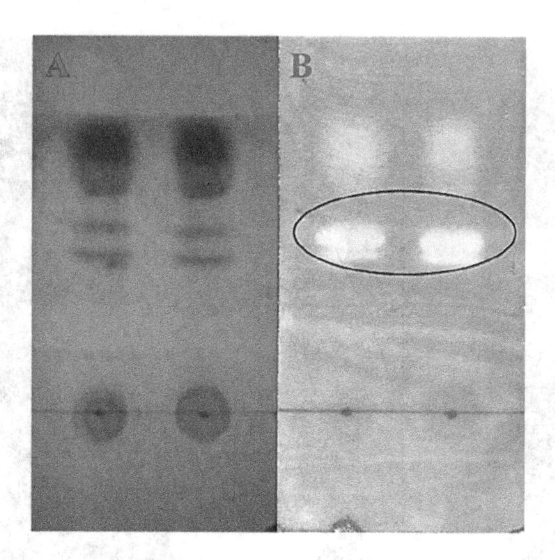

Figure 2. Display of (A) TLC plate loaded with extract under UV light and (B) TLC plate developed with CAS Dye.

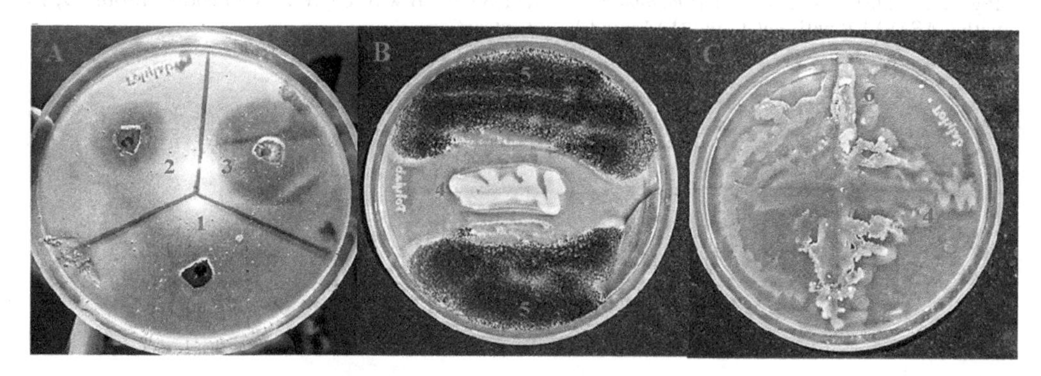

Figure 3. Displays (A) Anti- microbial Assay (1) Control (Sterile DW) (2) Siderophore extract (20 mm) (3) Standard antibiotic – Ciprofloxacin (36 mm) (B) Antifungal activity of *Pseudomonas* sp. (4) against *Aspergillus niger* (5) and (C) Compatibility test of *Pseudomonas* sp. (4) against *Staphylococcus aureus* (6).

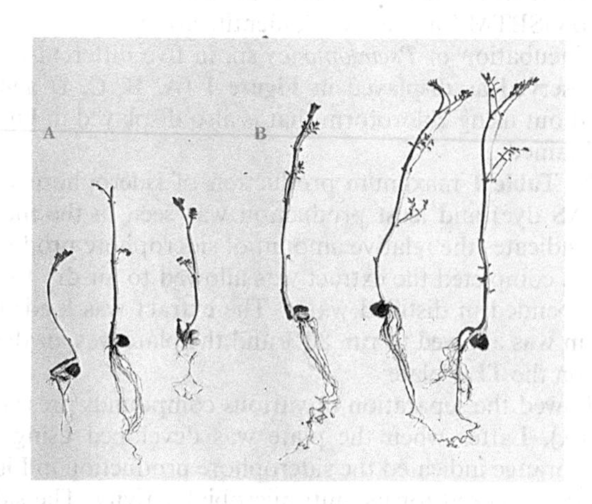

Figure 4. Display of (A) Control plants and (B) Treated plants of *Cicer aritenum* after 14 days of germination.

Table 2. Morphological measurements of plants.

Plant	Root Length (in mm)	Shoot Length (in mm)	Total Length (in mm)	Fresh Weight (in g)	Dry Weight (in g)	No. of Leaves	No. of Branches
Control	63.67 ± 16.07	95.33 ± 8.08	159.00 ± 23.39	0.47 ± 0.06	0.11 ± 0.01	19.00 ± 5.20	2.33 ± 0.58
Treated	111.67 ± 11.85	169.00 ± 12.29	280.67 ± 13.32	0.72 ± 0.08	0.13 ± 0.01	37.67 ± 1.53	3.67 ± 0.58

activity was also assessed against *Aspergillus niger* and biocompatibility was checked with *Staphylococcus aureus* as portrayed in **Figure 3 (A, B and C)** respectively.

Pseudomonas spp. are also known for its PGP activity and here we checked the efficacy in *Cicer aritenum* seeds (**Figure 4**). The seeds treated with *Pseudomonas* sp. showed better growth and on analysing the variance P = 0.0043 (**Table 2**) which showed that the data is very significant.

Certain associates of the genus *Pseudomonas* are considered to be an imperative phytopathogens, while several other strains and species divulge bioremediation and biocontrol abilities [11]. Under the current research we have isolated a strain of *Pseudomonas* sp. and sent it for its 16S rRNA gene sequencing and we claimed the point of proving high resourcefulness of this strain making it beneficial as a PGPR and Bio-control agent.

Siderophores are iron-specific compounds which are secreted under low iron stress and we found that production of siderophores in the media was less in $FeCl_3$ and was more in media containing CAS dye which acted as inducer. Moreover, the siderophore compound showed good anti-microbial property, similar zone of inhibition was obtained compared to the standard antibiotic and it also showed good antagonist activity against fungal pathogen i.e *Aspergillus niger.* Alone with these *Pseudomonas* sp. was compatible with the other normal flora.

The PGP activity of *Pseudomonas* sp. is well observed on *Cicer aritenum* plants. Figure 4 and Table 2 shows that *Pseudomonas* sp. has shown almost two-fold increase in early days of plant growth. PGP activity of *Pseudomonas* sp. is due to production of Indole acetic acid, siderophore and antibiotics [12]. Thus, *Pseudomonas* sp. can be used as bio-fertilizer.

4 CONCLUSION

Pseudomonas sp. was isolated and assessed for siderophore production using different media and they were extracted as well as quantified. Siderophore showed good anti-microbial properties (bacteria and fungi). *Pseudomonas* sp. showed good PGP activity on *Cicer aritenum* plants promising us to be used as biofertilizer.

ACKNOWLEDGMENT

Authors are thankful to the Gujarat State Biotechnology Mission (GSBTM) for providing the funding under FAP 2017-18 GSBTM/MD/PROJECTS/SSA/5041/2017-18 project and St. Xavier's College Ahmedabad for providing necessary facilities.

REFERENCES

[1] Jurkevitch, E., Hadar, Y., & Chen, Y. (1986). The remedy of lime-induced chlorosis in peanuts by Pseudomonas sp. siderophores. *Journal of Plant Nutrition*, 9(3-7), 535-545.

[2] Saha, R., Saha, N., Donofrio, R. S., & Bestervelt, L. L. (2013). Microbial siderophores: a mini review. *Journal of basic microbiology*, 53(4), 303-317.

[3] Gupta, G., Parihar, S. S., Ahirwar, N. K., Snehi, S. K., & Singh, V. (2015). Plant growth promoting rhizobacteria (PGPR): current and future prospects for development of sustainable agriculture. *J Microb Biochem Technol*, 7(2), 096-102.

[4] Dimkpa, C. O., McLean, J. E., Britt, D. W., Johnson, W. P., Arey, B., Lea, A. S., & Anderson, A. J. (2012). Nanospecific inhibition of pyoverdine siderophore production in *Pseudomonas chlororaphis* O6 by CuO nanoparticles. *Chemical research in toxicology*, 25(5), 1066-1074.

[5] Saleem, M., & Shah, M. (2017). Detection of Siderophore production in Uropathogenic *Escherichia. coli* in patients with Type 2 Diabetes Mellitus. *Int J Med Microbiol Trop Dis*, 3, 176-177.

[6] Sattiraju, K. S., Kotiyal, S., Arora, A., & Maheshwari, M. (2019). Plant Growth-Promoting Microbes: Contribution to Stress Management in Plant Hosts. In *Environmental Biotechnology:* For *Sustainable Future* (pp. 199-236). Springer, Singapore.

[7] Leipert, J., Bobis, I., Schubert, S., Fickenscher, H., Leippe, M., & Tholey, A. (2018). Miniaturized dispersive liquid-liquid microextraction and MALDI MS using ionic liquid matrices for the detection of bacterial communication molecules and virulence factors. *Analytical and bioanalytical chemistry*, 1-12.

[8] Goswami, D., Patel, K., Parmar, S., Vaghela, H., Muley, N., Dhandhukia, P., & Thakker, J. N. (2015). Elucidating multifaceted urease producing marine *Pseudomonas aeruginosa* BG as a cogent PGPR and bio-control agent. *Plant growth regulation*, 75(1), 253-263.

[9] Patel, D., Patel, A., Vora, D., Menon, S., Vadakan, S., Acharya, D., & Goswami, D. (2018). A resourceful methodology to profile indolic auxins produced by rhizo-fungi using spectrophotometry and HPTLC. *3 Biotech*, 8(10), 413.

[10] Patel, P. R., Shaikh, S. S., & Sayyed, R. Z. (2018). Modified chrome azurol S method for detection and estimation of siderophores having affinity for metal ions other than iron. *Environmental Sustainability*, 1(1), 81-87.

[11] Alberoni, D., Gaggìa, F., Baffoni, L., & Di Gioia, D. (2016). Beneficial microorganisms for honey bees: problems and progresses. *Applied microbiology and biotechnology*, 100(22), 9469-9482.

[12] Munawar, T. M., Rao, R. S. V., & Raj, H. B. (2017). screening and invitro characterization of siderophore producing *Pseudomonas fluorescens*.

Biotechnology and Biological Sciences – Sen et al. (Eds)
© 2020 Taylor & Francis Group, London, ISBN 978-0-367-43161-7

Parametric optimization for production of chromate reductase from *Bacillus* sp. under solid state fermentation using by-products of food and feeds

Bhagwat Prasad Rath*
Department of Biotechnology, College of Engineering and Technology, Biju Patnaik University of Technology, Techno-Campus, Ghatikia, Bhubaneswar, India

Achutya Nanda Acharya
Department of Chemistry, College of Engineering and Technology, Biju Patnaik University of Technology, Techno-Campus, Ghatikia, Bhubaneswar, India

Hrudayanath Thatoi
Department of Biotechnology, North Orissa University, Sriram Chandra Vihar, Takatpur, Baripada, Mayurbhanj, India

ABSTRACT: Solid State Fermentation (SSF) or Solid Substrate Cultivation (SSC) is envisioned as a prominent bioconversion technique to transform natural raw materials into a wide variety of chemical as well as bio-chemical products. This process involves the fermentation of solid substrate medium with microorganism in the absence of free flowing water. Recent developments and concerted focus on SSF enabled it to evolve as a potential bio-technology as an alternative to the traditional chemical synthesis. However, agricultural and food-industry residues constitute a major proportion (almost 30%) of worldwide agricultural production. These wastes mainly comprise lignocellulosic materials, fruit and vegetable wastes, sugar-industry wastes as well as byproducts. Agro-residues are rich in many bioactive and nutraceutical compounds, such as polyphenolics, carotenoids and dietary fiber among others. Technologies available for protein enrichment of these wastes include solid substrate fermentation, ensiling, and high solid or slurry processes. Technologies to be developed for the reprocessing of these wastes need to take account of the peculiarities of individual wastes and the environment in which they are generated, reprocessed, and used. Physico chemical and environmental factors such as inoculum type, moisture and water activity, pH, temperature, substrate, particle size, aeration and agitation, nutritional factors plays a major role in enzyme production. The advantages of SSF over Submerged Fermentation (SmF) are indicated. The need for adopting SSF technology in bioremediation of toxic compounds using by-products of food and feeds; agro-products and residues is emphasized. Among the tested by-products, rice bran supported more chromate reductase production. In all the studies, along with chromate reductase production; variation of protein content and chromate reductase activity activity were also observed. Attempts were made to explain the effects and also gauge their implications for large-scale production.

Keywords: *agro-residues*, *Bacillus* sp, *Cr(VI) reductase*, solid-state fermentation (SSF), *bioioremediation*

*Corresponding author: bpr.srath@gmail.com

1 INTRODUCTION

Solid state fermentation (SSF) refers to the microbial fermentation, which takes place in the absence or near absence of free water, thus being close to the natural environment to which the selected microorganisms, especially fungi, are naturally adapted. The current status of SSF research globally was discussed in terms of articles publication. This was followed by discussion of the advantages of SSF and the reason for interest in SSF as a notable bioprocessing technology to be investigated and compared to submerged fermentation (SmF) for the production of various added-value products. SSF also proved to be a potential technology to treat solid waste produced from food and agricultural industry and to provide environmental benefits with solid waste treatment. Therefore it is of great importance to discover the new chromate reductase producing microbial strains and optimize their enzyme production conditions in order to meet this increasing demand in both submerged and solid state fermentations. However, it is necessary to reduce the high production cost of SmF and find alternative methods. At this point, solid-state fermentation (SSF) is a method to be considered as an attractive alternative.There are some important factors that should be optimized in order to maintain microbial growth and enzyme synthesis. The aim of optimization of the process is to increase the productivity and yields by building up a base for a large scale process. With this perspective, the selection of the type of substrate and formulation of the fermentation medium should be considered to support the growth of the chosen microorganism and synthesis of the target product. The conclusion on the type of microorganism and substrate should be followed by determination of the environmental factors that significantly affect the microbial growth and product formation (Raimbault, 1998). The agro-industrial residues used for the production of chromate reductase were mentioned and some of the most important factors affecting the SSF were thoroughly experimented. One of the originality of this current study is the production of chromate reductase having reasonably high activities with the use of only suitable solid substrate and distilled water. For this purpose, the most abundant agro-wastes residues of the food industry were screened. In order to improve the chromate reductase production by SSF process, the most important factors were optimized.

Table 1. Advantages and disadvantages of the Solid state fermentation.

Advantages	Disadvantages
• The culture media are simple. Some substrates can be used directly as a solid media or enriched with nutrients • The product of interest is concentrated, that which facilitates its purification • The used inoculum is the natural flora of the substrates, spores or cells • The low humidity content and the great inoculum used in a SSF reduce vastly the possibility of a microbial contamination • The quantity of waste generated is smaller than the SmF • The enzymes are low sensive to catabolic repression or induction	• The used microorganisms are limited those that grow in reduced levels of humidity • The determination of parameters such as humidity, pH, free oxygen and dioxide of carbon, constitute a problem due to the lack of monitoring devices • The scale up of SSF processes has been little studied and it presents several problems

Table 2. Basic differences in solid state and submerged fermentation.

Factors	Submerged Fermentation	Solid Fermentation
Medium	Medium free flowing	Medium is not free-flowing
Substrates	Soluble Substrates (sugars)	Polymer Insoluble Substrates: Starch Cellulose Pectines Lignin
Aseptic conditions	Heat sterilization and aseptic Control	Vapor treatment, non sterile conditions
Water	High volumes of water consumed and effluents discarded	Limited Consumption of Water; low Aw. No effluent
Metabolic Heating	Easy control of temperature	Low heat transfer capacity
Aeration	Limitation of by soluble oxygen High level of air required	Easy aeration and high surface exchange air/substrate
pH control	Easy pH control	Buffered solid substrates
Mechanical agitation	Good homogenization	Static conditions preferred
Scale up	Industrial equipments Available	Need for Engineering & New design Equipment
Inoculation	Easy inoculation, continuous process	Spore inoculation, batch
Contamination	Risks of contamination for single strain bacteria	Risk of contamination for low rate growth fungi
Energetic consideration	High energy consuming	Low energy consuming
Volume of Equipment	High volumes and high cost technology	Low volumes &low costs of equipments
Effluent & pollution	High volumes of polluting effluents	No effluents, less pollution
Concentration of Products	30-80 g/1	100/300g/1
Size of the fermentor volume and depth	Large volume and grater depth	Less and Shallow depth
Microbial distribution	Uniformly distributed	Bacterial and yeast cells adhere to solid and grow
Size of Inoculam volume	Large	Less
Phase of fermentation	2 phase	3 phases
Solubility of nutrient	Medium nutrients dissolved in water	Medium absorbs water

2 MATERIALS AND METHODS

2.1 *Chemicals*

All chemicals used in the Himedia® India Pvt. Ltd, SRL® India Pvt. Ltd, Fisher Scientific® India Pvt. Ltd and Merck® India Pvt. Ltd. A stock of 10000 μM of Cr(VI) was prepared by dissolving $K_2Cr_2O_7$ (Fisher Scientific® India Pvt. Ltd) in microbiological grade water (Millipore India), sterilizing and adding to medium before incubation to obtain the desired initial concentration of Cr(VI). 1, 5- Diphenylcarbazide 0.25 (%, w/v) was freshly prepared by dissolving it in acetone.

2.2 *Glasswares*

All glasswares (Conical flask, Measuring cylinders, Beakers, Petriplates, Test tubes, etc.) were procured from Borosil India Pvt. Ltd.

2.3 *Enzyme extraction*

At the end of the fermentation, appropriate amount of fermented sample in 250 ml Erelenmeyer flask was added with 50 ml phosphate buffer (pH 7.0) in the ratio of 1:10 (sample amount: phosphate buffer solution). The content was dispersed with a glass burette until a homogeneous mixture was obtained and shaken at 100 rpm and 25°C for 30 minutes. Afterwards, the pH of the flask content was measured with a pH meter and the content was squeezed through muslin cloth. Again, the mixture was filtered through Whatman filter paper No. 1 and the suspension was then centrifuged at 7,000g at 4°C for 10 min and the supernatant was used for enzyme assay.

3 RESULTS AND DISCUSSION

A total of 10 different substrates were selected as solid substrate; locally available and screened for chromate reductase activity and protein concentration. Out of the 10, rice husk showed the highest activity of 19 U/mg and protein concentration of 33 mg/ml. From the production of this agricultural products, a plenty of derivatives in terms of rice husks, rice straw and ashes have been yielded as by-products. Rice straw is considered to account for the largest portion of available biomass feedstock in the world, i.e. 7.31 x 1014 kg of dry rice straw per year, and Asia contributes about 90% of the annual global production. Rice husk has been proven to provide a valuable source of carbon of about 70% and used as significant solid substrate support for enzyme production.

Initial Cr(VI) concentration with varying range (10, 20, 50, 75, 100, 200, 500, 750, 1000 and 2000 μM) were studied and it was found that 100 μM of Cr(VI) concentration was found to be the most prominent for chromium reductase production (11 U/ml). At higher initial concentration, Cr(VI) has been tend to decrease the growth rate of bacterial cells, accompanied by morphological changes which is due to toxicity and chromate stress. If the Cr(VI) concentration exceeds above the toxicity limit of the bacteria, the bacterial growth become stop and hence no chromate reductase production occurs.

Chromate reductase production by *Bacillus* sp. was determined by changing the incubation temperatures from 25-50°C. The *Bacillus* sp. was able to grow at temperatures ranging from 30-50°C. But the maximum enzyme production (22 U/mg) was achieved at 35°C at pH-7.0 under 3 d incubation period. The cell growth is also significantly affected by temperature because the variation in temperature affects the viability of the cells. Increase in temperature affects proteins by causing thermal denaturation. Thermal denaturation of proteins can cause loss of chromate reductase function.

pH of the culture media is one of the most important factor for enzyme biosynthesis. The bacterium, *Bacillus* sp. can tolerate wide range of initial medium pH and significant amount of enzyme production (18.33 U/ml) was noted at pH-7.0 at 35°C under 3 d incubation period. The decrease in enzyme activity was less in alkaline condition (with respect to optimum pH) as compared to acidic condition. The optimum pH for the growth of *Bacillus* sp. was 8.0.

For optimal conditions, the substrate Cr(VI) should saturate the enzyme, which minimizes the time between the completion of one reaction and the start of the next one (Uhlig, 1998). As bacterial Cr(VI) reduction is enzyme mediated, rate of this enzyme catalysed reaction increases with the increase in the number of active collisions, as Cr(VI) occupies more enzyme active sites. Consequently, Cr(VI) increases reflecting increased enzyme activity. The rise in enzyme activity continues until the active sites are occupied i.e. until Cr(VI) concentration saturates the enzyme (Narayan and Shetty, 2013).

4 CONCLUSION AND FUTURE ASPECTS

The enhanced production of chromate reductase by *Bacillus* sp. has been examined by using rice husk in solid state fermentation process. The highest chromate reductase activity of 3.99 U/g was produced after 48 h of incubation time. For the results of incubation temperature, rice straw yielded 19.0 U/g at 35°C. Solid-state fermentation has gained remarkable attention in recent years with respect to its low cost and potential of value added product formation. This study includes the production of chromate reductase from *Bacillus* sp. by solid-state fermentation technique using the agro-industrial residues as the substrate. A detailed literature survey on the solid-state fermentation technique, chromate reductase production by solid-state fermentation technique was stated. The experimental results performed for the investigation of the optimization parameters for production of chromate reductase from *Bacillus* sp. by solid-state fermentation technique were presented.

REFERENCES

Acuna-Arguelles, M.E.; Gutierrez-Rojas, M.; Viniegra-Gonzales, G.; FavelaTorres, E. Production and Properties of Three Pectinolytic Activities Produced by *Aspergillus niger* in Submerged and Solid-state Fermentation. *Applied Microbiology Biotechnology* 1995, 43, 808–814.

Acuna-Arguelles, M.E.;Gutierrez-Rojas, M.;Viniegra-Gonzales, G.; FavelaTorres, E. Effect of Water activity on Exo-pectinase Production by *Aspergillus niger* CH4 on Solid State Fermentation. *Biotechnology Letters* 1994, 16, 23–28.

Alazard, D.; Raimbault, M. Comparative Study of Amylolytic Enzymes Production by *Aspergillus niger* in Liquid and Solid-state Cultivation. *European Journal of Applied Microbiology and Biotechnology* 1981, 12, 113–117.

Bellon-Maurel, V.; Orliac, O.; Christen, P. Sensors and Measurements in Solid State Fermentation: A Review. *Process Biochemistry* 2003, 38, 881–896.

Bhargav, S.; Panda, B.P.; Ali, M.; Javed, S. Solid-state Fermentation: An Overview. *Chem. Biochem. Eng. Q.* 2008, 22, 49–70.

Blandino, A.; Iqbalsyah, T.; Pandiella, S.; Cantero, D.; Webb, C. Polygalacturonase Production by *Aspergillus awamori* on Wheat in Solid-state Fermentation. *Appl. Microbiol. Biotechnol.* 2002, 58, 164–169.

Camilios-Neto, D.; Bugay, C.; Santana-Filho, A.P.; Joslin, T.; de Souza, L.M.; Sassaki, G.L.; Mitchell, D.A.; Krieger, N. Production of Rhamnolipids in Solidstate Cultivation using a Mixture of Sugarcane Bagasse and Corn Bran Supplemented with Glycerol and Soybean Oil. *Appl. Microbiol. Biotechnol.* 2011, 89, 1395–1403.

Chen, L.; Yang, X.; Raza, W.; Luo, J.; Zhang, F.; Shen, Q. Solid-state Fermentation of Agro-industrial Wastes to Produce Bioorganic Fertilizer for the Biocontrol of Fusarium Wilt of Cucumber in Continuously Cropped Soil. *Bioresource Technology* 2011, 102, 3900–3910.

Couto, S.R.; Sanroman, M.A. Application of Solid-state Fermentation to Ligninolytic Enzyme Production. *Biochemical Engineering Journal* 2005, 22, 211–219.

Díaz, A.B.; de Ory, I.; Caro, I.; Blandino, A. Enhance Hydrolytic Enzymes Production by *Aspergillus awamori* on Supplemented Grape Pomace. *Food and Bioproducts Processing* 2011, 90, 72–78.

Freitas, P.; Martin, N.; Silva, D.; Silva, R.; Gomes, E. Production and Partial Characterization of Polygalacturonases Produced by Thermophilic *Monascus* sp. N8 and by Thermotolerant *Aspergillus* sp. N12 on Solid-state Fermentation. *Braz. J. Microbiol.* 2006, 37, 302–306.

Gasiorek, E. Effect of Operating Conditions on Biomass Growth during Citric Acid Production by Solid-state Fermentation. *Chemical Papers* 2008, 62, 141–146.

Grajek, W.; Gervais, P. Influence of Water Activity on the Enzyme Biosynthesis and Enzyme Activities Produced by *Trichoderma viride* TS in Solid-state Fermentation. *Enzyme Microb. Technol.* 1987, 9, 658–662.

Hölker, U.; Lenz, J. Solid-state Fermentation - Are There any Biotechnological Advantages?. *Current Opinion in Microbiology* 2005, 8, 301–306.

Jay, J. *Modern Food Microbiology* 6th Ed.; Aspen Publishers Inc.: Maryland, USA, 2000; pp. 38–56.

Kammoun, R.; Naili, B.; Bejar, S. Application of a Statistical Design to the Optimization of Parameters and Culture Medium for A-amylase Production by *Aspergillus oryzae* CBS 81972 Grown on Gruel (wheat grinding by-product). *Bioresource Technology* 2008, 99, 5602–5609.

Krishna, C., Solid-state Fermentation Systems-An Overview. *Crit. Rev. Biotechnol.* 2005, 25, 1–30.

Kumar, Y.S.; Varakumar, S.; Reddy, O.V.S. Production and Optimization of Polygalacturonase from Mango (*Mangifera indica* L.) Peel using *Fusarium moniliforme* in Solid State Fermentation. *World J. Microbiol. Biotechnol.* 2010, 26, 1973–1980.

Lee, C.K.; Darah, I.; Ibrahim, C.O. Production and Optimization of Cellulase Enzyme using *Aspergillus niger* USM AI 1 and Comparison with *Trichoderma reesei* via Solid State Fermentation System. *Biotechnology Research International* [Online] 2011, Article ID 658493, 6 pages.

·Lonsane, B.K.; Saucedo-Castaneda, G.; Raimbault, M.; Roussos, S.; ViniegraGonzalez, G.; Ghildyal, N.P. Scale-up Strategies for Solid-state Fermentation. *Process Biochem.* 1992, 27, 259–273.

Lu, W.; Li, D.; Wu, Y. Influence of Water Activity and Temperature on Xylanase Biosynthesis in Pilot-scale Solid-state Fermentation by Aspergillus sulphureus. Enzyme and Microbial Technology 2003, 32, 305–311.

Martin, N.; Souza, R.S.; Silva, R; Gomes, E. Pectinase Production by Fungal Strains in Solid-state Fermentation using Agro-industrial Bioproduct. *Braz. Arch. Biol.*
Nigam, P.; Singh, D. Solid-state (substrate) Fermentation Systems and their Applications in Biotechnology. *J. Basic Microbiol.* 1994, 34, 405–423.
Ooijkaas, L.P.; Weber, F.J.; Buitelaar, R.M.; Tramper, J.; Rinzema, A. Defined Media and Inert Supports: Their Potential as Solid-state Fermentation Production Systems. *Trends in Biotechnology* 2000, 18, 356–360.
Pal, A.; Khanum, F. Production and Extraction Optimization of Xylanase from *Aspergillus niger* DFR-5 through Solid-state Fermentation. *Bioresour. Technol.* 2010, 101, 7563–7569.
Pandey, A. Solid-state Fermentation. *Biochem. Eng. J.* 2003, 13, 81–84.
Pandey, A.; Ashakumary, L.; Selvakumar, P.; Vijayalakshmi, K.S. Influence of Water Activity on Growth and Activity of Aspergillus niger for Glycoamylase Production in Solid-state Fermentation. World Journal of Microbiology & Biotechnology 1994, 10, 485–486.
Patil, S.; Dayanand, A. Optimization of Process for the Production of Fungal Pectinases from Deseeded Sunflower Head in Submerged and Solid-state Conditions. Bioresource Technol. 2006a, 97, 2340–2344.
Pérez-Guerra, N.; Torrado-Agrasar, A.; López-Macias, C.; Pastrana, L. Main Characteristics and Applications of Solid Substrate Fermentation. Electron. J. Environ. Agric. Food Chem. 2003, 2, 343–350.
Qureshi AS, Khushk I, Ali CH, et al. Coproduction of protease and amylase by thermophilic Bacillus sp. BBXS-2 using open solid-state fermentation of lignocellulosic biomass. Biocatal Agric Biotechnol. 2016;8: 146–151.
Raghavarao, K.; Ranganathan, T.; Karanth, N. Some Engineering Aspects of Solid-state Fermentation. Biochem. Eng. J. 2003, 13, 127–135.
Raimbault, M. General and Microbiological Aspects of Solid Substrate Fermentation. Electronic Journal of Biotechnology 1998, vol 1, no 3.
Ramachandran, S.; Patel, A.K.; Nampoothiri, K.M.; Francis, F.; Nagy, V.; Szakacs, G.; Pandey, A. Coconut Oil Cake—A Potential Raw Material for the Production of A-amylase. Bioresource Technology 2004, 93, 169–174.
Rathakrishnan P, Nagarajan P. Red gram husk: a potent substrate for production of protease by Bacillus cereus in solid-state fermentation. Int J ChemTech Res. 2011;3:1526–1533.
Ray, B. Fundamental Food Microbiology 3rd Ed.; CRC Press LLC: Boca Raton, 2004; pp. 67–80.
Rodriguez-Leon, J.A.; Soccol, C.R.; Pandey, A.; Roddguez, D.E. Factors Affecting Solid-state Fermentation. In Current Developments in Solid-state Fermentation; Pandey, A., Larroche, C., Soccol C.R., Eds.; Springer: Delhi, 2008; pp. 230–252.
Sangeetha R, Geetha A, Arulpandi I. Pongamia pinnata seed cake: a promising and inexpensive substrate for production of protease and lipase from Bacillus pumilus SG2 on solid-state fermentation. Indian J Biochem Biophys. 2011;48:435–439.
Sangeetha, P.T.; Ramesh, M.N.; Prapulla, S.G. Production of Fructosyl Transferase by Aspergillus oryzae CFR 202 in Solid-state Fermentation using Agricultural Byproducts. Appl. Microbiol. Biotechnol. 2004, 65, 530–537.
Santos, M.M.; Rosa, A.S.; Dal'Boit, S.; Mitchell, D.A.; Krieger, N. Thermal Denaturation: Is Solid-state Fermentation Really a Good Technology for the Production of Enzymes?. Bioresource Technology 2004, 93, 261–268.
Schutyser, M.A.I.; de Pagter, P.; Weber, F.J.; Briels, W.J.; Boom, R.M.; Rinzema, A. Substrate Aggregation due to Aerial Hyphae during Discontinuously Mixed Solid-state Fermentation with Aspergillus oryzae: Experiments and Modeling. Biotechnol. Bioeng. 2003, 83, 503–513.
Shivasharana CT, Naik GR. Production of alkaline protease from a thermoalkalophilic *Bacillus sp.* JB-99 under solid state fermentation. Int J Pharm Biol Sci. 2012;3:571–587.
Silva, D.; Martins, E.; Silva, R.; Gomes, E. Pectinase Production by *Penicillium viridicatum* RFC3 by Solid State Fermentation Using Agricultural Wastes and Agro-industrial By-products. *Braz. J. Microbiol.* 2002, 33, 318–324.
Singhaniaa, R.R.; Patel, A.K.; Soccol, C.R.; Pandey, A. Recent Advances in Solidstate Fermentation. *Biochemical Engineering Journal* 2009, 44, 13–18.
Smits, J.P.; Rinzema, A.; Tramper, J.; van Sonsbeek, H.M.; Hage, J.C.; Kaynak, A.; Knol, W. The Influence of Temperature on Kinetics in Solid-state Fermentation. *Enzyme and Microbial Technology* 1998, 22, 50–57.
Vijayaraghavan P, Lazarus S, Vincent SGP. De-hairing protease production by an isolated *Bacillus cereus* strain AT under solid-state fermentation using cow dung: Biosynthesis and properties. Saudi J Biol Sci. 2014;21: 27–34.

Biotechnology and Biological Sciences – Sen et al. (Eds)

Evaluations of the VITEK 2 BCL card for identification of biosurfactant producing bacterial isolate SPS1001

Varsha Singh & Padmini Padmanabhan
Department of Bio-Engineering, Birla Institute of Technology, Mesra, Ranchi, Jharkhand, India

Sriparna Saha
Department of Computer Science and Engineering, Indian Institute of Technology, Patna, Bihar, India

ABSTRACT: The objective of the present study is to evaluate the VITEK 2 Bacillus Identification Card (BCL) method for the biochemical identification of biosurfactant producing environmental bacteria, by employing reference strains. In the present study, bacterial isolate SPS1001 isolated from crude oil contaminated soil obtained from Haldia oil refinery, Haldia, India was found to be aerobic, gram positive, spore forming and large rod- shaped bacteria. The characterization of bacterial isolate SPS1001 was performed by VITEK 2 BCL card, which contains micro-wells of different identification substrates and antimicrobials. Bacterial strain SPS1001 in incubation time of 14.25 hours showed good confidence level with 90 percent probability to that of reference strain *Bacillus megaterium,* with a contraindicating typical bio pattern(s) APPA(29), AspA(6), PSCNa(10), AlaA(94), PyrA(79). The state-of-art technology VITEK 2 BCL card method performs both qualitative and quantitative test, hence provides the accurate and rapid identification of *Bacillus* species from environmental samples.

Keywords: Bacteria identification, BCL card, VITEK 2

1 INTRODUCTION

Today, researchers have access to biological mini-factories in the form of bacteria, fungi and yeast to manufacture industrial products such as biosurfactant. Biosurfactants are composed of hydrophilic and hydrophobic portions and are classified based upon the nature of the hydrophilic "head groups", which is either ionic (cationic or anionic), non-ionic or amphoteric, and the hydrophobic tails which is usually a hydrocarbon chain (Silva et al. 2014). Amphiphilic biomolecules biosurfactant reduces the surface tension and the interfacial tension at the air-water and oil-water interfaces (Satpute et al. 2010; Banat et al. 2010). Biosurfactants have potential for applications in various industries such as cosmetics, food, petroleum and pharmaceuticals (Sachdev & Cameotra, 2013). Many of the biosurfactant-producing microorganisms enhance the bioavailability of polycyclic aromatic hydrocarbon were studied for their biodegradation of hydrophobic compounds (Siegmund &Wagner 1991).

Biochemical characterizations generally failed in identification of microbes isolated from environmental samples, as there is large diversity of microorganisms in these habitats (Torsvik et al. 2002). Several studies reveal that VITEK 2 has been used as an efficient method to characterize and identify isolates originated from particular ecosystems.

2 MATERIALS AND METHODS

2.1 *Environmental bacterial isolate*

Biosurfactant producing environmental bacterial isolate SPS1001 were selected for analysis in this study. Bacterial strain SPS1001 were isolated by enrichment method on modified mineral salt medium as described by Patowary et al. (2016) from crude oil contaminated soil obtained from Haldia oil refinery, Haldia, India.

2.2 *VITEK 2 identification*

BCL Card was selected to run identification analysis according to the strains to be tested. Moreover, the metabolic properties such as enzymatic reactions and sugar assimilation were studied by VITEK 2 system (bioMérieux) using BCL card according to the guidelines. BCL card contains forty six substrates to measure utilization of carbon sources, enzyme activities, inhibition by 6.5% NaCl and resistance to the antibiotics namely kanamycin, oleandomycin and polymyxin B (Atlas 1993, Claus & Berkeley 1986, Gordon et al. 1973, Logan & Berkeley 1984).

2.3 *Gene sequencing*

A VITEK 2 result discrepancy was resolved by molecular method 16S rRNA gene sequencing. If the homologous rate noted above 99% than the results would be considered valid.

3 RESULTS & DISCUSSION

3.1 *Environmental bacterial isolates*

A pure bacterial colony of biosurfactant producing strain SPS1001 was found to be aerobic, gram-positive, large rod-shaped (Figure 1) and spore-forming bacteria. Colony characteristics are listed in Table 1.

3.2 *Strain identification VITEK 2 system ID-BCL card*

VITEK 2 Systems identification method is based on the data characteristics and biochemical test being analyzed. VITEK 2(bioMérieux) BCL reagent card contains micro-wells of individual test substrate and measures various biochemical tests like acidification, alkalization, growth in presence of inhibitory substances and hydrolysis of enzyme. The list of test substrates of BCL card for biochemical test is shown in Table 3.

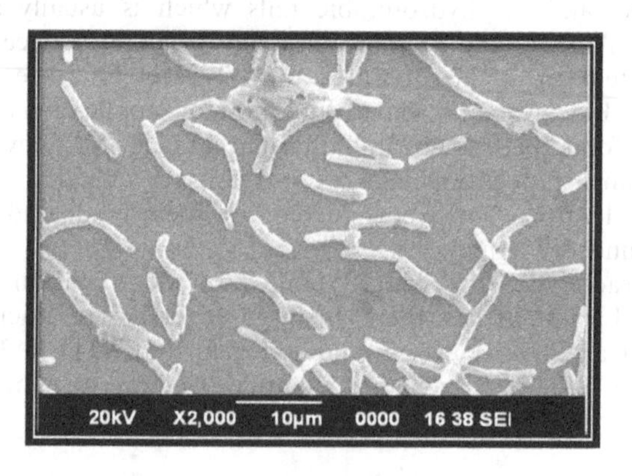

Figure 1. Scanning Electron Microscopy (SEM) image of gold sputter coated bacterial strain SPS1001 at 5000X magnification.

Table 1. Colony features of isolated bacterial strain SPS1001.

Colony features	Results
Configuration	Circular
Margin	Entire
Elevation	Raised
Pigment	Creamish
Density	Opaque
Gram nature	Positive
Shape	Long Bacilli
Spore formation	Positive

Table 2. Identification with VITEK 2 BCL card of designated bacterial strain SPS1001.

Identification levels	Result
Identity with VITEK 2	*Bacillus megaterium*
% Probability	90
Bionumber	4356510545456320
Confidence level	Good Identification
Contraindicating Typical Biopattern(s)	APPA(29), AspA(6), PSCNa(10), AlaA(94), PyrA(79)

Table 3. Biochemical details of designated bacterial strain SPS1001 using VITEK/BCL card.

Well	Test Substrate	Mnemonic	Amount/ well	SPS1001 Result
1	BETA XYLOSIDASE	BXYL	0.0324 mg	-
3	L-Lysine ARYLAMIDASE	LysA	0.0228 mg	-
4	L-Aspartate ARYLAMIDASE	AspA	0.024 mg	-
5	Leucine -ARYLAMIDASE	LeuA	0.0234 mg	+
7	Phenyalanine ARYLAMIDASE	PheA	0.0264 mg	+
8	L-Proline ARYLAMIDASE	ProA	0.0234 mg	-
9	BETA-GALACTOSIDASE	BGAL	0.036 mg	+
10	L-Pyrrolydonyl- ARYLAMIDASE	PyrA	0.018 mg	-
11	ALPHA-GALACTOSIDASE	AGAL	0.036 mg	-
12	Alanine ARYLAMIDASE	AlaA	0.0222 mg	-
13	Tyrosine ARYLAMIDASE	TyrA	0.0282 mg	+
14	BETA-N-ACETYL-GLUCOSAMINIDASE	BNAG	0.0408 mg	(+)
15	Ala-Phe-Pro ARYLAMIDASE	APPA	0.0384 mg	+
18	CYCLODEXTRIN	CDEX	0.3 mg	-
19	D-GALACTOSE	dGAL	0.3 mg	-
21	GLYCOGEN	GLYG	0.1875 mg	+
22	myo-INOSITOL	INO	0.3 mg	-
24	METHYL-A-D-GLUCOPYRANOSIDE acidification	MdG	0.3 mg	-
25	ELLMAN	ELLM	0.03 mg	-
26	METHYL-D-XYLOSIDE	MdX	0.3 mg	-
27	ALPHA MANNOSIDASE	AMAN	0.036 mg	-
29	MALTOTRIOSE	MTE	0.3 mg	+
30	Glycine ARYLAMIDASE	GlyA	0.012 mg	-
31	D-MANNITOL	dMAN	0.3 mg	+
32	D-MANNOSE	dMNE	0.3 mg	-
34	D-MELEZITOSE	dMLZ	0.3 mg	-
36	N-ACETYL-D-GLUCOSAMINE	NAG	0.3 mg	(+)

(Continued)

Table 3. (*Continued*)

Well	Test Substrate	Mnemonic	Amount/ well	SPS1001 Result
37	PAI ATINOSE	PLE	0.3 mg	+
39	L-RHAMNOSE	IRHA	0.3 mg	-
41	BETA-GLUCOSIDASE	BGLU	0.036 mg	+
43	BETA-MANNOSIDASE	BMAN	0.036 mg	-
44	PHOSPHORYL CHOLINE	PHC	0.0366 mg	-
45	PYRUVATE	PVATE	0.15 mg	-
46	ALPHA-GLUCOSIDASE	AGLU	0.036 mg	+
47	D-TAGATOSE	dTAG	0.03 mg	-
48	D-TREHALOSE	dTRE	0.03 mg	+
50	INULIN	INU	0.12 mg	-
53	D-GLUCOSE	dGLU	0.03 mg	+
54	D-RIBOSE	dRIB	0.03 mg	-
56	PUTRESCINE assimilation	PSCNa	0.201 mg	+
58	GROWTH in 6.5% NaCl	NaCl 6.5%	1.95 mg	+
59	KANAMYCIN RESISTANCE	KAN	0.006 mg	-
60	OLEANDOMYCIN RESISTANCE	OLD	0.003 mg	-
61	ESCULIN hydrolysis	ESC	0.0225 mg	+
62	TETRAZOLIUM RED	TTZ	0.0189 mg	-
63	POLYMIXIN_B RESISTANCE	POLYB_R	0.00093 mg	-

* Reactions result that are shown in parentheses "(–)" or "(+)" indicate weak reactions that are near to the threshold values.
** Note: Other well numbers not designated in this table are empty.

A transmittance optical system in VITEK 2 allows interpretation of test reactions using different wavelengths in the visible spectrum. The change in transmission of light is compared with a preset threshold value for each well. Values greater than or equal to threshold values are considered to be positive reactions, whereas the values less than the threshold considered to be negative for test substrates (Table 3). Biochemical reactions required an average of 14.25 hours of incubation for completion of test. Several factors have an effect on the results obtained by an automated biochemical identification system such as age of the culture, card lots, cell suspension density, medium, saline diluent concentration, pH and the algorithm and database of the machine (Vitek 2 bioMerieux).

Percent Probability

As part of the biochemical identification process, in VITEK 2 system the software compares the test set of reactions to the expected set of reactions of each organism, which can be identified by the product. The percent probability, a quantitative value is calculated and relates to how well the observed reactions compare to the typical reactions of each organism. Values closer to 99 shows a closer match to the typical pattern for the given organism. When the reaction pattern is not sufficient to discriminate between two to three organisms, the percent probabilities reflect this ambiguity. In this present study bacterial strain SPS1001 showed good confidence level with 90 percent probability (Bionumber 4356510545456320) to that of reference strain *Bacillus megaterium,* with a contraindicating typical biopattern(s) APPA(29), AspA(6), PSCNa(10), AlaA(94), PyrA(79) (Table 2).

4 CONCLUSIONS

The VITEK 2 system shows accuracy for automated biochemical identification for a range of test substrates with biosurfactant producing bacterial strain SPS1001 isolated from crude oil contaminated soil obtained from Haldia Oil Refinery, Haldia. In VITEK 2 (bioMérieux) BCL card performs both qualitative and quantitative test, thus bacterial isolates are identified more accurately and efficiently by this method. Rapid identification, reduced waste, robust, updated and error free database of bacterial strains makes this method of identification useful in environmental microbiology laboratories.

ACKNOWLEDGEMENT

The authors would like to thank Department of Science and Technology (Government of India) Women Scientist Scheme (WOS-A) for providing financial support, Project file no. SR/LS-154/2018 to carry out this research work. We also like to acknowledge Birla Institute of Technology, Mesra, Jharkhand, India for allowing us to undertake this project.

REFERENCES

Atlas, R.M., 1993. Handbook of microbiological media. CRC press. *Boca Raton/London.*

Banat, I.M., Franzetti, A., Gandolfi, I., Bestetti, G., Martinotti, M.G., Fracchia, L., Smyth, T.J. and Marchant, R., 2010. Microbial biosurfactants production, applications and future potential. *Applied microbiology and biotechnology*, *87*(2), pp.427-444.

Claus, D., 1986. The genus Bacillus. *Bergey's manual of systematic bacteriology*, *2*, pp.1105-1139. Edited by P. H. A. Sneath, N. S. Mair, M. E. Sharpe & J. G. Holt. Baltimore: Williams & Wilkins.

Gordon, R.E., Haynes, W.C., Pang, C.H.N. and Smith, N.R., 1973. The genus bacillus. *US Department of Agriculture handbook*, (427), pp.109-126.

Logan, N.A. and Berkeley, R.C.W., 1984. Identification of Bacillus strains using the API system. *Microbiology*, *130*(7), pp.1871-1882.

Patowary, K., Patowary, R., Kalita, M.C. and Deka, S., 2016. Development of an efficient bacterial consortium for the potential remediation of hydrocarbons from contaminated sites. *Frontiers in microbiology*, *7*, p.1092.

Sachdev, D.P. and Cameotra, S.S., 2013. Biosurfactants in agriculture. *Applied microbiology and biotechnology*, *97*(3), pp.1005-1016.

Satpute, S.K., Banpurkar, A.G., Dhakephalkar, P.K., Banat, I.M. and Chopade, B.A., 2010. Methods for investigating biosurfactants and bioemulsifiers: a review. *Critical reviews in biotechnology*, *30*(2), pp.127-144.

Siegmund, I. and Wagner, F., 1991. New method for detecting rhamnolipids excreted by Pseudomonas species during growth on mineral agar. *Biotechnology Techniques*, *5*(4), pp.265-268.

Silva, R., Almeida, D., Rufino, R., Luna, J., Santos, V. and Sarubbo, L., 2014. Applications of biosurfactants in the petroleum industry and the remediation of oil spills. *International journal of molecular sciences*, *15*(7), pp.12523-12542.

Torsvik, V., Øvreås, L. and Thingstad, T.F., 2002. Prokaryotic diversity–magnitude, dynamics, and controlling factors. *Science*, *296*(5570), pp.1064-1066.

Biotechnology and Biological Sciences – Sen et al. (Eds)
© 2020 Taylor & Francis Group, London, ISBN 978-0-367-43161-7

In-vitro study of the efficacy of ethanolic extract of Zingiber officinale on biofilm formed by pharyngeal isolate

Dibyajit Lahiri & Moupriya Nag
Department of Biotechnology, University of Engineering and Management Kolkata

Bandita Dutta
Department of Biotechnology, University of Engineering and Management Kolkata
Department of Biotechnology, MaulanaAbulKalam Azad University of Technology, West Bengal

Indranil Mukherjee, Shreyasi Ghosh & Soumik Dey
Department of Biotechnology, University of Engineering and Management Kolkata

Sudipta Dash & Rina Rani Ray
Department of Biotechnology, MaulanaAbulKalam Azad University of Technology, West Bengal

ABSTRACT: Biofilm are the colony of sessile group of organisms that remain adhered to biotic and abiotic surface with the help of pilli and Extracellular Polymeric Substances (EPS). This is composed of carbohydrates, protein, nucleic acids and several other minerals that not only provides nutrition but also prevents the penetration of antibiotics and other drugs rendering antibiotic resistance. The overuse of antibiotics has resulted in the development of genotypic changes among the organism which has also been a great concern to fight against the resistant group of organisms. Thus our re-search focuses on antibiofilm activity of *Zingiber officinale* (Tuber) and to compare its efficacy with Azithromycin on the biofilm producing organisms isolated from pharyngeal throat infection. The organism obtained were characterised by 16S rRNA sequencing followed by sequence homology and the organism was identified to be *Bacilliusparamycoides*. The ethanolic extract of *Zingiberofficinale* showed Minimum Inhibitory Concentration (MIC) of 0.2514 μg/ml and its bioactive compounds showed inhibition of the viable cells of the organism at a greater extent than azithromycin. This was further studied with Scanning Electron Microscopy (SEM) which also showed that maximum reduction of the biofilm forming cells occurred due to the challenge of the ethanolic extract of *Zingiberofficinale*. The binding energy obtained from *in-silico* studies showed that the bioactive compounds tightly bind with the biofilm forming protein and thus inhibits the further proliferation of biofilm. Thus the present study concludes that the phytocompounds has efficiency to act against the biofilm forming bacterial cells to which antibiotic fails to act.

1 INTRODUCTION

Antibiotics have been widely used as a conventional treatment against pathogens since the discovery of penicillin in 1940s [1]. But at the end of 1980s, rapid use of antibiotics over various bacteria influenced the development of antibiotic resistant strains of bacteria leading to a new threat to human health. One of the major reasons of this resistance is the formation of biofilms by microorganisms. Biofilms are adherent group of sessile communities characterized by cells that adhere to foreign body or a mucosal surface irreversibly using pili and extracellular polymeric substances (EPS) [2]. This development of EPS makes the biofilm impermeable to antibiotics and drugs resulting in the development of multidrug resistant organisms [3]. The genetic modifications within the bacterial cells also resulted in the development of acquired drug resistance that has made this issue a global concern. Recently, it has been observed that throat infections are becoming more resistant to antibiotic treatment leading to failure in eradicating the diseases such as chronic rhinosinusitis [CRS], chronic otitis media [COM] and otitis media with effusion (OME).

Fortunately, it has been seen that the traditional medicinal plants having wide range of bioactive compounds such as alkaloids, terpenoids, saponin, quercetin, anthocyanin, flavonoids, essential oils etc can effectively treat the bacterial infection [4]. Moreover, these herbal treatments, unlike synthetic medicines, do not cause severe side effects. This work focuses on the comparative analysis to study the efficacy of the ethanolic extract of *Zingiberofficinale* to treat throat infection with an objective to find out the alternate strategy of the plant extracts having the potential to act effectively in comparison to standard antibiotic for throat infection.

2 MATERIALS AND METHODS

2.1 *Microorganisms*

The strain forming oral biofilm used in this study was identified as *Bacilliusparamycoides* by sequencing of 16S r-RNA.

2.2 *Phytoextract*

The ethanolic extract of the pulverized rhizome of *Zingiberofficinale* was used.

2.3 *Chemicals: The chemicals used were all research grade*

2.4 *Determination of Minimum Inhibitory Concentration (MIC) and Minimum Biofilm Eradication Concentration (MBEC)*

The strain forming oral biofilm used in this study was identified as *Bacilliusparamycoides* by sequencing of 16S r-RNA.Theworking strains were inoculated at semi agar plates at 4°C with 15% glycerol and was preserved at -70°C until further use. The ethanolic extract of *Zingiberofficinale* and azithromycin was chosen for this study. The MIC was determined using CLSI guidelines (2006). The comparative susceptibility of the plant extract and the antibiotic was determined by disc dilution method and the MBEC was determined using with 3-[4,5-dimethyl-2-thiazolyl]-2, 5-diphenyl-2H-tetrazolium bromide (MTT dye) to quantify the viability of the bacterial cells in the presence of the challenge using ELISA plate reader (2018 GEN-NET)

2.5 *Determination of biofilm development and viability count*

The oral isolate was made to develop biofilm upon chitin flakes in accordance to the method of Ander *et. al (2009)*. The bacterial cultures were grown at Luria BertaniAgar (LB Agar) slants at 37°C. The overnight grown culture was inoculated in LB Broth to get an OD value of 0.05 at 540 nm indicating a cell concentration of $1*10^6$ CFU/mL.Such cells were then inoculated at sterile LB broth containing chitin flakes and was incubated at 37°C for a period of 72 hours in a shaking incubator [5]. The biofilm formedthus upon the chitin flakes was washed with 0.9% saline water to remove the planktonic group of cells. The biofilm forming cells were spotted upon LB agar culture plates and were incubated at 37°C. The biofilm formed upon these flakes were assessed by scanning electron microscope(Model- ZEISS EVO-MA 10) by standard protocol [6]. The results thus obtained were expressed as mean ± SE and two tailed unpaired t-test was performed to determine the statistical significance.

2.6 *Determination of docking interaction*

The docking interaction was analyzed using software Schordinger to calculate ligand dependent protein docking using the Lamarckian Genetic Algorithm (LGA) method which deals with ligands having more degree of freedom. Standard docking settings were applied and the energetically most favourable binding poses (lowest docked energy) were taken to obtain the best conformation [7].

3 RESULTS

The MIC of 72 hours isolate of planktonic and the sessile group of cells were compared. It was observed that the MIC of the sessile group of cells were more than the planktonic cells which increased from 0.25 μg/mL to 9 μg/mL for ethanolic extract of *Zingiberofficinale* and 0.32 μg/mL from 12 μg/mL for azithromycin. The cell viability assay (MBEC) showed that 72 hours old biofilm was reduced by 67.5 ± 2.13% in the presence of the challenge of the phytoextract which was considerably more than that of the antibiotic. It was observed that the planktonic cells were effectively killed by the antibiotic but it ceases to act upon the cells which has already formed the biofilm thus it implies that antibiotic can be effective before the time of conversion of cells from planktonic to sessile form. On the other hand the phytoextract showed its efficacy on both the sessile and planktonic group of cells. The Figures 1 and 2 shows that the CFU/ml of the biofilm forming cells reduced from 7.542 ± 0.02 to 7.112 ± 0.05 with respect to control in the presence of the ethanolic extract of *Zingiberofficinale* whereas the colony count reduced from 7.542 ± 0.02 to 7.164 ± 0.01 in the presence of the antibiotic azithromycin. The revival studies showed that maximum revival occurred to the organisms that were challenged with antibiotic and minimum with the phytoextract. All data were statistically significant with respect to control ($p < 0.05$). The scanning electron microscopic images (Figure 3) shows that the phytoextract causes considerable removal of the biofilm with respect to the control. This work was further validated using *in-silico* (Figure 4) studies that the major bioactive compound like furyl hydroxyl methyl ketone when docked with the biofilm forming target protein showed a binding constant value of $\Delta G = -26.996$ Kcal/mole respectively which was better than normally used antibiotics like azithromycin.

4 DISCUSSION

The results of this work provides us the information that the regular group of antibiotics like azithromycin is not sufficient to combat with the biofilm associated infection being present upon the oral cavity as these organisms has the tendency of reviving back to a larger extent.

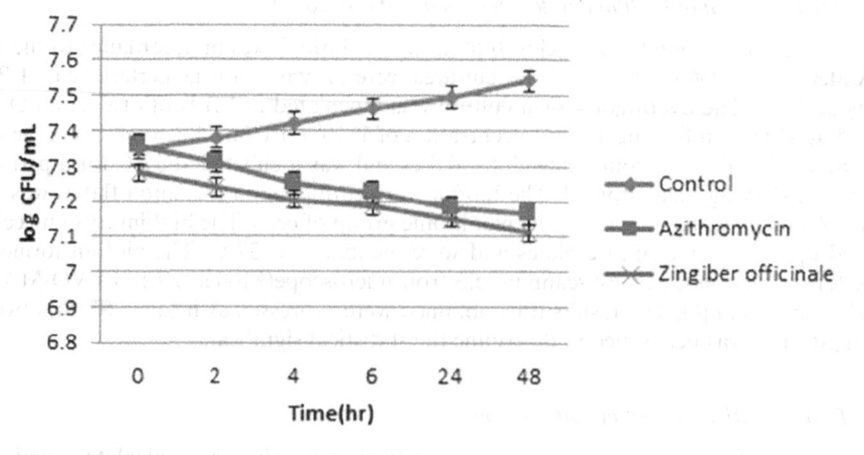

Figure 1. The viability of the biofilm forming cells in the presence of the ethanolic extract of *Zingiberofficinale* and azithromycin on *Bacillus paramycoides*.

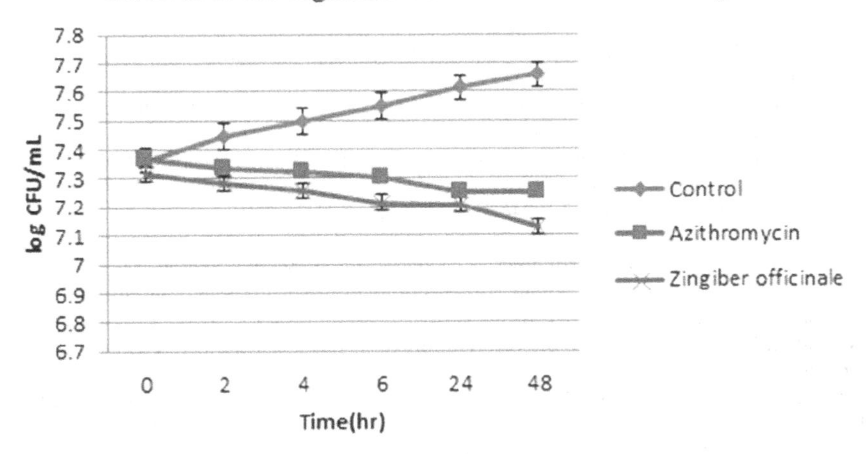

Figure 2. The revival of biofilm forming cells after the challenge of the ethanolic extract of *Zingiberofficinale* and azithromycin on *Bacillus paramycoides* was removed.

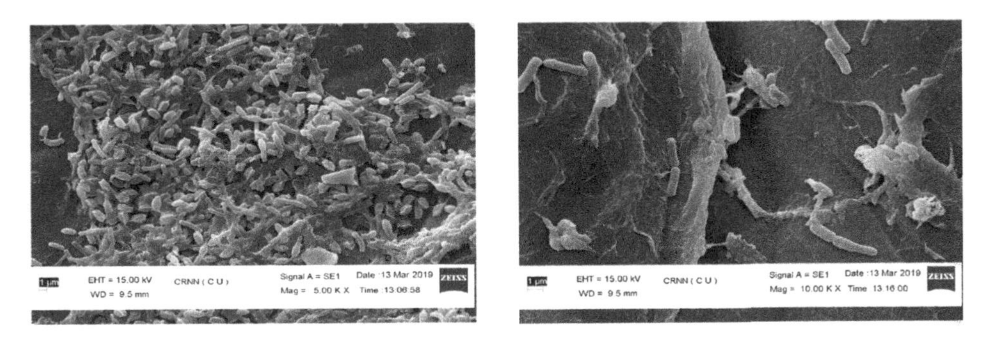

Figure 3. Shows the SEM view of the biofilm upon the chitin flakes that shows that after treating with the phyto extract there was maximum reduction of biofilm.

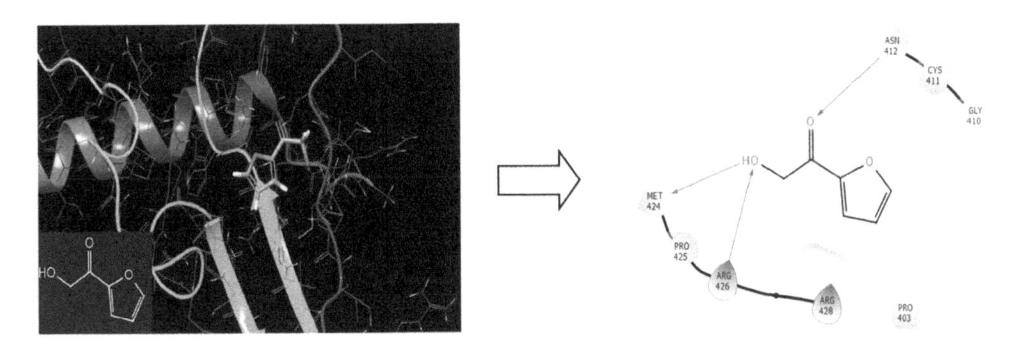

Figure 4. Oral Biofilm forming Protein vs furyl hydroxyl methyl ketoneΔG:-26.996 Kcal/mol.

Hence the use of the phytoextract from *Zingiberofficinale* provided an alternative model to treat the biofilm associated infection of throat. Thus this drift from the conventional treatment would provide an alternative strategy without cytotoxic effect upon person suffering from throat infection.

REFERENCES

1. Andersson, S. (2009). *Characterization of bacterial biofilms for wastewater treatment* (Doctoral dissertation, Kungliga TekniskaHögskolan).
2. Low -density polyethylene degradation by *Pseudomonas sp. AKS2* biofilm, Prosun Tribedi & Alok K. Sil DOI 10.1007/s11356-012-1378-y
3. Biodegradation of polyethylene by the thermophilic bacterium Brevibacillus borstelensis D. Hadad, S. Geresh and A. Sivan Department of Biotechnology Engineering, Ben-Gurion University of the Negev, Beer Sheva, Israel 2004/0825: received 15 July 2004, revised 22 October 2004 and accepted 23 October 2004.
4. Rummel, C. D., Jahnke, A., Gorokhova, E., Kühnel, D., & Schmitt-Jansen, M. (2017). Impacts of biofilm formation on the fate and potential effects of microplastic in the aquatic environment. *Environmental Science & Technology Letters*, *4*(7), 258–267.
5. J. N. Anderl, M. J Franklin & P. S Stewart. Antimicrob. Agents Chemother. 44(2000), 1818–1824.
6. F.Gomes, P. Teixeira, H. Ceri & R. Oliveira. Indian J.Med. Res. 135(2012) 542–547.
7. Morris, G. M., Goodsell, D. S., Halliday, R. S., Huey, R., Hart, W. E., Belew, R. K., & Olson, A. J. (1998). *Automated docking using a Lamarckian genetic algorithm and an empirical binding free energy function* to compute its bulk and surface properties. Journal of Computational Chemistry, 19(14), 1639–1662.

Biotechnology and Biological Sciences – Sen et al. (Eds)
© 2020 Taylor & Francis Group, London, ISBN 978-0-367-43161-7

Management of polythene waste through biodegradation by using few species of Aspergillus: An environment friendly approach

Nitesh Verma & Sharmita Gupta*
Botany Department, Dayalbagh Educational Institute, Agra, Uttar Pradesh

ABSTRACT: Polythenes are highly inert plastic material made of long chain monomer of ethylene. Polythene are light weight, stable, durable and low cost product. These properties make this synthetic polymer valuable in domestic, industrial, and environmental applications. These endless applications of polythene have subsequently led to the generation of large quantities of waste in the environment. In this study four strains of *Aspergillus* sp. were screened using synthetic medium for degradation, these *Aspergillus* sp. show remarkable ability to degrade polythene. The biodegradation experiment was performed for 9 months in soil and one month in synthetic medium in shake flasks test. The FTIR, % weight loss and the SEM were used to determine structural and surface change on polythene. The efficacy of the microorganism in the degradation of polythene was analyzed in shake flask biodegradation test. Among the *Aspergillus* species, *Aspergillus fumigatus* degraded 26% of polythene in one month shake flask test. In soil among the *Aspergillus* species *Aspergillus fumigatus* degraded 31% of polythene in 9 months.The results showed that the microorganisms were capable of degrading polythene. This reflects that these microbes can be used both in-vivo and in-vitro for the attainment of polythene degradation.

Keywords: *Aspergillus* spp, Polythene biodegradation, Shake flask, FTIR etc

1 INTRODUCTION

Polythene plays very important role in our everyday life. Polythene are petroleum based linear hydrocarbon polymers consisting of long chain of the ethylene monomers **[10]**. Polythene is classified into different types such as low density polythene (LDPE), high density polythene (HDPE) etc [9]. Today we use different kind of plastic goods is increasingly rapidly from domestic use to industrial use. Polythene waste is compounding annually at 22%. All the type of plastic, consumption of polythene bag is much more i.e. about 3.27 million tonnes in 2006-07 [8].

Polythene wastes engender intimidating challenge in their control. Polythene is semi –rigid, translucent, water proof, good chemical resistance,low cost and easily processed by most methods. Due to these qualities there is drastic rise in use of polythene. As a result, plastic waste accumulating in the environment and creates an ecological threat to land, water and air **[3]**. Number of solutions have been proposed for polythene waste management such as incineration, dumping of plastic waste in land and recycling etc. but none of these methods are up to marks, every methods have some drawbacks [13]. To overcome such problems, biodegradation of plastics is an ecofriendly and cost effective method. *A.terreus*, *A. flavus*, *A. japonicus* and *Mucor* species were isolated and identified as polythene degrading fungi [4]. In the present study 15 different types of fungi were screened for polythene degradation on synthetic medium. Degradation of polythene was determined by weight loss method, FTIR and SEM analysis.

*Corresponding author: drsharmitagupta123@gmail.com

2 MATERIAL AND METHOD

2.1 *Sample preparation & screening*

2.1.1 *Preparation of polythene powder*

Polythene samples taken were commercially available in local market used as carry bags of 20 micron thickness HDPE. Polythene film were shredded into small strips and dipped into the xylene.This was later boiled for half an hour. Xylene dissolved the polythene and the residue was crushed with mixer grinder. Residual xylene was removed from polythene powder by washing with ethanol which was later evaporated. Obtained polythene powder was oven dried at 60°C for 12-15 hours

2.1.2 *Screening of polythene degrading fungi*

Fungal strains were procured from the Botany department's Microbiology Lab.15 fungal cultures were screened for polythene degradation on synthetic medium. Ingredients of synthetic medium in (gram/litre) areNH_4NO_3:1.0,$MgSO_4.7H_2O$:0.2,K_2HPO_4:1.0,KCl:0.15, Agar: 20, 1% polythene powder, 1% Tween20 [2].

2.1.3 *Preparation of fungal suspension*

Fungi were cultured on SDA slants and incubated at 30°C. After completion of sporulation, spore of fungi on slants were washed with 6ml physiological saline solution. The spore mixture was placed in a small beaker on stirrer and stirred for 10 minutes aseptically. Added 3ml of this spore mixer in known volume of distilled water. Fungal spore were counted by haemocytometer.

2.2 *Biodegradation experiment*

2.2.1 *Soil burial treatment*

Pre-weighted polythene strips (3 cm × 2 cm) were disinfected with 70% ethanol for 30 minutes, and then transferred to benzene for 30 minutes to remove the plasticizer. Later it was placed in distilled water for 10 minutes and dried for 15 minutes in laminar air flow chamber. This disinfected film is buried in pots containing sterilized soil and inoculated with fungal suspension.

2.2.2 *Shake flask test (in liquid synthetic medium)*

In this biodegradation experiment 2 ml of fungal suspension was added to flask containing 150 ml of liquid synthetic medium(composition of synthetic medium is same as screening medium except agar). 3X2 cm sterilized film was immersed in flask and kept in orbital shaker for 1 month incubation at 30°C/120 rpm.

2.2.3 *Film harvest*

After incubation period nine months in soil and one month in liquid synthetic medium polythene strips were harvested and washed with running water. After this 70% ethanol used to remove as much as cell mass from the residual film as possible. The film was dried 24 hours in oven.

2.3 *Assessment of degradation of polythene*

2.3.1 *Weight loss method*

Sample weight loss was determined by an analytical balance (SHIMADZU CORPORATION TYPE AY220).

2.3.2 *FTIR*

Infrared spectra of polythene film were recorded on Cary 630 FTIR (Agilen technologies) over a range of 4000cm^{-1}-800cm^{-1}. Sampleswere powdered and analyzed. For monitoring, formation and disappearance of carbonyl and double bond FT-IR is necessary to elucidate the mechanism of the biodegradation process.

2.3.3 *SEM*

The surface morphology and microstructure of the polyethylene strip due to biodegradation were analyzed through scanning electron microscopy [7]. The polythene film were examined by JSM 6490 LV (JEOL JAPAN)

3 RESULTS AND DICUSSION

3.1 *Assessment of polythene degradation through weight loss*

Out of 15 fungal cultures, 4 fungal cultures *viz Aspergillus niger. A.fumigatus, A.flavus, A.candidus,* show growth on synthetic medium in screening. These 4 fungi were taken for further biodegradation test. Degrading ability of polythene carry bag of HDPE through *Aspergillus niger, Aspergillus fumigatus, Aspergillus flavus, Aspergillus candidus,* has been shown in Tables 1 and 2. The degradation of polythene was determined by calculating the percentage of weight loss of polythene. The percentage of weight loss is shown in Tables 1 and 2.

These 4 fungal species were separately allowed to degrade the polythene under the soil burial method and in shake flask method in liquid synthetic medium. Among the fungal species *Aspergillus fumigatus* was found to be most active degrading 31% in soil and 26% in liquid synthetic medium.

3.2 *Assessment of polythene degradation through FTIR*

The FTIR spectra of untreated HDPE film and HDPE film incubated with *Aspergillus* species for 30 days in liquid synthetic medium and 90 days in soil containing polythene as sole carbon source are shown in Figure 1. Depiction of transmittance and wavelength were interpreted using books "Spectrometric identification of organic compound" [12] and "Interpretation of infrared spectra, a practical approach" [4].

In the present study FTIR spectra showed the native band at 1463 cm^{-1} was decreased to 1461 to 1459 cm^{-1} which show methylene C-H bend asymmetric/symmetric on HDPE film inoculated with fungi *A.niger, A.candidus, A.fumigatus, A.flavus* [5]. New absorption band at 2100 $cm-^{1}$-1800 $cm-^{1}$ and 2140 $cm-^{1}$- 2348 $cm-^{1}$ of the spectra were observed in fungi treated HDPE polythene film.

Table 1. Degradation of polythene in shake flask test(synthetic medium) within one month.

Fungi	Initial Weight(mg)	Final Weight(mg)	Weight of P.E. degraded(mg)	Percentage of degraded P.E.
Aspergillus niger	7.0	5.7	1.3 ± 0.52	20%
Aspergillus flavus	7.0	5.8	1.2 ± 0.06	16%
Aspergillusfumigates	7.0	5.5	1.5 ± 0.05	26%
Aspergillus candidus	7.0	6.4	0.6 ± 0.40	7%

Table 2. Degradation of polythene in soil within nine months.

Fungi	Initial Weight(mg)	Final Weight(mg)	Weight of P.E. degraded(mg)	Percentage of degraded P.E.
Aspergillus niger	7.0	5.3	1.7 ± 0.15	22%
Aspergillus flavus	7.0	5.4	1.6 ± 0.92	12%
Aspergillusfumigatus	7.0	5.0	2.0 ± 0.26	31%
Aspergillus candidus	7.0	5.6	1.4 ± 0.2	20%

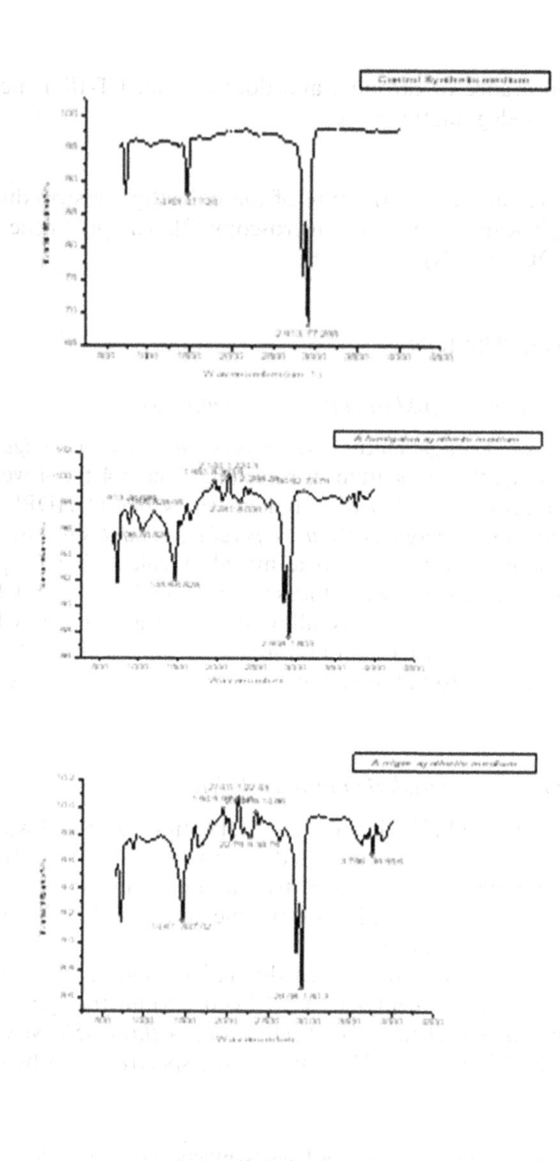

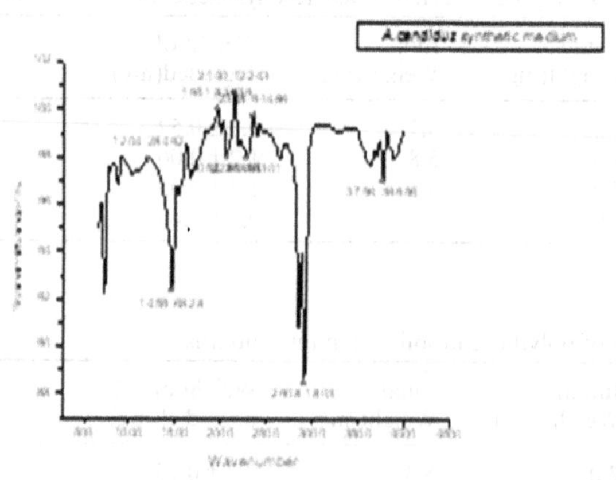

Figure 1. Graphs of FTIR showing degradation of polythene through breakdown and stretching of bond between hydrogen and carbon in liquid synthetic medium.

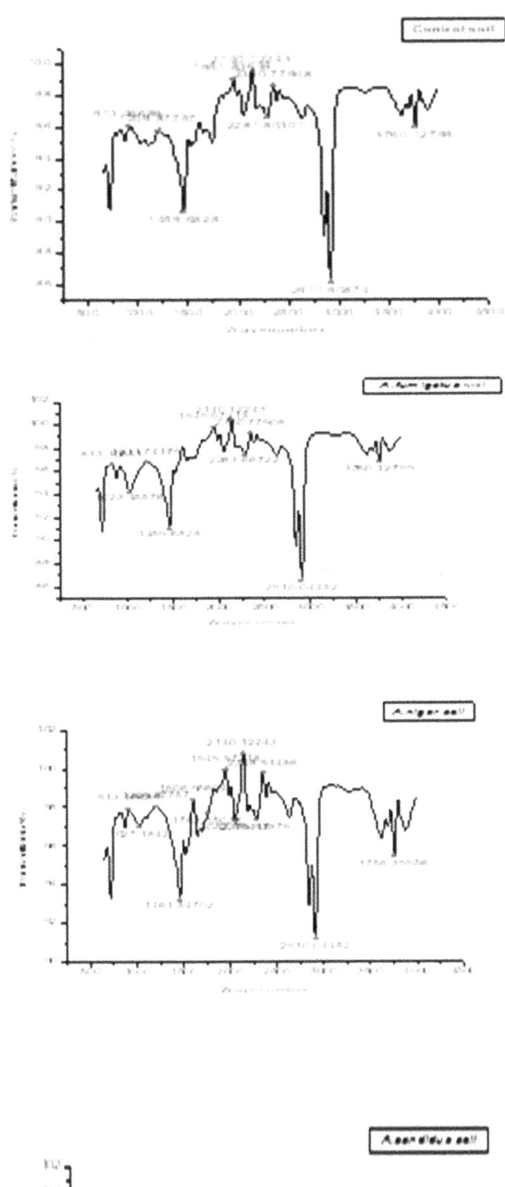

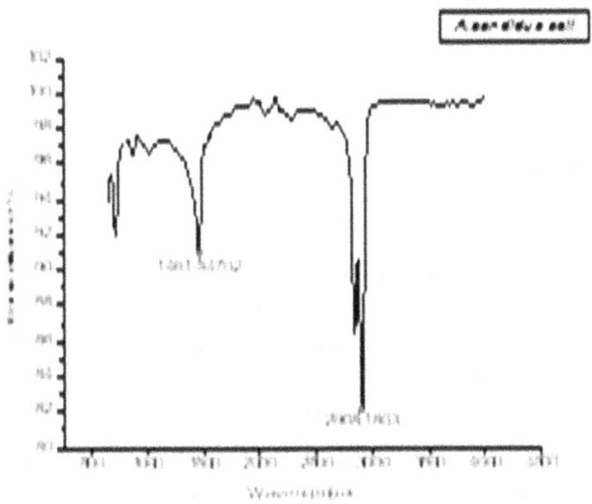

Figure 2. Graphs of FTIR showing degradation of polythene through breakdown and stretching of bond between hydrogen and carbon and in soil.

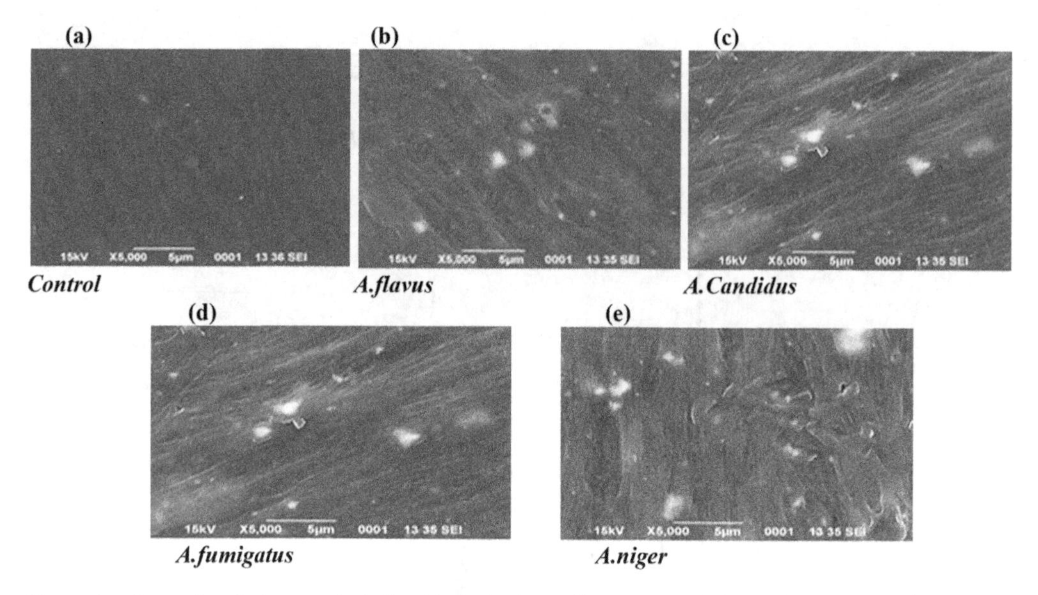

Figure 3. Treated polythene strip in liquid synthetic medium a.control, b. *A.flavus*,c. *A.candidus*. d.*A. fumigatus*, e. *A.niger*.

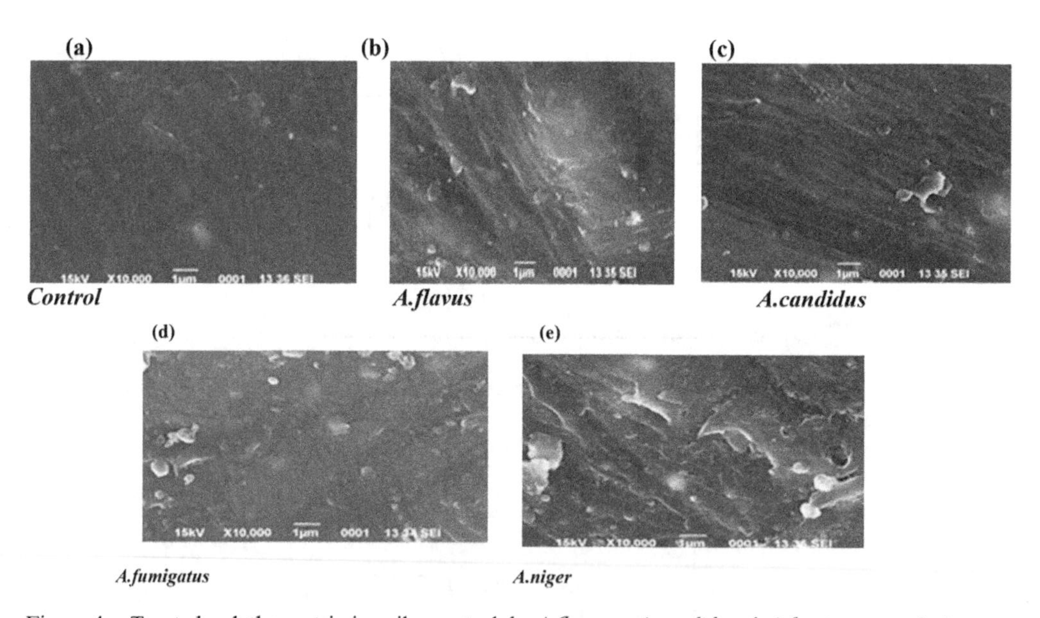

Figure 4. Treated polythene strip in soil a.control, b. *A.flavus*,c. *A.candidus*. d.*A.fumigatus*, e. *A.niger*.

This is possibly due to formation of transition metal carbonyl group frequency. In addition the bands at 913 cm^{-1} and 911 cm^{-1} also appeared by the action of *A.fumigatus* in both the degrading medium (soil and liquid synthetic medium) due to C-H bend (vinyl C-H out of plane bend). *A.flavus* in soil HDPE polythene strip show band at 1202 cm^{-1} which show C-C vibration. In all the fungi treated HDPE polythene film the new band at 2140 cm^{-1} appeared which exhibit C≡C stretch except *A.fumigatus* and *A.candidus* in soil medium. The native band 2913 cm^{-1} is decreased to 2910 cm^{-1} in in soil medium by *A.fumigatus* and *A.niger* and 2908 cm^{-1} to 2903 cm^{-1} by *A.candidus, A.flavus, A.fumigatus* in liquid synthetic medium due to strong stretch in methyne C-H group and also show broadening of band in these ranges [1]. Additionally new absorption bands between 3800 cm^{-1} -3100 cm^{-1} of the spectra were observed in the HDPE film and this is due to formation of carboxylated compounds.

FTIR results, when compared with control revealed alteration in peaks. New peaks were formed, certain peaks vanished and for some there was transformation in the peak range. Variation in FTIR are considered as assessing parameter, which occur owing to changes on polythene surface due to fungal activity [10]. The change in the peak values of almost all functional groups support the conformational changes on polythene surface

3.3 *Assessment of polythene degradation through SEM*

Scanning electron micrograph show different kinds of surface alterations. Controls are used to establish the baseline degradation for comparison. The polythene film buried in soil and in synthetic medium showed the holes and cracks on its surface. It was also observed that the control show no such changes [6].

SEM results exhibit the surface of the polythene film becomes pitted and eroded. SEM pictures of polythene film revealed that the film treated with *A.fumigatus* show more cracks and holes and polythene film treated with *A.flavus* show less cracks and pits in soil degradation. In shake flask test *A.fumigatus* also show more holes and cracks and *A.candidus* show less degradation.

4 CONCLUSION

Polythene is probably the polymer used frequently in the world, because it is easily processed, stable and low cost product. In order to get rid of polythene waste with an ecofriendly way there is only one solution to exploit microorganism to degrade polythene. In present study biodegradation of HDPE polythene film analyzed for 90 days in soil burial and 30 days by shake flask test in liquid synthetic medium. The surface of polythene film become pitted. Owing to fungal action, there was considerable reduction in weight of polythene film. The degradation was evaluated by different methods including percent weight loss, FTIR and SEM analysis. The present work concludes that these 4 species of *Aspergillus* fungi are able to degrade polythene in natural environment as well as lab conditions. Among all the fungi *A.fumigatus* show high rate of degradation in soil as well as by shake flask test in liquid synthetic medium.

REFERENCES

1. Ambika devi K., P. V. Ratnasari, B. K. M. Lakshmi, and K. P. J. Hemalatha (2014) Isolation of polythene degrading bacteria from marine waters of Viskhapatnam IndiaInt J of Curr Micro and App Sci 310: 269–283.
2. Esmaeili A., A. A. Pourbabaee, H. A. Alikhani, F. Shabani, and L. Kumar (2014) Colonization and Biodegradation of Photo-Oxidized low-Density Polyethylene(LDPE) by New Strains of *Aspergillus* sp. and *Lysinibacillus* sp. Biorem J 18: 213-226.
3. Gilan(Orr), Y. Hadar, and A. Sivan. (2004) Colonization,biofilm formation and biodegradation of polythene by a strain of *Rhodococcus* ruber App Micro and cell physio 65: 97-104.
4. John Coates (2000) Interpretation of Infrared Spectra, A Practical Approach Encyclopedia of Analytical Chemistry R.A. Meyers Ed. Copyright John Wiley and Sons Ltd.
5. Mahalaxmi V., A. Siddiq, and S. N. Andrew (2012) Analysis of Polythene degrading potential of Microorganism Isolated From Compost Soil Int J of Phar & Bio Arc 35: 1190-1196.
6. Mona Gouda, Azza E., Swellam and Sanaa H. omar (2012) Biodegradation of synthetic polyester (BTA &PCL) with natural flora in soil burial and pure cultures under ambient temperature Res J of Env and Earth Sci 43: 325–333.
7. Gnanavel G., Dr. Marimuthu, Thirumarimurugan and V. P.Mahana Jeya Valli (2016) Biodegradation of oxopolyethylene: An approach using soil compost degraders Int J of Adv Eng Tech 7: 140-144
8. National Plastic Waste Management Task Force Report 2008.
9. Pramila R. and Ramesh K. V. (2015) Potential biodegradation of low density polyethylene (LDPE) by Acenitobacter baumannii AfrJ of Bact Res 7: 24–28

10. R. Kavitha, Anju K. Mohan and Bhuvaneswari (2014) Biodegradation of low density polythene by isolated from oil contaminated soil Int J of Plant Ani and Env 4: 601-610.
11. Raaman R., N. Rajitha, A. Jayshree, and R. Jegadeesh (2012) Biodegradation of plastic by *Aspergillus spp.* isolated from polythene polluted sites around Chennai. Acad Indus Res16: 313-316.
12. Robert M., Francis X. Webster,David J. Kiemle (2005) Spectrometric identification of organic compounds 7th Ed. Copyright John Wiley and Sons Ltd.
13. T. Mumtaz, M.R. Khan and M.A. Hassan (2010) Study of environmental biodegradation of LDPE films in soil using optical and scanning electron microscopy Micron ELS.

Biotechnology and Biological Sciences – Sen et al. (Eds)
© 2020 Taylor & Francis Group, London, ISBN 978-0-367-43161-7

Protoplast fusion of yeast strains for strain improvement to enhance mixed substrate utilization range

S. Sharma & A. Arora*
Division of Microbiology, ICAR-Indian Agricultural Research Institute, New Delhi,

D. Paul
Amity Institute of Biotechnology, Amity University, Noida, India,

ABSTRACT: Protoplast fusion, a part of evolutionary engineering has a great potential for genetic analysis and strain improvement [3]. It breaks down the barriers to genetic exchange imposed by conventional mating systems. It can serve the purpose for developing a strain with mixed substrate fermentation abilities. *Saccharomyces cerevisiae* strains are suitable hexose fermenters but pentose fermentation is a setback for these strains. Although recombinant strains have been engineered but with limited success and certain constraints. *Pichia stipitis* strains are better pentose fermenters. Therefore, in this study, *S. cerevisiae* (LN) (~80% fermentation efficiency) and *P. stipitis* NCIM 3498 exhibiting xylose assimilation were taken. Firstly, selection markers were identified for the selection of fusants. *S. cerevisiae* LN was sensitive to hygromycin B (50 ppm) while *P. stipitis* NCIM 3498 was sensitive towards cycloheximide (5 ppm). Protoplast formation was optimized using glucanex enzyme (Sigma). 10 mg mL^{-1} enzyme concentration yielded highest protoplast frequency after 72 h. Fusion was carried out using 35% PEG and protoplasts incubated at 30 °C for ~30 h. Fusants were observed under microscope after 16 h of incubation but they could not regenerate well in the regeneration medium.

Keywords: Protoplast fusion, evolutionary engineering, *S. cerevisiae*, *P. stipitis*

1 INTRODUCTION

Metabolic engineering has paved way for improvement of microbial strains through various tools and techniques. These techniques include evolutionary engineering/adaptive evolution, medium engineering, random mutagenesis and/or protoplast fusion. Among these techniques, protoplast fusion technique has a great possibility for genetic analysis as well as for strain improvement. This technique is generally applied for developing inter specific, intra specific and inter generic, intra generic supra hybrids with higher capability. It is a significant tool for genetic manipulation as it resolves the barrier to genetic exchange imposed by conventional mating systems. It is particularly useful for industrially important microorganisms [4].

This technique offers several advantages over other conventional mating systems. These include transfer of relatively large segments of genomic DNA [6] and several genomes can be recombined simultaneously without prior knowledge of genomic information. Therefore, several gene mutations can randomly and efficiently occur. This generates a large number of fusants/mutants to be screened.

Protoplast fusion has three main steps: protoplast formation, fusion of protoplasts and then regeneration of cells. Disadvantages of this technique include instability of hybrids and reversion into parental strains which is more pronounced in distantly related species than closely related species [1].

*Corresponding author: E-mail: anjudev@yahoo.com

In this study, two intergeneric strains have been fused namely, *Saccharomyces cerevisiae* LN and *Pichia stipitis* NCIM 3498. *Saccharomyces* being glucose fermenter and *Pichia* being xylose fermenter have been chosen for the study. Similar studies have been conducted by several other researchers to obtain a high ethanol yielding strain [2, 3, 5].

2 MATERIAL AND METHODS

2.1 *Microbial strains*

Saccharomyces cerevisiae LN, hexose fermenter isolated from fruit juices along with *Pichia stipitis* NCIM 3498, a xylose assimilator and fermenter was selected for protoplast fusion. Both the cultures were grown on MGYP medium (3 g L^{-1} Malt Extract, 10 g L^{-1} Glucose, 3 g L^{-1} Yeast Extract and 5 g L^{-1} Peptone) at 30 °C and stored at 4 °C as slants.

2.2 *Marker selection for screening fusants*

Both cultures were screened for antibiotic tolerance (Hygromycin B, Cycloheximide and Nystatin) using the concentration range 1-50 ppm, temperature tolerance (4°C- 40 °C), ethanol tolerance (4%-14%).

2.3 *Enzyme used for protoplast formation*

Glucanex enzyme (Sigma Aldrich) was used for formation of protoplasts of both the cultures. It is a cost effective, yeast lytic enzyme.

2.4 *Protoplast formation*

Overnight grown cultures with 1 OD_{600nm} were used for protoplast formation. 5 mL culture was taken in tubes and centrifuged at 8,000 rpm for 10 min. Culture supernatant was discarded and pellet was washed with protoplast solution (0.6 M KCl, 10 mM β-mercaptoethanol, 50 mM phosphate buffer). Different concentrations of enzyme (10 mg mL^{-1} to 50 mg mL^{-1}) prepared in protoplast solution were added to the culture pellets and incubated for 72 h at 30 °C and 150 rpm. Protoplast formation was monitored after 24 h interval by staining cells with lactophenol cotton blue dye under bright field microscope.

2.5 *Protoplast fusion*

Obtained protoplasts were washed with protoplast buffer. Both the culture protoplasts were mixed in equal proportion. Protoplasts were resuspended in fusion buffer containing PEG (35%), $CaCl_2$ and sorbitol. Protoplasts were then incubated at 30 °C, 150 rpm for 72 h.

2.6 *Regeneration of protoplasts*

Cells were washed with protoplast buffer and added to 10 ml regeneration medium (containing 50 ppm Hygromycin B, 5 ppm Cycloheximide and 1 mg mL^{-1} Kanamycin and 1% xylose in minimal medium: 1 g L^{-1}KH2PO4, 5 g L^{-1}MgSO4, 5 g L^{-1}(NH4)SO4, 1 g L^{-1}Yeast extract). Cultures were incubated at 30 °C for 72 h and 150 rpm.

3 RESULTS AND DISCUSSION

3.1 *Strain and marker selection*

Strains were selected on the basis of their properties. Since this study aims at creating a new strain capable of mixed substrate utilization from ligocellulosic biomass, *Saccharomyces* and *Pichia* strains were used for intergeneric fusion to obtain an improved strain.

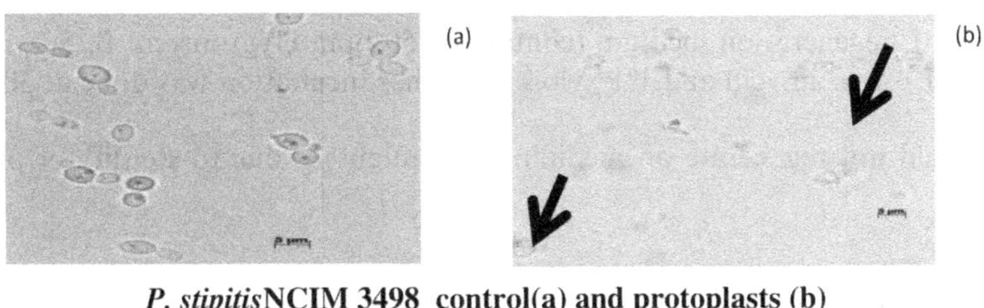

***P. stipitis*NCIM 3498 control(a) and protoplasts (b)**

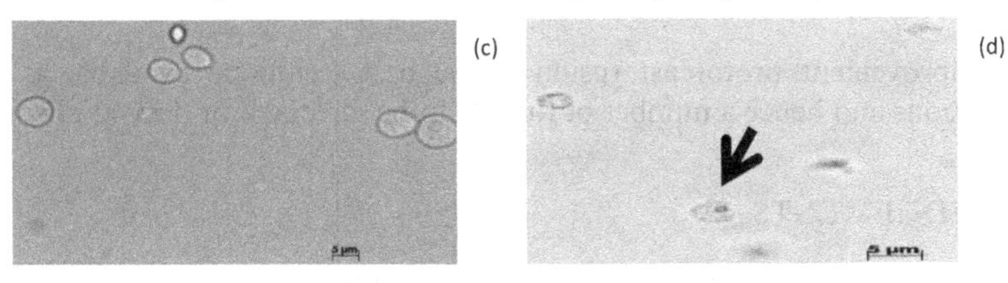

***S. cerevisiae*LN control (c) and protoplasts (d)**

Figure 1. Control (a, c) and Protoplasts (b, d) of *P. stipitis* NCIM 3498 and *S. cerevisiae* LN respectively as observed under microscope.

To differentiate fusants from the parent strains, a set of three parameters were chosen; antibiotic resistance, temperature tolerance and ethanol tolerance. Krishnamoorthy et al. 2010 [3] chose cycloheximide as a marker since *S. cerevisiae* LN was sensitive while *P. stipitis* NCIM 3498 was efficiently growing on the same antibiotic at 5 ppm concentration. However, *P. stipitis* was susceptible for Hygromycin B (50 ppm) and *S. cerevisiae* was efficiently growing on it. Both cultures exhibited similar profile for temperature and ethanol tolerance. Therefore, Hygromycin B and Cycloheximide were selected as selection markers.

3.2 *Protoplast formation with glucanex enzyme*

Cells in logarithmic phase were used for protoplast formation by glucanex enzyme as they are younger cells and easily susceptible to enzymatic lysis [6]. Therefore, cultures with 1 OD were used. Highest frequency protoplasts were noticed after 72 h [Figure 1].

Protoplasts obtained at higher concentrations of enzymes were aggregated and clumped. Different yeast strains exhibit different rates of protoplast release because of variation in thickness and composition of cell wall layers [6].

3.3 *Protoplast fusion*

Fusants were observed after 16-17 h of incubation [Figure 2]. PEG was removed by centrifugation. Cells were washed with protoplast buffer (0.1 M phosphate buffer, 0.8 M sorbitol) and

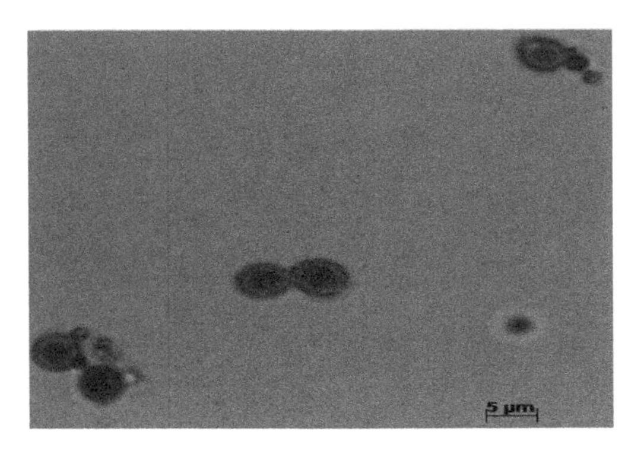

Figure 2. Fusants formed after incubation (16-17 h) with PEG.

added to 10 mL regeneration medium (containing 50 ppm Hygromycin B, 5 ppm Cycloheximide, 1 mg mL^{-1} Kanamycin and 1% xylose). Further incubation was done at 30 °C for 72 h, 150 rpm.

Fusants could not regenerate on medium. This might be due to sterility of protoplasts to regenerate.

4 CONCLUSION

For strain improvement protoplast fusion is the best technique available as it generates random mutations and hence a number of fusants to be screened for desired phenotypes.

ACKNOWLEDGEMENTS

This work was performed at ICAR-Indian Agricultural Research Institute, New Delhi, India. Funding was provided by ICAR-AMAAS (12-124).

REFERENCE.

1. Attfield P.V., Bell PJ, (2003) Genetics and classical genetic manipulations of industrial yeasts. In Functional genetics of industrial yeasts, (17-55), Springer Berlin Heidelberg.
2. Jingping G, Hongbing S., Gang S., et al (2012) A genome shuffling-generated *Saccharomyces cerevisiae* isolate that ferments xylose and glucose to produce high levels of ethanol. J Ind Microbiol Biotechnol 39(5): 777-787.
3. Krishnamoorthy R.N., Vijila K., Kumutha K. (2010) Intergeneric protoplast fusion of yeast for high ethanol production from cheese industry waste Whey. J Yeast Fungal Res, 1(5): 81-87.
4. Muralidhar R.V., Panda T. (2000) Fungal protoplast fusion-a revisit. Bioprocess Biosyst Eng 22(5): 429-431.
5. Shi D.J., Wang C.L., Wang K.M. (2009) Genome shuffling to improve thermotolerance, ethanol tolerance and ethanol productivity of *Saccharomyces cerevisiae*. J Ind Microbiol Biotechnol 36(1): 139-147.
6. Thomson ISI, Fawzia JS, Noor AI et al (2016) Asian J Biol Sci 9(1-2): 10-18.

Biotechnology and Biological Sciences – Sen et al. (Eds)
© 2020 Taylor & Francis Group, London, ISBN 978-0-367-43161-7

Green synthesis and characterization of silver nanoparticles using aqueous extract of microalga *Desmococcus Olivaceus* and its biological activities

K. Sangeetha & Uma Ramaswamy*
PG & Research Department of Biochemistry, Dwaraka Doss Goverdhan Doss Vaishnav College, Chennai

ABSTRACT: The present work was carried out for the synthesis and characterization of silver nanoparticles of green microalga *Desmococcus olivaceus* and to evaluate its antioxidant, antimicrobial and anti proliferative activities. The maximum synthesis of silver nanoparticle-(AgNP) was obtained in 5% algal extract and 1mM silver nitrate within 48 hours at 37°C. AgNP's were characterized by using UV-visible spectroscopy, Scanning electron microscopy EDAX and Fourier transform spectroscopy(FTIR). Silver Surface Plasmon Resonance (SPR) occurred at 430 nm for 1mM AgNP's respectively.SEM EDAX analysis showed the presence of polydispersed, spherical AgNP's . FTIR spectra revealed the presence of reducing groups in the extract responsible for AgNP synthesis .The synthesized AgNP exhibited superoxide and nitric oxide radical scavenging activities. Microalgal silver nanoparticles demonstrated anti-bacterial activity against gram positive and negative bacteria which includes *Staphylococcus aureus,Bacillus subtilis, Klebsiella pneumoniae* and *Escherichia coli* . The *invitro* anti prolifera-tive activity was confirmed by MTT assay in the cell lines of cervical cancer Hela cells and vero cell lines showed IC_{50} value of silver nanoparticles at 1.56 & 7.04 μg/ml respectively.

Keywords: silver nanoparticle, *Desmococcus olivaceus*, Hela cells, antibacterial

1 INTRODUCTION

Metal nanoparticles possess distinct physical properties differing from that of the bulk phase. Metal nanoparticles were widely used in the fields like imaging, drug delivery, medicine and biosensing. Algae a diverse group of plant kingdom, contains different bioactive compounds. The bioactive substances produced by actively growing cells of algae includes proteins, fats, lipids, carbohydrates, vitamins, free amino acids, enzymes, growth regulators, pigments, toxins and antibiotics(Becker, 1994). *Desmococcus* is a genus of green algae belong to the family Chaetophoraceae and are rich in secondary metabolites and widely used in phycoremediation.

2 MATERIALS AND METHODS

2.1 *Collection and cultivation of microalgae*

Freshwater green microalgae, *Desmococcus olivaceus* were obtained from the Phycospectrum environmental research lab, Anna nagar, Chennai-40. *D.olivaceus* were grown in improvised CFTRI medium and the pH was maintained at 10.

*Corresponding author: umaramesh.rg@gmail.com

2.2 *Preparation of silver nitrate solution*

0.168g of silvernitrate was dissolved in one litre of distilled water and stored in amber bottle (1mM silver nitrate solution).

2.3 *Preparation of aqueous extract of* desmococcus olivaceus

Wet weight of algal biomass of 3.0 g was dissolved in 60 ml of double distilled water for 30 minutes at 100°C in an conical flask. After boiling, the mixture was cooled and centrifuged at 5,000 rpm for 15 min. Supernatant was collected and was stored at 4°C for further analysis.

Effect of different Concentration of the extract on the synthesis of AgNP: The different volumes of the algal extract (5%) 0.5,1, 2, 3, 4 ml was added to 9.5, 9, 8, 7, 6 ml of 1 mM $AgNO_3$ aqueous solution labeled as (A1-A5), kept at 37°C for 48hrs. The time of addition of extract in to aqueous silver nitrate solution was considered as the start of the reaction.

2.4 *Characterization of silver nanoparticle*

UV-Visible Spectroscopy: UV-Visible Spectroscopy of the silver nanoparticles was recorded using Thermoscientific Evolution 201 UV-Visible Spectrophotometer. About 2 ml of the reaction mixture was taken and the optical density was measured from a wavelength of 200-800 nm.

Scanning Electron Microscopy (SEM): A drop of the sample was placed on an aluminium foil and allowed to evaporate in a vacuum dessicator. The foil containing the sample was placed on a carbon coated copper grid and SEM (Scanning electron Microscopy) observation of the AgNP .

2.5 *Biological activities of the AgNPs of* Desmococcus olivaceus

Antioxidant scavenging activity: The effect of nanoparticle on superoxide (SOD) and nitric oxide(NO) radical were estimated using the alkaline DMSO method (Govindarajan,2003) and Sreejayan,1996;Sreejayan 1997.

Antibacterial activity: Antibacterial activity of biologically synthesized Agnes of *Desmococcus.olivaceus* was determined by Disc diffusion method (Bauer *et al.*, 1966).

Cytotoxic activity -MTT assay: The cytotoxicity of samples (AgNP's) on Hela cells was determined by MTT assay(3-(4, 5-dimethyl-2 thiazolyl)-2,5-diphenyl-tetrazolium bromide).The color developed was measured at 570 nm [Mossmann T.,1983].The effect of the samples on the proliferation of human colorectal cancer cells was expressed as the % cell viability, using the following formula:

% Cell viability = Optical density of treated cells/Optical density of control cells × 100.

3 RESULTS AND DISCUSSIONS

UV-Visible spectrum analysis: Optimum intensity of UV-Vis spectra peak or SPR (surface Plasmon resonance band) centered between 400 - 450 nm occurred for 0.5-2.5 ml of the extract + 9.5-7.5 ml of 1 mM $AgNO_3$ at 37°C for 48hour (Figure 1 A1 –A5). However the intensity of the peak increased with respect to the volume of the extract (0.5 -2.5ml) concentration (5%) and it is stable even after 48 hours.

SEM analysis: Figure 2 shows the SEM image of AgNP's of *D.olivaceus* .The AgNP formed were spherical in shape, agglomerated and polydispersed. The AgNP diameter varied from 18-97 nm (Table 1).

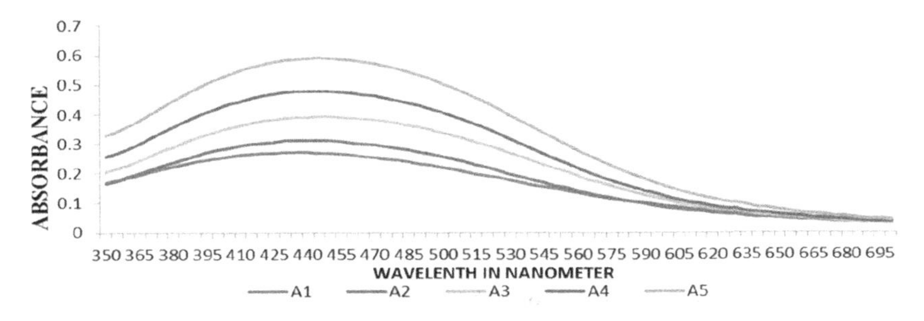

Figure 1. UV-Visible spectra recorded as a function of concentration of aqueous extract of *Desmococcus olivaceus* reaction with 1 mM $AgNO_3$ at 37°C (A1-0.5 ml extract+9.5 ml 1 mM$AgNO_3$; A2-1.0 ml extract +9.0 ml 1 mM$AgNO_3$; A3- 1.5 ml extract+8.5 ml 1 mM$AgNO_3$; A4-2.0 ml extract+8.0 ml 1 mM$AgNO_3$; A5-2.5 ml extract+7.5 ml 1 mM$AgNO_3$).

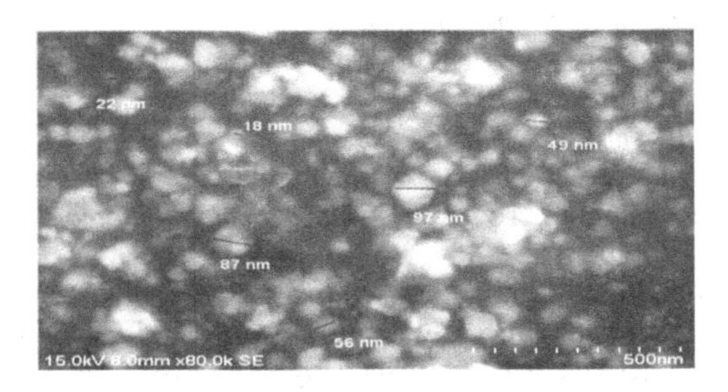

Figure 2. SEM Analysis of a *D.olivaceus* extract + 1 mM $AgNO_3$ at 37°C.

Table 1. EDAX Spectra of *desmococcus olivaceus*-AgNP.

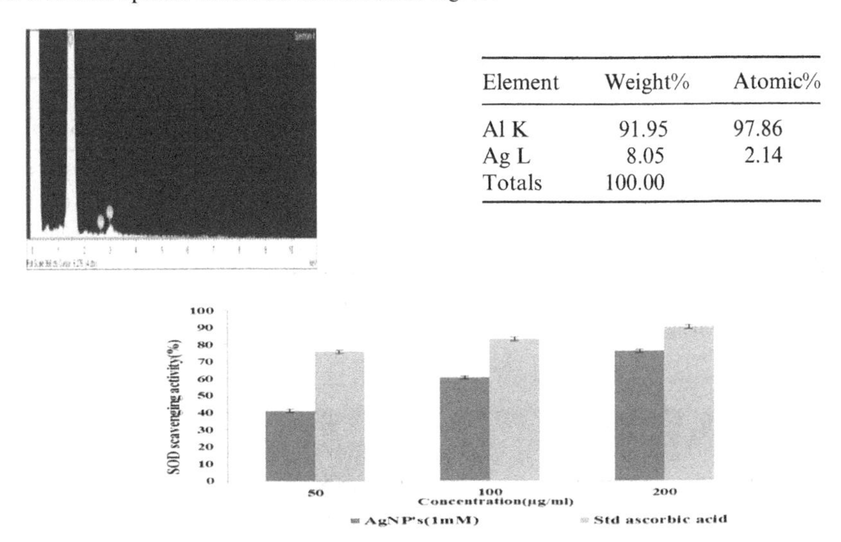

Element	Weight%	Atomic%
Al K	91.95	97.86
Ag L	8.05	2.14
Totals	100.00	

Figure 3. SOD Scavenging activity of AgNP's of *D.olivaceus* Anti-oxidant Activity.

Antioxidant activities: Silver nanoparticles of *Desmococcus olivaceus* showed significant free radical antioxidant scavenging activity in a dose dependent manner. SOD scavenging activity of algal silver nanoparticles was maximum of about 76% when compared to the NO scavenging activity (46.1%). Silver nanoparticles of *D.olivaceus* showed potent NO & SOD scavenging activities when compared to the standard ascorbic acid. (Figure 3 & 4)

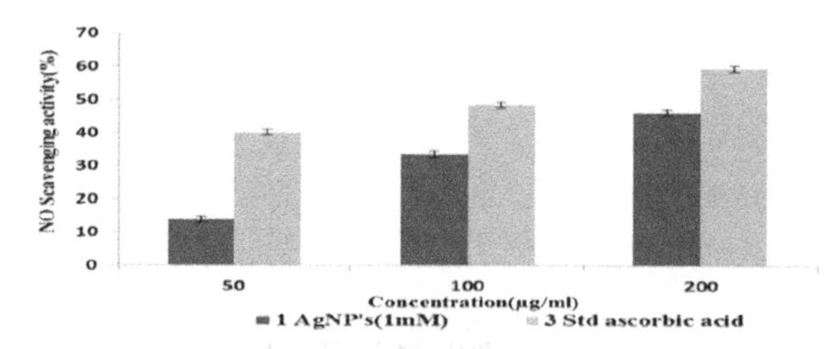

Figure 4. Nitric oxide Scavenging activity of AgNP's of *D.olivaceus.*

Antimicrobial activities: The synthesized AgNP inhibit the growth of micro organism. Growth inhibition zone values are obtained in mm for the synthesized Ag/*D.olivaceus* nanoparticle against four micro organism which includes *Escherichia coli, Bacillus subtilis, Klebsiella pneumoniae* and *Staphylococcus aureus* were compared with positive control (streptomycin) and negative control Dimethyl sulphoxide (Table 2 & Plate I). Silver nanoparticles of *D.olivaceus* shown 7.1, 6.5,7.8 and 6.2 mm inhibition respectively against E. *coli,, Bacillus subtilis, klebsiella pneumonia and S. aureus*. AgNP of *D.olivaceus* shows maximum activity on *E.coli* and *K.pneumoniae*. Biosynthesis and characterization of silver nanoparticles using micro algae *Chlorococcum humicola* and its anti microbial activity (Jayashree *et al.*, 2013).

Anti proliferative activities: AgNP's of *D.olivaceus* exhibited anti proliferative activity against Hela cell lines (Cervical cancer) with IC $_{50}$ value of 1.56 µg/ml. (Figure 5 & 6). The *invitro* anti proliferative activity was confirmed by MTT assay in the vero cell lines showed IC_{50} value of 7.04µg/ml respectively. Antiproliferative activity was reported in *D.olivaeus* against Hep2 cell lines (Lung carcinoma cell line (Uma *et al.*, 2011).

Table 2 . Antibacterial activity of synthesized AgNPs'(1mM) of *D.olivaceus.*

S. no	*Micro organisms*	12 µl	16 µl	20 µl	24 µl	Streptomycin
1	*Escherichia coli*	5.8 ± 0.62	6.4 ± 0.3	6.8 ± 0.4	7.1 ± 0.2	18.2 ± 0.5
2	*Bacillus subtilis*	5.4 ± 0.5	5.8 ± 0.5	6.2 ± 0.3	6.5 ± 0.5	16.2 ± 0.4
3	*Klebsiella pneumoniae*	6.6 ± 0.16	7.0 ± 0.2	7.4 ± 0.4	7.8 ± 0.4	15.3 ± 0.6
4	*Staphylococcus aureus*	5.0 ± 0.4	5.4 ± 0.5	5.8 ± 0.3	6.2 ± 0.4	17.4 ± 0.3

Values are expressed as Mean±SD (n=3): Volume of the standard streptomycin:10 µl/disc(10 µg):Volume of the algal extracts;10 µl/disc (100 µg)

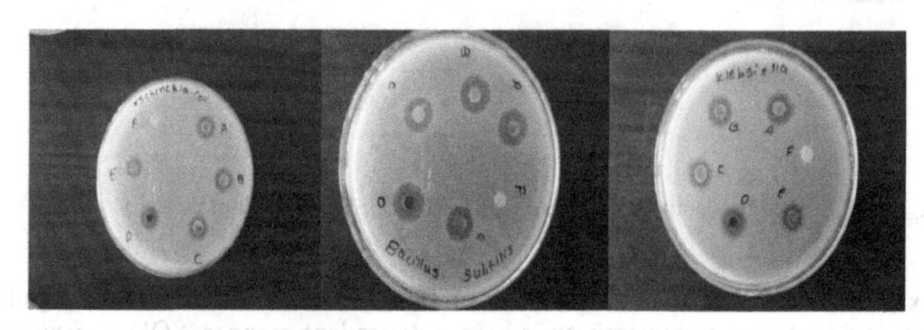

Plate I:. *Invitro* antibacterial activity of silver nanoparticles of *Desmococcus olivaceus* [A:12 µl ; B:16 µl; C:20 µl; D:24 µl; E:1mMAgNO3, F:DMSO]

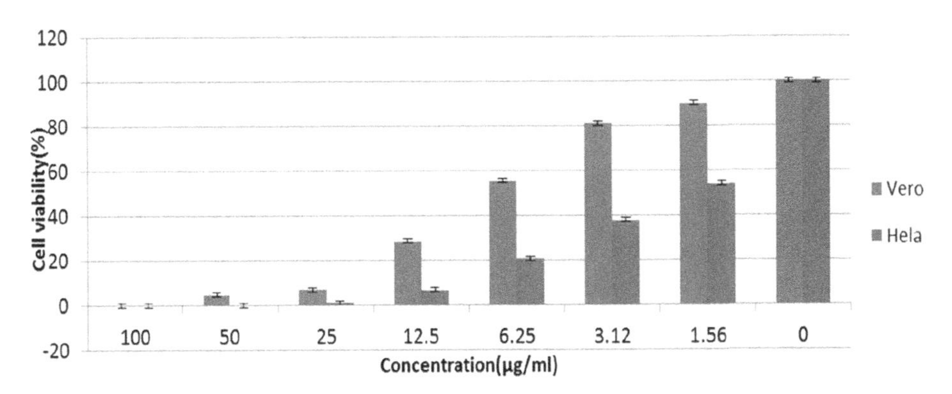

Figure 5. Invitro cytotoxic activity of silver nanoparticles of *D.olivaceus* against vero and Hela cell lines.

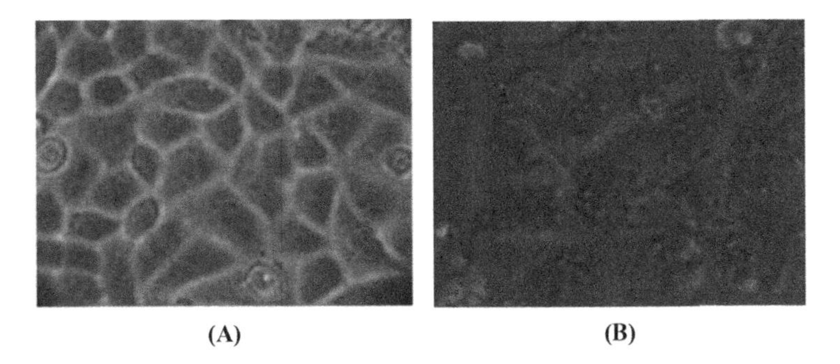

Figure 6.(A) Hela control cell line. Figure 6. (B) AgNP treated cell line(1.56μ g/ml).

4 CONCLUSION

In this study a simple, ecofriendly, pollutant free and economic biological procedure has been developed to synthesis the AgNPs of *Desmococcus olivaceus*. The AgNP's were characterized by UV-visible spectra and SEM. AgNP's synthesized by the present method are of spherical shape and poly dispersed. Algal derived silver nanoparticles exhibited potent antioxidant scavenging,antibacterial and anti proliferative activities against cervical cancer (HeLa) cell line.

REFERENCES

Bauer AN., Kirby WMM., Sherries, JC., Truck M., 1966, Antibiotic susceptibility testing by standardised single disk method, *Am.j. Clin. Pathol*, 45: 493–496.

Becker EW.Micro algae.Cambridge Univ.press.Cambridge, Newyork,1994.

Govindarajan R., M. Vijaya Kumar, A.K.S. Rawat, S. Mehrotra, 2003, Free radical scavenging potential of *Picrrorhiza kurroa* Royle ex Benth., *Indian J. Exptl. Biol.*, 41, pp. 875.

Jayashree Jena, Nilotpola Pradhan, Bisnu Prasad Dash, Lala Behari Sukla, Prasanna Kumar panda, 2013, *International Journal of nanomaterials and biostructures*, 3(4), 1–8.

Mossmann T. 1983, Rapid colorimetric assay for cellular growth and survival: application to proliferation and cytotoxicity assays. *J Immunol Meth* ,65 (1-2), 55–63.

Sreejayan, Rao MN. Nitric oxide scavenging by curcuminoids. *J Pharm Pharmacol* 1997; 49: 105–107.

Sreejayan, M.N.A. Rao, 1996, Free radical scavenging activity of curcuminoids, *Drug Res*, 46, pp. 169.

Uma R, Sivasubramanian V and Niranjali Devaraj S. "Evaluation of *in vitro* antioxidant activities and anitproliferative activity of green micro algae of *Desmococcus olivaceus* and *Chlorococcum humicola*. Jl of Algal Biomass Utilization, 2011, 2 (3), 82–93.

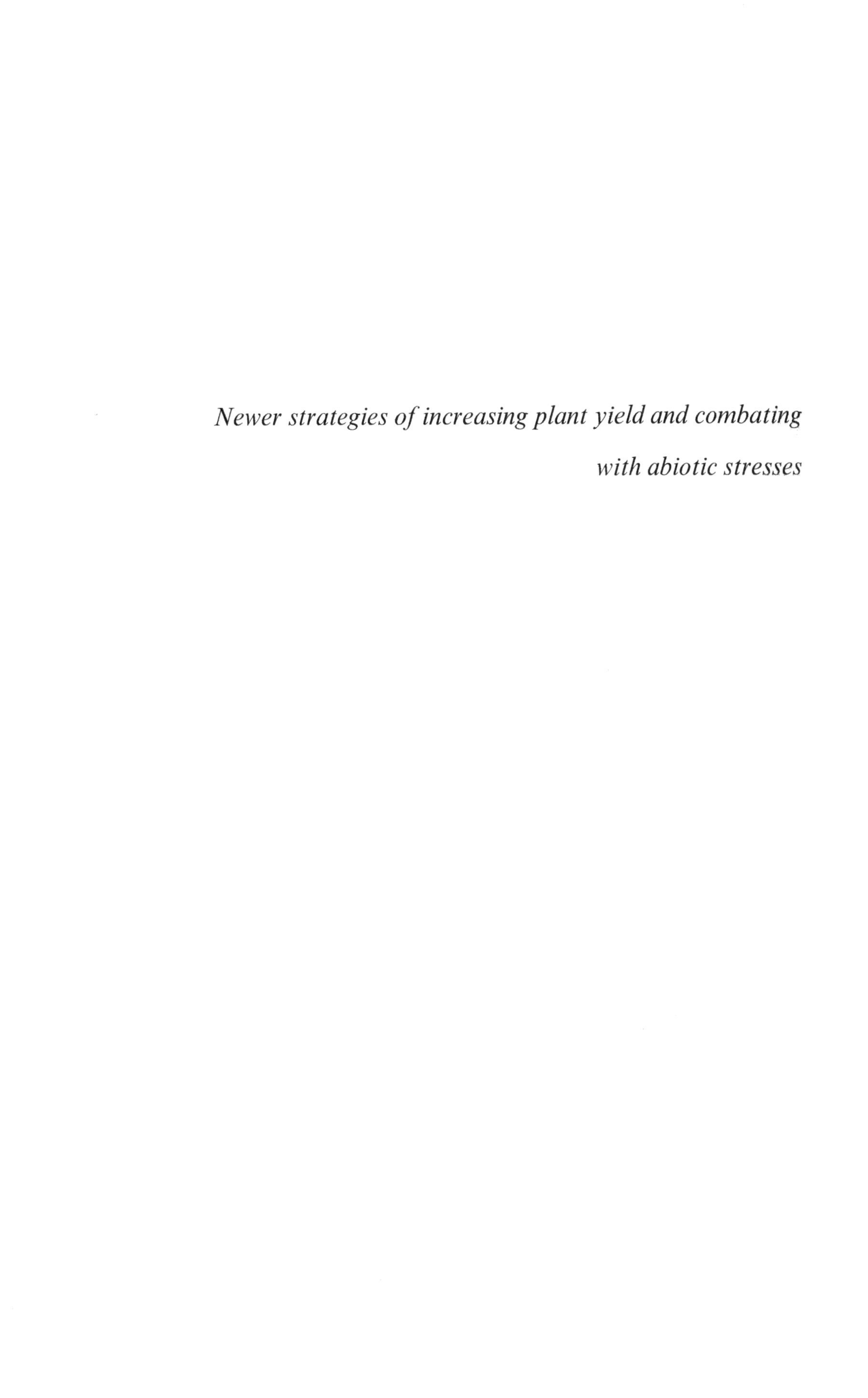

Newer strategies of increasing plant yield and combating with abiotic stresses

Biotechnology and Biological Sciences – Sen et al. (Eds)
© 2020 Taylor & Francis Group, London, ISBN 978-0-367-43161-7

Rapid mass propagation and conservation of *Rhododendron leptocarpum* Nutt

Mithilesh Singh
G.B.Pant National Institute of Himalayan Environment and Sustainable Development, Kosi-Katarmal, Almora, Uttarakhand, India

Osin Rai & K.K. Singh
G.B.Pant National Institute of Himalayan Environment and Sustainable Development, Sikkim Regional Centre, Pangthang, Gangtok, Sikkim, India

ABSTRACT: An efficient protocol for micropropagation of *Rhododendron leptocarpum* Nutt., has been standardized using nodal segment explants. Among various tested media combinations, Anderson's Medium (AM) supplemented with 5 μM 2-isopentenyladenine (2-iP) induced the highest multiple shoot proliferation from the explants. On AM + 2-iP (5 μM), an average of 5.00 shoots per explants were proliferated within 5 weeks in 85 % cultures. The shoots grew well and each shoot has attained an average height of 3.1 cm. Regenerated shoots rooted spontaneously with 100 % frequency in AM-liquid medium supplemented with IBA (Indole-3-Butyric Acid). *In vitro* grown plantlets were acclimatized successfully with 87 % survivability. The present study, thus, provides a fast, efficient technique for repetitive multiplication of *R. leptocarpum* without damaging its natural populations. This method can be used for large scale propagation and conservation of this ecologically and economically important threatened species.

1 INTRODUCTION

The genus *Rhododendron* (family Ericaceae), renowned for its beautiful flower, is represented by more than 1025 species in the world (Chamberlain et al., 1996). From India, 135 taxa of *Rhododendron* have been reported, of which 132 taxa are found in the north eastern region between upper temperate and sub-alpine zone (1600-3600 m above msl). Arunachal Pradesh comprises the highest number i.e. 119 taxa, whereas Sikkim is abode of 42 taxa (Mao et al., 2017). The genus has immense ecological and ethnobotanical significance. Some *Rhododendron* species are reported to possess medicinal properties such as estrogenic, antimicrobial, anti-inflammatory, hepatoprotective, immunomodulatory, adaptogenic, antidiarrheal, anticancer and antidiabetic activities (Rawat et al., 2017). These virtues are mainly due to the presence of secondary metabolites for instance alkaloids, terpenoids, steroids, saponins, phenolic acid and flavonoid (Nisar et al., 2013; Shrestha et al., 2017).

In spite of significant importance, this genus is very less explored for research purposes. Recently, decrease in the number of *Rhododendron* species and drastic changes in their natural habitat, as well, as in some cases, complete population extinction in the wild has become a strong issue of concern. It has been reported that if proper management and conservation initiatives are not taken up, the species may be wiped out in near future from the biota. Considering this, in the present study, strong efforts have been made to propagate and conserve a threatened *Rhododendron* species of Eastern Himalaya. *Rhododendron leptocarpum,* commonly known as *Slender-fruited Rhododendron* or *Jhinophale Gurans*, have become rare and critically endangered (Singh et al., 2003). It is an evergreen epiphytic shrub that grows up to

2-3 m in height. Biogeographically this species is restricted only in Sikkim, Bhutan and South-East China. It is fairly rare in Sikkim and represented by few records at ca. 3000 m amsl.

2 MATERIALS & METHODS

The seeds of *R. leptocarpum* were collected from West Sikkim, India during the month of October, 2013. After collection, the seeds were cleaned, dried for a week and then wetted with detergent solution (Tween-20) for 15 minutes. After washing with distilled water for 10 minutes, the seeds were sterilized with Bavistin (0.2%, 20 min) and mercuric chloride (0.1%, 3 min). For germination, seeds were inoculated in basal MS (Murashige and Skoog, 1962) medium having 0.8 % w/v agar and 3% sucrose. Subsequently, pH was adjusted at 5.8 with 1N HCL/1N NaOH and the cultures were maintained in culture room at 21±2 °C with photoperiod 16 h light/8 h dark of 60 $\mu Mm^{-2}s^{-1}$ fluorescent light.

After 6 weeks, nodal segments were carefully excised from seedlings and cultured on freshly prepared MS, ½ MS and AM (Anderson medium, 1984) basal medium supplemented with sucrose (3%, w/v) and vitamins, viz., myo-inositol (100 mg/l), nicotinic acid (0.5 mg/l), pyridoxine (0.5 mg/l), glycine (2 mg/l) and thiamine (0.1mg/l). Different auxins and cytokinins alone and in combinations were tested for shoot proliferation.

For rooting experiments, the micro-shoots of 3-4 cm (approx.) length were carefully separated from 5 weeks old shoot clump and transferred to MS and ½ MS liquid nutrient medium containing different concentration of auxins viz., NAA (α-naphthalene acetic acid), IAA (indole-3-Acetic Acid), IBA (indole-3-butyric acid).

3 RESULTS & DISCUSSION

On the MS basal medium 75% seeds of *R. leptocarpum* germinated within 20-25 days of inoculation. *In vitro* axenic nodal segments, excised from 6-week old *in vitro* grown seedlings, were cultured in MS, ½ MS and Anderson's media supplemented with different concentrations of three cytokinins viz. BAP, 2-iP and Kinetin. Direct shoot proliferation from explants were achieved on MS, ½ MS media having 2.0, 4.0, 6.0 µm concentrations of BAP and AM medium supplemented with 5.0, 10.0 and 15.0 µm concentrations of 2-iP. There was no positive response noticed on medium devoid of growth hormone and in Kinetin supplemented medium. Shoot multiplication was achieved within 5 weeks.

Among the different responded media combinations, the frequency of multiple shoot was highest (85%) in Anderson's medium. On AM + 2-iP (5 µM), an average of 5.00 shoots per explants were proliferated within 5 weeks in 85 % cultures. The shoots grew well and each shoot has attained an average height of 3.1 cm (Figure 1a-c). Followed to this was MS + BAP (4 µM), in which 75% cultures responded with multiple shoot proliferation.

It was observed that the shoot growth rate as well as the rate of multiplication of shoots were decreased simultaneously with increase in concentration of 2-iP and BAP (Figure 2).

To further improve shoot proliferation response, auxins (IAA and NAA) were also incorporated in cytokinin containing medium. But, auxin incorporation could not enhance shoot proliferation/multiplication response.

In accordance with this study, earlier works on propagation and conservation of important *Rhododenron* species report that 2-iP is better than other cytokinins for *Rhododendron in vitro* shoot proliferation and growth (Singh 2008; Mao et al., 2011; Singh et al., 2016, Mao et al., 2018).

Root initiation is an important part of *in vitro* regeneration of plants. The 3 weeks old individual micro-shoots of *R. leptocarpums* with normal morphology were transferred to AM-liquid nutrient medium containing different concentration (1.0, 2.0 and 3.0 µm) of auxins viz. NAA, IAA and IBA. Root induction was observed after 2 weeks but the complete root development took 5 weeks. Rooting was observed in all the three auxins, but IBA was found more

Figure 1. *In vitro*-propagation of *R. leptocarpum* (a & b) Shoot proliferation from nodal segments on AM medium supplemented with 2-iP (5 μM). (c); Profuse shoot proliferation on AM medium supplemented with 2-iP (5 μM) after 5 weeks of culture.

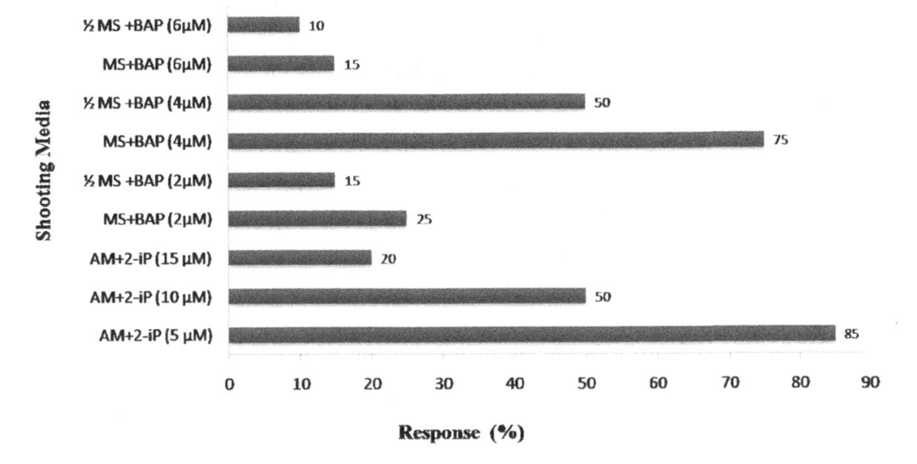

Figure 2. Effect of different media and growth regulators on shoot proliferation from nodal segments of *R. leptocarpum*. Growth period: 5 weeks.

suitable for root induction than NAA and IAA. IBA at 1.0 μM concentration induced rooting in 100 % *in vitro* shoots of *R. leptocarpum* (Figure 3a & b).

Higher concentration of IBA (2.0 and 3.0 μM) inhibited both percent rooting response and number of roots (Table 1). Callusing from the base of shoots was also observed in 2.0 and 3.0 μM

Figure 3. Rooting and acclimatization of *R. leptocarpum* (a); Root induction on AM-liquid medium supplemented with IBA (1.0 μM). (b); Well established rooted plantlets after 8-9 weeks of culture on AM-liquid medium containing 1.0 μM IBA (c); Acclimatization of plants in pot.

Table 1. Effect of different auxins incorporated in Anderson medium on rooting of regenerated shoots of *R. leptocarpum* after five weeks.

Auxins (μM)	Rooting (%)	No. of roots (no. ± SD)
Control	22	1.50 ± 0.57
IBA 1	100	7.00 ± 1.00
2	55	3.66 ± 0.58
3	35	3.00 ± 3.00
IAA 1	50	5.66 ± 0.58
2	30	5.00 ± 1.00
3	25	2.00 ± 2.64
α-NAA 1	25	3.33 ± 0.57
2	20	2.66 ± 2.15
3	15	2.00 ± 1.00

IBA containing medium. In IAA and NAA containing medium less than 50 % rooting response was observed. In concurrence to the present study, IBA has been found to be the best for rooting in other *Rhododendron* species viz. *R. griffithianum, R. maddenii, R. dalhousiae* var. rhabdotum, *R. elliottii, R. johnstoneanum* and *R. watti* (Singh 2008; Mao et al., 2011; Singh et al., 2016, Mao et al., 2018).

In vitro full plants were acclimatized and transferred in normal garden soil in the greenhouse under high temperature and relative humidity (80 %). The regenerated plants showed 87 % survival inside greenhouse (Figure 3c).

4 CONCLUSION

We report here, a reproducible and rapid micropropagation protocol for *R. leptocarpum*. Using this method, large number of *R. leptocarpum* plants can be produced year round in limited time and space. The developed efficient tissue culture protocol can be used for conservation of this important *Rhododendron* species.

ACKNOWLEDGMENT

The author is grateful to the Director, GBPNIHESD for providing necessary facilities. Department of Biotechnology, New Delhi is highly acknowledged for financial support. Mr. Mohan Thapa is thanked for his assistance.

REFERENCES

Anderson, W.C. 1984. A revised tissue culture medium for shoot multiplication of Rhododendron. Journal *of the American Society for Horticultural Science* 109:343-347.

Chamberlain, D., Hyam, R., Argent, G., Fairweather, G., Walter, K.S. 1996. The genus Rhododendron: its classification and synonymy. UK, Edinburgh: Royal Botanic Garden Edinburgh.

Mao, A.A., Dash, S.S., Singh, P. 2017. Rhododendrons of North East India: a pictorial handbook. Published by Director Botanical Survey of India, Kolkata, India.

Mao, A.A., Kaliamoorthy, S., Ranyaphi, R.A., Das, J., Gupta, S., Athili, J., Yumnam, Y., Chanu L.I. 2011. In vitro micropropagation of three rare, endangered, and endemic rhododendron species of Northeast India. *In Vitro Cellular & Developmental Biology Plant* 47:674-681.

Mao, A.A., Vijayan, D., Nilasana Singha R.K., Pradhan, S. 2018. In vitro propagation of *Rhododendron wattii* Cowan-a critically endangered and endemic plant from India. *In Vitro Cellular & Developmental Biology Plant*, https://doi.org/10.1007/s11627-017-9869-7.

Murashige, T., Skoog, F. 1962. A revised medium for rapid growth and bioassays with tobacco cultures. *Physiologia Planta*, 15: 473–497.
Nisar M, Ali S, Qaisar M, Gilani SN, Shah MR, Khan I, Ali G. Antifungal activity of bioactive constituents and bark extracts of *Rhododendron* arboreum. Bangladesh J Pharmacol. 2013;8:218–222.
Rawat, P., Rai, N., Kumar, N. and Bachheti, R.K. 2017. Review on *Rhododendron arboreum*-a magical tree. *Oriental Pharmacy and Experimental Medicine* 17:297-308.
Shrestha, A., Said, I.H., Grimbs, A., Thielen, N., Lansing, L., Schepper, H., Kuhnert, N. 2017. Determination of hydroxycinnamic acid present in Rhododendron species. *Phytochemistry* 144: 216-225.
Singh, K.K., Kumar, S., Rai, L.K., Krishna. A.P. 2003. Rhododendron Conservation in Sikkim Himalaya. *Current Science* 85(5):602-606.
Singh K.K. 2008. In vitro plant regeneration of an endangered Sikkim Himalayan Rhododendron (R. *maddeni* Hook. f.) from Alginate- Encapsulated shoot tips. *Biotechnology* 7:144-148.
Singh, K.K., Singh, M., Chettri, A. 2016. *In vitro* propagation of *Rhododendron griffithianum* Wt. an endangered Rhododendron species of Sikkim Himalaya. *Journal of Applied Biology and Biotechnology* 4(02): 072-075.

Biotechnology and Biological Sciences – Sen et al. (Eds)
© 2020 Taylor & Francis Group, London, ISBN 978-0-367-43161-7

In vitro micropropagation of rose apple and assessment of plantlets through biochemical analysis and DNA fingerprinting

P.K Das
Department of Agricultural Biotechnology, Ramakrishna Mission Vivekananda Educational and Research Institute, Narendrapur, Kolkata, India

A. Chakraborty
YP-I, Department of Agricultural Biotechnology, Ramakrishna Mission Vivekananda Educational and Research Institute, Narendrapur, Kolkata, India

S. Chakraborty
Department of Biological Sciences, IISER Kolkata, Kalyani, India

ABSTRACT: True to type mass propagations were attempted using nodal explant and shoot tip meristems in rose apple, a nutritionally rich fruit of India. For shoot multiplication from nodal explant and shoot tip, best response was in 4mg/l BAP in MS or ½ MS and 3mg/l kinetin in ½ MS respectively. Shoot multiplication from nodal explant took lesser period as compared to that of shoot tips. Rooting in shootlets of nodal explant recorded in 6mg/l IBA in half MS while shootlets derived from shoot tips induced roots in 7mg/l IBA in MS. *In vitro* raised plantlets from nodal explant took almost 90 days but 120 days from shoot tips. 1year conserved plantlets compare to immediately raised, recorded higher level of soluble protein content and peroxidase but lower catalase, depicting a variable pattern during growth and development. DNA fingerprinting using RAPD and ISSR markers did not show any DNA polymorphism, confirming the genetic stability.

1 INTRODUCTION

Rose apple, a high value nutritionally rich fruit (Orwa *et al.*,2009) belonging to family Myrtaceae is grown extensively in South East Asia including India. It is traditionally propagated through seed which produces heterozygous plants. In order to achieve time neutral mass micropropagation, *in vitro* direct organogenesis was attempted using nodal and shoot tip explants. One-year old plantlets conserved *in vitro* were compared with immediately *in vitro* raised for different biochemicals such as soluble protein, peroxidase and catalase to profile their status during growth and development. Simultaneously these plantlets (conserved and immediately raised) were assessed for the genetic fidelity using different molecular markers which would help to study DNA polymorphism. The significant observations and findings are presented and discussed.

2 MATERIALS AND METHODS

Extracted seeds from collected rose apple fruits were soaked overnight in water followed by surface sterilization with 1% (v/v) savlon + 1% (v/v) Tween 20 for 30 minutes, washed 4 times in distilled water, treated with 1% (w/v) Bavistin for 10 minutes, washed for 4 times in distilled water and were treated with 0.1% (w/v) Mercuric Chloride ($HgCl_2$) for 10 minutes and washed for 4 times with distilled water. Surface sterilized seeds were cultured in MS media with 30g/l sucrose, 100 mg/l myoinositol without any PGR. Nodal explants of 1.5-2 cm length were excised from *in vitro* regenerated 70 days old plants. Shoot tips (0.5-1 cm) collected from 3 weeks old plants grown *in vitro*.

For shoot growth and multiplication nodal explants were cultured on MS and ½ MS supplemented with different concentrations of BAP (0, 1, 2, 4, 6 and 8 mg/l). shoot tips were cultured on MS and ½ MS supplemented with different concentrations of BAP (0.5, 1 mg/l), TDZ (0.25, 0.5 mg/l) and kinetin (1.5, 3 mg/l).

For root induction, shootlets of nodal explant were cultured on ½MS with 6 and 8 mg/l IBA. Also, for rooting shootlets of shoot tips were cultured on MS supplemented with IBA and NAA having concentrations of 5, 6 and 7 mg/l for both. During these *in vitro* operations, temperature was set to 25°±1 C under standard cool fluorescent tubes providing 60 $\mu mol/m^2/s$ irradiance at a 16-hour photoperiod.

One-year old in vitro conserved plantlets and immediately raised plantlets were analysed for soluble proteins (Lowry's method, 1951), catalase (Aebi *et al*, 1984) and peroxidise (Pine et al, 1984) from leaf extracts.

Genetic fidelity study was done to ensure the genetic purity of the conserved as well as immediately raised *in vitro* by employing 10 RAPD markers and 7 ISSR markers (Table 1). For this purpose genomic DNA was extracted from leaf using the protocol of Allen *et al.* PCR amplification was done using the said primers and thermal profile was set to- 94°C for 2 min (initial denaturation), (94°C for 1 min, 35–50°C [variable] for 1 min, 72°C for 2 min) 35 cycles and final extension at 72°C for 10 min. Aliquots of 5µl PCR products, along with 100 bp and 50 bp DNA ladder were resolved by 1.5% agarose gel in 1x TAE buffer, stained with Ethidium Bromide (EtBr) (0.5µg/ml).

3 RESULTS AND DISCUSSIONS

3.1 *In vitro micropropagation*

3.1.1 *Nodal explant*

It took almost 17-18 days of culture for bud breaking. Maximum shoot length with highest number of leaves was recorded within 50-55 days of bud break. 4 mg/l BAP was found to be the best irrespective of MS strength as it produces the longest shoot (2 cm) and maximum number of leaves (30). But shoot survivality during multiplication is another important factor as polyphenol exudates causes browning of media that ultimately affect shoot survivality.

Table 1. List of primers.

Number of primers	Name of primers	Nucleotide sequence	Annealing temperature
1	UBC 810	GAGAGAGAGAGAGAGAT	45.4
2	UBC 813	CTCTCTCTCTCTCTCTT	45.4
3	UBC 814	CTCTCTCTCTCTCTCTA	44.7
4	UBC 815	CTCTCTCTCTCTCTCTG	46.8
5	UBC 822	TCTCTCTCTCTCTCTCA	47.0
6	UBC 823	TCTCTCTCTCTCTCTCC	48.1
7	UBC 824	TCTCTCTCTCTCTCTCG	48.5
8	OPA-2	TGCCGAGCTG	40.7
9	OPA-4	AATCGGGCTG	35.1
10	OPA-7	GAAACGGGTG	33.2
11	OPA-8	GTGACGTAGG	31.1
12	OPA-10	GTGATCGCAG	33.1
13	OPA-17	GACCGCTTGT	35.7
14	OPA-20	GTTGCGATCC	33.5
15	OPB-1	GTTTCGCTCC	35.8
16	OPB-2	TGATCCCTGG	32.2
17	OPB-4	GGACTGGAGT	32.2

Reduction of MS strength is proved to be effective to solve this problem as shootlets survival was significantly higher in half MS (86.67%) than MS (20%). Thus, from these observations, it can be concluded that for shoot multiplication from nodal explant half MS with 4 mg/l BAP is the best choice. Profuse root can be grown in shootlets within 17-18 days of culture in root induction media. It was found that both 6 mg/l and 8 mg/l IBA in half MS produced 12-15 robust roots of about 1 cm length. No significant differences between these two concentrations were found. It can be suggested that 6 mg/l IBA is good enough to induce root *in vitro* which supports the earlier observation of Prashantha *et al* (2003). Different series of events during plantlet production from nodal explant were depicted in Figure 1.

3.1.2 *Shoot tip meristem*

Almost 10 days were required for shoot induction in shoot tip meristem culture. It took almost 80 days for proper shoot growth. The best combination was found to be half MS with 3 mg/l kinetin which produces longest shoot (1.73 cm) with maximum leaves (22). Root induction was little slower in this case which took almost 30 days and best result was found in MS with 7 mg/l IBA, producing 12-17 number of healthy well growing roots. Different series of events during plantlet production from shoot tip meristem were depicted in Figure 2.

Both nodal explant and shoot tip meristem were found to be successful for multiplication in *in vitro* condition. But shoot multiplication from shoot tip meristem was little slower than multiplication from nodal explant. This may be due to the application of Kinetin in shoot tip culture but BAP in nodal explant, suggesting differential response of PGRs during growth and differentiation. The comparison between these two is shown in Figure 3.

4 CONSERVED PLANTLETS ASSESSED

Plantlets of nodal explant were subcultured and conserved for 1 year in half MS with 4 mg/l BAP while shoot tip derived plantlets were maintained in half MS with 3 mg/l kinetin. To assess the conserved plantlets biochemical and molecular study were done.

Biochemical study showed higher level of soluble protein content and higher peroxidase activity but lower catalase concentration level in conserved plantlet while comparing with immediately raised (Figure 4). This indicates variable pattern during

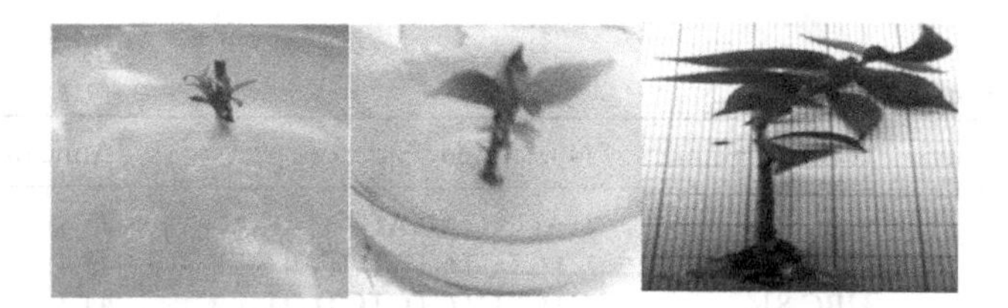

Figure 1. Series of events during shoot multiplication from nodal explant. From left to right- bud break, shootlet and plantlet with roots.

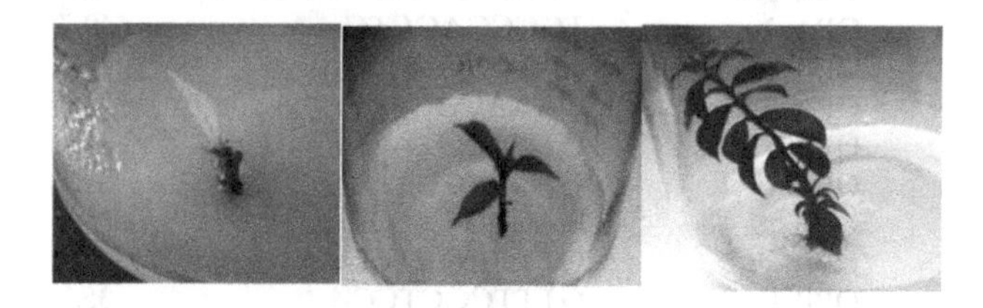

Figure 2. Series of events during shoot multiplication from shoot tip explant. From left to right-shoot tip inoculation, shootlets, plantlets with roots.

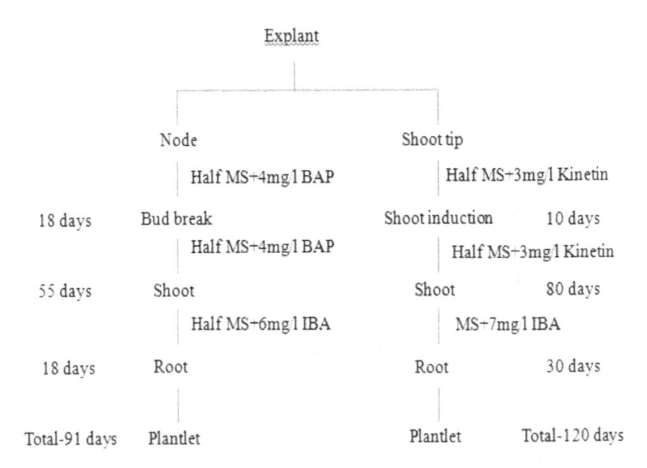

Figure 3. Flow diagram showing *in vitro* micropropagation.

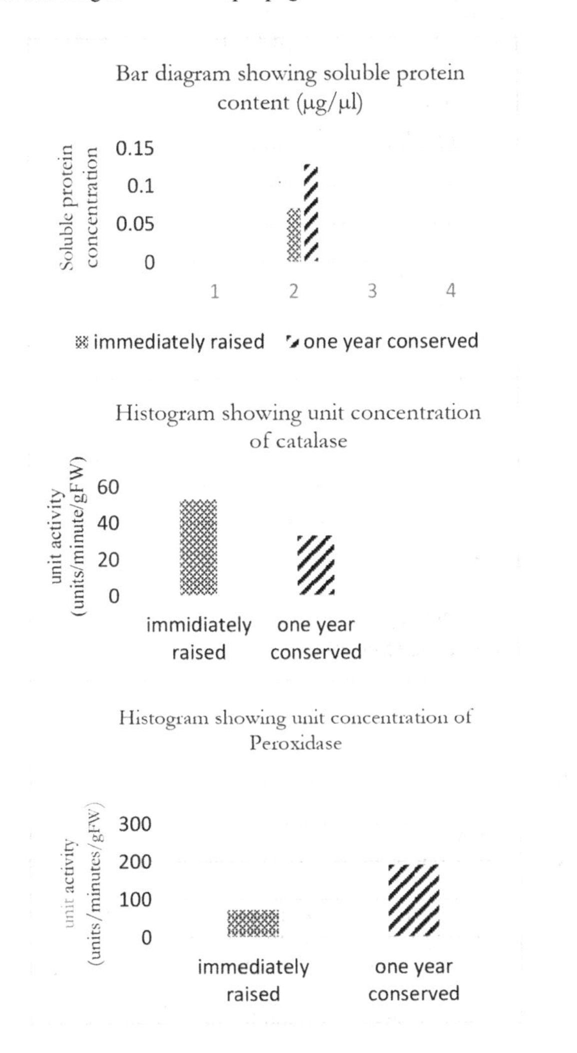

Figure 4. Assessment of biochemical characteristics between one-year-conserved and immediately raised leaves- from top to bottom- assessment of soluble protein, assessment of soluble protein, assessment of isozyme catalase and assessment of isozyme peroxidase.

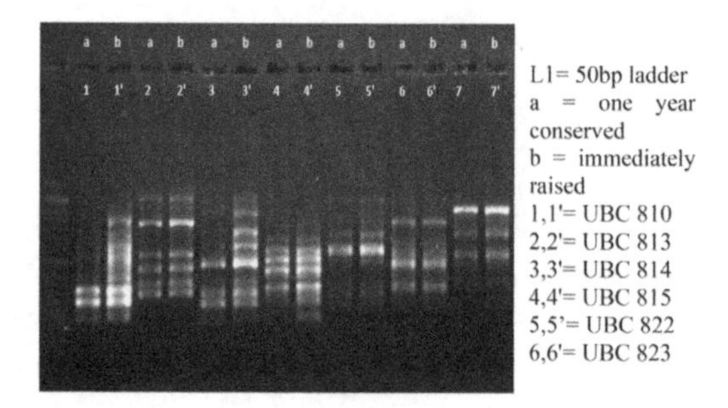

Figure 5. Genetic fidelity study using molecular marker.

growth and differentiation with stress adaptation due to aging which supports the observation of Middlebrook (1954).

DNA fingerprinting using molecular markers did not show any polymorphism either in RAPD or in ISSR markers used (Figure 5). This confirms genetic stability of *in vitro* raised plantlets using molecular markers as reported by Rani *et al* (2000). The very significant finding in this case is that even after one-year conservation *in vitro* genetic fidelity is faithfully maintained. This would suggest the possibility of medium-term in vitro conservation which would help steady supply of propagules in rose apple.

5 CONCLUSIONS

Following conclusions are derived:

Shoot multiplication from either nodal explant or shoot tip can be an alternative approach for mass propagation of rose apple. But nodal explant would be better choice as it took lesser time to produce a complete plantlet.

During growth and development with the advance of age, plantlets develop different biochemical outfit to face the challenge of acclimatization. Genetic fidelity test confirms that medium term *in vitro* conservation is feasible without any changes in genetic material.

ACKNOWLEDGEMENT

We thank the facilities provided by the Ramakrishna Mission Vivekananda Educational and Research Institute to conduct the experiment.

REFERENCES

Aebi, H. (1984). [13] Catalase in vitro. In *Methods in enzymology* (Vol. 105, pp. 121–126). Academic Press.

Allen G.C, Flores-Vergara M.A, Krasynanski S, Kumar S & Thompson W F. (2006). A modified protocol for rapid DNA isolation from plant tissues using cetyltrimethylammonium bromide, *Nature protocols*, *1*(5), 2320–2325.

Lowry, O.H., Rosebrough, N.J., Farr, A.L., & Randall, R.J. (1951). Protein measurement with the Folin phenol reagent. *Journal of biological chemistry*, *193*, 265–275.

Middlebrook, G. (1957). Isoniazid-resistance and catalase activity of tubercle bacilli. *Am Rev Tuberc*, *69*, 471–472.

Murashige T., & Skoog F, (1962). A revised medium for rapid growth and bio assays with tobacco tissue cultures. Physiologia plantarum, 15(3), 473–497.

Orwa., C, Mutua, A., Kindt,R., Jamnadass, R., & Simons, A., (2009), Agroforestry database: a tree species reference and selection guide version 4.0. World Agroforestry Centre ICRAF, Nairobi, KE.

Pine, L., Hoffman, P.S., Malcolm, G.B., Benson, R.F., & Keen, M.G. (1984). Determination of catalase, peroxidase, and superoxide dismutase within the genus Legionella. *Journal of clinical microbiology*, *20* (3), 421–429.

Prashantha, K.G., Sathyanarayana, B.N., Mathew, D. & Sondur, S.N. (2003). In vitro callus induction and plant regeneration in rose apple (*Syzygium jambos* L.). J. Plant Biol. 30: 99–102.

Rani V & Raina SN. (2000). Genetic fidelity of organized meristem-derived micropropagated plants: a critical reappraisal. *In Vitro Cellular & Developmental Biology-Plant*, *36*(5), 319–330.

Biotechnology and Biological Sciences – Sen et al. (Eds)
© 2020 Taylor & Francis Group, London, ISBN 978-0-367-43161-7

Antioxidant and nutritional properties of *Hydrocotyle javanica* Thunb

Manab Mandal, Debabrata Misra & Vivekananda Mandal*
Department of Botany, University of Gour Banga, Malda, India

ABSTRACT: *Hydrocotyle javanica* Thunb. represent one of the most commonly used plant in traditional systems of Indian medicines.To study the meatbolites (primary and secondary) and *in-vitro* free radical scavenging activity of aerial parts of *H. javanica.* Meatbolite content of the aerial parts of *H. javanica* and the *in-vitro* free radical scavenging activities were evaluated in 80% methanol, ethanol and water extract using standard protocols. It was found that most of the secondary metabolites like flavonoid, steroid, terpenoid etc were found in 80% methanol extract of *H. javanica* when compared with other solvent extraction. It also showed better inhibition potential in various *in-vitro* scavenging models when compared to the standard. Hence, the present study reveals that the aerial parts of *H. javanica* might be used as antioxidant in herbal drugs.

Keywords: Antioxidant, Nutrients from plants, Free radicals, *Hydrocotyle javanica* Thunb.

1 INTRODUCTION

The plant chemicals are classified as primary or secondary metabolites. The primary metabolites are important for the growth and development of plant. The carbohydrates, fats and proteins are referred to as the proximate principles and form the major portion of diet of herbivores and omnivores [Seal, 2011]. On the other hand, secondary metabolites are produced as by-products of metabolic pathways and not essential for the survival of the plant, are important in their defence system. A variety of secondary metabolites has been isolated from plants viz. alkaloids, coumarins, flavonoids, polyphenols, quinines, saponins, tannins and terpenoids. These major groups represent classes of structurally and chemically diverse groups of compounds that exert strong physiological effects in humans. Their therapeutic properties have been utilized since long and research is still in progress to explore their applications as medicines [Briskin, 2000]. All of these secondary metabolites compounds have been reported as scavengers of free radicals and also have been considered as good therapeutic candidates for free radical related pathologies [Mradu et al., 2012]. Nowadays, there is an increasing focus on the search of anti-oxidants (non-synthetic) from medicinal plants such as carotenoids, ascorbic acid (vitamins), phenolic and flavonoids. Anti-oxidative activity is a measure of capability of compound to scavenge free Hydroxyl groups, Nitrogen group and Oxygen species. So, presence of novel antioxidant is very important property of medicinal plants. Free radicals are causative agents for different diseases such as cancer. Therefore, antioxidant properties are an index of antioxidant potential against reactive species (free radicals) [Mradu et al., 2012]. Our early studies showed that aerial parts of *H. javanica* Thumb. are rich source of phytochemicals including flavonoids, phenolics, alkaloids, glycosides and minerals like K, Ca, Fe and P [Mandal et al., 2016, 2017]. Therefore this study was conducted with the aim to investigate phytoconstituents present in aerial parts of *H. javanica* Thumb. as potent antioxidant candidates.

*Corresponding author: vivekugb@gmail.com

2 MATERIALS AND METHODS

2.1 *Preparation of plant extracts*

Organic extracts of leaves are prepared by using three different solvents (80% methanol, ethanol and water). Dried plant powder weighed carefully and used for extract preparation through Soxhlet apparatus at respective temperature. The extract obtained was filtered and concentrated in rotary evaporator.

2.2 *In - vitro antioxidant assays*

The assessment of antioxidant activities of the bioactive fraction were carried out by evaluating different radical (1, 1-diphenyl-2-picryl hydrazyl, Hydroxyl, Nitric oxide) scavenging activities and total antioxidant capacity by Phosphomolybdanum antioxidative power assay using the standard protocols as mentioned below.

DPPH radical scavenging activity was measured according to the method of Blois, 1958. The OH scavenging ability was measured by the method of Yu et al. (2004). NO scavenging activity was determined according to the method of Garratt, 1964. The PAP assay was worked out by the method described by Prieto et al., (1999). In all the radical scavenging tests ascorbic acid (0.1 to 1 mg/mL) was used as standard and scavenging activity was calculated using the following formula.

Scavenging % = $(A_0\text{-}A_S) / A_0 \times 100$, where A_0 = OD of Blank and A_S = OD of Sample

In PAP assay mg equivalent ascorbic acid per g of dry weight (dw) was calculated from the standard curve of ascorbic acid at the concentration of 0.1 to 1mg/mL. Butylated hydroxyl toluene (BHT) was taken as positive control.

2.3 *Extraction and estimation of chlorophyll pigments and primary metabolites*

Chlorophyll a and Chlorophyll b were estimated using the method of Lichtenthaler and Welburn, 1983. The content of reducing sugar was estimated using DNS reagent and standard curve of maltose at the concentration of 20 to 100 g/mL according to the Miller (1959) method. Starch content was estimated using anthrone method and from the standard curve of starch at the concentration of 20 to 100 g/mL as per the method given by Hansen and Mollar, 1975. Protein was estimated by Bradford method, (1976). The protein content was expressed as g BSA equivalent (BSAE) per 100 g of dry weight (dw). Total fat was estimated according to the method of Uraku et al, 2015. Ascorbic acid was estimated following standard method [Sadasivam & Manickam, 1996]. The ascorbic acid content was expressed as mg ascorbic acid per g of dry weight (dw). Fresh leaves as well as dried leaf powder were taken as the test samples for comparison.

2.4 *Quantification of secondary metabolites*

Flavonoid was extracted with 80% methanol from 5 g leaf powder and content was estimated as mg quercetin equivalent (QE) per g of dry weight (dw) following the method of Chang et al., 2002. The phenol content was estimated using standard curve of quercetin prepared at a concentration range of 0.02 to 0.20 mg/mL in methanol and the flavonoid content was expressed as mg quercetin equivalent (QE) per g of dry weight (dw) by the Harborne method (1998). The flavonol estimation was evaluated by [Boham and Kocipai, 1994] with minor modification it was expressed as mg flavonol per g of dry weight (dw). The estimation of alkaloids was done by method of Harborne, 1988. The alkaloid content was expressed as mg alkaloid per g of dry weight (dw). Glucoside content was determined from the standard curve of D- Glucose in chloroform - methanol at the concentration of 0.02 to 0.20 mg/mL and the content was expressed as mg D-glucose equivalent (DGE) per g of dry weight (dw). Saponin was estimated following the method of Obadoni and Ochuko, 2001. The saponin content was expressed as mg saponin per g of dry weight (dw).

2.5 *Statistical analysis*

All the analyses were performed in triplicate and the results were statistically analyzed and expressed as mean (n = 3) ± standard deviation (SD).

3 RESULTS

In DPPH radical scavenging activity methanol extract showed the highest DPPH radical scavenging activity among the other sample in terms of IC_{25} value which was 0.375 mg/mL while the value was 0.23 mg/mL for ascorbic acid which was taken as standard which was shown in the Figure 1(A). In OH radical scavenging activity the lowest IC_{50} value was 0.835 mg/mL in the methanol extract while the value was 0.63 mg/mL in ascorbic acid which was taken as standard as shown in the Figure 1(B). In NO radical scavenging activity water extract showed the highest scavenging activity among the other solvents with the IC_{50} value 0.61 mg/mL while the values in standard were 0.65 mg/mL. as shown in the Figure 1(C). In PAP assay the ethanol extract showed the maximum antioxidant activity with the value of 0.121 mg equivalent ascorbic acid/g dry extract while the standard antioxidant BHT showed the value 0.18 mg. Water and methanol extracts had the value of 0.054 mg/g and 0.048 mg/g dry extract respectively as shown in the Figure 1(D).

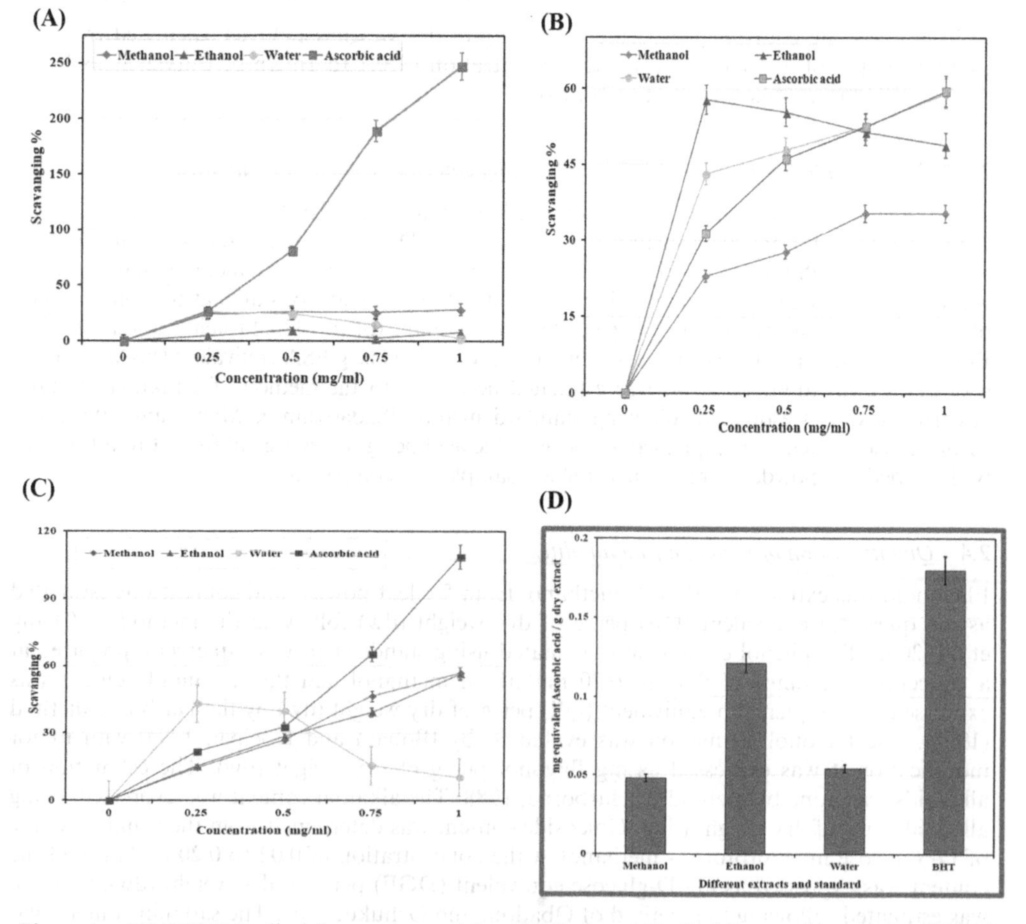

Figure 1. (A) DPPH Radical Scavenging activity. (B) Hydroxyl Radical Scavenging activity. (C) NO Radical Scavenging activity. (D) Phosphomolybdanum Antioxidative Power Assay

The chlorophyll a content was 5.32 and 0.33 µg/g in fresh leaves and dried powder sample, respectively and chlorophyll-b contents were 1.92 and 0.31 µg/g in fresh leaves and dried powder sample, respectively as shown in Figure 2(A). The content of different primary metabolite viz. reducing sugar, soluble carbohydrate, starch, total carbohydrate, fat and protein were 1.11, 0.271, 3.32, 4.588, 0.125 and 4.2 respectively in the fresh leaves and 2.18, 0.304, 3.52, 4.195, 0.094 and 7.8 g% in the dried leaf sample as shown in the Figure 2(B). The content of different secondary metabolite like ascorbic acid, alkaloid, flavonoids, flavonol, glucoside, phenol and saponins were 0.095, 0.414, 0.017, 0.161, 0.173, 0.112 and 0.1476 mg/g respectively in the fresh leaves sample and 0.063, 0.4219, 0.006, 0.161, 0.154, 0.001 and 0.08398 mg/g in the dried powder sample as shown in the Figure 2(C).

4 DISCUSSION

4.1 *DPPH radical scavenging activity*

The IC_{50} value is the concentration of sample at which the inhibition percentage reaches 50%. IC_{50} value is negatively related to the antioxidant activity, as it expresses the amount of

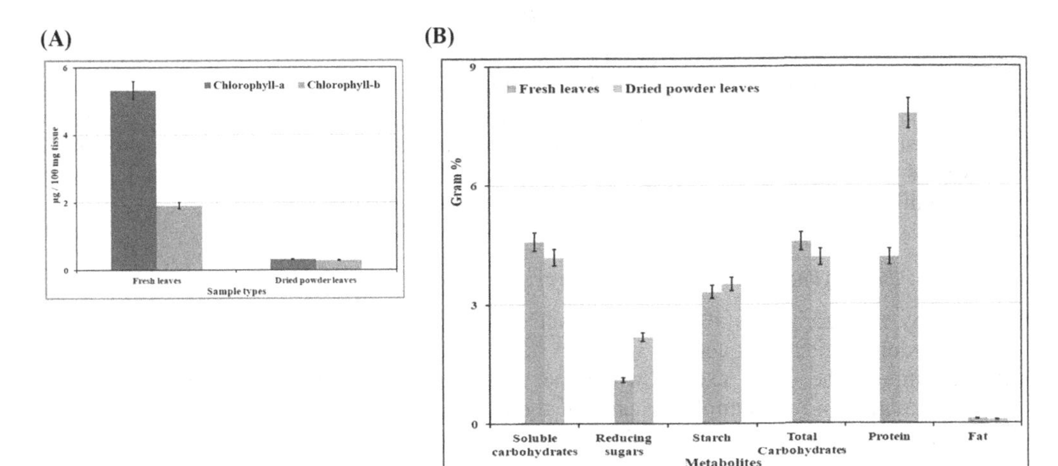

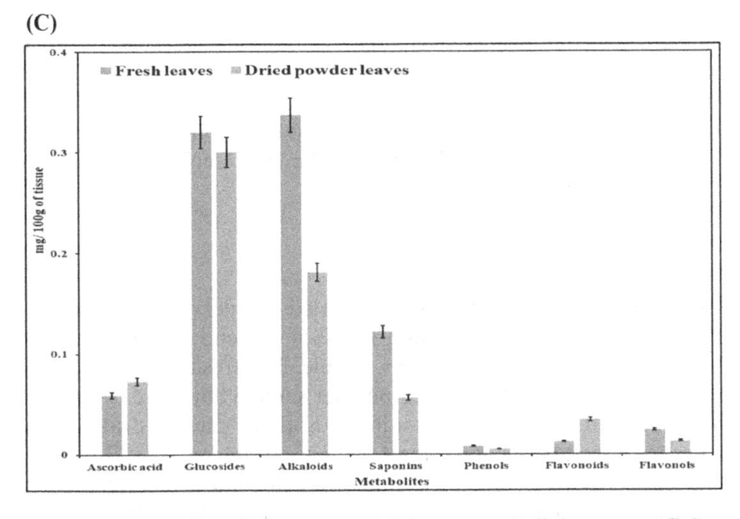

Figure 2. (A) Chlorophyll content. (B) Primary metabolites content. (C) Secondary metabolites content.

antioxidant needed to decrease its radical concentration by 50%. The lower the IC_{50} value the higher is the antioxidant activity of the tested sample. In DPPH radical scavenging activity the methanol extract showed the highest DPPH radical scavenging activity among the different samples in terms of IC_{25} value. The IC_{25} value was 0.375 mg/mL while the value of ascorbic acid taken as standard was 0.23 mg/mL while the ethanol and water extract failed to show such radical scavenging activity.

4.2 *Hydroxyl (OH) radical scavenging activity*

Hydroxyl radical (OH^-) is produced by the reaction of O_2 and H_2O_2 and extremely toxic and attacks biological molecules such as DNA, proteins and lipids [Ranjan et al., 2014]. Thus, OH^- scavenging activity of antioxidants can be accomplished through direct scavenging or preventing of OH^- formation. Among the different sample the lowest IC_{50} value was 0.835 mg/mL in the methanol extract while the value was 0.63 mg/mL in ascorbic acid as standard. The ethanol extract showed the IC_{50} value was 0.90 mg/mL while the water extract failed to show such radical scavenging activity.

4.3 *Nitric Oxide (NO) radical scavenging activity*

In human system excess concentration of (NO) causes several disease viz. hypertension, cardiovascular dysfunction and asthma. NO and their derivatives may have a genotoxic effect and play a vital role in carcinogenesis process [Parul et al., 2013]. The water extract showed the highest scavenging activity among the other extract with the IC_{50} value 0.61 mg/mL while the values in standard was 0.65 mg/mL. The methanol extract showed the scavenging activity 0.85 mg/mL while the ethanol extract failed to show such activity.

4.4 *Phosphomolybdanum antioxidative power (PAP) assay*

PAP assay evaluate the total antioxidant activity of the extract. The ethanol extract showed the maximum antioxidant activity with the value of 0.121 mg equivalent ascorbic acid/g dry extract among the other extract while the standard antioxidant BHT showed the value 0.18 mg. Water and methanol extracts had the low activity with the equivalent ascorbic acid value of 0.054 mg/g and 0.048 mg/g dry extract, respectively.

Previous literature survey reported that the different *Hydrocotyle* species contained total phenolic compounds, flavonoids and flavonols, which were associated with a broad spectrum of biological activities including radical scavenging properties. It was also reported that the phenolic compounds are powerful chain breaking antioxidants [Wagh et al., 2012] and the present study material also contained secondary materials viz. phenolic compounds, flavonoids and flavonols in high quantity. The methanol extract of dry seeds of few apiaceae plants (*Foeniculum vulgare, Anethum graveolens, Coriandrum sativum, Carum carvi*) have been reported to have high phenolic and flavonoid content with high antioxidant activities [Bagdassarian et al., 2014] and the methanol extract of test sample also showed antioxidant activity.

4.5 *Photosynthetic pigments and metabolites*

The study showed that the sample variation (fresh leaves and dried powder) had the effect on the pigments and primary metabolites contents. Chlorophyll-a content was higher than Chlorophyll-b content in both the fresh and dried powder sample. The fresh leaf sample contained seven time higher of Chlorophyll-a than the dried powder sample but there is little difference in the Chlorophyll-b content in both the samples. Soluble carbohydrates, total carbohydrates were higher in the fresh leaves sample than the dry powder sample. But starch was comparatively higher in the dry powder sample. Reducing sugar and protein content in the dry powder sample were almost double of the content in the fresh leaf sample but fat content was more or less same in both the sample. Hence sample variation had a direct impact on the content variability of primary metabolites in the plant under study. The study also showed that the sample variation had the effect on the secondary metabolites content also. The content of alkaloids, ascorbic acid, flavonoids, glucoside and saponins were higher in the fresh leaf sample than the dried powder sample. The phenol content was almost same in the both the sample, whereas flavonoids content in the dried powder sample was almost double of the content in the fresh leaves sample. Hence, the secondary metabolites production in the fresh leaves sample was higher than the dried powder sample because there may be a possibility of secondary metabolites synthesis in a fresh plant to tolerate or avoid drought, defence and different types of stress.

5 CONCLUSION

On the basis of the results obtained from different antioxidant activity assays, the 80% methanolic extract of *H. javanica* Thunb. has shown a significant total antioxidant, and reduction capacity of free radical scavenging activity. The reducing capacity depends on phenolic contents which were present in the 80% methanol extract in high quantity and this activity might be due to the synergistic effect of the active compounds. Keeping in view its high antioxidant property of the methanolic extract of this plant can also be used alone or in combination in the form of different herbal formulations to protect the body from deleterious effects of free radicals. Therefore, the individual compounds responsible for such activities can be purified, isolated and characterized for their further study.

ACKNOWLEDGEMENT

We are thankful to DST, Government of India (SERB/F/5369/2013-'14) for partial financial help of the study. We are also grateful to the people of Darjeeling hills (West Bengal, India) for their kind information and cooperation.

REFERENCES

Bagdassarian, V.L.C., Bagdassarian, K. S., Atanassova, M. S., Ahmad, A., Comparative analysis of total phenolic and total flavonoid contents, rutin, tannins and antioxidant capacity in Apiaceae and Lamiaceae families, *Indian Horti. J.* 4 (3/4): (2014), 131-140.

Blois, M.S., Antioxidant determinations by the use of a stable free radical. *Nature*, 181, (1958), 1199-1200.

Boham, B. A., and Kocipai, A. R., Flavonoids and condensed Tannins from leaves of Hawaiian *Vaccinium vaticulatum* and *V. calycinium*. Paci Sc, 48, (1994), 458–463.

Bradford, M.M., A rapid and sensitive method for the quantitation of microgram quantities of protein utilising the principle of protein- dye binding, Anal Biochem, 72, (1976), 248-254.

Briskin, D.P., Medicinal plants and phytomedicines. Linking plant biochemistry and physiology to human health. *Plant Physiology*, 124(2): (2000), 507–514.

Chang, C. C., Yang, M. H., Wen, H. M., Chern, J. C., Estimation of total flavonoid content in propolis by two complementary colorimetric methods. *J of Food and Drug Analysis*, 10, (2002), 178-182.

Garrat, D. C., The Quantitative analysis of drugs. Chapman and Hall Ltd., Japan, 3,(1964), 456-458.

Hansen, J., and Moller, I., Percolation of starch and soluble carbohydrates from plant tissue for quantitative determination with anthrone. *Anal Biochem*, 68, (1975), 87-94.

Harborne, J.B., Textbook of phytochemical methods. A guide to modern techniques of plant analysis, 5th edition, Chapman and Hall, London, (1988), 21-72.

Laware Shankar, L., Sequential extraction of plant metabolites. *Int. J. Curr. Microbiol. App. Sci.* 4(2): (2015), 33-38.

Lichtenthaler, H.K., and Welburn, A.R., Determination of total carotenoids and chlorophyll a and b of leaf extracts in different solvents. *Biochem. Soc. Transac.* 11, (1983), 591-592.

Loo, A.Y., Jain, K., Darah, I., Antioxidant activity of compounds isolated from the pyroligneous acid, *Rhizophora apiculata. Food Chem.*, 107, (2008), 1151–1160.

Loo, C. K., McFarquhar, T. F., Mitchell, P. B., A review of the safety of repetitive transcranial magnetic stimulation as a clinical treatment for depression. *Int J NeuropsychoPharmacol*, 11(1): (2008),131-147.

Mandal, M., Misra, D., Ghosh, N. N., Mandal, V., Physicochemical and elemental studies of *Hydrocotyle javanica* Thunb. for standardization as herbal drug. *Asi Pac J of Trop Biomed.* 7(11): (2017), 979–986.

Mandal, M., Paul, S.,Uddin, M. R., Mandal, M. A., Mandal, S., Mandal, V., *In vitro* antibacterial potential of *Hydrocotyle javanica* Thunb. *Asi Pac J of Trop Disease*, 6,(2016), 54-62.

McCready, R.M., Guggols, J., Silviers, V., Owen, H.S., Determination of starch and amylase in vegetable. *Ann. Chem*, 22, (1950), 1156-1158.

Miller, G. L.,Use of Dinitrosalicylic acid reagent for determination of reducing sugar. *Anal Chem.*, 31(3): (1959), 426–428.

Mradu, G., Saumyakanti, S., Sohini, M., Arup, M., HPLC profiles of standard phenolic compounds present in medicinal plants. *Int J PharmacogPhytochem Res*, 4, (2012), 162-167.

Obadoni B. O., and Ochuko P. O., Phytochemical studies and comparative efficacy of the crude extracts of some haemostatic plants in Edo and Delta States of Nigeria. *Global J of Pure and Applied Sc*, 8(2), (2002) 203–208.

Parul, B. B., Sharma, B., Jain, U., Virulence associated factors and antibiotic sensitivity pattern of *Escherichia coli* isolated from cattle and soil. *Veterinary World*, 7(5): (2014), 369-372.

Prieto, P., Pineda, M., Aguilar, M., Spectrophotometric quantitation of antioxidant capacity through the formation of a phosphomolybdenum complex: Specific application to the determination of vitamin E. *Anal Biochem*, 269, (1999), 337–341.

Ranjan, N., Chaudhary, U., Chaudhry, D., Ranjan, K. P., Ventilator-associated pneumonia in a tertiary care intensive care unit: Analysis of incidence, risk factors and mortality, *Ind J Crit Care Med.* 18, (2014), 200-204.

Sadasivam, S., and Manickam, A., Biochemical methods, (2nd ed.), NAI(P) Ltd, Coimbatore, (2004).

Seal, T., Determination of nutritive value, mineral contents and antioxidant activity of some wild edible plants from Meghalaya state, India. *Asi J Applied Sci*, 4(3):(2011), 238-246.

Uraku, A. J., Nutritional potential of *Citrus sinensis* and *Vitis vinifera* peels, *J of advan in Med and Life Sc*, 3(4), (2015), 2348–294X.

Wagh, S. S., Jain, S. K., Patil, A.V.,Vadnere, G.P., *In -vitro* free radical scavenging and antioxidant activity of *Cicer arietinum* L. (Fabaceae) *Int. J. Pharm Tech. Res.*, 4 (1): (2012), pp. 343-350.

Yu, W., Zhao, Y., Shu, B., The radical scavenging activities of *Radix puerariae* isoflavoniods: A chemi luminescence study. *Food chemistry*, 86, (2004), 525-529.

Biotechnology and Biological Sciences – Sen et al. (Eds)
© 2020 Taylor & Francis Group, London, ISBN 978-0-367-43161-7

Production and characterization of xanthan gum by *Xanthomonas campestris* using sugarcane bagasse as sole carbon source

Jignesh Prajapati & Rajvi Panchal
Department of Biochemistry and Biotechnology, St. Xavier's College (Autonomous), Ahmedabad, India

Dhavalkumar Patel
Department of Biochemistry and Biotechnology, St. Xavier's College (Autonomous), Ahmedabad, India
Department of Biotechnology and Microbiology, Parul Institute of Applied Science and Research, Parul University, Ahmedabad, India

Dweipayan Goswami
Department of Biochemistry and Biotechnology, St. Xavier's College (Autonomous), Ahmedabad, India
Department of Microbiology and Biotechnology, School of Sciences (SoS), Gujarat University, Ahmedabad, India

ABSTRACT: Xanthan gum is a polysaccharide that is widely used as stabilizer and thickener in many industrial applications. The sugarcane bagasse is an abundant residue obtained during the sugar processing and its disposal in the environment causes several drawbacks. The bioconversion of this by-product in valuable products is an important alternative to overcome this environmental problem. In this work, the feasibility of using sugarcane bagasse as carbon source for xanthan gum production was investigated using *Xanthomonas campestris*. In addition, sucrose as carbon source was investigated as a control. A production of 6.76 ± 0.09 gL^{-1} and 9.79 ± 0.11 gL^{-1} was obtained when sucrose and sugarcane bagasse was used respectively. The solutions of xanthan formed was subjected to rheological analyses. The findings indicated that these polysaccharide solutions has pseudoplastic properties. Based on a preliminary analysis, the use of sugarcane bagasse for xanthan gum production has the potential to be a cost-effective supplemental carbon source to produce non-food grade xanthan gum.

Keywords: *Xanthomonas campestris*, sugarcane bagasse, xanthan, pseudoplasticity

1 INTRODUCTION

Xanthan gum is a microbial exopolysaccharide produced by *Xanthomonas campesteris* through aerobic fermentation [1]. Xanthan gum was discovered in the late 1950s by a US scientist, and is the first biopolymer produced industrially. In 1969, the Food and Drug Authority (FDA) authorized the use of xanthan gum in food products, marking the introduction of the first industrially produced biopolymer to the food industry without any specific quantity limitations [2,3]. This polysaccharide employed in food, cosmetics, water-based paints, pharmaceutical and petroleum industries as emulsifying and thickening agent because of its unique physico-chemical properties including high shear stability, pseudo-plastic features and stability on a wide variety of temperatures and pH values [4-6]. Xanthan is relatively expensive, since commercial production of xanthan gum uses glucose or sucrose as the substrate. However providing cheap carbon source is the main challenge. For example sugar beet molasses [7], cheese whey [8], tapioca pulp [9], palm date juice [1°], sugarcane broth [11], and whey permeate [12] are used as substrates for xanthan production.

In this study, the authors postulate that the bagasse, which is abundantly available in nature as a waste of sugarcane harvesting, can be used as a cheap substrate for xanthan gum production. The properties of xanthan obtained from these substrate were tested and compared with xanthan obtained from sucrose. The viscosity and shear thinning property of solutions was analyzed.

2 MATERIALS & METHODS

Microorganism and growth conditions *Xanthomonas campestris* (NCIM 2954) was obtained from National Collection of Industrial Microorganisms (NCIM), Pune, India. The microorganism was maintained on MGYP agar (Hi-media) slant containing malt extract 3 gL^{-1}; Glucose 10 gL^{-1}; Yeast extract 3 gL^{-1}; Peptone 5 gL^{-1} with 2% Agar adjusted to pH 7 as recommended by NCIM. It was allowed to grow for 24 h at 30°C and then stored at 4°C. This was sub-cultured every two weeks.

Sugarcane Bagasse hydrolysate preparation Sugarcane bagasse was obtained from nearby sugarcane juice merchants at Ahmedabad, India. It was then Sun dried and ground to fine size using mixer. The substrate was hydrolyzed at 121°C for 20 min in distilled water. The hydrolysate was filtered using Whatman filter paper No. 1 to remove suspended particles and stored at 4°C.

Inoculum preparation One loop of cells from 72 h old agar slant were inoculated into the 50 mL MGYP broth medium prepared using tap water in 250 mL conical flask and this was used as the inoculum for fermentation. 24 h culture was used as seed for the fermentation process [13].

Fermentation process For evaluation of xanthan production, three different types of broth medium were used. The media constituents were (in gL^{-1}) as follows: Medium XGL: Sucrose – 40 and Luria Bertani broth – 20. Medium XGS: Sucrose – 40; KH_2PO_4 – 5; Yeast extract – 5; Citric acid – 2; $MgSO_4{\cdot}7H_2O$ – 0.2; $CaCO_3$ – 0.02 and H_3BO_3 – 0.006. Medium XGB: KH_2PO_4 – 5; Yeast extract – 5; Citric acid – 2; $MgSO_4{\cdot}7H_2O$ – 0.2; $CaCO_3$ – 0.02; H_3BO_3 – 0.006 and hydrolysate with 40 gL^{-1} initial sugar. The medium pH was adjusted to 6.8. Batch fermentation was carried out with 100 mL of production media in 250 mL conical flask as triplicates for all medium. All flasks were inoculated with 5% *Xanthomonas campestris*. The flasks were incubated at room temperature for 96 h in rotary shaker at 150 rpm.

Determination of xanthan gum Xanthan gum was determined in culture supernatants as the method described by Rottava et al. (2009) by some modification. Samples withdrawn from the fermentation broth after 72 h were centrifuged at 7500 rpm, 30°C for 20 min to remove bacterial cells. Ice-cold acetone at 8°C was added to the cell free supernatant in the ratio of 3:2 (v/v) and the mixture was left for overnight at 8°C for precipitation of xanthan. Precipitated xanthan was collected by centrifugation (15 min, 7500 rpm) and dried at room temperature until constant weight was obtained. The dried gum was ground with mortar-pestle and stored in a sealed container. Xanthan concentration was evaluated by measuring the weight of dry product in grams per litre of fermented broth. Analysis of Variance (ANOVA) was carried out using triplicate value to identify significance difference in yield of xanthan gum (gL^{-1}) between various media. Mean values were compared at significance levels of 1% LSD [14].

Determination of Rheological properties of the xanthan gum Polysaccharide solutions of 0.5 gL^{-1} were prepared in distilled water. For complete dissolution of sample, solutions were kept under agitation at 150 rpm for 18 h. Viscosity measurement were performed by Searles Viscometer. Apparent viscosity, shear stress and shear rate was calculated by following formulas.

$$\eta = \frac{1}{\omega}\frac{M}{4\pi h}\frac{Ro^2 - Ri^2}{Ro^2 Ri^2} \tag{1}$$

$$\omega = \frac{2\pi}{T} \tag{2}$$

$$M = Dmg \tag{3}$$

$$\gamma = \frac{Ro^2 + Ri^2}{Ro^2 - Ri^2}\left(\frac{2\pi}{T}\right) \tag{4}$$

$$\tau = \frac{Ro^2 - Ri^2}{Ro^2 Ri^2}\left(\frac{1}{4\pi h}\right) \tag{5}$$

Where; η is the apparent viscosity in pascal-seconds (Pa·s), γ is the shear rate in reciprocal seconds (s^{-1}), τ is the shear stress in Pascals (Pa), ω is the angular velocity in radians per - second, M is the torque produced by the weights, h is the height of immersion in centimeters of the inner cylinder in the liquid media (cm), Ro is the radius in centimeters of the outer cylinder (cm), Ri is the radius in centimeters of the inner cylinder (cm), D is the radius of the spool in centimeters (cm) and m is the total mass in grams (g) [15].

3 RESULTS & DISCUSSION

Determination of xanthan gum Xanthan concentration was evaluated by measuring the weight of dry product in grams per litre of fermented broth. Results obtained in present study support the previously reported data [16] that medium formulation affects xanthan biosynthesis [16]. As displayed in Figure 1, maximal xanthan production (9.79 ± 0.11 gL^{-1}) was obtained in medium XGB followed by medium XGS (6.76 ± 0.09 gL^{-1}). Media XGB and XGS contains Bagasse hydrolysate and Sucrose as sole carbon source respectively.

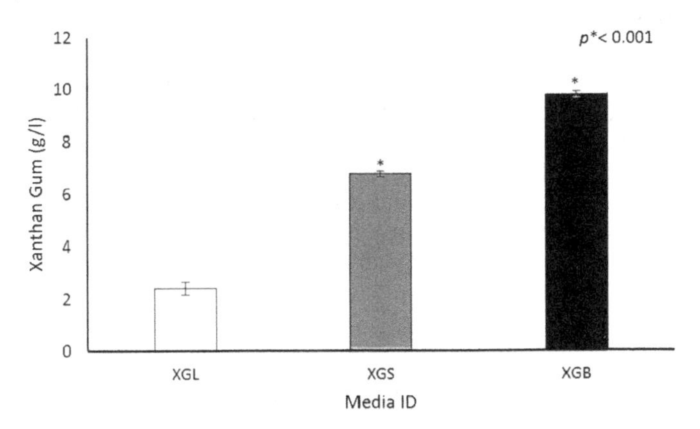

Figure 1. Graphical presentation for produced xanthan gum (gL^{-1}) in various media.

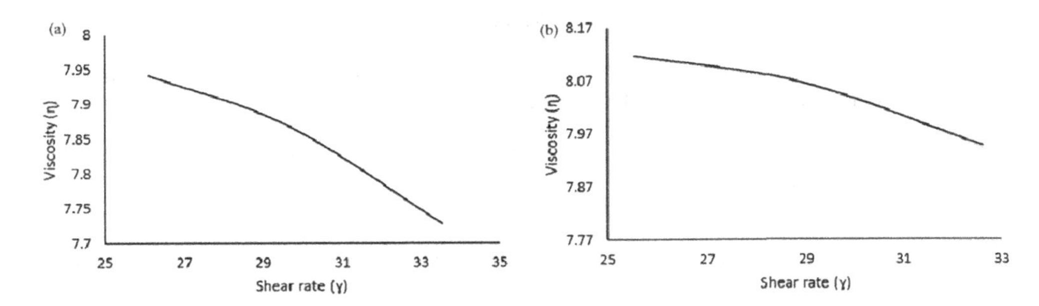

Figure 2. Graphical presentation for Shear Thinning behavior of xanthan from medium (a) XGS and (b) XGB.

Table 1. Viscosity of water and xanthan gum solutions.

Solution	m	T	h	τ	γ	η
Water	40	0.47	3	103.68	22.87	5.98
Water	60	0.6	3	155.52	22.54	6.09
Water	80	0.44	3	207.36	34.83	5.95
XGS	80	0.59	3	207.36	26.11	7.94
XGS	90	0.52	3	233.28	29.64	7.87
XGS	100	0.46	3	259.20	33.54	7.72
XGB	80	0.6	3	207.36	25.54	8.12
XGB	90	0.53	3	233.28	28.91	8.07
XGB	400	0.47	3	259.20	32.60	7.95

We got higher xanthan production in XGB media than XGS media, this is may be due to Sugarcane Bagasse hydrolysate provide simple sugar like mannose which is easy for assimilation and direct integration in xanthan biosynthesis. All points on graphs are the mean of three individual flasks.

Rheological Properties of xanthan solution The viscosity of xanthan gum solution produced by *X. campestris* in XGS and XGB medium was 7.94 Pa·s and 8.12 Pa·s respectively at 0.5 gL^{-1} concentration as shown in Table 1. As applied shear rate and shear stress increased, Viscosity of xanthan gum solution was decreased.

Shear rate against Viscosity graph is displayed in Figure 2 and decrease in viscosity was observed as shear rate is increases, which confirms the shear thinning property of xanthan gum.

4 CONCLUSION

Overall, the findings indicated that these polysaccharide solutions has pseudoplastic properties. The present study indicate that the Sugarcane Bagasse presented in the current study seems to have strong potential and promising properties that can open new pathways for the production of efficient and cost-effective xanthan gum. It can, therefore, be considered as a strong candidate for future industrial and commercial applications related to xanthan gum. The results are, in fact, promising in that they suggest that xanthan gum production can be industrially extended and maximized through the use of this low cost carbon source.

REFERENCES

[1] Kerdsup, P., Tantratian, S., Sanguandeekul, R., & Imjongjirak, C. (2011). Xanthan production by mutant strain of *Xanthomonas campestris* TISTR 840 in raw cassava starch medium. *Food and bioprocess technology, 4*(8), 1459-1462.

[2] Garcıa-Ochoa, F., Santos, V. E., Casas, J. A., & Gomez, E. (2000). Xanthan gum: production, recovery, and properties. *Biotechnology advances, 18*(7), 549-579.

[3] Katzbauer, B. (1998). Properties and applications of xanthan gum. *Polymer degradation and Stability, 59*(1-3), 81-84.

[4] López, M. J., Vargas-Garcıa, M. C., Suarez-Estrella, F., & Moreno, J. (2004). Properties of xanthan obtained from agricultural wastes acid hydrolysates. *Journal of Food Engineering, 63*(1), 111-115.

[5] Sutherland, I. W. (1998). Novel and established applications of microbial polysaccharides. *Trends in biotechnology, 16*(1), 41-46.

[6] Palaniraj, A., & Jayaraman, V. (2011). Production, recovery and applications of xanthan gum by *Xanthomonas campestris. Journal of Food Engineering, 106*(1), 1-12.

[7] Kalogiannis, S., Iakovidou, G., Liakopoulou-Kyriakides, M., Kyriakidis, D. A., & Skaracis, G. N. (2003). Optimization of xanthan gum production by *Xanthomonas campestris* grown in molasses. *Process Biochemistry, 39*(2), 249-256.

[8] Niknezhad, S. V., Asadollahi, M. A., Zamani, A., Biria, D., & Doostmohammadi, M. (2015). Optimization of xanthan gum production using cheese whey and response surface methodology. *Food Science and Biotechnology*, *24*(2), 453-460.

[9] Gunasekar, V., Reshma, K. R., Treesa, G., Gowdhaman, D., & Ponnusami, V. (2014). Xanthan from sulphuric acid treated tapioca pulp: influence of acid concentration on xanthan fermentation. *Carbohydrate polymers*, *102*, 669-673.

[10] Salah, R. B., Chaari, K., Besbes, S., Ktari, N., Blecker, C., Deroanne, C., & Attia, H. (2010). Optimisation of xanthan gum production by palm date (*Phoenix dactylifera L.*) juice by-products using response surface methodology. *Food Chemistry*, *121*(2), 627-633.

[11] Faria, S., Vieira, P. A., Resende, M. M., Ribeiro, E. J., & Cardoso, V. L. (2010). Application of a model using the phenomenological approach for prediction of growth and xanthan gum production with sugar cane broth in a batch process. *LWT-Food Science and Technology*, *43*(3), 498-506.

[12] Savvides, A. L., Katsifas, E. A., Hatzinikolaou, D. G., & Karagouni, A. D. (2012). Xanthan production by *Xanthomonas campestris* using whey permeate medium. *World Journal of Microbiology and Biotechnology*, *28*(8), 2759-2764.

[13] Yoo, S. D., & Harcum, S. W. (1999). Xanthan gum production from waste sugar beet pulp. *Bioresource Technology*, *70*(1), 105-109.

[14] Rottava, I., Batesini, G., Silva, M. F., Lerin, L., de Oliveira, D., Padilha, F. F. & Treichel, H. (2009). Xanthan gum production and rheological behavior using different strains of *Xanthomonas sp. Carbohydrate Polymers*, *77*(1), 65-71.

[15] European Pharmacopoeia Commission, European Directorate for the Quality of Medicines & Healthcare. (2010). *European pharmacopoeia* (Vol. 1). Council of Europe.

[16] Lo, Y.M. Yang, S.T. & Min, D. (1997). Effects of yeast extract and glucose on xanthan production and cell growth in batch culture of *Xanthomonas campestris. Appl Microbiol Biotechnol*, 47(6), 689-694.

Biotechnology and Biological Sciences – Sen et al. (Eds)
 ISBN 978-0-367-43161-7

Development of bio-process strategy for production of *Pleurotus eryngii* mycelium with improved ergosterol content

Umesh Singh, Ashwani Gautam, Vikram Sahai & Satyawati Sharma
Centre for Rural Development & Technology, IIT DELHI, India

ABSTRACT: Ergosterol, converted to vitamin D2 when exposed to direct sunlight or irradiated with UV, is a steroid found in the fungal cell membrane of mushrooms. It is also considered a biomarker for fungal biomass. *Pleurotus eryngii* an edible nutraceutical mushroom could be a good source of ergosterol. Using higher biomass production strategies under shake flasks submerged culture conditions, ergosterol enhancement has been observed in *Pleurotus eryngii.* The highest dry biomass obtained in PDB medium after 12 days was 10 ± 1 g/l and ergosterol content was 1.5 ± 0.2 mg/g (dry cell mass).

Keywords: Ergosterol, mushroom, submerged culture, vitamin D2, UV-radiation

1 INTRODUCTION

Mushrooms are being used as a food from ancient times [1]. The edible mushrooms are the good source of vitamin D_2 after sun exposure or UV irradiation [2]. The cell membrane of the mushrooms contains ergosterol which converts in to vitamin D_2 when exposed to direct sunlight [3] or irradiated with UV light [4]. Ergosterol is the most abundant sterol present in the mushrooms and comprises more than 85% of total sterol content [5]. It helps in maintaining membrane fluidity [6], activity of membrane bound enzymes as well as it is vital for cellular growth rate [7].

The *Pleurotus eryngii* popularly known as King Oyester, a medicinal and edible variety of mushroom [8-12]. Cultivation of mushroom fruiting bodies takes about 90 to 100 days [13]. Since the growth of mushroom is also dependent on temperature it can only be grown during specific months of the year, if temperature not being controlled mechanically. These are the obvious limitations for growing of mushrooms fruiting bodies. The alternative route to overcome these limitation is the submerged cultivation of fungal mycelia that comparatively require short incubation time, labor and space. In literature limited study is available on production of *Pleurotus eryngii* mycelia under submerged cultivation focusing on ergosterol enhancement. Majority of the literature available on mushroom fruiting bodies and little study is available on ergosterol and its conversion to vitamin D in mycelial form of mushrooms. The focus of this study was to develop submerged cultivation strategies for obtaining higher fungal biomass with improved ergosterol content in fungal dried biomass.

2 MATERIALS & METHODS

2.1 *Culture strain*

The strain of *Pleurotus eryngii* was obtained from HAIC Murthal research centre, Haryana, India. The fungal mycelia were maintained on potato dextrose agar (PDA) plate at 28°C. The inoculum was develop by transferring an 8 mm culture plug into 50 ml of potato dextrose broth (PDB) and incubated for 96h at 28°C and 130 rpm. A 10% of inocula were used to inoculate production flask in 50 ml of PDB and incubated at 28°C and 130 rpm. The mycelia was harvested and washed thrice with deionized water after 12 days of incubation and lyophilized for further use.

2.2 *Growth profile*

The growth profile studies were performed in solid media PDA in petri plates (90 mm) and in PDB in flasks. The initial pH was adjusted to 6.0 and it was incubated at 28°C and 130 rpm. The growth on solid media was measured as mean radial growth. The sampling from submerged culture was done at regular intervals by harvesting one flask at a time. All the flasks contained equal volumes of media.

2.3 *Biomass estimation*

Fresh harvested mycelial biomass was filtered on pre-weighted Whatman no.1 filter paper and washed thrice with distilled water. It was then dried overnight at 50°C and weight was taken after constant weight achieved.

2.4 *Extraction and quantification of ergosterol from mycelia*

Lyophilized mushroom mycelia was powdered using liquid nitrogen. Ergosterol from mycelia extracted and analyzed according to the method of Mau et al. (1998), modified as given below. 0.5 g of crushed mycelia powder was accurately weighed into 250 ml round bottom flasks and mixed with 4 ml of sodium ascorbate solution (17.5 g of sodium ascorbate in 100 ml of 1 M NaOH), 50 ml of ethanol (95% pure, Merck), 10 ml of 50% potassium hydroxide (85% pure, Merck Chemicals). Ergosterol standards prepared using 99% pure ergosterol (Sigma chemicals).

The mixture was saponified under reflux at 80°C for 1 h, then, it was immediately cooled to room temperature and transferred into a separating funnel. The mixture was first extracted with 15 ml de-ionized water, followed by 15 ml ethanol and then with three-stages of n-hexane of volumes 50, 50 and 20 ml, respectively. The pooled organic layers were washed three times with 50 ml of 3% KOH in 5% ethanol and then finally with de-ionized water until neutralized. The organic layer was transferred into a round bottom flask. Solvent evaporated to dryness at 40°C using rotavapor, and immediately re-dissolved in 5 ml methanol using ultra-sonication for 5 min. After centrifugation (10 min, 10,000g, 4°C) and membrane filtration (0.22 μm) of the re-dissolved residue, the supernatant was used for analysis by high performance chromatography with a diode array detector (HPLC–DAD, Agilent 1260). The HPLC method was used similar to Jasinghe and Perera (2004).

3 RESULTS & DISCUSSION

3.1 *Growth profile of P. eryngii on solid medium (agar plates)*

The growth profile data on solid plate indicated that *P. eryngii* culture entered in to log phase of growth on day-2 of incubation [Figure 1 (F)] and covered the complete surface of the plate on day-10 of incubation [Figure 1 (A-E)]. On day-12 a cottony white growth was appeared.

3.2 *Growth profile of P. eryngii under submerged agitated conditionson*

3.2.1 *Growth profile and yield of mycelia*

The experiments were performed in triplicates. The flasks were incubated at 28°C, 120 rpm. Forty two flasks of 250 ml with a working volume of 50 ml were used in this

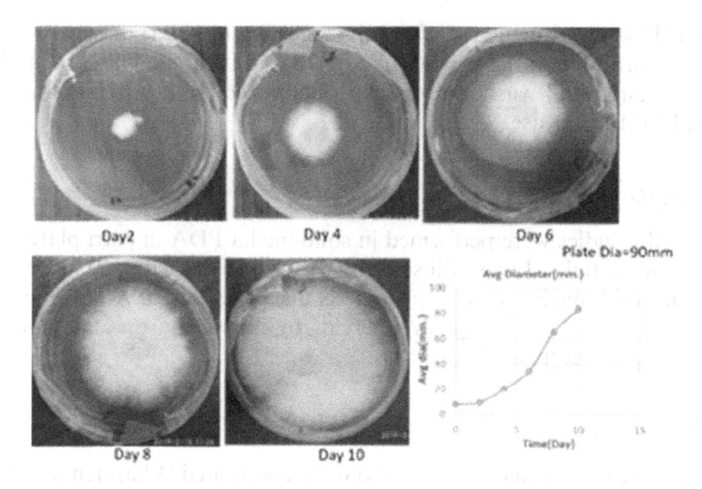

Figure 1. Growth studies of *P. eryngii* on potato dextrose agar plate. (A-E) shows radial fungal growth [A. Day-2, B. Day-4, C. Day-6, D. Day-8, E. Day-10 of incubation. (F). shows growth curve of *P. eryngii* (measured as average mycelia diameter). Note: Petri plate diameter is 90 mm.

Figure 2. Growth of *P. eryngii* mycelia on PDB medium at 28°C, 120 rpm at different days.

experiment. Sampling was done by harvesting one flask on alternate days (Destructive sampling) (Figure 2). Samples were analysed for dry biomass, ergosterol and residual glucose concentrations.

We were able to reduce the total cultivation time to 12 days from 21 days. Maximum biomass (~ 10 g/L) build on day-10 of incubation and there after culture entered into stationary phase (Figure 3). The plateau observed in biomass profile (Figure 3) could be due to the consumption of all the available N-source in PDB on day-10 as sufficient C-source was found to be available (~4 g/L). Ergosterol content estimated in day 12 sample was ~1.72mg/g of dried biomass (Figure 4).

3.2.2 *Scale up studies on growth up to 1L flask*

The data obtained at 250 ml shaking flask were validated till 1L shaking flask. Scale up results are shown in Table 1. Figure 5 shows the harvested mycelia from 1L flask after day 12.

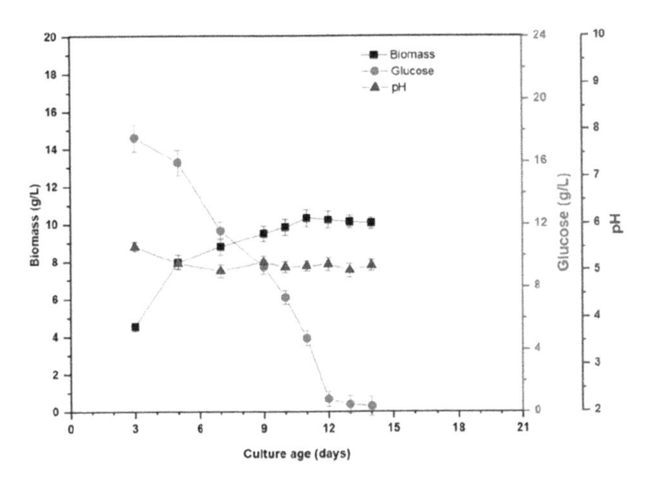

Figure 3. Growth of *P. eryngii* on PDB medium at 28°C at 120 rpm. The graph representing Dry biomass (g/L), pH, Residual substrate concentrations (g/L) during the process.

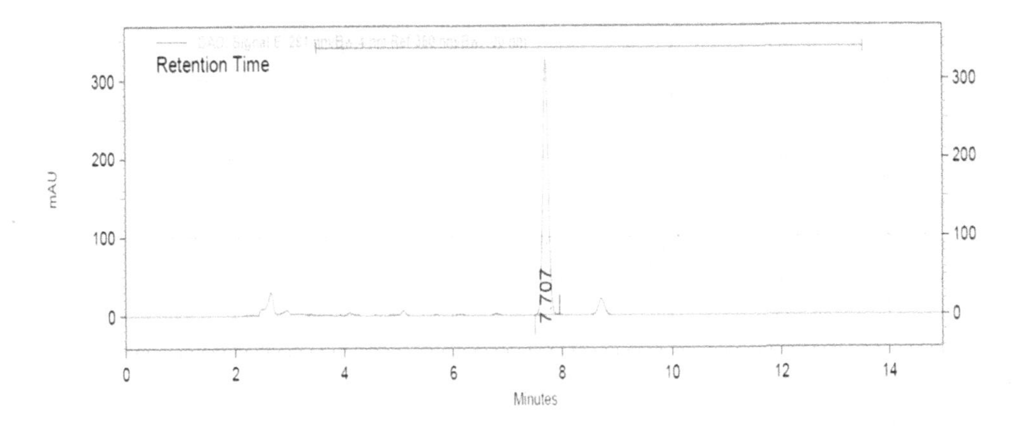

Figure 4. HPLC graph of ergosterol in mycelia sample.

Table 1. Scale up results from 250 ml flask to 1L flask.

Flask scale (mL)	Working volume (mL)	Number of days	RPM	Biomass (g/L)
250	50	12	120	10.18
500	100	12	120	10.125
1000	200	12	120	10.89

Figure 5. *P.eryngii* mycelia in pellet form.

4 CONCLUSIONS

Mycelial form of *P. eryngii* obtained using agitated submerged culture techniques cultivated in short duration can be used as an alternative to mushroom fruiting bodies for ergosterol production.

ACKNOWLEDGMENT

The funding from Ministry of Food Processing Industries (MoFPI), GOI for conducting this research is highly acknowledged.

REFERENCES

1. Wani, A., Bodha, R.H. Wani, A.H. (2010). Nutritional and medicinal importance of mushrooms, J. Med. Plants Res. 4 (24) 2598–2604.
2. Phillips, K.M., Ruggio, D.M., Horst, R.L., Minor, B., Simon, R.R., Feeney, M.J., Byrdwell, W.C., Haytowitz, D.B. (2011). Vitamin D and sterol composition of 10 types of mushrooms from retail suppliers in the United States. Journal of Agricultural and Food Chemistry 59, 7841–7853.
3. Havinga, E., Kock, R. J. D., & Rappoldt, M. P. (1960). The photo chemical interconversions of provitamin D, lumisterol, previtamin D and tachysterol. Tetrahedron, 11, 276–284.
4. Jasinghe, V. J., & Perera, C. O. (2005). Distribution of ergosterol in different tissues of mushrooms and its effect on the conversion of ergosterol to vitamin D2 by UV irradiation. Food Chemistry, 92 (3),541–546.
5. Mattila, P., Lampi, A.-M., Ronkainen, R., Toivo, J., Piironen, V. (2002). Sterol and vitamin D2 contents in some wild and cultivated mushrooms. Food Chemistry 76, 293–298.

6. Guan, X. L., C. M. Souza, H. Pichler, G. Dewhurst, O. Schaad, K. Kajiwara, H. Wakabayashi, T. Ivanova, G. A. Castillon, M. Piccolis, et al. (2009). Functional interactions between sphingolipids and sterols in biological membranes regulating cell physiology. Mol. Biol. Cell 20: 2083–2095.
7. Wollam J, Antebi A (2011). Sterol regulation of metabolism, homeostasis and development. Annu Rev Biochem 80:885–916. doi:10.1146/annurev-biochem-081308-165917
8. Yang Z, Xu J, Fu Q, Fu X, Shu T, et al. (2013). Antitumor activity of a polysaccharide from Pleurotus eryngii on mice bearing renal cancer. Carbohydr Polym 95: 615-620.
9. Chen P, Yong Y, Gu Y, Wang Z, Zhang S, et al. (2015). Comparison of antioxidant and anti-proliferation activities of polysaccharides from eight species of medicinal mushrooms. Int J Med Mushrooms 17: 287-295.
10. Xue Z, Li J, Cheng A, Yu W, Zhang Z, et al. (2015). Structure identification of triterpene from the mushroom Pleurotus eryngii with inhibitory effects against breast cancer. Plant Foods Hum Nutr 70: 291-296.
11. Krupodorova T, Rybalko S, Barshteyn V. (2014). Antiviral activity of basidiomycete mycelia against influenza type A (serotype H1N1) and herpes simplex virus type 2 in cell culture. Virol Sin 29: 284-290.
12. Wang SJ, Li YX, Bao L, Han JJ, Yang XL, et al., (2012). Eryngiolide A, a cytotoxic macrocyclic diterpenoid with an unusual cyclododecane core skeleton produced by the edible mushroom Pleurotus eryngii. Org Lett 14: 3672-3675.
13. Kapahi, Meena. (2018). Recent Advances in Cultivation of Edible Mushrooms. 10.1007/978-3-030-02622-6_13.

Biotechnology and Biological Sciences – Sen et al. (Eds)
© 2020 Taylor & Francis Group, London, ISBN 978-0-367-43161-7

Isolation of pectinolytic fungi from rotten tomato, brinjal and banana

U. Bhattacharyya, P. Paul, S. Mustafi, C. Sen, M. Gupta & B. Roychoudhury
Department of Biotechnology, Bengal Institute of Technology, Kolkata, India

ABSTRACT: Pectinase is the most widely used enzyme commercially. It is very important to explore this enzyme as its uses are many. Pectinase is an important enzyme in the field of food industries and is mainly used in fruit and vegetable extraction, oil extraction and fermentation of coffee, cocoa and tea. It is of both plant and microbial origin. In our present study, we have given emphasis on isolating pectinolytic fungi from rotten fruits and vegetables. The study mainly focused on characterizing the nature of the enzyme based on the clearance zone formed due to the activity of different fungi on pectin. For this, Czapek-Dox medium was used. Under our study, we found that the clearance zone formed by the medium containing fungal growth from tomato was of biggest diameter.

1 INTRODUCTION

Biotechnology is an application of living organisms and their components to industrial products and processes that requires less energy and is based on renewable raw materials (Awan 1993, Rolin 1993 & Ridley 2001). Microbial enzymes are used in many industrial purposes. Due to the increasing concern about pollution, there is an increased demand to replace the chemical processes with biotechnical processes involving microorganisms and few enzymes such as Pectinase.

Pectin is a generic name used for compounds which are acted upon by pectinolytic enzymes. They do not have a definite molecular weight unlike proteins, lipids and nucleic acids. However, their relative molecular masses range between 25 to 360 kDa (Reddy & Sreeramulu 2012). Pectins are acidic in nature that is they are negatively charged complex macromolecules. Pectinases are used for extraction and clarification of fruit juices, preparations of fruit cordials. It also enhances the quality of wine. It is also used for fermentation of cocoa, coffee and tea. Since pectinase are widely used enzymes for different industrial application, it is important to increase the pectinases production which also included selection of better pectinases production sources. Therefore in the present study we have attempted to isolate different types of pectinolytic fungi from rotten brinjal, tomato and banana and characterized those fungi using potato as the pectin source.

2 MATERIALS & METHODS

2.1 *Materials*

Rotten banana, tomato, brinjal.

2.2 *Methods*

In order to get the chief source of pectinase production three different types of rotten fruits and vegetables plated in pectin agar medium followed by pure culture at liquid broth medium and isolated on Czapek-Dox agar medium.

Table 1. Composition of Czapek-Dox medium.

Components	Quantity (g/l)
$NaNO_3$	3.0
K_2HPO_4	1.0
$MgSO_4,H_2O$	0.50
KCl	0.50
FeSO4	0.01
Sucrose	30.0
Agar	15.0

2.3 *Isolation of pectinolytic fungi*

Rotten waste vegetables and fruits were collected from market. The vegetables included tomato, brinjal and among fruits, banana was taken. The vegetables were kept in a warm place to enhance the growth of microbes on them. 1000 ml of Pectin agar media was prepared which contained 5 g pectin, 0.50 g potassium hydrogen phosphate (K_2HPO_4), 0.10 g magnesium sulphate heptahydrate ($MgSO_4.7H_2O$), 0.20 g sodium chloride (NaCl), 0.20 g calcium chloride dehydrate ($CaCl_2.2H_2O$), 0.01g ferric chloride hexahydrate ($FeCl_3.6H_2O$), 1.00 g yeast extract and 20 g agar. The agar was melted after mixing of all the ingrediients and then autoclaved for 30 mins at 121°C and 15psi pressure. The medium was next poured into Petri dishes inside the laminar flow to avoid any kind of contamination. After the pectin agar medium was set, a sterile spreader was vigorously rubbed on the waste vegetables and then spreaded along the entire surface of the pectin agar media on the dishes. The dishes were labelled according to the vegetable or fruit source and incubated at 30° C for 3-4 days.

The Petri dishes containing fungal growth were taken out from the incubator. Now, potato dextrose agar (PDA) media was prepared which contained 200ml potato extract, 20 g dextrose, 20g agar and 0.1% ampicillin. The pH was maintained at 5.6. The media was autoclaved at 121°C and 15psi pressure for 30 mins. Pure cultures (one strain per plate) of the previously grown different fungi strains on the pectin agar media were obtained by repeated subculturing on PDA plates. The PDA plates were incubated for 4-5 days and the process was continued to obtain pure culture.

Potato Dextrose Broth was next prepared which contained 200 g/l potato extract, 20 g dextrose. Its pH was maintained at 5.1. The potato dextrose broth was poured into six different conical flasks containing 50 ml in each flask. Fungi from the PDA medium was next poured into the flask and kept in the shaker incubator for 6-7 days at 31°C. A slimy layer of fungal growth was observed above the PDB medium. This layer was next filtered through a filter paper. The slimy fungal layer was collected and used for the next step. The pure fungal spores so produced were spreaded on Czapek-Dox media.

The isolated spores were next cultivated on Czapek-Dox agar medium. it contained pectin as the sole carbon source. The growth of fungal strains was observed and the pectinolytic activity of those individual strains was determined by formation of inhibiton zone by the strain. The clearance zone formed around colonies was determined using Gram iodine solution.

3 RESULTS & DISCUSSION

As the main objective of this study was to isolate pectinolytic enzymes from the cheap sources, spoiled fruits and vegetables were selected. The strain which produced maximally was cultured by changing the various parameters for the better output.

High pectinase producing strains were further screened semi quantitatively by plate assay method.

Table 2. Primary screening of Pectinolytic fungi isolated from different sources.

Serial No.	Source	Pectinolytic fungi
1	Tomato	Rhizopus sp., Penicillium sp., Aspergillus sp.
2	Banana	Penicillium sp., Rhizopus sp.
3	Brinjal	Aspergillus sp.

Pectinases are novel enzymes which are integral in fruit processing industry. In this study a group of pectinolytic fungi vegetables and *Penicilliumcitrinum* was found to be the potent source for pectinase.

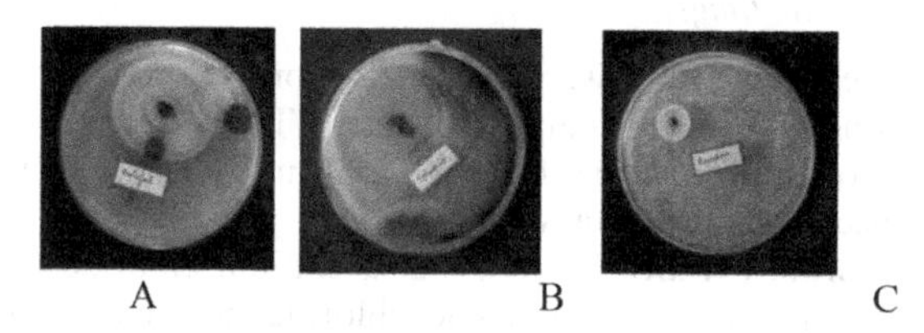

Figure 1. Clearance ring formed on Czapek-Dox Petri dish for A. Brinjal, B. Tomato & C. Banana Induction of fungal pectinase production.

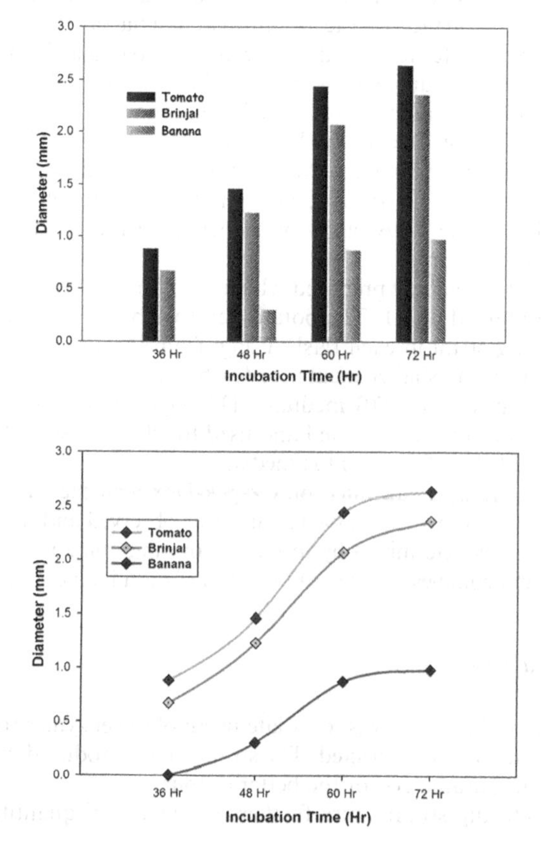

Figure 2. Bar diagram for representing the source of pectinase enzyme on the x-axis and the diameter of clearance zone on the y-axis.

Similar work has been done by Tobechuku c. ezike 2012. and the results that they have found are:

3.1 *Pectin extraction*

The percentage yield of pectin from ground orange peels is 15.5% at pH 2.2 and at a temperature of 700C. Pandharipande and Makode (2012) obtained 48% pectin at pH 1.0 from orange peels. The variations in the yield may be resulted from extraction technique, changes in pH, temperature and extraction time as observed by Pandharipande & Makode 2012 during pectin extraction.

The extracted pectin was used to induce pectinase production in three *Aspergillus* species: *A. niger*, *A. fumigatus* and *A. flavus*. Grenerally, the medium for the production of pectinase always contains pectic metarials as an inducer. From the three pectinolytic fungal species isolated from natural sources (*Aspergillus flavus*, *Aspergillus fumifatus* and *Aspergillus niger*), *Aspergillus niger* showed higher pectinase activity on a selective media containing orange pectin.

3.2 *Production of pectinase from aspergillus niger*

The pectin extracted from orange peels was used to induce pectinase production in *Aspergillus niger* which had more pectinase activity under submerged fermentation. The entire fermentation process was carried out at room temperature (30°C). The accumulation of maximal extracellular activity was observed after 4 days of fermentation.

The graph clearly shows that as the time proceeds the production of pectin increases, highest in banana.

4 CONCLUSION

The present study can further be extended by identifying the fungi strain which were responsible for the pectinase enzyme production.

Pectinases are the growing enzymes of the biotechnological sector. These are used in various industries like wine, fruit juice, coffee-tea etc and they are eco-friendly too. The enzyme activity can be enhanced by enzyme engineering. The strains which are used for its production can also be genetically engineered in order to get enzymes with higher potency.

REFERENCES

Awan, A.J. 1993. *Elements of food nutrition.* Faisalabad: Pakistan.

Ezike, T. C. 2012. Production and characterization of pectinases obtained from *Aspergillus niger* under submerged fermentation system using pectin extracted from orange peels as carbon source Doctoral dissertation, University of Nigeria Nsukka.

Pandharipande, S., & Makode, H. 2012. Separation of oil and pectin from orange peel and study of effect of pH of extracting medium on the yield of pectin. *Journal of Engineering Research and Studies*, 3(2),06–09.

Reddy, P. L., & Sreeramulu, A. 2012. Isolation, identification and screening of pectinolytic fungi from different soil samples of Chittoor district. *International Journal of Life Sciences Biotechnology and Pharma Research*, 1(3),1–10.

Ridley, B. L., O'Neill, M. A., & Mohnen, D. 2001. Pectins: structure, biosynthesis, and oligogalacturonide-related signaling. *Phytochemistry*, 57(6),929-967.

Rolin, C. 1993. *Pectin, Industrial Gums: Polysaccharides and Their Derivatives.* Academic press: New York.

Biotechnology and Biological Sciences – Sen et al. (Eds)
© 2020 Taylor & Francis Group, London, ISBN 978-0-367-43161-7

Phyto chemical abundance (in %) and *in silico* based molecular target interaction aptitude of essential oil components

Smaranika Pattnaik
Department of Biotechnology and Bioinformatics, Sambalpur University, Jyoti Vihar, Burla, India

Arka Kumar Das Mohapatra
Department of Biotechnology and Bioinformatics, Sambalpur University, Jyoti Vihar, Burla, India
Department of Business Administration, Sambalpur University, Jyoti Vihar, Burla, India

ABSTRACT: Plants are defined as the 'Solar powered Phytochemical factory' for synthesizing a myriad metabolites, both primary and secondary. The essential oils, being secondary metabolites, are made up of very diverse molecular assemblies with 'Chemical polymorphism'. In the literature and in most cases, only the major constituents of essential oils have been considered for the Chemotype determination. However, it is worth noting that minor compounds can play a very important role in establishing antibacterial activity of a given species.

This study was designed to put the chromatographic data of Lemongrass, Palmarosa and Citronella oil in one basket with their observed antibacterial activity, from which the abundance (in %) of essential oil components was inferred. From 'Components Abundance ' analysis (followed by a t- test of all components of essential oils available in database, a chemotype of entity respective essential oils were defined. There were varied opinion with respect to the presence of Limonene, Citronellol and Linalool as observed from the t-test. There was existence of a significant difference in the opinion with respect to the presence of Limonene, Citronellol, Linalool and Caryophyllene in Lemongrass, Palmarosa, and Citronella oils, as the significance value was less than 0.05 at 99% confidence level.

Further, *in silico* based docking simulation studies were carried out between the above mentioned components of essential oils with DNA gyrase, a molecular target using molecular docking tools. It was observed that there was no correlation between abundance (%) of components and respective molecular interaction aptitude of each active constituents considered for this study.

1 INTRODUCTION

1.1 *Plant is the solar powered phytochemical factory*

Phytochemical constituents of medicinal plants encompass a diverse space of chemical scaffolds which can be used for rational design of novel drugs (Mohanraj *et al.*, 2017) hence defined as a solar powered biochemical factory (Figure 1.1), synthesizing a myriad of secondary metabolites that are derived from central or primary metabolism (Aharoni and Galli, 2011). The large number of different structures arise from the various combinations of multiple hydroxyl groups, methyl groups, glycosides and acyl substituents (Piasecka, 2017) with wide range of commercial and industrial applications (Vikram *et al.*, 2014). It is worthwhile to study the variety of co-existing phytochemical constituents in the plant, which may be responsible for its unique bioactivity (Liu *et al.*, 2013) including antioxidant and antimicrobial activities (Bhat and Dahihan, 2014). Currently, data on the antimicrobial activity of numerous plants, so far considered empirical,

* Corresponding author: akdm.2002@gmail.com

have been scientifically confirmed, concomitantly with the increasing number of reports on pathogenic microorganisms resistant to antimicrobials (Silva and Fernades, 2010). Different authors have shown that medicinal and aromatic plants have bioactive compounds that could act alone or in synergy with antibiotics against bacterial isolates (Aires, 2016).

1.2 *Phytochemical profile of medicinal and aromatic plants (MAPs)*

This is pertinent to mention here that different varieties of the same MAP species may have different phytochemical profile and thus different types of effects (Pattnaik *et al.*, 1995; Pattnaik *et al.*, 1999).

The medicinal and aromatic plants constitute essential oils (Egamberdieva, 2016) as products of secondary metabolism. These are biosynthesized in specialized cells types, such as osmophores, glandular trichomes, ducts and cavities, present on different parts (flower, seeds, bark, root, leaves) or the whole plants (Rehman and Hanif, 2016). Figure 1 is a pictorial representation of position of oil glands present in a leaf lamina where the secreted essential oils are deposited. Further, EOs being secondary metabolites are made up of very diverse molecular assemblies with 'Chemical polymorphism', each having different properties raising the concept of 'Chemotype' or 'Chemical race' (Balint *et al.*,2016).

1.3 *Phytochemical polymorphism of essential oil components*

In the literature and in most cases, only the major constituents of essential oils have been considered for the Chemotype determination. Figure 2 depicts about onset of chemical polymorphism due to process of isomerization, rearrangement and dehydrogenation for conversion of Pinene into Limonene, Terpinene and p-Cymene respectively. In addition,

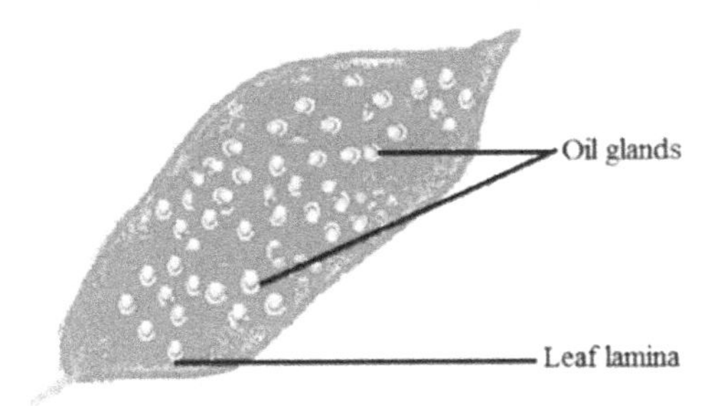

Figure 1. A diagrammatic sketch showing the presence of oil glands.

Figure 2. Inter conversion of Citronellol to Isopulegol showing the presence of oil glands a); Pinene to P-Cymene (b); interconversion of Citral to Citronellal, Geraniol, Citronellol (c); Geranyl acetate to Limonene (d).

conversion of Citral to Citronellal which in turn conversion into Citronellol and finally Geraniol and Nerol by the process of hydrogenation is also evidenced. It is also seen that Citral is converted into Geraniol and Nerol by the process of hydrogenation. Limonene is formed from geranyl pyrophosphate, via cyclization of a neryl carbocation (Mann, 1994). The chemo selective hydrogenation of the carbonyl group of citronellal leads to the unsaturated alcohol, Citronellol (Cahyono, *et al.*, 2010).

The distribution of compounds in various essential oils can play a role in characterizing MAP species regarding their bioactivity. There have been research to study the effectiveness of individual components (Pattnaik *et al.*,1997). However, in literature, there is a big gap in validating the distribution of chemical entities and degree of bioactivity. Analyses of distribution of compounds proffer a statistical validation also.Hence, this chapter is designed to corroborate the available essential oil chromatographic data of some selected plant essential oils (Lemongrass, Palmarosa and Citronella oil) belonging to MAP family Poaceae L. with potent antimicrobial activity from which the abundance (in %) of respective components in the essential oils was inferred. The 'Components Abundance' analysis (followed by 't' test analysis) of all components of essential oils were made. In addition,, efforts were made to carry out *in silico* based essential oil component-target enzyme molecular docking predictions. The test compounds were evaluated on the basis of degree of target enzyme interaction. An analysis was made to link the degree of bio efficacy with the pattern of distribution of essential components included in this study.

2 MATERIALS AND METHODS

2.1 *The distribution of components in various essential oils*

For the detection of the distribution (in %) of phyto constituents of essential oils in source plants belonging to Poaceae family, an extensive review of the literature was undertaken in different national and international scientific sources like Pub Med, Science direct, Research Gate etc. The authors reporting phytochemicals of respective essential oils of source plants belonging to said family are given in Table 1 and the Figure 3 depicts the three species of *Cymbopogon*.

2.2 *Distribution pattern and t- test*

The output of all the reports were systematically analysed; different components of essential oils were documented to observe that four compounds, namely, Linalool, Caryophyllene, Limonene and Citronellol were present in the said essential oils. The occurrences of specific components (discrete variables) of each essential oils were counted. Further, a simple

Table 1. List of authors studied about Phytochemistry of respective essential oils.

Names of plants	Authors names
Cymbopogon citratus	Salma *et al.*, 2016; Pinto *et al.*, 2015; Joga Rao *et al.*, 2015; Boukhatem *et al.*, 2014; Mohamed Hanna *et al.*, 2012; Vazirian *et al.*, 2012; Shah *et al.*, 2011; Aiemsaard *et al.*, 2011; Ganjewala, 2009; De Bona Da Silva *et al.*, 2008; Nigrelle and Gomes, 2007; Kasali *et al.*, 2001; Lewinsohn, 1998;Onawunmi, 1988;Onawunmi and Ogunlana, 1986
Cymbopogon martini	Kakaraparthi *et al.*, 2015; Lawrence *et al.*, 2012;Padalia *et al.*, 2011; Rajeswar, 2010; Akbar and Saxena, 2009; Rao *et al.*, 2009; Ganjewala, 2009; Randriamiharisoa and Gaydo, 2003; Raina *et al.*, 2003; Dubey *et al.*, 2000; Prashar *et al.*, 1987
Cymbopogon flexuosus	Singh and Kumar, 2017; Adukwu *et al.*, 2016; Jantamas *et al.*, 2016; Batubara *et al.*, 2015; Chong *et al.*, 2015; Gupta and Ganjewala, 2015;Pal *et al.*, 2011; Silva *et al.*, 2011; Jaroenkit *et al.*, 2011; Iruthythas *et al.*, 1977; Kakarla and Ganjewala, 2009; Nakahara *et al.*, 2003; Matsuda *et al.*, 1992

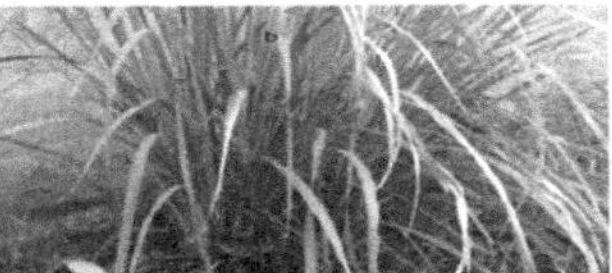

Figure 3. a: The Lemongrass plant (*Cymbopogon citratus*), b: Palmarosa plant (*Cymbopogon martinii*) and c: Citronella plant (*Cymbopogon flexuosus*) belonging to Poaceae family.

percentages calculations were made by dividing the number of occurrences of specific components (discrete count data) by total number of reports, expressed in percentages. Thus, t-test (Mishra and Das Mohapatra, 2010) was resorted to resolve such varied opinion.

2.3 *Molecular docking study with target enzyme, DNA gyrase subunit B*

When a molecular docking study was carried using Auto dock 1.5.6 and visualized in Discovery studio -2016 software offline (Pattnaik et al., 2017). The four components namely, Limonene, Citronellol, Caryophyllene and Linalool were taken for the docking study. The pdb files were retrieved from the Pubchem (www.ncbi.ac.in) data base. The DNA gyrase sub unit B (PDB ID 1aj6) of *Escherichia coli* O157:H7 was considered as the target enzyme (Reece and Maxwell, 1991). The best docking poses were predicted with highest docking score and formation of conventional Hydrogen bond.

3 RESULTS AND DISCUSSION

3.1 *Distribution pattern and t- test*

Table 2 is displaying the distribution (in %) of Linalool, Caryophyllene, Limonene and Citronellol in *C. citratus, C. martini* and *C. flexuosus* which was reported in literature. Figure 4 depicts about the pattern of distribution of the above four compounds in three different species of genera *Cymbopogon*. There was variation in distribution of oil components in different species of *Cymbopogon*. Citronellol and Limonene were found to be most abundant (in %) in Citronella essential oil. However, the abudance of Linalool in Citronella oil was least. Higher percentage of distribution of Linalool was found both in Lemongrass oil and Palmarosa oil. The distribution pattern of Caryophyllenene was similar both in Lemongrass oil and Citronella oil.

Table 3 displays the results of t-test carried out for Limonene, Citronellol, Caryophyllene and Linalool in Lemongrass, Palmaroasa and Citronella oils.

It is evident from Table 4 that there exists a significant difference in the opinion with respect to the presence of Limonene, Citronellol, Caryophyllene, Linalool in the three essential oils, namely, Lemongrass, Palmarosa and Citronella oil, as the significance value is less than 0.05 at 99% confidence level. Hence, it may be inferred that the natural distribution pattern of phytochemicals in all the essential oils remains the same which is also corroborated by the results of the t-test.

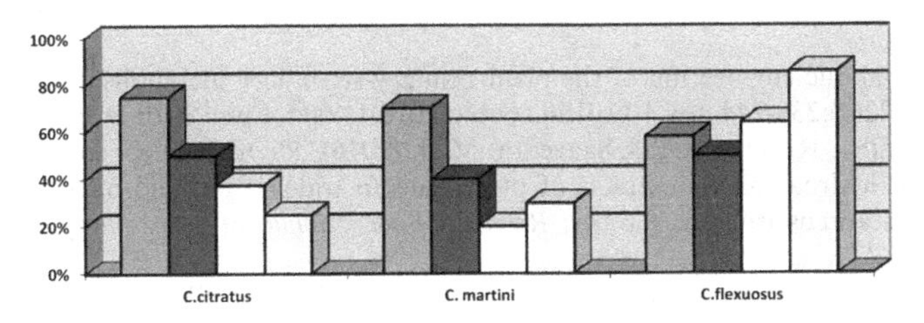

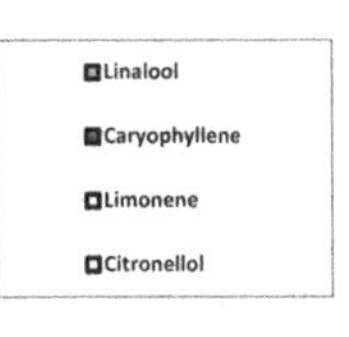

Figure 4. The percentage of distribution of Linalool, Caryophyllene, Limonene and Citronellol present in *Cymbopogon citratus* (Lemongrass oil), *Cymbopogon martini* (Palmarosa oil) and *Cymbopogon flexuosus* (Citronella oil).

Table 3. Result of t- test.

Names of Essential oils	Test Value = 0 t	df	Sig. (2-tailed)	Mean Difference
Lemon grass oil	4.392	3	0.022	46.87500
Palmarosa oil	3.703	3	0.034	40.00000
Citronella oil	8.356	3	0.004	64.50000

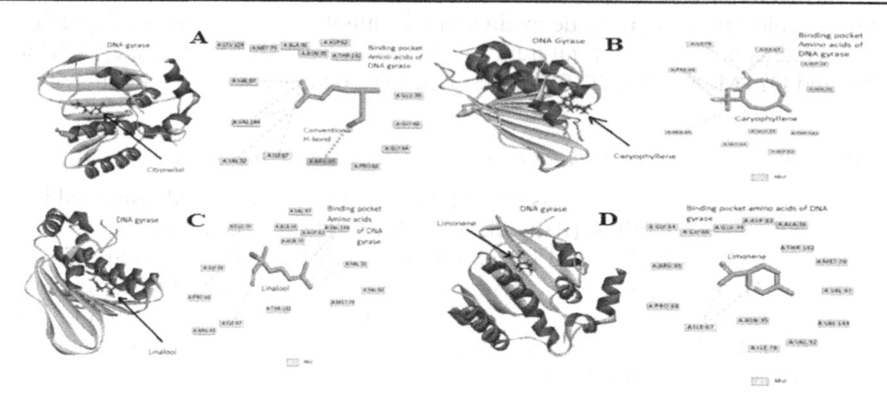

Figure 5. A: The molecular docking poses; A: DNA gyrase Sub unit B with Citronellol, B:with Caryophyllene, C: with Linalool and D: with Limonene.

3.2 *Molecular docking study with target enzyme, DNA gyrase subunit B*

From the molecular docking study, it was found that there were variation in bonding pattern with the bacterial target enzyme. Among the test essential oil components, Citronellol was found to be most potent component (Figure 5) to interact with the target DNA gyrase enzyme (followed by Caryophyllene, Linalool and Limonenene. It was forming a conventional Hydrogen bonding with Arginine 65. Formation of Hydrogen bonding is regarded as higher qualification of a lead molecule in context to its druggability (Pattnaik et al., 2017). It may therefore be concluded that the distribution pattern of individual components of essential oils, validated through t-test, has no correlation with the distribution of individual components for their bioactivity.This study comprising the distribution pattern analysis of Limonene, Citronellol, Caryophyllene and Linalool in *Cymbopogon citratus*, *Cymbopogon martini* and *Cymbopogon flexuosus*, validated with t- test have enough implications in the screening for active compounds present in essential oils for their respective bioactivity.

REFERENCES

Adukwu EC, Bowles M, Edwards-Jones V, Bone H. Antimicrobial activity, cytotoxicity and chemical analysis of Lemongrass essential oil (*Cymbopogon flexuosus*) and pure Vitral. Appl Microbiol Biotechnol. 2016;100(22):9619–9627.

Aharoni A, Galili G.2011. Metabolic engineering of the plant primary-secondary metabolism interface. *Curr Opin Biotechnol.* 2011. 22(2):239-244.doi: 10.1016/j.copbio.2010.11.004. Epub 2010 Dec 6.

Aires, A., Marrinhas, E., Carvalho, R., Dias, C., & Saavedra, M. J. (2016). Phytochemical composition and antibacterial activity of hydroalcoholic extracts of pterospartum tridentatum and mentha pulegium against staphylococcus aureus isolates. *BioMed Research International*, https://doi.org/10.1155/2016/5201879

Aiemsaard, J., Aiumlamai, S., Aromdee, C., Taweechaisupapong, S., & Khunkitti, W. 2011. The effect of lemongrass oil and its major components on clinical isolate mastitis pathogens and their mechanisms of action on Staphylococcus aureus DMST 4745. *Research in Veterinary Science*, *91*(3), 2011.

Akbar N. Saxena B.K. Isolation of Geraniol content from various essential oils. 2009; The Asian Journal of Experi mental Chemistry 4 (2), 1-4.

Bálint, M., Bahram, M., Eren, A. M., Faust, K., & Fuhrman, J. A. (2018). Millions of reads, thousands of taxa : microbial community structure and associations analyzed via marker genes. *FEMS Microbiology Reviews*,40 (5), 686–700.

Batubara I, Suparto IH, Sa'diah S, Matsuoka R, Mitsunaga T. (2015).Effects of inhaled citronella oil and related compounds on rat body weight and brown adipose tissue sympathetic nerve. *Nutrients*.;7 (3):1859–1870.

Boukhatem, M. N., Ferhat, M. A., Kameli, A., Saidi, F., & Kebir, H. T. (2014). Lemon grass (*Cymbopogon citratus*) essential oil as a potent anti-inflammatory and antifungal drugs. Libyan Journal of Medicine, 9.

Bhat, R. S., & Al-Daihan, S. (2014). Phytochemical constituents and antibacterial activity of some green leafy vegetables. *Asian Pacific Journal of Tropical Biomedicine*, *4*(3), 189–193.

Cahyono, E., & Pranowo, H. D. (2010). *Cyclisation-Acetylation Kinetic Of (R) - (+) -Citronellal By Zn 2 + – Natural. 10*(2), 196–201.

Chong DJW, Latip J, Hasbullah SA, Sastrohamidjojo H. 2015. Optimal extraction and evaluation on theoil content of citronella oil extracted from *Cymbopogon nardus*. Malaysia J Anal Sci. 19(1):71–76.

Dubey VS, Mallavarapu GR, Luthra R.2000.Changes in the essential oil content and its composition during Palmarosa (*Cymbopogon martinii*) inflorescence development. Flavour Fragr J.15(5):309.

Egamberdieva, D., Ovidi, E., Tiezzi, A., & Craker, L. (2017). *Phytochemical and Pharmacological Properties of Medicinal Plants from Uzbekistan : A Review*. *5*(2), 59–75. https://doi.org/10.7275/R5571969

Ganjewala, D. (2009). Cymbopogon essential oils : Chemical compositions and bioactivities. *International Journal of Essential Oil Therapeutics*, *3* (APRIL2009), 56–65.

Grace O. Onawunmi (1989) Evaluation of the Antifungal Activity of Lemon Grass Oil, International *Journal of Crude Drug Research*, 27:2, 121-126, DOI: 10.3109/13880208909053950

Gomes, M. S., Das, M., Cardoso, G., Soares, M. J., Batista, L. R., Machado, S. M. F., Rodrigues, L. M. A. (2014). Use of Essential Oils of the Genus Citrus as Biocidal Agents. *American Journal of Plant Sciences*, *5*(February), 299–305. https://doi.org/10.4236/ajps.2014.53041

Gupta AK, Ganjewala D. 2015.A study on developmental changes in essential oil content and composition in *Cymbopogon flexuosus* cultivar Suvarna. *Acta Biol Szeged*. 59(2):119–125.

Jantamas S, Matan N, Matan N, Aewsiri T. Improvement of antifungal activity of Citronella oil against *Aspergillus flavus* on Rubberwood (*Hevea brasiliensis*) using heat curing. J Trop For Sci. 2016;28(1):39–47.

Jaroenkit P, Matan N, Nisoa M.2011. *In* vitro and *in vivo* activity of Citronella oil for the control of spoilage bacteria of semi dried round scad (*Decapterus maruadsi*). 1(3):234–239.

Hanaa, A. R. M. (2012). *Lemongrass (Cymbopogon citratus)* essential oil as affected by drying methods. 57, 113–116.

Iruthayathas EE, Herath H, Wijesekera R, Jayewardene1977. Variations in the Composition of Oil in Citronella. J Natn Sci Coun Sri Lanka. 2(5):133–146.

Kakaraparthi PS, Srinivas KVNS, Kumar JK, Kumar AN, Rajput DK, Anubala S. 2015. Changes in the essential oil content and composition of Palmarosa (*Cymbopogon martini*) harvested at different stages and short intervals in two different seasons. *Ind Crops Prod*. 69:348–354.

Kakarla S, Ganjewala D. 2009.Antimicrobial Activity of Essential Oils of Four Lemongrass (*Cymbopogon flexuosus* Steud) Varieties. Med Aromat Plant Sci Biotechnol. (1): 107–109.

Kasali, A.A., Oyedeji, A.O., Ashilokun, AO., 2001. Volatile leaf oil constituents of *Cymbopogon citratus* (DC) Stapf, Flavour and Fragrance journal, 16 (5). DOI:10.1002/ffj.1019

Lawrence R, Lawrence K, Lawrence R, Lawrence K, Srivastava R, Gupta D. 2015. Antioxidant activity Of Lemon grass Essential Oil (*Cympopogon citratus*) grown in North Indian plains.

Lewinsohn E, Dudai, N., Tadmor, Y., Katzir, I., Ravid, U., Putievsky, E., & Joel DM 1998. *Histochemical Localization Of Citral Accumulation In Lemongrass Leaves* (*Cymbopogon Citratus (Dc .) Stapf ., Poaceae)*, 35–39.

Liu, Z., Liu, Y., Liu, C., Song, Z., Li, Q., Zha, Q., Lu, C., Wang, C., Ning, Z., Zhang, Y., Tian, C. & Lu, A. E. 2013. The chemotaxonomic classification of Rhodiola plants and its correlation with morphological characteristics and genetic taxonomy. *Chemistry Central Journal*. https://doi.org/10.1186/1752-153X-7-118

Mann, S. 1994, Molecular tectonics in biomineralization and biomimetic materials chemistry. *Nature* 365, 499-505.

Matsuda BM, Surgeoner GA, Heal JD, Tucker AO, Maciarello MJ.1992. Essential oil analysis andfield evaluation of the citrosa plant "*Pelargonium citrosum*" as a repellent against populations of *Aedes mosquitoes*. J Am Mosq Control Assoc. 12:69–74.

Mohanraj, K., Karthikeyan, B. S., Chand, R. P. B., Aparna, S. R., Mangalapandi, P., & Samal, A. (2018). OPEN IMPPAT : A curated database of Indian Medicinal Plants, Phytochemistry And Therapeutics. *Scientific Reports*, (February), 1–17. https://doi.org/10.1038/s41598-018-22631-z

Mishra, P. S., & A. K. Das Mohapatra. (2010). Relevance of Emotional Intelligence for Effective Job Performance: *Vikalpa: The Journal for Decision Makers, 35*(1), 53–61.

Nakahara K, Alzoreky NS, Yoshihashi T, Nguyen HTT, Trakoontivakorn G. 2003.Chemical Composition and Anti fungal Activity of Essential Oil from *Cymbopogon nardus* (Citronella Grass). Japan Agric Res Q. 2003;37(4):249–252.

Silva., NCC & A, F. J. (2010). Biological properties of medicinal plants : a review of their antimicrobial activity. *16*(3), 402–413.

Onawunmi, G. O., & Ogunlana, E. O. (1986). *A Study of the Antibacterial Activity of the Essential Oil of Lemon Grass (Cymbopogon citratus (DC .) Stapf).*

Onawunmi, G. O. (1988). In vitro studies on the antibacterial activity of phenoxyethanol in combination with lemon grass oil. *Pharmazie, 43*(1), 42–44.

Padilla-gonzalez GF, Antunes F. 2016.Critical Reviews in Plant Sciences Sesquiterpene Lactones : More than Protective Plant Compounds With High Toxicity Sesquiterpene Lactones;2689.

Pattnaik S, Behera SK, Mohapatra N. 2017.Bioinformatics Homology modeling of FtsZ proteinfrom virulent bacteri al strains and its interaction with Eucalyptol : An *In silico* approach for therapeutics.1:24870.

Pattnaik, S., Subramanyam, V. R., Kole, C. R., & Sahoo, S. (1995). Antibacterial activity of essential oils from Cymbopogon: inter- and intra-specific differences. *Microbios, 84* (341).

Pattnaik, S., Subramanyam, V. R., Bapaji, M., & Kole, C. R. (1997). Antibacterial and antifungal activity of aromatic constituents of essential oils. *Microbios, 89*(358).

Prashara A, Hili P, Veness RG, Evans CS. 2003. Antimicrobial action of palmarosa oil (*Cymbopogon marti nii*) on *Saccharomyces cerevisiae*. Phytochemistry. 2003;63(5):569–575.

Pinto, Z. T., Sánchez, F. F., Santos, A. R. Dos, Amaral, A. C. F., Ferreira, J. L. P., Escalona-Arranz, J. C., & Queiroz, M. M. D. C. (2015). Chemical composition and insecticidal activity of Cymbopogon citratus essential oil from Cuba and Brazil against housefly. *Revista Brasileira de Parasitologia Veterinária, 24*(1), 36–44.

Rajeswara Rao BR, Rajput DK, Patel RP, Purnanand S.2010 Essential oil yield and chemical compositionchanges during leaf ontogeny of Palmarosa (*Cymbopogon martinii* var. motia). Nat Prod Commun.2010;5(12):1947–1950.

Raina VK, Srivastava SK, Aggarwal KK, Syamasundar K V., Khanuja SPS.2003. Essential oil composition of *Cymbopogon martinii* from different places in India. *Flavour Fragr J.*;18 (4):312–315.

Randriamiharisoa RP, Gaydou EM.1987. Composition of Palmarosa (*Cymbopogon martinii*) Essential Oilfrom Madagascar. J Agric Food Chem.;35(1):62–66.

Rao, H. J., Kalyani, G., & King, P. (2015). Isolation Of Citral From Lemongrass Oil Using Steam Distillation : Statistical Optimization By Response Surface Methodology. 13(3),1305–1314.

Rehman, R., & Asif Hanif, M. (2016). Biosynthetic Factories of Essential Oils: The Aromatic Plants. *Natural Products Chemistry & Research, 04*(04). https://doi.org/10.4172/2329-6836.1000227

Reece, R. J., & Maxwell, A. (1991). DNA Gyrase: Structure and Function. Critical Reviews in Biochemistry and Molecular Biology, 26(3-4), 335–375.doi:10.3109/10409239109114072

Shah, G., Shri, R., Panchal, V., Sharma, N., Singh, B., & Mann, A. 2011. Scientific basis for the therapeutic use of Cymbopogon citratus, stapf (Lemon grass). *Journal of Advanced Pharmaceutical Technology & Research, 2*(1), 3. https://doi.org/10.4103/2231-4040.79796

Salma, M., Abdellah, F., El Houssine, A., Kawtar, B., & Dalila, B. (2016). Comparison of the chemical composition and the bioactivity of the essential oils of three medicinal and aromatic plants from Jacky Garden of Morocco. *International Journal of Pharmacognosy and Phytochemical Research, 8*(4), 537–545.

Singh A, Kumar A. 2017. Cultivation of Citronella (*Cymbopogon winterianus*) and evaluation of its essential oil, yield and chemical composition in Kannauj region. Int J Biotechnol Biochem .; 13(2):139–146.

Selim, S. A. (2011). Chemical composition, antioxidant and antimicrobial activity of the essential oil and methanol extract of the Egyptian lemongrass *Cymbopogon proximus* Stapf. 62(1),55–61.

Vazirian, M., Kashani, S. T., Ardekani, M. R. S., Khanavi, M., Jamalifar, H., Fazeli, M. R., & Toosi, A. N. (2012). Antimicrobial activity of lemongrass (*Cymbopogon citratus* (DC) Stapf.) essential oil against food-borne pathogens added to cream-filled cakes and pastries. *Journal of Essential Oil Research, 24*(6), 579–582.

Vikram, P., Chiruvella, K. K., Ripain, I. H. A., & Arifullah, M. (2014). A recent review on phytochemical constituents and medicinal properties of kesum (Polygonum minus Huds.). *Asian Pacific Journal of Tropical Biomedicine, 4*(6), 430–435. https://doi.org/10.12980/APJTB.4.2014C1255

Zouari, N. (2013). Essential Oils Chemotypes: A Less Known Side. *Medicinal & Aromatic Plants, 02*(02), 2167.

Biotechnology and Biological Sciences – Sen et al. (Eds)
© 2020 Taylor & Francis Group, London, ISBN 978-0-367-43161-7

Toxicity assessment of naphthalene on *Anabas testudineus*

Susri Nayak, Dipti Raut & Lipika Patnaik*
Department of Zoology, Environmental Science Laboratory, Centre of Excellence in Environment and Public Health, Ravenshaw University, Cuttack

ABSTRACT: Naphthalene is a Polycyclic Aromatic Hydrocarbon (PAH) commonly used to make mothballs, insecticides, cleaning solutions and is an important component of petroleum and its by-products. PAHs enter the aquatic ecosystem through various routes and exert its effects on aquatic biota. In the present study naphthalene doses of varying concentration was tested for a period of 72 hours on *Anabas testudineus* and its toxicity was examined through biochemical and histological investigations. LC_{50} was calculated to be 5.4mgl^{-1} using probit and graphical method. Experimental fishes were exposed to naphthalene dose concentration of 4.4 mgl^{-1}, 4.6 mgl^{-1}, 4.8 mgl^{-1} and 5 mgl^{-1} for a period of 72 hours and a parallel control set was maintained. Liver tissue was used to observe the changes incorporated as a result of naphthalene exposure. The glycogen content in muscle tissue of *Anabas testudineus* decreased with increase in the toxicant dose concentration in comparison to the control set. Decrease in the glycogen content is indicative of more carbohydrate utilization as a result of stress induced by the toxicant. Similar results were obtained for acid and alkaline phosphatases in experimental fishes. Reduction in ACP and ALP enzyme concentration in experimental fishes might be due to endoplasmic reticulum damage as a result of exposure to naphthalene. Different histopathological changes like vacuolization, necrosis, formation of intercellular spaces, swelling of hepatocytes and shifting of the nucleus towards periphery has been noticed in the liver tissue of experimental fishes. Change in the structural configuration of liver confirms and supports decrease in the concentration of glycogen and other marker enzymes like Acid phosphatase (ACP) and Alkaline phosphatase (ALP) on exposure to naphthalene.

1 INTRODUCTION

Pollution of aquatic environment has become a major problem which is brought about by increasing anthropogenic activities (Abalaka 2015). One such pollutant of concern is polycyclic aromatic hydrocarbons and its toxic effects on aquatic animals (Castle et al., 2005). In 2001 PAHs were ranked as ninth most threatening compound with both carcinogenic and mutagenic properties (Hossain et al., 2010). They are regarded as persistent organic pollutants (POPs) and are also included in the European Union and United States Environmental Protection Agency (EPA) because of their ubiquitous presence in the environment (Hossain et al., 2010; Baklanov et al., 2007; Latimer et al., 2003). PAHs damages the cell membrane structure and enzymes associated with it (Rengarajan et al., 2015). The United States Environmental Protection Agency (EPA) and WHO have identified sixteen PAHs and one of them is Naphthalene (Tuvikene 1995). Naphthalene has been one of the extensively studied PAH due to its high toxicity, high persistence in water and lower molecular weight compound (Palanikumar et al., 2013). Naphthalene is a two ring PAH and is rampantly used as mothballs which serve

*Corresponding author: E-mail: lipika_pat@yahoo.co.uk

the purpose of fumigant repellants and toilet deodorant blocks (Santos et al., 2011). Fishes are excellent tools to evaluate the health condition of the aquatic ecosystem (Jeheshadevi et al., 2014). Fish contribute about 6% of world supply of protein (Al-Halani et al., 2018) and is being for aquatic ecotoxicological studies (Latif et al., 2014). Aquatic fishes can absorb lipophilic organic compounds and it can concentrate in liver prior to excretion (Ranasingha and Pathiratne, 2015). Histopathological changes in animal tissues act as powerful indicators of change induced by any xenobiotic which may be the result of any enzymatic, biochemical and physiological changes in organisms (Reddy and Waskale, 2013). The aim of the present study was to assess the toxic effects of Naphthalene in liver tissues. The histological analysis of liver was performed to confirm the structural damage. Liver glycogen was determined as the stress indicator of the fish health. Enzymes like Alkaline Phosphatases (ALP) and Acid Phosphatases (ACP) are predominantly found in liver and were assessed for any change induced by the toxicant.

2 MATERIALS AND METHODS

Fresh water fish, *Anabas testudineus* were collected from Central Institute of Freshwater Aquaculture (CIFA) and were acclimatized in laboratory conditions for 10-14 days. Fishes were stocked in glass aquariums and were given dry commercial feed having 45% protein. Water was cleaned every day to remove excess feed and feacal matter. At the end of acclimatization period fishes of length 10-12cm and weight of 10-12gms were selected for biochemical, enzymatic and histopathological studies. Twelve aquariums of 50 L capacity were taken and divided into six sets. Six sets were labeled as S-1, S-2, S-3, S-4, S-5 and S-6 and their duplicates were parallelly maintained. Prior to naphthalene administration the fishes were starved for a day. S-1 served as the control (no dose) while S-2, S-3, S-4, S-5, S-6 were given dose concentrations of 4.2mgL^{-1}, 4.4mgL$^{-1,}$ 4.6mgL^{-1}, 4.8mgL^{-1} and 5 mgL^{-1} respectively (LC_{50} – 5.4 mgL^{-1}). Six fishes were kept in each aquarium and after 72 hours all the fishes were harvested and the liver tissue was removed for the estimation of ACP and ACP enzymes. The liver tissue was removed, cleaned and homogenized in chilled 0.9% ice cold saline solution. The homogenate was further centrifuged at 3000 rpm for 10 minutes and the supernatant was used for glycogen, ACP and ALP enzyme assays.

2.1 *Estimation of glycogen*

In a clean test tube 1 ml of supernatant was taken and 1 ml of 10N KOH was added. The test tube was boiled for one hour and 0.5 ml of glacial acetic acid was added to it after cooling. The volume was made up to 10ml by addition of distilled water. In another test tube 2 ml of anthrone reagent was taken and 1 ml of the aliquot from the first test tube was added to it. The test tube was heated in water bath for 10 minutes, cooled and the optical density was read at 650nm in UV-spectrophotometer. The values obtained was plotted on a standard glucose graph.

2.2 *Estimation of acid phosphatase*

In a clean test tube 0.5 ml of supernatant was taken and 0.5ml of the substrate solution (p-nitrophenyl phosphate) and 0.5 ml of 0.1 N citrate buffer were added. The test tube with the above solution was kept in a water bath maintained at 37°C for 30 minutes. After completion of 30 minutes, the reaction was arrested in the extracts by adding 3.8 ml of 0.1N sodium hydroxide. The colour formed at the end of the reaction was read at 415 nm in UV spectrophotometer. The values were expressed in μ moles of phenol liberated / min / 100 mg of protein.

2.3 *Estimation of alkaline phosphatase*

In a clean test tube 0.5 ml of supernatant, 0.5 ml of the substrate solution (p-nitrophenyl phosphate) and 0.5 ml of glycine buffer was taken. The test tube was kept in a water bath maintained at 37°C for 30 minutes. After completion of 30 minutes, the reaction was arrested in the extract by adding 10 ml of 0.02N sodium hydroxide. The colour developed was read at 415 nm in UV spectrophotometer and the values were expressed in μ moles of phenol liberated / min / 100 mg of protein.

2.4 *Histopathological studies*

Tissue specimens were collected from both control and experimental fishes and fixed in aqueous Bouin's solution for 24 hours, washed under tap water for 24 hours, processed through graded series of alcohols, cleared in xylene and embedded in paraffin wax. Sections were cut at 7μ thickness using Weswox Rotary microtome. Sections were spread on a clean glass slide coated with Mayer's albumin. The slides were subjected to hydration process through graded alcohol series (100%, 90%, 70%, 50%, 30% and distilled water). Then they were stained with haematoxylin and eosin. After staining they were subjected to dehydration (30%, 50%, 70%, 90% and 100%) and then cleared in xylene. The cleared slides were mounted in DPX. Finally, the prepared sections were examined and photographed under 10x and 40x magnifications using a light microscope.

3 RESULT AND DISCUSSION

The results of the present study are given below. Depletion of glycogen content was noticed in experimental fishes as compared to control fishes. Decline in glycogen content (Table 2) of experimental liver tissues ranged from 18.80% to 37.93% from the control liver tissues. Statistical differences among the means of control and experimental tissues were calculated using ANOVA. The p value of < 0.05 was considered as significant against control. Liver ALP and ACP activity

Table 1. Glycogen values are given in mg/g, ACP and ALP values are expressed as μ moles of PNP liberated/hour/mg of protein. Values within brackets indicate % decline and are Mean ± S.D (n=6).

Dose	Glycogen	ACP	ALP
Control	0.319 ± .011	6.60 ± .165	6.99 ± .394
4.2mgL^{-1}	0.254 ± .005 (-18.80%)	5.88 ± .160 (-10.90%)	5.86 ± .118 (-16.16%)
4.4mgL^{-1}	0.219 ± .017 (-31.34%)	5.85 ± .121 (-11.36%)	5.99 ± .020 (-14.30%)
4.6mgL^{-1}	0.221 ± .003 (-30.72%)	5.61 ± .046 (-15.15%)	5.64 ± .031 (19.31%)
4.8mgL^{-1}	0.219 ± .001 (-31.34%)	5.13 ± .040 (-22.27%)	4.70 ± .162 (-32.76%)
5 mgL^{-1}	0.198 ± .001 (-37.93%)	5.16 ± .115 (-21.81%)	4.51 ± .149 (-35.47%)

Table 2. Glycogen values are given in mg/g, ACP and ALP values are expressed as μ moles of PNP liberated/hour/mg of protein. (Values are Mean ± SE n = 6).

Dose	Glycogen	ACP	ALP
Control	0.319 ± .006	6.60 ± .095	6.99 ± .227
4.2mgL^{-1}	0.254 ± .003	5.88 ± .093	5.86 ± .068
4.4mgL^{-1}	0.219 ± .010	5.85 ± .070	5.99 ± .012
4.6mgL^{-1}	0.221 ± .002	5.61 ± .026	5.64 ± .094
4.8mgL^{-1}	0.219 ± .001	5.13 ± .023	4.70 ± .094
5 mgL^{-1}	0.198 ± .0003	5.16 ± .066	4.65 ± .086

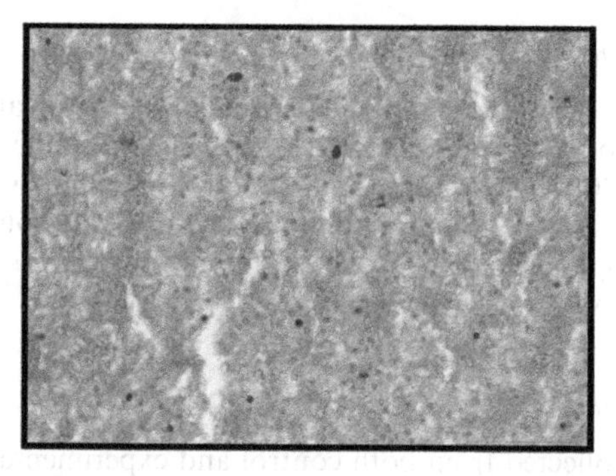

Figure 1. Microphotograph of normal structure of liver.

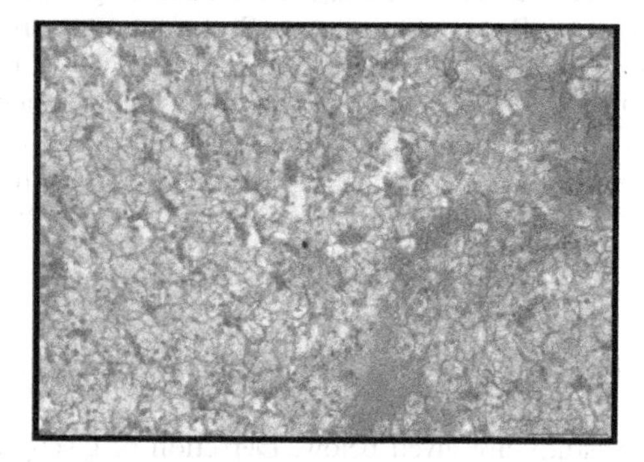

Figure 2. Microphotograph of exposed liver of fish *Anabas testudineus* showing swelling of hepatocytes of control fish *Anabas testudineus.*

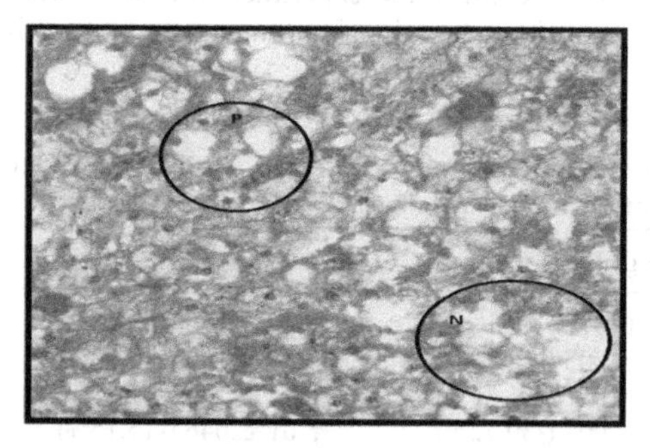

Figure 3. Microphotograph of exposed liver of fish *Anabas testudineus* (P - Nucleus towards the periphery, N – Necrosis).

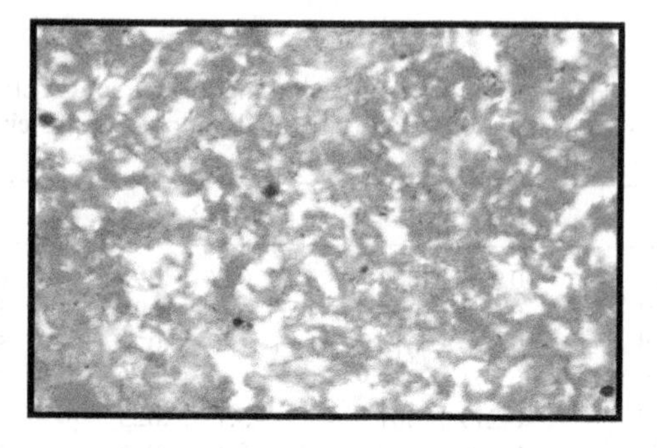

Figure 4. Microphotograph of exposed liver of fish *Anabas testudineus* showing fatty cysts.

inhibited significantly compared to control values at the end of exposure (Table 1). The inhibition of ALP enzyme activity in liver tissues of experimental *Anabas* varied from 16.16% to 35.47%. The acid phosphatase activity was found to be decreased significantly by 10.90% - 22.27% as compared to control.

The liver histology of control fishes showed normal cells and were regularly arranged. The control hepatic cells showed distinct cell membrane and spherical nucleus. However, the hepatocytes of experimental fishes showed changes like vacuolization, swelling, necrosis and structural distortion.Liver glycogen is concerned with storage and transport of hexose units for maintenance of blood glucose. A fall in the glycogen level clearly indicates its rapid utilization to meet the enhanced energy demands in fish exposed to toxicant through glycogenolysis (Somaiah et al., 2014). Naphthalene brought about a significant change in glycogen metabolism. There was a significant decline of 18.80% to 37.93% from the control fish liver. Similar condition of glycogen depletion was noticed in *Sarotherodon mossambicus* when exposed to naphthalene (Dange and Masurekar, 1982). Reduction in glycogen content in the liver indicates glycogenolysis to meet the stress demands induced by naphthalene (Nagarjuna and Rathnamma, 2018). Naphthalene was also seen to induce hypoxic conditions in crab *Scylla tranquebarica* due to which glycogen levels depleted resulting in extra expenditure of carbohydrates (Vijayavel et al., 2009). *Oncorhynchus mykiss* when treated with PAHs like β-Naphthoflavone and benzo(a)pyrene also showed decreased glycogen content to overcome stress conditions (Tintos et al., 2008).Enzymes like ALP and ACP are predominantly found in liver and serve as indices for the diagnosis of tissue damage. Alkaline phosphatase is a hydrolase enzyme which removes various phosphate esters at an alkaline pH and mediates membrane transport whereas acid phosphatase is a hydrolytic enzyme which frees attached phosphate groups from other molecules at an acidic pH. Decrease in ALP and ACP activities in the liver of *Anabas testudineus* exposed to naphthalene could be either due to their possible leakage from the cytosol into general blood circulation through damaged plasma membrane or decrease in their synthesis as a result of organ dysfunction. The decrease could be the symptom of oxidative stress caused by naphthalene (Sunmonu et al., 2015). A decreasing trend in phosphatases activity was also noticed in estuarine crab *Scylla tranquebarica* from naphthalene contaminated habitats (Vijayavel et al., 2009). Necrosis of liver tissue and uncoupling of oxidative phosphorylation might be responsible for the decrease and inhibition of acid phosphatase (Magar and Shaikh 2013). Decrease in ALP enzyme activities probably would facilitate the increased activity of phosphorylation enzyme in the tissue of fish and cause subsequent break down of glycogen for energy release during toxic stress (Suresh et al., 2016). Similar reports in *Clarias gariepinus* indicates fall in the rate of synthesis of glycogen resulting from low metabolism demand and decrease in metabolic transport due to decreases in ALP (Evelyn et al., 2013). Our results are in complete agreement with studies made on changes in ACP and ALP in the liver and kidney of *Labeo rohita* exposed to high concentrations of heavy metal (Mir et al., 2016). Decrease in ALP enzyme activity may result in altered transport, inhibitory effect on the cell growth and proliferation whereas decrease in ACP may be due to increased glycogenolysis or changes in the mitochondrial membrane function (Sreekala and Zutshi, 2010). Decrease in ALP and ACP enzyme activities have also been noticed in *Channa punctatus* when exposed to chromium trioxide for different time periods (Pal and Trivedi, 2016). Decrease in ALP enzyme activity was also noticed in *Heterobranchus bidorsalis* exposed to another PAH anthracene (Sunmonu et al., 2009).Histological studies help to understand the pathological state of the fish. The liver is the target organ of all toxic agents as it deals with detoxification and biotransformation of toxic agents (Montaser et al., 2010). The results of this study showed that *Anabas* when exposed to naphthalene induced histological alterations like hepatic vacuolization, pyknosis and cellular swelling. Formation of fatty cysts was also noticed. Vacuole formation is a cellular defensive mechanism of hepatocytes against toxic compounds and accumulating them inside the vacuole. This prevents the toxic materials from interfering with the physiological activities of the hepatocytes (Asifa and Chitra, 2018). The large vacuole also pushes the nuclei to the periphery. Similar changes were noticed when *Channa punctatus* were exposed to 4-Nonylphenol (Sharma et al., 2018). The liver cells of exposed fish also showed necrosis and congestion. Similar results were also noticed when

Anabas was exposed to copper (Joseph and Asha, 2018). The histological results support the biochemical and enzymatic changes accounting for the functional disruption on the activity of the organ due to cellular damage.

4 CONCLUSION

The present study clearly indicates inhibition of enzyme activity in the liver tissue of *Anabas testudineus* by naphthalene. Alteration in the structure of hepatocytes might be due to naphthalene accumulation. Thus, Enzymatic and histopathological studies can be used biomarkers for monitoring naphthalene toxicity in aquatic ecosystem.

REFERENCES

Abalaka, S.E., 2015. Heavy metals bioaccumulation and histopathological changes in *Auchenoglanis occidentalis* fish from Tiga dam, Nigeria. ***Journal of Environmental Health Science & Engineering***. 13:67.

Al-Halani, A.A., 2018. Effect of Seasonal Changes on Physiological and Histological Characteristics of Nile tilapia (*Oreochromis niloticus*) Inhabited Two Different Freshwater Habitats. ***International Journal of Modern Biology and Medicine***. 9(1).

Asifa, K.P., Chitra, K.C. 2018. Evaluation of acute toxicity of octylphenol in the cichlid fish, *Pseudetroplus maculatus* (Bloch, 1795). ***International Journal of Applied Research***. 4(3): 197-203.

Baklanov, A., Hänninen, O., Slordal, L.H., 2007. Integrated systems for forecasting urban meteorology, air pollution and population exposure. ***Atmos Chem Phys***. 7, 855-874.

Castle, D.M., Montgomery, M.T., Kirchman, D.L., 2006. Effects of naphthalene on microbial community composition in the Delaware estuary. ***Federation of European MicrobiologicalSocieties***. 56.

Dange, A.D., Masurekar, V.B. 1982. Naphthalene-induced changes in carbohydrate metabolism in *Sarotherodon mossambicus* Peters (Pisces: Cichlidae). ***Hydrobiologia***, 94(2): 163-172.

Evelyn, E.N., Bebe, Y., Chukwuemeka, N., 2013. Changes in liver and plasma enzymes of *Clarias gariepinus* exposed to sublethal concentration of diesel. African Journal of Biotechnology. 12 (4): 414-418.

Hossain, M.A., Yeasmin, F., Rahman, S.M.M., Rana, S., 2010. Naphthalene, a polycyclic aromatic hydrocarbon, in the fish samples from the Bangsai river of Bangladesh by gas chromatograph–mass spectrometry. ***Arabian Journal of Chemistry***. 7, 976-980.

Jeheshadevi, A.K., Ramya, T.M, Sridhar, S., Chandra, J.H., 2014. Histological Alterations on the Muscle and Intestinal Tissues of *Catla Catla* Exposed to Lethal Concentrations of Naphthalene. ***International Journal of Applied Engineering Research***. 9(2).

Joseph, M. L., Asha, K. R. K. 2018. Heavy Metal Pollution Induced Histopathological Changes in *Anabas testudineus* collected from Periyar River at Ernakulam district and the Recovery Responses in Pollution free water. ***Journal of Pharmacy***. 8(1): 27-37.

Latif, F., Khalid, M., Ali, M. 2014. An Assessment of Naphthalene Stress on Renal and Hepatic Functional Integrity in Labeo rohita. ***International Journal of Current Engineering and Technology***, 4(1): 366-372.

Latimer, J., Zheng, J., 2003. The sources, transport, and fate of PAH in the marine environment. ***An ecotoxicological perspective***. New York: John Wiley and Sons Ltd.

Magar, R.S., Shaikh, A., 2013. Effect of malathion on acid phosphatase activity of fresh water fish *Channa punctatus*. International Journal of Pharmaceutical, Chemical and Biological Sciences. 3(3), 720-722.

Montaser, M., Mahfouz, M.E., El-Shazly, S.A.M., Abdel-Rahman, G.H., Bakry, S. 2010. Toxicity of Heavy Metals on Fish at Jeddah Coast KSA: Metallothionein Expression as a Biomarker and Histopathological Study on Liver and Gills. ***World Journal of Fish and Marine Sciences***. 2(3): 174-185.

Nagaraju, B., Rathnamma, V.V. 2018. Fluctuations of certain Biochemical constituents as a consequence of Chlorantraniliprole toxicity in the Mud Fish, *Channa punctatus* (Bloch, 1974). ***Journal of Zoological Research***, 2(3): 11-16.

Pal, M., Trivedi, S.P., 2016. Effect of Chromium trioxide on Liver biochemistry of freshwater fish, *Channa punctatus* (Bloch). Archives of Applied Science Research. 8, 1-3.

Palanikumar, L., Kumaraguru, A.K., Ramakritinan, 2013. Biochemical and genotoxic response of naphthalene to fingerlings of milkfish ***Chanoschanos.Ecotoxicology***.

Ranasingha, R.A.T.C.S., Pathiratne, A., 2015. Histological alterations and polycyclic aromatic hydrocarbon exposure indicative bile fluorescence patterns in fishes from Koggala lagoon, Sri Lanka. ***Journal of the National Science Foundation of Sri Lanka***. 43(1).

Reddy P.B., Waskale K., 2013. Using histopathology of fish as a protocol in the assessment of aquatic pollution. ***Journal of Environmental Research And Development***. 8(2).

Rengarajan, T., Rajendran, P., Nandakumar, N., Lokeshkumar, B., Rajendran, P., Ikuo, Nishigaki., 2015. Exposure to polycyclic aromatic hydrocarbons with special focus on cancer. ***Asian Pacific Journal of Tropical Biomedicine***. 5(3),182-189.

Samal, D., Sethy, J., Sahu, H.K. 2016. Ichthyofauna diversity in relation to physico-chemical characteristics of Budhabalanga River, Baripada,Mayurbhanj, Odisha. ***International Journal of Fisheries and Aquatic Studies***, 4(1): 405-413.

Santos, T.C.A., Gomes, V., Passos, M. J. A. C. R., Rocha, A. J. S., Salaroli, R. B., Ngan, P. V., 2011. Histopathological alterations in gills of juvenile Florida pompano *Trachinotus carolinus* (*Perciformes, carangidae*) following sublethal acute and chronic exposure to naphthalene. ***Pan-American Journal of Aquatic Sciences***. 6(2),102-120.

Sharma, M., Chada, P., Borah, M. K. 2018. Histological alterations induced by 4-Nonylphenol in different organs of fish, *Channa punctatus* after acute and sub chronic exposure. ***Journal of Entomology and Zoology Studies***. 6(4): 492-499.

Somaiah, K., Satish, P. V. V., Sunita, K., Nagaraju, B., Oyebola, O.O. 2014. Toxic Impact of Phenthoate on Protein and Glycogen Levels in Certain Tissues of Indian Major Carp *Labeo rohita* (Hamilton). ***Journal of Environmental Science, Toxicology and Food Technology***, 8(1): 65-73.

Sreekala, G., Zutshi, B., 2010. Acid and alkaline phosphatase activity in the tissues of *Labeo rohita* from freshwater lakes of Bangalore. ***The Bioscan***. 2, 365-372.

Sunmonu, T.O., Owolabi, O.D., Oloyede, O.B., 2009. Anthracene – Induced enzymatic changes as stress indicators in African Catfish, *Heterobranchus bidorsalis*. ***Research Journal of Environmental Science***. 3 (6),677-686.

Suresh, V., Bose, M.J., Deecaraman, M., 2016. Toxic effects of cadmium on the acid and alkaline phosphatase activity of female fiddler crab, (*Uca annulipes*). World Journal of Zoology. 11(2): 123-130.

Tintos, A., Gesto, M., Miguez, J. M., Soengas, J. L. 2008. β-Naphthoflavone and benzo(a)pyrene treatment affect liver intermediary metabolism and plasma cortisol levels in rainbow trout *Oncorhynchus mykiss*. ***Ecotoxicology and Environmental Safety***, 69, 180-186.

Tuvikene. A., 1995. Responses of fish to polycyclic aromatic hydrocarbons (PAHs). ***Annales Zoologici Fennici***. 32.

Vijayavel, K., Anbuselvam, C., Balasubramanian, M.P., Samuel, V.D., Gopalakrishnan, S. 2006. Assessment of biochemical components and enzyme activities in the estuarine crab *Scylla tranquebarica* from naphthalene contaminated habitants. ***Ecotoxicology***, 15: 469-476.

Biotechnology and Biological Sciences – Sen et al. (Eds)
© 2020 Taylor & Francis Group, London, ISBN 978-0-367-43161-7

In-vitro study of antifungal activity of *Lentinus edodes* mushroom extract against *Alternaria triticina*

Puja Kumari* & Sunil Kumar Choudhary
P.G.Department of Biotechnology, T.M. Bhagalpur University, Bhagalpur

Arvind Kumar
Regional Director, Agricultural Research Institute, Patna

ABSTRACT: The present study was aimed to evaluate the antifungal activity of solvent based extracts of *Lentinus edodes* mushroom on the phytopathogen A*lternaria triticina*, the causual agent of leaf blight disease in wheat plant by Poisoned Food Technique. Result showed that the aqueous extracts of this mushroom was not so much effective while methanolic extract significantly reduced their growth. Only ethanolic extract of *Lentinus edodes* mushroom was found to inhibit the growth of this pathogen completely. These studies suggest that *Lentinus edodes* mushroom contain potential compound which may be used for controlling leaf blight disease of wheat and it is profitable for economy of the country.

Keywords: *Lentinus edodes*, *Alternaria triticina*, Ethanol and Methanol

1 INTRODUCTION

Lentinus edodes mushroom commonly called shiitake mushroom belongs to class Basidiomycetes. It is the second most important edible cultivable mushroom in the world (Royse, 2001) on the basis of customer choice (Bisen et al., 2010). It is of wood decaying nature of white rot fungus which naturally inhabits on dead wood or wooden logs of hardwood tree under different climatic condition (Ohga, 1992). Phytocompounds such as b-D-glucan, heteroglucan, xylomannan, lentinan and eritadenine; mannose, mannitol, trehaloseand glycerol; vitamins (B2, B12, D2) and dietary fibre are present (Hobbs, 2000) which shows the medicinal properties including antitumor, antimicrobial, antifungal, antibacterial, and hypoglycemic and antioxidant properties (Kitzberger et al., 2007). Lentin, Lentinan and Lentinamycin are the compounds present in the *Lentinus edodes* which makes it liable to inhibit the growth of pathogens of *Trichoderma* sp (Tokimoto et al., 1987) and Gram positive bacteria (Komemushi et al., 1996) and Gram negative bacteria, filamentous fungi, yeast and viruses (Sasaki et al., 2001). Antifungal protein was isolated from *Lentinus edodes* mushroom such as lentinamicin (octa-2,3- diene-5,7 diyne-1-ol), β-ethyl phenyl alcohol and lentin and it shows the inhibition against fungi *Physalospora piricola, Botrytis cinerea*, and *Mycosphaerella arachidicola* (Ngai and Ng, 2003).

Alternaria triticina is a cosmopolitan's fungus which causes leaf blight disease on wheat and firstly identified in India in 1962 (Prasada and Prabhu, 1962). It can causes losses of up to 2.72-36% in grain yield in India. The symptoms appear on the leaf surface with small lesions, gets enlarged and colasscess with another lesion potentially cover entire area. Leaf colour becomes dispigmented and appears as burnt like (Prabhu and Prasada, 1966).

*Corresponding author: E-mail: pkumari402@gmail.com

2 MATERIALS AND METHODS

Isolates of *Alternaria triticina*, Potato dextrose agar media and *Lentinus edodes* mushroom; Distilled water, ethanol and methanol were taken.

2.1 *Preparation of mushroom extract*

The fresh shiitake mushroom was broken down into small pieces and dried under the shade. It was grinded properly in powder form. Different types of solvent like distilled water, ethanol and methanol were properly mixed into prepared mushroom powder in 1:2 ratio separately and placed on Rotatry shaker for 6-7 days for making its aqueous, ethanolic and methanolic extract. Finally extract was filtered through filter paper and further air dried. Extracts were stored in bottles with proper labeling and kept in a refrigerator at 40°C±1 until use.

2.2 *Antifungal activities of mushrooms extract against Alternaria triticina in vitro condition by Poisoned Food Technique*

The experiment trial was conducted *in vitro* conditions to evaluate the efficacy of mushroom extract Shiitake against *Alternaria triticina*. Aqueous, ethanolic and methanolic extracts of prescribed material were evaluated by using Poisoned Food Technique (Nene and Thapliyal, 1979).Required concentration of extracts were mixed with PDA media and poured into the petridishes. After 12 hours duration, mycelial disc of particular microbe or mold *Alternaria triticina* was placed at the centre of the petridishes with the help of forceps and incubated at 27° C for a week. The PDA media plate without extracts acts as a media control. The PDA media plate with only mycelial disc of organism acts as an organism control. Each treatment was replicated thrice. All the petriplates were incubated for a week at 25°C- 27°C. After completion of incubation period, colony diameter or radial growth of the microbes were measured. The data obtained during the trials were subjected to statistical analysis.

2.3 *Result and discussion*

All the extract reduced the mycelial growth of *Alternaria triticina*. At 0.5% of concentration *Alternaria triticina* showed the similar radial growth in aqueous, ethanolic and methanolic extracts containing PDA media. Aqueous extract of this mushroom was not so much effective in comparison to ethanolic and methanolic extracts. But ethanolic extract showed the complete inhibition of *Alternaria triticina* at 5%, 10%, 20% and 40% concentration. It also showed 37.51% of inhibition at 2% of extracts concentration. Methanolic extract of 20% and 40% of concentration of mushroom extract also showed 100% of inhibition. Least concentration of 5% of *Lentinus*

Table 1. Antifungal activity of *Lentinus edodes* mushroom extracts against *Alternaria triticina* by Poisoned Food Technique.

Treatments	Concentration	Aqueous (Radial growth)		Ethanolic (Radial growth)		Methanolic (Radial growth)	
		(cm)	**P.I (%)**	**(cm)**	**P.I (%)**	**(cm)**	**P.I (%)**
T1	Control	5.6	0.00	5.6	0.00	5.6	0.00
T2	0.5%	5.3	5.32	5.1	8.91	5.3	5.32
T3	1%	5.3	5.34	5.0	10.70	5.1	8.85
T4	2%	5.2	7.16	3.5	37.51	5.0	10.70
T5	5%	5.0	10.61	0.0	100	3.3	41.07
T6	10%	5.0	10.70	0.0	100	2.5	55.35
T7	20%	4.6	17.82	0.0	100	0.0	100
T8	40%	4.6	17.93	0.0	100	0.0	100

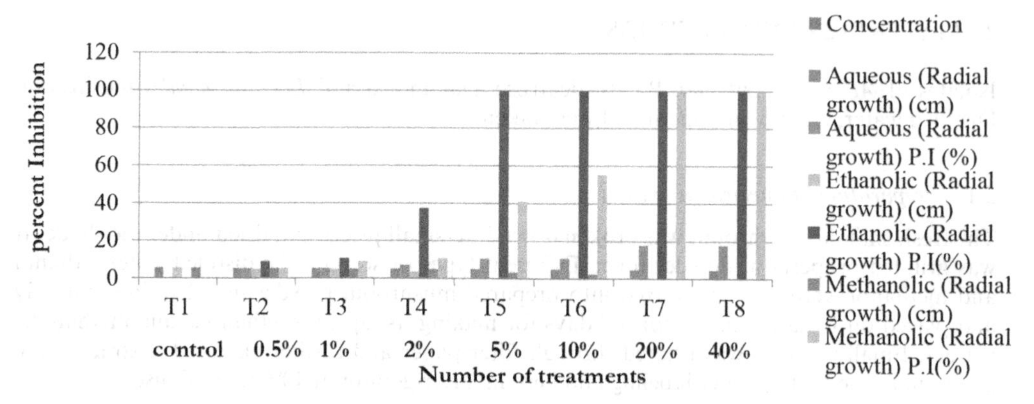

Figure 1. Graphical representation of Antifungal activity of *Lentinus edodes* mushroom extracts against *Alternaria triticina* by Using Poisoned Food Technique.

edodes mushroom extract was the most suitable for further trials. The sesquiterpene rufuslactone (Lindequist et al., 2005) showed activity against some phytopathogenic fungi such as *Alternaria alternata*, *Alternaria brassicae*, *Botrytis cinerea* and *Fusarium grammearum.* Other sesquiterpenes, enokipodim F,G and I (3a-c), isolated from *Flammulina velutipes* mycelium presented low activity against *Aspergillus fumigates* with Ic50 value 229.1±3.6, 233.4±3.8 and 235.1±4.2 respectively. Cloratin A (Quang et al.,2006), derivative of benzoic acid, was isolated from *Xylaria intracolarata* and showed activity against *Aspergillus niger* (IZD = 15mm) and *Candida albicans* (IZD = 17mm); similar to the control (nystatin; with a IZD = 17mm). Piccinin, (2000) reported that aqueous extracts of *Lentinus edodes* reduces the growth of *C. sublineolum.* (Khan and Hsiang, 2003) reported that the inhibition of spore germination and appressorium formation are very important for controlling phytopathogen.

3 CONCLUSION

In the present study *Lentinus edodes* has been found possessing phytocompound which makes capable of conferring defense against fungal phytopathogens by inhibiting mycelial growth. The present findings may be further explored *in vivo* models to substantiate the therapeutic use of *Lentinus edodes* for phytopathogens.

For controlling the phytopathogen and protection of loss in yield, application of chemical fungicides is common practice. As it controls disease, it has some negative effects on our environment including human beings. It necessitates the need of scientific research for seeking an alternative of chemical fungicides to control plant diseases and the present study is an attempt in this direction dealing with the *in vitro* antifungal activity of *Lentinus edodes* mushroom extract against phytopathogen *Alternaria triticina.*

ACKNOWLEDGEMENTS

The author likes to acknowledge Dr.Sunil Kumar Choudhary, Head, University Department of Botany and Dr.Prabhat Kumar Roy, Head, University Department of Biotechnology, Tilka Manjhi Bhagalpur University Bhagalpur for providing lab and library facilities.

REFERENCES

Bisen P S, Baghel R K, Sanodiya B S, Thakur G S and Prasad G. B. 2010 *Lentinus edodes*: A macro fungus with pharmacological activities. Current Medicinal Chemistry 17(22): 2419–30.

Hobbs, C. Medicinal value of *Lentinus edodes* (Berk.) Sing. 2000 A literature review. International Journal of Medicinal Mushroom 2: 287–02.

Khan, A; Hsiang, T. 2003. The infection process of *Colletotrichum graminicola* and relative aggressiveness on four turf grass species. Canadian Journal Microbiology, Ottawa, 49(7): 433-442.

Kitzberger, C.S.G; Smania, A.; Pedrosa, R.C. et al. 2007. Antioxidant and antimicrobial activities of shiitake (*Lentinulaedodes*) extracts obtained by organic solvents and supercritical fluids. Journal of Food Engineering **80**, 631–638.

Komemushi, S., Yamamoto, Y., Fujita. T. 1996. Purification and identification of antimicrobial substances produced by *Lentinus edodes*. Journal of Antibacterial and Antifungal Agents, 24, 21–25.

Lindequist, U., Niedermeyer, T.H.J., Julich, W.D.2005. The pharmacological potential of mushrooms Electronic centralised aircraft monitor 2, 285–299.

Luo, D-Q: Wang, F; Bian, X-Y; Liu, J.-K.2005. Rufuslactone, a new antifungal sesquerterpene from the fruiting bodies of the Basidiomycete Lactarius rufus. Journal of Antibiotics 58, 456-459.

Nene, Y.L. and B.W. Thapliyal 1979. Fungicides in plant disease control. Oxford & IBH Publisher house New Delhi. 425.

Ngai, Phk; Ng, Tb. (2003), Lentin, a novel and potent antifungal protein from shitake mushroom with-inhibitory effects on activity of human immunodeficiency virus-1 reverse transcriptase and proliferation of leukemia cells. *Life Sci.*, **73**, 3363–3374.

Ohga, S. 1992 Adaptability of *Lentinu sedode* strains to a sawdust-based cultivating procudere. *Mokuzai-Gakkaishi*, 38, 301-309.

Piccinin, E. 2000 Potencial de preparações do cogumelo comestível "shiitake" (Lentinula edodes) no controle de fitopatógenos fúngicos, bacterianos e virais em sorgo, maracujá e fumo162 p. Tese (Doutorado em Fitopatologia) – Escola Superior de Agricultura "Luiz de Queiroz", Universidade de São Paulo, Piracicaba.

Prabhu AS, Prasada R, 1966. Pathological and epidemiological studies on leaf blight of wheat caused by *Alternaria triticina*. Indian Phytopathology, 19:95-111.

Prasada R, Prabhu AS, 1962. Leaf blight of wheat caused by a new species of Alternaria. Indian Phytopathology, 15:292-293.

Punja, Z.K.; Utkhede R.S. Using fungi and yeasts to manage vegetable crop diseases. Trends Biotechnology, Phyladelphia, v. 21, p.400-407, 2003.

Quang, D.N., Bach, D.D., Hashimoto,T., Asakawa, Y.Chemical constituents of the Vietnamese inedible mushroom Xylaria intracolorata Nat.Prod.Res,2006,20,317-321.

Royse, D.J. (2001), *Cultivation of Shiitake on Synthetic and Natural Logs*. College of Agricultural Sciences,Cooperative Extension, Pennsylvania State University, University Park, PA, USA, 12p.

Sasaki, S.H.; Linhares, R.E.C.; Nozawa, C.M. et al. (2001), Strains of *Lentinulaedodes*suppress growth of phytopathogenic fungi and inhibit Alagoas serotype of vesicular stomatitis virus. *Braz. J.Microbiol.*, **32**, 52-55.

Tokimoto, K., Fujita, T., Takaishi, Y. (1987), Increased or induced formation of antifungal substance in cultures of *Lentinusedodes*by the attack of *Trichoderma ssp. Proc. Japan Acad. Series B. Phys.and Biol. Sciences*, **63**, 277-280.

Biotechnology and Biological Sciences – Sen et al. (Eds)
© 2020 Taylor & Francis Group, London, ISBN 978-0-367-43161-7

In-vitro antioxidant, LC-MS analysis for bioactive compounds of *Artemisia nilagirica*

P. Parameswari
Sathyabama Institute of Science and Technology, Chennai

R. Devika
Aarupadai Veedu Institute of Technology, Paiyanoor

ABSTRACT: *Artemisia nilagirica* (Clarke) Pamp were collected from Theni district. In the present investigation, the leaf extract of Artemisia nilagirica (clarke) pamp was subjected to column chromatographic separation and the fractions were subjected to thin layer chromatography. The Rf value was authenticated to be 0.71 for flavonoids from *Artemisia nilagirica*. In our work study, petroleum ether, ethyl acetate, methanol and aqueous extracts and standard L. Ascorbic acid of *Artemisia nilagirica* (Clarke) Pamp have been evaluated for in vitro antioxidant activity at different concentration of 20, 40, 60, 80, 100 respectively. The IC50 value for DPPH inhibition of methanol (81.68 μg/ml), ethyl acetate (69.73 μg/ml), Aqueous (77.37 μg/ml), petroleum ether (65.41 μg/ml) and L Ascorbic acid (85.39 μg/ml) respectively. In LC-MS of the active compound of quercetin gave (M-H)- at M/Z 301. For showing a highest peak and molecular weight to be 302. The results of this study also provide a scientific support to medicinal uses of *Artemisia nilagirica* in Theni. The In Vitro antioxidant activitites could be accredited to the highest peak of active components identified in *Artemisia nilagirica* leaves by LC-MS analysis.

Keywords: *Artemisia nilagirica*, Column Chromatography, Antioxidant activity, LC-MS Analysis

1 INTRODUCTION

Artemisia nilagirica (Clarke) Pamp alluded to locally as "Indian Wormwood", has a spot with the family Asteraceae. It is a helpful plant that has been used for more than ten years to treat sickness and manifestations, for instance, jungle fever, irritation, diabetes, stress, melancholy, diabetes, malignancy and numerous other microbial ailments [1]. Cell reinforcements are significant in the counteractive action of human maladies. Normally happening cell reinforcements in verdant vegetables and seeds, for example, ascorbic corrosive, flavonoids mixes have the capacity to lessen the oxidative harm related with numerous ailments, including malignancy, cardiovascular illness, waterfalls, atherosclerosis diabetes, joint pain, and insusceptible inadequacy sicknesses and maturing [2-6]. Free radicals are synthetics species which contain at least one unpaired electrons because of which they are exceedingly precarious and cause harm to different atoms by extricating electrons from them so as to achieve steadiness [7]. Cancer prevention agent assumes a noteworthy job in protecting our body from ailment by decreasing the harm to cell segment brought about by ROS [8]. Fluid Chromatography/Mass spectrometry (LC/MS) is quick creating and it's the favored apparatus of fluid chromatographers. Fluid chromatography-mass spectroscopy (LC-MS/MS) is a strategy that utilizations fluid chromatography (or HPLC) with the mass spectroscopy. It is a diagnostic science system that consolidates the physical detachment abilities of fluid chromatography with mass examination capacities of mass spectroscopy. LC-MS is usually utilized in research facilities for the subjective and quantitative examination of medication substances, sedate items and organic examples [9].

2 MATERIALS AND METHODS

2.1 *Column chromatography*

The crude methanolic leaf extract (20g) was mixed (adsorbed) with silica gel (100-200 mesh) (Merck) and chromatographed on a silica gel initially eluting with continuous suitable solvent system and gradually increasing the polarity mixture of solvent (Hexane, Hexane: Chloroform, Chloroform, Chloroform: Ethyl acetate, Ethyl acetate, Ethyl acetate: Ethanol). The fractions were eluted using TLC and similar TLC pattern were pooled in to major fraction to obtain (126) a pale greenish amorphous powder. 20 grams of the Sample were chromatographed over silica gel column (100 – 200 mesh). The mixture was packed on a silica gel column (Merck, India) and eluted with 100% hexane and then solvents were added in the increasing order of their polarity namely Chloroform,Ethyl acetate and Ethanol in the ratios of 90:10, 80:20,70:30, 60:40 and 50:50. Based on TLC profile, the eluates were pooled into some fractions. Column fraction126 (Ethyl acetate:Ethanol – 50:50) gave a solid which was crystallized from methanol to yield a pale greenish amorphous powder (Yield: 125mg).

2.2 *Thin layer chromatography*

Thin-layer chromatography (TLC) is the simplest and cheapest method of detecting plant constituents [15]. The thin layer chromatography was developed in twin through chamber with silica gel 60 F254 Pre coated aluminium plate of 0.2 mm thickness using ethyl acetate: methanol (1:1) as the developing solvent system. Rf values were calculated and visualized by dipping the plate in vanillin sulphuric acid (1%) and heated on 105° C and the colour of the spot appeared distinctly under visible light, short UN 245 nm and long UV 365 nm.

2.3 *Liquid Chromatography – Mass Spectroscopy (LC-MS)*

Electrospray is produced by applying a strong electric field to a liquid passing through a capillary tube with a weak flux. This produces charged large droplets which is then subjected to solvent evaporation. The mass spectra operated in the positive turbo ion spray (ESI) mode. Chromatographic separation was achieved on a phenomenex C18 reversed phase column with an ID of 5 μm and dimension of 50 m x 4.68 mm. 10 micro litre samples were injected using a Shimadzu auto-sampler fitted with a Hamilton 100-μl syringe. Different gradient of mobile phase composition of 0.1% formic acid in water and acetonitrile at a flow rate of 1 minute were used. The column oven temperature was operated at room temperature. By using ESI high mass sample, non-volatile molecules, liquids can be ionized and disadvantage of this source of ionization is poor sensitivity, low fragmentation and source is instable [10].

2.4 *In vitro antioxidant activity*

The reaction mixture (3.0ml) consisting 1.0 ml DPPH in methanol, ethyl acetate, petroleum ether and aqueous (0.1mM), 1.0ml and 1.0 ml different concentration of the Crude fraction (20, 40, 60, 80, and 100 μg/ml) was incubated in dark for 10 min, and the OD was measured at 517nm against blank. For control, 1.0 ml of methanol, ethyl acetate, petroleum ether and aqueous was used in place of crude fraction. L- Ascorbic acid was used as positive control [11]. Percentage inhibition of DPPH was calculated using the formula.

Inhibition (%) = OD of Control – OD of Experiment/OD of Control x 100

The data of antioxidant were subjected to statistical analysis and the mean and SE for five individual observations was calculated. The significance of the sample mean was tested by Two Way ANOVA using SPSS software. The differences were considered s significant at $p<0.05$ level [12].

3 RESULT AND DISCUSSION

3.1 *Liquid Chromatography – Mass Spectroscopy (LC-MS)*

The methanolic extracts was analysed by LC-MS techniques which allowed the identification of active compound of quercetin. The bioactivity guided fractionation of the methanol extract on silica gel column. Gave the active fraction eluted with chloroform: Methanol [7:3] (Figure 1). The active fraction contains the flavonoid quercetin. Quercetin has been reported to inhibit the oxidation of other molecules and is therefore classified as an antioxidant. Quercetin contains a polyphenolic chemical structure that stops oxidation by eliminating the free radicals responsible for oxidative chain reactions [13]. TLC of the fraction over silica gel with ethyl acetate: Methanol 4:1 gave a single spot (light yellow) to observe in visible light, long and short UV, Rf = 0.7. The spot turned bluish green with 0.1% alcoholic ferric chloride. LC-MS of the compound in negative mode gave (M- H^-) at M/Z 301.0.

3.2 *In vitro antioxidant activity*

The DPPH activity of aqueous extract, ethyl acetate, methanol and petroleum ether compared with L-Ascorbic acid (Standard) and the data are presented in Table 1. The DPPH inhibition was recorded as 14.06 % and -10.24 % at 20µg/ml concentration of methanol and aqueous. Similarly, The DPPH inhibition was recorded as -11.30 % and -10.54 % at 20 µg/ml concentration of ethyl acetate and Petroleum ether. As the concentration increased DPPH inhibition increased to -73.65 % and -59.69 % at 100µg/ml of methanol and aqueous. Similarly, the concentration increased DPPH inhibition also increased to -57.14 and -73.43 at 100µg/ml ethyl acetate and petroleum ether. The percent inhibition of DPPH was directly proportional to the concentration. Likewise, DPPH activity of standard ascorbic acid showed a gradual inhibition from -22.52 % to 75.43 % as the concentration of L-Ascorbic acid increased from 20µg/ml to 100µg/ml. The inhibition of DPPH in methanol fraction was more or less equal to that of the standard L Ascorbic acid [14].

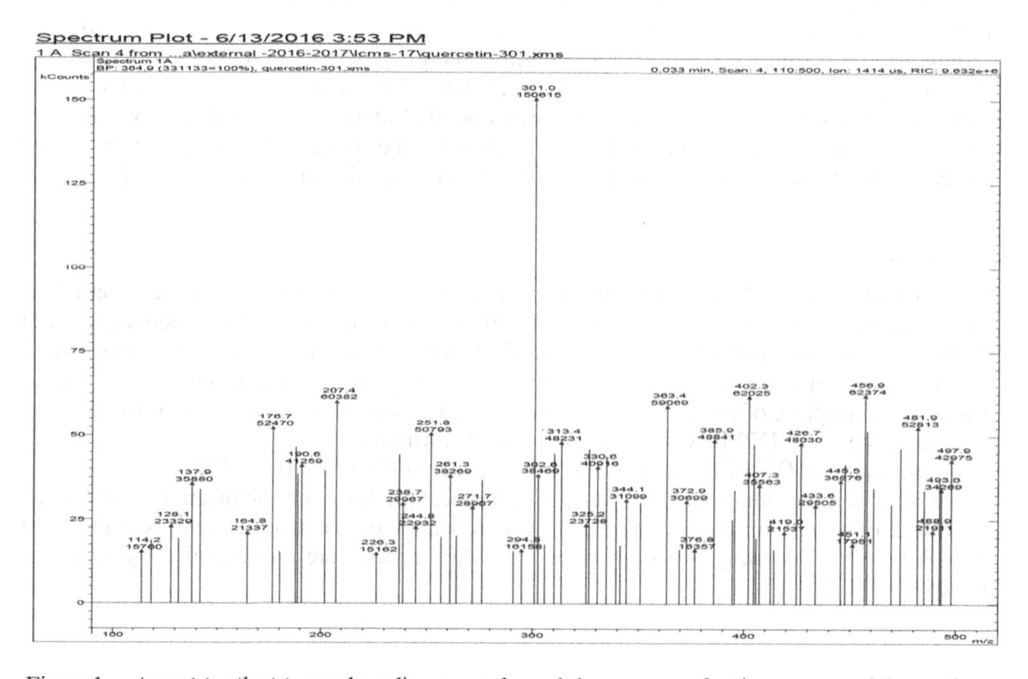

Figure 1. *Artemisia nilagirica* methanolic extract showed the presence of active compound Quercetin.

Table 1. DPPH activity of *Artemisia nilagirica* methanolic extract of various solvent.

Concentration μg/ml	Methanol	Aqueous	Ethyl acetate	Petroleum ether	Standard (L- Ascorbic Acid)
Control (0)	1.212±0.000	1.212±0.000	1.212±0.000	1.212±0.000	1.212±0.000
20	0.914±0.005 (-14.06)	0.955±0.002 (-10.24)	0.975±0.004 (-11.30)	1.005±0.003 (-10.54)	0.924±0.003 (-22.52)
40	0.877±0.007 (-26.97)	0.872±0.004 (-17.98)	0.846±0.005 (-20.45)	0.835±0.004 (-12.06)	0.813±0.004 (-32.98)
60	0.722±0.006 (-41.72)	0.704±0.004 (-33.77)	0.524±0.003 (-31.89)	0.521±0.003 (-22.77)	0.792±0.003 (-44.29)
80	0.641±0.004 (-53.82)	0.594±0.004 (-44.17)	0.604±0.003 (-43.23)	0.483±0.005 (-35.80)	0698±0.002 (-57.89)
100	0.359±0.008 (-73.65)	0.425±0.004 (-59.69)	0.456±0.003 (-57.14)	0.527±0.005 (-50.43)	0.482±0.003 (-75.43)

Values are mean ± SE of five individual observations.
Values in parenthesis are per cent change over standard
+ denotes per cent increase over standard.
*denotes values are significant at $p<0.05$.

Table 2. Extract IC_{50} value of DPPH activity of *Artemisia nilagirica* methanolic extract.

S.No	Samples	Concentration (μg/ml)
1.	L- Ascorbic Acid	85.39
2.	Methanol	81.68
3.	Aqueous	77.37
4.	Ethyl acetate	69.73
5.	Petroleum ether	65.41

Statistical significance as verified by two ways ANOVA revealed that the data were significant at 5% level. The median inhibitory concentration (IC_{50}) of Methanol, ethyl acetate, aqueous and Petroleum ether and L Ascorbic acid was found to 81.68 μg/ml, 69.73 μg/ml, 77.37 μg/ml, 65.41μg/ml and 85.39 μg/ml respectively. The results clearly indicate that methanol extract – based biosynthesized *Artemisia nilagirica* Showed profound DPPH reducing activity against stable free radicals as shown in Table 1 and 2 respectively.

4 CONCLUSIONS

Subsequently it tends to be inferred that the *Artemisia nilagirica* (Clarke) Pamp could be pharmaceutically exploited for antioxidant properties by LC-MS analysis. For future work to concentrate an *In vitro* anti-inflammatory activity and anticancerous by utilizing a Colo320 DM cell lines for inflammatory bowel disease (IBD).

REFERENCES

Andlauer, W & Furst, P. 1998. Antioxidative power of phytochemicals with special reference to cereals. *Cereal Foods World* 43(2): 356–359.

Andrew, M. 2007. High-lights in the evolution of phytochemistry. 50 years of the Phytochemical *Society of Europe* 68(22): 2786–2792.

Dario, M, Cabezas, B, Diehl, W.K & Mabel, C.T. 2013. Emulsifier and antioxidant properties of by-products obtained by enzymatic degumming of soyabean oil. *European Journal of Lipid Science and Technology* 15(1): 659–667.

Huda, A.W, Munira, M.A, Fitrya, S.D & Salmah, M. 2009. Antioxidant activity of Aquilaria malaccensis (Thymelaeaceae) leaves. *Pharmaceutical Research* 1(2): 270–273.

Jayapraksha, G.K, Selvi, T & Sakariah, K.K. 2003. Antibacterial and antioxidant activities of grape seed extract. *Food Research International* 36(3): 177–122.

Kumar, P.R, Dinesh, S.R & Rini, R. 2016. LC-MS- a Review and a recent update. *World Journal of Pharmacy and Pharmaceutical Sciences* 5(5): 377–391.

Lee. K.G, Mitchell, A.E & Shibamoto, T. 2000. Determination of antioxidant properties of aroma extracts from various beans. *Journal of Agricultural Food Chemistry* 48(1): 4817–4820.

Leong. C.N, Tako, M, Hanashiro, I & Tamaki, H. 2008. Antioxidant Flavonoids glycosides from the leaves of Ficus pumila L. *Food Chemistry* 109(2): 415–420.

Middleton, E, Kandaswamy, C & Theoharides, T.C. 2000. The effects of plant flavonoids on mammalian cells: implications for inflammation, heart disease and cancer. *Pharmacological Review* 52(2): 673–751.

Nishanthi, M, Vijay Aanandhi, M, Azhagesh, Raj K & Vijaya kumar, B. 2012. Evaluation of In Vitro Anti-inflammatory activity of methanolic leaf extract of Vigna radiate (L). Wileze. *International Journal of Pharmacological Screening methods* 2(2): 89–90.

Paramita, B & Camelia M. 2016. *In Vitro* Antioxidant activities and polyphenol contents of seven commercially available fruits. *Pharmacognosy Research* 8(4),256–264.

Petra, T, Blaz, C, Tomaz, P, Janez, H & Tomaz Pozrl. 2016. LC-MS analysis of phenolic compounds and antioxidant activity of buck wheat at different stages of malting. *Food chemistry* 210(2): 9–17.

Pietta, P, Simonetti, P & Mauri, P. 1998. Antioxidant activity of selected medicinal plants. *Journal of Agricultural Food Chemistry* 46(2): 4487–4490.

Subramani, P, Anish, R, Subramani, B, Selvadurai, M, Kalaimani Jayaraj, K &Venugopal, V. 2014. An overview of Liquid Chromatography- Mass Spectroscopy instrumentation. *Pharmaceutical Methods* 5 (2): 47–55.

Walter, H.L, Memory Elvin, L.P.F. 2013 Medicinal Botany, 2nd (edn) pp. 345-350. Parameswari: Devika.

Biotechnology and Biological Sciences – Sen et al. (Eds)
© 2020 Taylor & Francis Group, London, ISBN 978-0-367-43161-7

Screening, evaluation and *in silico* modelling and docking studies of isolated fungal cellulases for enhanced saccharification of lignocellulosic grass (*Pennisetum* sp.) biomass for biofuel production

Hrudayanath Thatoi
Department of Biotechnology, North Orissa University, Baripada, India

Sonali Mohapatra
Department of Biotechnology, College of Engineering and Technology, Biju Patnaik University of Technology, Bhubaneswar, India

Manish Paul
Department of Biotechnology, North Orissa University, Baripada, India

ABSTRACT: Cellulases are a group of enzymes that catalyzes the hydrolysis of cellulose by breaking the 1, 4-β-glycosidic bonds in between the cellulose chain of lignocellulosic biomass the most abundant natural biopolymer on earth to smaller sugar components such as glucose subunits which can be microbially fermented to alcohol for biofuel production. The enzyme cellulases are generally produced by several types of microorganisms including bacteria and fungi. Cellulase is expensive and contributes only 50% to the overall cost of hydrolysis during biofuel production. Besides, this enzyme has wide application in food, brewery, wine, animal feed textile, laundry, paper, and pulp industries. In the recent years, there has been much research aimed at obtaining new microorganisms capable of producing cellulase enzymes with higher specific activities and greater efficiency. Keeping these in view, in, the present study eleven cellulolytic fungi were isolated from the lignocellulose waste containing soil samples collected from four different areas of Bhubaneswar, Odisha using fungal basal medium containing 1% CM cellulose following dilution plate techniques. These fungi were screened for their CM Case and FPase activity. A fungal strain was found to exhibit high cellulase activity was identified as *Aspergillus fumigatus*. Further *in silico* studies like docking was undertaken to analyse the activity of cellulase enzyme.

Keywords: Cellulase, Fungi, CMC, FPase, Glucose, Protein estimation, Molecular docking, Binding energy

1 INTRODUCTION

Grass is the world's cheapest lignocellulosic biomass which has potential for use in biofuel production. In general, grasses are perennial crop that can grow throughout every region of the world and throughout the seasons and can be cultivated in marginal lands. There has been increasing interest for producing bioethanol from grass feedstock because of high yields, low costs, good suitability for low quality land, and low environmental impact (Rabelo et al., 2008). Cellulose is the main structural constituent in plant cell walls including grass species and is found in an organized fibrous structure present in both crystalline and amorphous forms. The celluloses present in lignocellulosic biomass are hydrolysed by enzyme cellulase to produce sugars which fermented into bioethanol using yeasts/bacteria (El-Naggar *et al.*, 2014). Cellulases are a group of hydrolyzing enzymes produced primarily by fungi and bacteria are employed for hydrolyzing cellulose present in lignocellulosic biomass to smaller molecule like glucose. Among these, the fungi are reported to produce maximum cellulase *in vitro* (Rana and Kaur, 2012). However, for cellulose production, pretreatment is an essential step which can be carried out through different physical, chemical and biological approaches.

Enzymatic (cellulose) hydrolysis is a major cost driver in bioethanol production due to the high cost of enzyme production (Klein-Marcuschamer, 2012). Thus research has been focused in enzyme production from potent cellulose producing microorganisms. Keeping these in view, attempt has been made to isolate cellulase producing fungi from soil samples and screened for their cellulase activity. Based on the screening test, an efficient cellulase producing fungus *Aspergillus fumigates* was identified and both its CM cellulase and FPase activities were evaluated. Further, an i*n silico* molecular docking approach has been adopted to evaluate the mode of interaction and affinity between cellulase from the *A. fumigatus* and a disaccharide substrate obtained from lignocellulosic biomass.

2 MATERIALS AND METHODS

2.1 *Isolation and screening of cellulolytic fungi*

The soil samples were collected from soils containing lignocellulosic wastes from different locations in Bhubaneswar which comprises soil containing decomposed plant materials, soil containing decomposed kitchen waste, Soil containing decomposed lignocellulosic waste from grass biomass and soil containing decomposed potato wastes. The soil samples were serially diluted with sterile tap water. The screening of the potent cellulolytic fungus was done using two step methods. In the first part all the twenty isolated fungi were evaluated for their zone of clearance in 1 % CMC agar plates. After seven days of inoculation, the culture plates were flooded with aqueous solution of 1% Congo red for 15 minutes at room temperature and then the excess dye was washed gently with NaCl. The observation was made to note the substrate utilization zone around the colony. Further, the micro-organisms showing the zone of clearance were screened for their cellulase activity (CMCase and FPase).

2.2 *Morphological and molecular identification*

The isolated fungi, was identified through the morphological studies carried out by using lactophenol cotton blue solution as per the protocol of Leck (1999) (Himedia, India). For molecular identification, genomic DNA was isolated from fruiting body of fungus following the CTAB method (with minor modifications) as described by Zhou *et al.*, (2008). Further, DNA quantification was carried out by taking 30 ng DNA/μL. Amplification of 25S~28S region of the isolated DNA was carried out by using universal primers LROR-(ACCCGCTGAACTTAAGC) and LR3 (GGTCCGTGTTTCAAGACGG).

2.3 *Fermentation for cellulase production*

For evaluating the cellulase activity, seven fungal isolates were inoculated in sterilized fungal basal medium (2.0 g KH_2PO_4, 0.3 g urea, 0.3 g $MgSO_4.7\ H_2O$, 0.3 g $CaCl_2$, 5 mg $FeSO_4.\ 7\ H_2O$, 1.6 mg $MnSO_4.H_2O$, 1.4 mg $ZnSO_4.7\ H_2O$ and 1.5 mg $CoCl_2.6H_2O$ in 1000 ml distilled water).

2.4 *Determination of CMCase and FPase activity*

CMCase activity was assayed using DNS method. For determination of the quantity of cellulolytic enzyme required to hydrolyze lignocellulosic biomass obtained from *Pennisetum* sp., Filter Paper Activity/Assay (FPA) was used.

2.5 *Docking of cellulase from* Aspergillus fumigatus *with disaccharide*

Molecular docking between the cellulase from *Aspergillus fumigatus*and disaccharide (cellulose) was done by knowledge based grid information obtained from the crystal structure (PDB ID: 4V20) (Moroz *et al.*, 2015).

2.6 *Binding energy calculation*

The binding free energy between the *A. fumigatus* cellulase and its disaccharide ligand was calculated using ACFIS server with the algorithms such as Poisson-Boltzmann surface area (MMPBSA), molecular mechanics generalized born surface area (MMGBSA) and linear interaction energy (LIE).

3 RESULTS AND DISCUSSION

3.1 *Isolation and pure culture of cellulose degrading fungi*

The isolation cellulolytic fungi was carried out from soil samples containing lignocellulosic wastes using fungal basal medium containing 1% CM cellulose following dilution plate techniques. A total of eleven fungal colonies were isolated and pure cultures were prepared (Figure 1).

3.2 *Screening of fungal isolates in CMC agar plates*

As a preliminary qualitative study, the isolated fungal cultures were screened for its cellulase activity by Congo red assay in terms of formation of zone of clearance in 1 % CMC agar plates. The fungi were evaluated for their cellulase producing ability based on the diameter of the zone of clearance. Out of the eleven isolates only seven fungal isolates showed zone of clearance in 1% CMC agar as shown in Figure 2. The fungi were named cellulose degrading fungi (CDF) and rated to be negative, good and very good for their as per their cellulase degrading capacity as shown in Table 1. Negative sign indicates that the organism did not showed any zone of clearance while good and very good indicates the ability of the fungus to hydrolyse the cellulose.

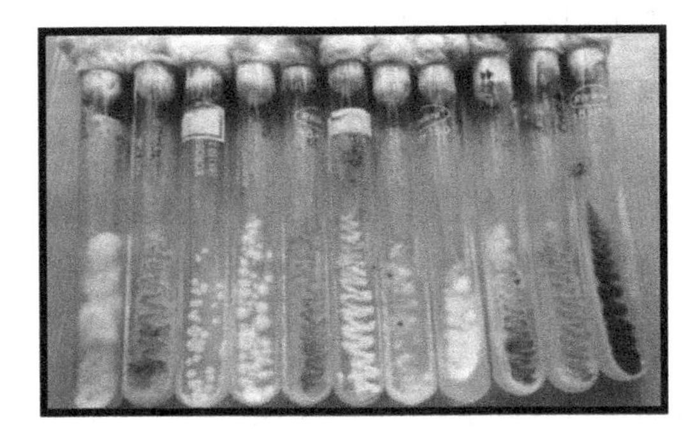

Figure 1. Pure cultures of the isolated fungi.

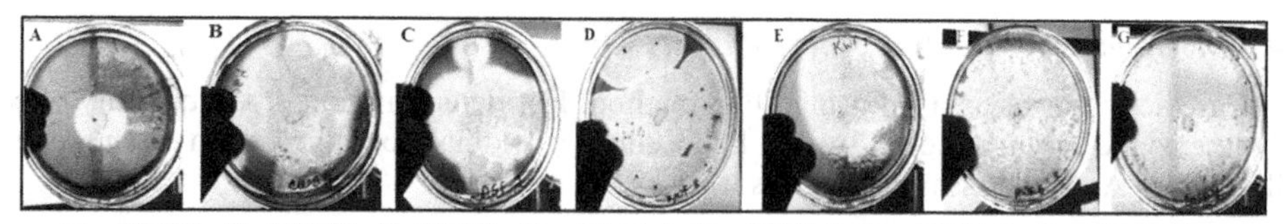

(A–CDF1, B–CDF2, C-CDF3, D– CDF 4, E– CDF 5, F– CDF 6, G– CDF 7).

Figure 2. Zone of clearance in CMCase agar plates for the seven fungal isolates.

Table 1. CMCase activity profile of seven fungal isolates.

Isolated microorganism	Day-2	Day-4	Day-6	Day-8	Day-10
CDF-2	0.08 ± 0.003	0.155 ± 0.15	0.41 ± 0.14	0.33 ± 0.32	0.29 ± 0.09
CDF -6	0.002 ± 0.002	0.162 ± 0.012	0.88 ± 0.26	0.66 ± 0.36	0.48 ± 0.023
CDF -7	0.088 ± 0.02	0.4 ± 0.03	0.59 ± 0.13	0.37 ± 0.15	0.21 ± 0.12
CDF -8	0.002 ± 0.001	0.21 ± 0.23	0.2 ± 0.43	0.59 ± 0.27	0.4 ± 0.25
CDF -9	0.002 ± 0.001	0.17 ± 0.05	0.5 ± 0.22	0.28 ± 0.32	0.11 ± 0.12
CDF -4	0.001 ± 0.03	0.14 ± 0.04	0.42 ± 0.12	0.72 ± 0.28	0.38 ± 0.03
CDF -7	0.103 ± 0.09	0.894 ± 0.23	0.97 ± 0.37	0.76 ± 0.25	0.66 ± 0.14

3.3 *Lactophenol staining*

Morphological identification of seven fungal strains was done microscopically by staining the culture with lactophenol cotton blue dye as shown in Figure 3. The fungal strains exhibited different morphological characteristics. The strains had morphological characteristics that were similar to *Aspergillus, Penicellum* and *Trichoderma* sp. as given in Table 3. The highly branched and lateral side branches that may be paired or not were observed. Further in some strains rebranch of the main branch with the secondary branches often paired with one another was observed. Majority of *Aspergillus, Fusarium, Alternaria, Rhizopus, Penicillum* and *Trichoderma* isolates were found to possess cellulytic activity. A wide range of *Aspergillus* species have been identified to possess all component of cellulase enzyme system (Vries and Visser, 2001).

3.4 *CMCase activity of the cellulolytic fungi*

The CMCase or exoglucanase activity was performed for seven fungal isolates (Table 1) and only two isolates (CDF 4 and CDF 6) were found to produce measurable CMCase in submerged fermentation. The isolate CDF 7 showed the highest CMCase activity (0.97 U/mL) after 6th day of incubation. It was followed by CDF-6 which had an exoglucanase activity of 0.88 U/mL. Another strain CDF 4 showed an exoglucanase activity of 0.72 U/mL on 8th day

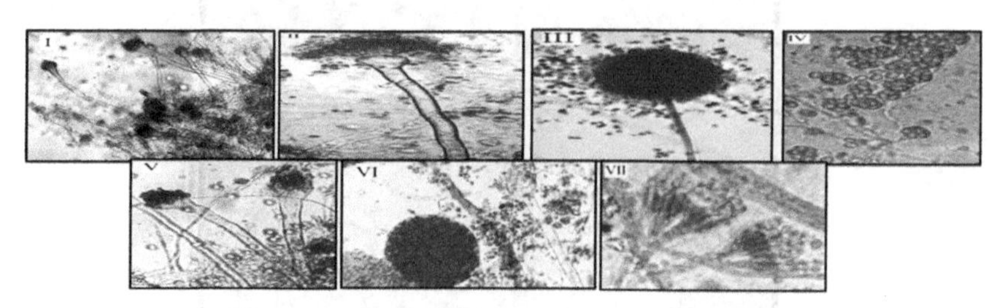

Figure 3. Microphotographs of seven fungal isolates identified as capable of cellulose utilization I) *Aspergillus fumigates,* II) *Aspergillus oryzae,* III) *Aspergillus niger,* IV) *Neosartoria fischeri* V) *Aspergillus flavus,* VI). *Eurotium amestaldomi,* VII) *Penicillium chrysogenum.*

of incubation.The CMCase activity of the other four strains was not evident as compared to these three strains. Philippidis (1994), stated that maximum enzyme activity in fungal species was noted after 5 days of incubation.

3.5 *FPAse activity of the cellulolytic fungi*

The FPase or endoglucanase activity was performed for seven fungal isolates (Table 2) and similar to CMCase activity. The isolates CDF 4 was found to produce highest FPase activity in submerged fermentation. The isolate CDF 4 showed FPase activity of 0.4U/mL after 6th day of incubation after which a decline in the activity was observed. The FPase activity was followed by CDF-2, CDF-6 and KWF-5 also had similar endoglucanase activity of 0.3 U/mL. All the strains showed a decline in the trend after 6th day except strain CDF-1 which showed a slight increase in the activity on the 8th day of incubation. In the study of Kale and Zanwar (2016) decrease in the CMCase activity in case of four fungal species belonging to two genera i.e. *Trichoderma* and *Aspergillus* was observed after the sixth day and the reason they described might be due to the depletion in carbon source.

Saccharification of alkali pretreated lignocellulosic biomass such as *Pennisetum* species (Dinanath) yielded 297.4 mg/g of gluocose and 98.5 mg/g of xylose using partial purified CMCase and FPase from *A. fumigatus* in equal concentration (1:1) against 504.4 mg/g glucose and 157.6 mg/g xylose when a commercial enzyme, Palkonal MBW (Map Enzyme Pvt. Ltd, Gujurat) was used.

3.6 *Three dimensional structure of disaccharide bound* A. fumigatus *cellulase*

Structure prediction of the cellulase of *A. fumigatus* conducted in the present study shown that the enzyme retain a typical α/β structure. The disaccharide substrate in this cellulase enzyme shown to bind with a site consists of mostly charged residues. One hexose sugar ring remains in well packed under the core region of the enzyme while the other ring remains projected outside the binding groove (Figure 4). Moroz et al. (2015) crystallized the three-dimensional

Table 2. FPase activity profile of seven fungal isolates.

Isolated microorganism	Day-2	Day-4	Day-6	Day-8	Day-10
CDF-1	0.0004 ± 0.009	0.074 ± 0.017	0.1 ± 0.021	0.1 ± 0.034	0.22 ± 0.024
CDF-2	0.001 ± 0.001	0.07 ± 0.02	0.39 ± 0.21	0.3 ± 0.11	0.07 ± 0.002
CDF-3	0.001 ± 0.12	0.037 ± 0.22	0.2 ± 0.18	0.1 ± 0.21	0.06 ± 0.018
CWS-4	0.001 ± 0.001	0.25 ± 0.18	0.4 ± 0.33	0.24 ± 0.25	0.1 ± 0.022
CDF-5	0.001 ± 0.001	0.18 ± 0.05	0.3 ± 0.24	0.1 ± 0.22	0.1 ± 0.18
CDF-6	0.004 ± 0.25	0.12 ± 0.21	0.32 ± 0.32	0.1 ± 0.23	0.01 ± 0.033
CDF-7	0.007 ± 0.001	0.22 ± 0.001	0.2 ± 0.006	0.1 ± 0.008	0.08 ± 0.003

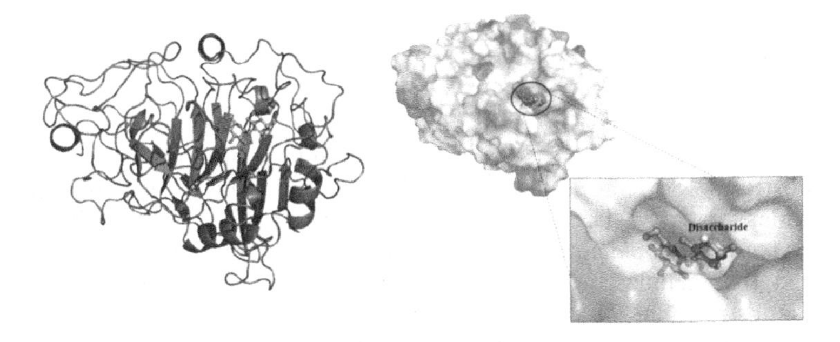

Figure 4. Three dimensional structure of disaccharide bound cellulase from *Aspergillus fumigatus* (Left). Electrostatic surface potential of the complex between *Aspergillus fumigatus* cellulase and disaccharide and the binding site of substrate (right).

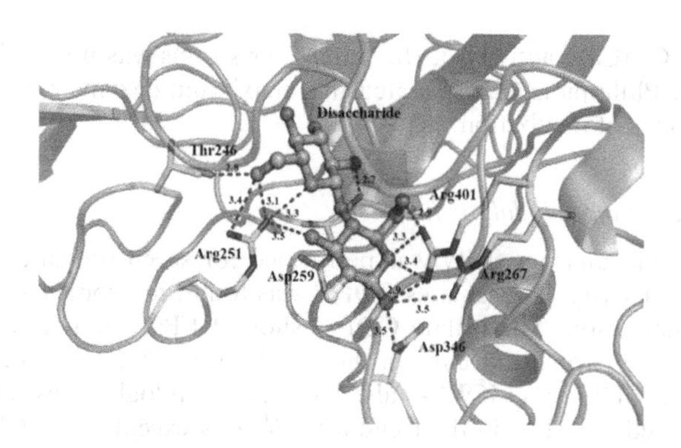

Figure 5. Molecular interaction between *Aspergillus fumigatus* cellulase and disaccharide.

structure of the enzyme cellobiohydrolase Cel7A from *A. fumigatus* both in the form of substrate unbound and bound forms at 1.8 Å and 1.5 Å resolutions respectively.

3.7 *Molecular interaction between* Aspergillus fumigatus *cellulase and disaccharide*

Analysis of molecular interactions between *Aspergillus fumigatus* cellulase and disaccharide revealed that there are six residues, Thr246, Arg251, Asp259, Arg267, Asp346, and Arg401 in the enzyme which mainly interacts with the ligand disaccharide. Moroz et al. (2015) in their study showed one difference between the substrate unbound and bound structure is that a loop movement containing the disaccharide binding residue Thr246. In their study, it was reported that the main chain in the region Thr246-Tyr247-Ser248 of this loop has an alternative conformation, partially moving away upon ligand binding.

3.8 *Binding energy between* Aspergillus fumigatus *cellulase and disaccharide*

Binding affinity between *Aspergillus fumigatus* cellulase and disaccharide was evaluated according to different energetic parameter like enthalpy (ΔH), entrophy ($-T\Delta S$), Gibbs free energy (ΔG), and ligand efficiency (LE). The value of enthalpy, entrophy, Gibbs free energy, and ligand efficiency of the enzyme-substrate complex was obtained as -45.010, 20.947, -24.063 Kcal/mol and 1.956 respectively (Table 7). From these different energetic values it can be concluded that disaccharide substrate has strong affinity with *Aspergillus fumigatus* cellulase. Sambasivarao et al. (2014) in their study identified the enzymatic mode of action for different family of cellulase enzymes with substrates by means of docking calculations. Comparative results from this docking calculation revealed that the average binding energies for cellobiose bound conformations at reducing end sub-sites are lower in Cel7B (PDB: 1EG1) compared to the non-reducing sub-sites in other Cel7B enzymes (PDB: 1EG1 and 2A39).

4 CONCLUSION

Fungi are considered as the primary decomposers of soils and they possess a number of advantages than bacteria and yeast. They are capability of utilising a wider variety of cellulosic substrates. Seven cellulase producing microorganisms isolated and evaluated for their cellulase activity. The fungal culture exhibiting highest cellulolytic activity was identified as *Aspergillus fumigatus*. Molecular docking study and binding energy analysis also suggested that cellulase of *Aspergillus fumigatus* have significant interactions with the cellulose substrate like disaccharide. Therefore it can be inferred that *Aspergillus* sp. has great potential for cellulolytic degradation and further improvement in the strain by mutation can be helpful for its effective use in enhanced saccharification of lignocellulosic biomass.

REFERENCES

Rabelo, S.C., Filho, R.M., Costa, A.C. (2008). A comparison between lime and alkaline hydrogen peroxide pretreatments of sugarcane bagasse for ethanol production. Applied Biochemistry and Biotechnology, 144(1):87–100.

El-Naggar, N.E., Deraz S. and Khalil, A. (2014). Bioethanol Production from Lignocellulosic Feedstocks Based on Enzymatic Hydrolysis: Current Status and Recent Developments. Biotechnology, 13: 1–21.

Rana, S., Kaur, M. (2012). Isolation and screening of cellulose producing microorganisms from degraded wood. IJPBSF. 2(1):10–15.

Klein-Marcuschamer, D., Oleskowicz-Popiel, P., Simmons, B.A., Blanch, H.W. (2012). The Challenge of Enzyme Cost in the Production of Lignocellulosic Biofuels. Biotechnology and Bioengineering, 1–5.

de Vries R.P., Visser J. (2001). Aspergillus enzymes involved in degradation of plant cell wall polysaccharides. Microbiol Mol Biol Rev. 65(4):497–522.

Philippidis, G.P. (1994). Cellulase Production technology. In: Enzymatic Conversion of Biomass for Fuel Production. (Eds): M. E. Himmel et al., ACS symposium series 566.

Kale R.A., Zanwar P.H. Isolation and Screening of Cellulolytic Fungi. IOSR Journal of Biotechnology and Biochemistry (IOSR-JBB). ISSN: 2455-264X, 2 (6) Part: II (2016), PP 57–61.

Leck, A. (1999). Preparation of Lactophenol Cotton Blue Slide Mounts. Community Eye Health, 12, 24.

Zhou, X., Wheeler, M.M., Oi F.M., Bennett, G.W., and Scharf, M.E. (2008). RNA interference in the termite Reticulitermes flavipes through ingestion of double-stranded RNA. Insect Biochemistry and Molecular Biology. 38, 805–815.

Moroz, O.V., Maranta, M., Shaghasi, T., Harris, P.V., Wilson, K.S., and Davies, G.J. (2015). The Three-Dimensional Structure of the Cellobiohydrolase Cel7A from Aspergillus Fumigatus at 1.5 A Resolution. Acta Crystallography., Section. F, 71, 114.

Sambasivarao, S.V., Granum, D.M., Wang, H., and Maupin, C.M. (2014). Identifying the Enzymatic Mode of Action for Cellulase Enzymes by Means of Docking Calculations and a Machine Learning Algorithm. AIMS Molecular Science, 1(1),59–80.

Biotechnology and Biological Sciences – Sen et al. (Eds)
 ISBN 978-0-367-43161-7

Attempts for genetic transformation of cotton through ternary transformation system in *Agrobacterium* mediated transformation- a strategy for improvement of gene transfer method

Nandini Bandyopadhyay
Department of Botany, Dum Dum Motijheel College, India

ABSTRACT: Cotton tissue in culture is a recalcitrant plant system. It was thus necessary to make available of a suitable protocol, efficient enough to cause Agrobacterium mediated transformation. An efficient gene transfer system should have the capacity to result high transformation incidence. The transformation frequency in any plant species is influenced by T- DNA transfer efficiency by the virulent Agrobacterium tumifaciens strain. A critical step in the development of robust Agrobacterium mediated transformation system in recalcitrant cotton is the establishment of optimal conditions for efficient T-DNA delivery into target tissue and recovery of transgenic plants. A dramatic increase in efficiency of T-DNA delivery was achieved by constitutive expression of additional vir genes, a mutant of virG, virGN54D in resident pCAMBIA vector in Agrobacterium strain LBA4404. Taking advantage of this phenomenon, we designed a strategy for ternary transformation system of a plant transformation by providing an additional plasmid containing virGN54D along with standard binary system of Agrobacterium mediated transformation. To start with, the influence of ternary system of transformation was tested by monitoring the expression of gus A gene, as marker. Analysis of stably transformed cell lines showed that, although the T-DNA transfer frequency is greatly enhanced and incidence of higher level of transformation in case of ternary system in comparison to binary system of transformation was observed. Agrobacterium mediated transformation of cotton via somatic embryogenic mode of plant regeneration remains by far the most dependable method of choice. This is due to the reliability and efficiency by which transgenic plants are recovered. Independent transformants showed stable expression of the transformed gus A gene and its seed transmission to the next filial generation.

Keywords: Agrobacterium, transformation; gus A gene, virGN54D, recalcitrant cotton, Ternary transformation system

1 INTRODUCTION

The transgenic approach of plant genetic engineering provides access to an unlimited gene pool for the transfer of desired genes between heterologous organisms, irrespective of their evolutionary or taxonomic relationship. Gene transfer through genetic transformation method, one can transfer any alien gene to any conceivable target crop plant. This facilitates production of agronomically desirable crops that exhibits increased resistance to pests, herbicides, pathogens, environmental stress and enhancement of qualitative and quantitative crop traits. Thus, gene transfer approach carries immense potential in bringing in improvement with additional genetic traits that does not exist in cotton gene

*Corresponding author: banerjee.nandini@gmail.com

pool. Thus, suitable method for gene transfer in cotton figures as the basic requirement for transgenic approach for improvement.

Agrobacterium-mediated transformation of cotton via passage through somatic embryogenesis remains, by far, the most common method of choice. This is due in large part to the reliability and efficiency by which transgenic plants are recovered from the culture, and the fact that this method was the first to be developed and widely adopted in both the private and public sectors (Firoozabady *et al.*, 1987; Umbeck *et al.*, 1987; Rajasekaran *et al.*, 2000). Although limited success in generation of transgenic cotton lines could be achieved, transformation of most commercial strain of cotton proved to be difficult. Recalcitrant to *in vitro* plant regeneration of all of the cotton cultivars turned out to be the major obstacle.

Cotton tissue in culture is a recalcitrant plant system. It was thus necessary to make available of a suitable protocol, efficient enough to cause *Agrobacterium* mediated transformation. An efficient gene transfer system should have the capacity to result high transformation incidence. The transformation frequency in any plant species is influenced by T-DNA transfer efficiency by the virulent *Agrobacterium tumifaciens* strain. Release and transfer functions of T-DNA are mediated by vir proteins encoded by several virulence (*vir*) genes that are present in the Ti/Ri plasmids. High expression of *vir* genes is expected to allow transferring T-DNA more profusely resulting enhanced transformation possibilities. Inducible virulence (*vir*) genes are under the control of a two component regulatory system. In response to environmental factors (phenolic compounds, sugars, pH), virA protein phosphorylates *virG*, which in turn induces with the promoters of other *vir* genes causing their functional activities. A mutation (Hansen *et al*, 1994) of *virG*, *virG*N54D (which codes for an Asn-54 to Asp amino acid change in the product) caused constitutive expression of other *vir* genes independent of *vir*A, the sensor of the two-component system (Pazour *et al.*, 1992). It was observed earlier that T-DNA transfer to tobacco and corn was enhanced in presence of *virG*N54D supplementation.

2 MATERIALS AND METHODS

2.1 *The 2x35S*-gusA-*nos chimeric gene construct:*

A 2 × 35S promoter element was constructed in the laboratory as a *Hind*III/*Bam*HI fragment. This promoter element was fused at the 5′-end of a *gusA* reporter gene present in pCAMBIA 1381 vector that contains a *nos* terminator at *Eco*RI/*Sac*I site of its multiple cloning site. The CAMBIA 1381 inherently contained a selection marker in the form of hygromycin phosphotransferase (*hpt II*) gene under 35S promoter in the T-DNA region.

2.2 *Preparation of competent cells for transformation of* Agrobacterium*:*

Transformation of pCAMBIA 1381 vector constructs into *Agrobacterium* strain LBA 4404 containing the *virG*N54D was carried out following the method of An *et al.*, (1988). The cells were aliquoted into 200μl eppendorf tubes and stored at -70°C freezer.

2.3 *Transfer of plasmid vector containing 2X35S*-gusA-nos *into* Agrobacterium*:*

Agrobacterium transformation was carried out using freeze-thaw method. About 6μg of the plasmid DNA (2X35S-*gusA-nos*) was layered on the top of a frozen 200μl aliquot of LBA4404/VirGN54D cell and gently flicked for mixing. The pellet was suspended in 100μl of broth and spread ontoYEP solid plates, supplemented with appropriate antibiotics together with 50mg/l kanamycin (Sigma) and incubated at 28°C for 3d.

2.4 *Mini plasmid isolation from* Agrobacterium *culture:*

Overnight grown bacterial culture was centrifuged in an eppendrof tube at 5000rpm for 6 min. The plasmid DNA was checked in a 0.8% agarose gel.

2.5 *Verification of the presence of the plasmid:*

Presence of the desired plasmid was checked by digesting the mini DNA of different colonies with *Hind*III/*Eco*RI that dropped the full gene cassette. The DNA was then transferred to the nitrocellulose membrane by overnight capillary transfer. The hybridization and prehybridization protocols were adopted as per Sambrook *et al.*, (1989). Autoradiograms were developed and colonies which signaled for positive hybridization were selected for further studies.

2.6 Agrobacterium *mediated transformation:*

2.6.1 *Bacterial strains and molecular vectors used:*
Agrobacterium strain LBA 4404 supplemented with a constitutive *virG* mutant gene (*virG*N54D) harbored in a binary plasmid pCAMBIA 1381 was used as a vector system for plant transformation. Hence, the chimeric gene construct pCAMBIA1381/BREF/GUS was produced.

2.6.2 *Culture of* Agrobacterium *strain:*
The LBA 4404 *virG*N54D *Agrobacterium* strain was cultured overnight in AB medium (Chilton *et al.*1974) in an incubator shaker (150rpm) at 28°C. The medium was supplemented with 20mg/l of rifampicin and 75 mg/l chloramphenicol. Culture of *Agrobacterium* strains were maintained in YEP medium supplemented with 1.6% agar in presence of corresponding antibiotics.

Table 2.1. Summary of the media used in *Agrobacterium* culture:.

Medium	Composition
YEP(solid)	10gm/l Bactopeptones, 10gm/l Yeast extract, 5gm/l NaCl and 1.6% agar(w/v).
YEP(liquid)	10gm/l Bactopeptones,10gm/l Yeast extract, 5gm/l NaCl
AB minimal media(liquid)	5gm/l glucose, 25 ml AB salt &25 ml AB buffer for one litre.

2.7 *Molecular analysis of the transformants:*

2.7.1 *PCR screening:*
For initial screening of the putative transgenic cotton lines selected on the basis of hygromycin resistance were subjected to PCR analysis. For this genomic DNA was isolated from the leaves of the lines following CTAB method

2.7.2 *Southern blot analysis of GUS transformed cotton lines:*
Some of the PCR positive plants were thereafter subjected to southern blot analysis. For this, 10μg of genomic DNA was isolated from the PCR positive plants.

2.7.3. β-Glucuronidase *(GUS) histochemical assay:*
Expression of GUS activity in different parts of the regenerated cotton plants was histochemically monitored by blue coloration due to the presence of GUS enzyme that converts the substrate 5-bromo-4-chloro-3-indonyl β-D glucuronide (X-Gluc) into an insoluble precipitate.

2.7.4 *Inheritance pattern of gus gene in T1 generation:*
Self-pollinated seeds of different T0 cotton lines transformed with *gus* gene were allowed to germinate in MS medium containing 3% sucrose (w/v). After 3 days, the germinated seeds were transferred to MS medium supplemented with 30mg/l of hygromycin. Resistance to hygromycin was scored after 7 days by counting the number of surviving plantlets amongst the total number of seeds sown.

3 RESULTS

3.1 *Attempts for genetic transformation through* Agrobacterium *infection*

3.1.1 *Ternary transformation system in* Agrobacterium *mediated transformation - a strategy for improvement of gene transfer method:*

We designed a strategy for a ternary transformation system by providing an additional plasmid containing *virG*N54D along with the standard binary system of *Agrobacterium* mediated transformation. This was done in order to augment the efficacy of transfer of T-DNA potential. This was done to develop additional transforming tools in addition to the use of standard LBA4404 *Agrobacterium* strain (Hoekama *et al.*, 1983) as the helper plasmid.

The binary vector pCAM/BREF/GUS was transferred to begin with the *Agrobacterium* strain LBA4404 to generate LBA4404: pCAM/BREF/GUS (**Figure 3.1**). This formed the basic binary system for transformation. Additionally a ternary vector system was created to constitute the ternary system by transferring *virG*N54D. In order to create this plasmid, the mutant *virG* gene which expresses constitutively *virG*N54D, was cloned as *Sac*I/*Hin*dIII fragment from PRAL6308 into plasmid pBBR1MCS (Kovach *et al.*, 1994).

The influence of *virG*N54D in facilitating transformation frequency was evaluated on tobacco leaf disc by monitoring expression of the transferred *gus* gene. The leaf discs were allowed to be infected for a specified regime of cocultivation with *Agrobacterium* strain. *Agrobacterium* infection was eliminated from the plated tobacco leaf discs in plant growth medium with the help of antibiotics, following the standard protocol. **Table 3.1** documents the results of our experimentation.

3.1.2 *Estimation of transformation frequency in cotton:*

The hypocotyl explants were cut into 3-5mm long pieces and were inoculated with *Agrobacterium tumifaciens* strain LBA4404/*virG*N54D harboring pCAMBIA 1381 carrying a *2x35S-gus-nos* gene construct. Infected hypocotyls were co-cultivated for 48 hours at 25°C incubation, where both the explants and *Agrobacterium* will grow in its acceptable environmental condition.

The transformant lines showing *gus* expression were further reconfirmed by searching for simultaneous presence of *hptII* gene. Since pCAMBIA1381/BREF/GUS construct had *hptII* gene as selection marker linked with the *gus* gene, presence of *hptII* gene along with *gus* gene was expected to be present. PCR analysis confirmed the presence of the *hptII* gene in all cases (**Figure 3.2**).

The plantlets that could be generated through selection at the early stage of infection were found to be PCR positive for the presence of the *hptII* gene. Randomly selected 6 of the hygromycin resistant plants were additionally searched for presence of the *gus* gene. All of them show the GUS-expression (**Figure 3.4** and **Figure 3.5**).

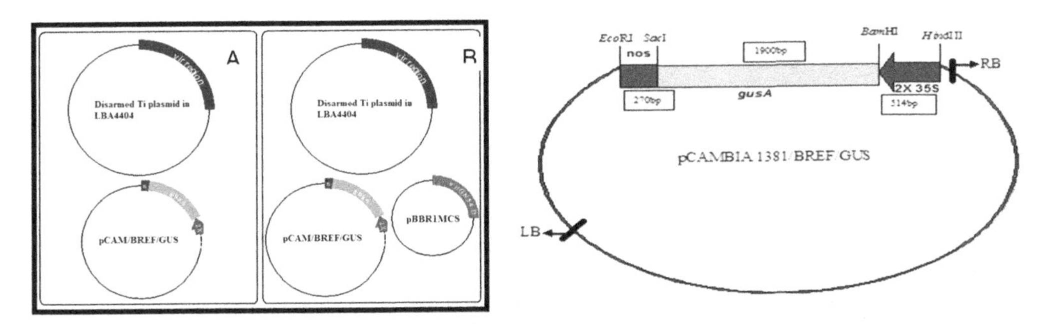

Figure 3.1. Schematic representation of *Agrobacterium* strain LBA4404 showing binary vector system(A) and a ternary system supplemented with a constitutive *virG* mutant gene(*virG*N54D) on a compatible plasmid(B). C shows a diagrammatic view of the chimeric gene construct *2x35S-gus-nos*.

Table 3.1. Addition of a *virG*N54D copy increases T-DNA transfer and stable transformation frequencies in tobacco leaf discs were cocultivated with *A.tumifaciens* strain LBA4404: pCAM/BREF/GUS with and without a copy of *virG*N54D:.

Agrobacterium strain	Experimental set up	No. of leaf discs inoculated	Total no. of regenerated plantlets	Total no. of GUS positive plantlets	Transformation frequency (%)
LBA4404: pCAM/	1	20	275	12	4.36
BREF/GUS.	2	25	304	15	4.93
	3	18	242	12	4.95
	4	23	284	14	4.92
	5	28	325	15	4.61
LBA4404: pCAM/	1	25	310	34	10.96
BREF/GUS.	2	19	251	27	10.75
pBBRvirGN54D	3	23	280	30	10.71
	4	28	330	36	10.90
	5	30	340	37	10.88

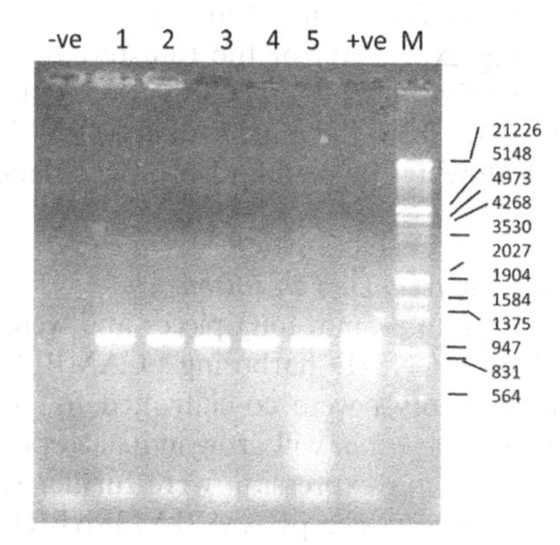

Figure 3.2. A representative electrophoregram of PCR amplified DNA showing band (~1026 bp) indicated the presence of the *hptII* gene. The untransformed plant representing for the negative control line does not show the presence.
Lane -ve: Untransformed control plant
Lane 1-5: Hygromycin resistant plant
Lane +ve: A_4 plasmid (product size-1026bp)
Lane M: Marker, λ *Eco*RI/*Hin*dIII digestion.

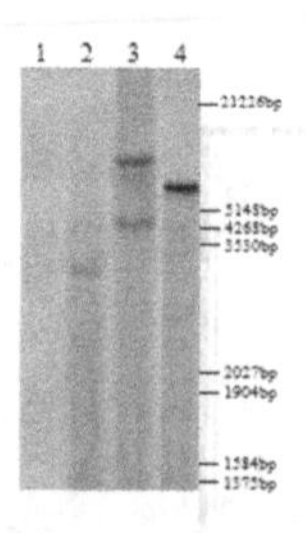

Figure 3.3. Southern analysis of PCR positive plants using *gus* gene as probe.
Lanes: 1-Untransformed control digested with *Hin*dIII
Lane: 2- DNA from PCR positive lines #G_1 digested with *Hin*dIII
Lane: 3- DNA from PCR positive line #G_5 digested with *Hin*dIII
Lane: 4- A_4 plasmid DNA (positive control)

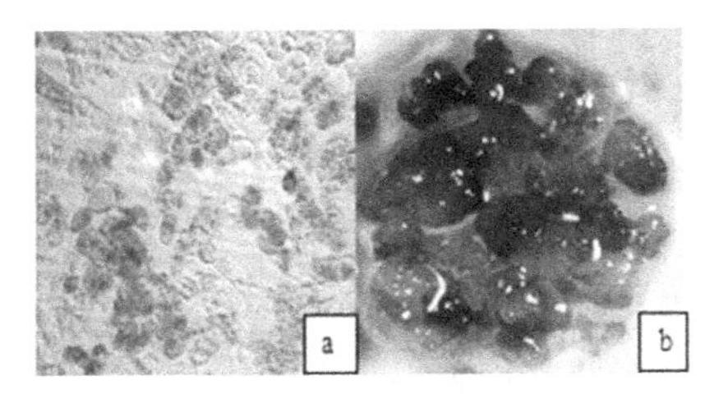

Figure 3.4. Expression of gus activity in MST_1 and MST_3 calli.
a. MST_1calli (x1000); b. Mass of embryogenic calli in MST_3 (x400).

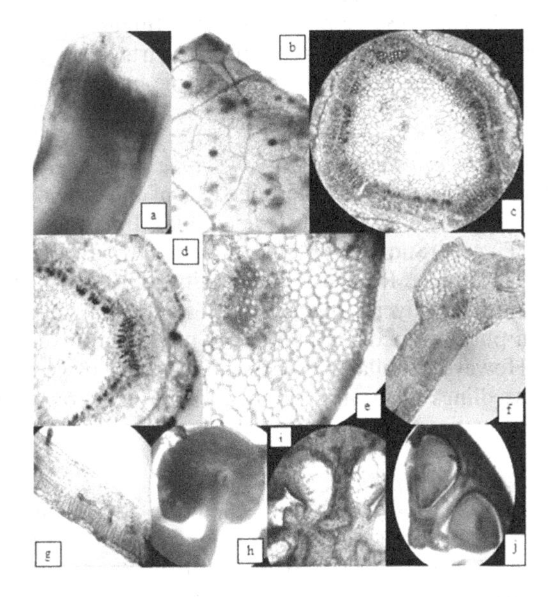

Figure 3.5. Histoenzymatic detection of the stable *gusA* gene expression in different plant parts and their transverse section; a. Portion of petiole(x100); b. Portion of leaf(x100); c. Section of stem portion showing pith and vascular bundles(x200);.
d. Section of portion of stem showing epidermis and vascular bundles(x200);
e. Section of petiole showing vascular bundle and epidermis;
f. Leaf section(x100);g. Leaf section showing palisade parenchyma(x200);
h. Anther lobe and stalk showing expression at cut stalk(x200);
i. Ovary chamber(x200); j. Ovule in ovary chamber(x200).

3.1.3 *PCR screening for the presence of the hygromycin gene:*
All plantlets derived from hygromycin resistant cell lines in culture provided evidence for the presence of *hptII* gene as gene specific primer resulted amplification of the predicted 1026bp DNA of the *hptII* gene. There after these plants were further analyzed for determining the copy number of the transferred gene. No amplification was observed in untransformed plant line. Fig. 3.1.5 shows presence of PCR amplified *hptII* gene in genomic DNA of the transformants.

3.1.4 *Southern blot analysis of PCR positive plants*
Southern hybridization was carried out for determining the nature of integration of the transferred DNA. For analysis radiolabelled *gusA* gene was used as probe (584bp). The nuclear DNA of the experimental plant lines were cleaved with *Hin*dIII digestion. It was observed that the transformed plant lines #$G_1$was found to have a single copy of the transgene near 3.5kb region while #G_5 was having 2 copies of the transforming DNA one in 4.5kb and another near 10kb approximately.

Fig. 3.6. Test of progenies for resistance to hygromycin. Diagram showing a number of resistant and sensitive seedlings in solid media (with phytagel) supplemented with 6mg/l hygromycin.

3.1.5 *Histochemical GUS assay.*

Histochemical observation for *gus* expression of T_0 transformants was carried out. Expression of *gus* activity in calli of MST_1, MST_3 could be detected (**Figure 3.4**). Stable expression of *gus* gene was detected by histochemical analysis of different parts of the transformed plants grown in greenhouse. Young leaves from the putative transgenic lines showed consistant *gus* expression (**Figure 3.5**).

3.1.6 *Seed transmission of the transferred* gusA *gene in the T_1 filial generation*

Seed transmission of the transgene was determined from the T1 population generated from selfed seeds of each of the two primary transformed lines. Inheritance pattern of hygromycin resistance gene in the T_1 generation was tested through growth of germinating T_1 seeds in presence of hygromycin (25mg/l). Resistant and sensitive seedlings could be distinguished in solid medium containing hygromycin (**Figure 3.6**). Germinated seeds from non-transformed control plant line when put into media containing 25mg/l hygromycin were characteristically etiolated after 7 days. However, similarly processed seeds of transgenic lines showed presence of green and etiolated seedlings in varying numbers. Table 3.3 shows segregation of such germinating seedlings into sensitive and resistant plantlets.

3.1.7 *Southern analysis of T_1 transgenics:*

The segregating populations of T_1generation of $G_1$and G_5 transgenic lines were checked for stability of transmission of the *gus* gene through seed. Randomly sampled 3 individual plants of T_1progenies of G_1 and G_5 events were tested. All plants under test showed presence of the transferred gene. Maintaining the same pattern of the T_0 plant. Figure 3.7A. shows the autoradiogram of the southern analysis. Lanes 2, 3 and 4 represent for T_1plants $G_{1\text{-}A}$, $G_{1\text{-}B}$, $G_{1\text{-}C}$.

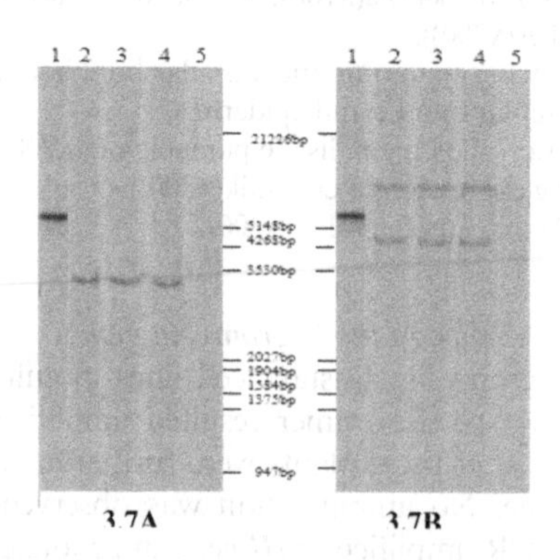

Figure 3.7. Southern analysis of hygromycin positive plants using *gus* gene as probe.
3.7A: T_1southern of G_1 transgenic line.
Lanes: 1- A_4 plasmid DNA (positive control).
Lane: 2-4 - DNA from PCR positive lines digested with *Hin*dIII (#$G_{1\text{-}A}$, #$G_{1\text{-}B}$, #$G_{1\text{-}C}$)
Lane: 5- Untransformed control plant.
3.7B: T_1southern of G_5 transgenic line.
Lanes: 1- A_4 plasmid DNA (positive control).
Lane: 2-4 - DNA from PCR positive lines digested with *Hin*dIII (#$G_{5\text{-}A}$, #$G_{5\text{-}B}$, #$G_{5\text{-}C}$)
Lane: 5- Untransformed control plant.

All of them provide evidence for the characteristic pattern of *Hin*dIII digested banding pattern. A similar evidence for three G_5 plants was also observed (Figure 3.7B).

4 DISCUSSION

Introduction of plant transgenic approach in biotechnology of many crop plants is severely constrained in situations where plant regeneration is difficult to achieve. It is well realized that transformation and regeneration of cotton via somatic embryogenesis has not been a simple proposition by any means and cotton tissues remain to be one of the major recalcitrant plant system to manipulate its growth pattern to result plant regeneration in culture (Wilkins *et al.*, 2000). Thus, the necessity to bring in improvement in the methodology to effect high incidence of *Agrobacterium* mediated transformation has remained as acute as ever. Based on this, an experimental strategy for adoption of *Agrobacterium* mediated gene transfer with the help of ternary transformation system was adopted in place of the usual binary system of *Agrobacterium* mediated transformation. The ternary system of transformation is known to enhance transformation frequency. To start with, the influence of ternary system of transformation was tested on tobacco leaf discs by monitoring the expression of the *gus*A gene, as the marker. Incidence of higher level of transformation in case of ternary system in comparison to binary system of transformation was observed **(Table 3.1)**. Based on number of plantlet regeneration, the transformation frequency was found to be **(Table 3.2)** higher than the earlier claims.

Agrobacterium-mediated transformation of cotton via somatic embryogenic mode of plant regeneration remains by far the most dependable method of choice. This is due to the reliability and efficiency by which transgenic plants are recovered from culture (Rajasekaran *et al.*, 1996, 2000). *Agrobacterium* mediated transformation remains to be the most efficient means for producing transgenic cotton plants at the present time. Although limited success in generation of transgenic cotton lines could be achieved in a non commercial cotton strain Coker, transformation of all commercial strains of cotton have still proved to be difficult.

An elaborate study on improvement and modification of cotton transformation protocol through *Agrobacterium* mediation had been conducted by Wilkins *et al.* (2004). Till date, *Agrobacterium* mediated transformation methods remained restricted to Coker and a few indiscriminate claims of success in certain other strains of cotton. A number of investigations have been conducted on the development of transgenic cotton by means of *Agrobacterium*

Table 3.2. Determination of transformation frequency of transformed plantlets.

Geno- -type	Exp. No.	Total no. of explants infected	Total no. of embryogenic calli produced	Total no. of regenerated plantlets	Total no. of GUS positive plants	Transformation frequency (%)	Mean transformation frequency (%)
Coker-	1	20	46	106	8	7.54	**7.4**
312	2	19	43	98	7	7.14	
	3	23	52	120	9	7.50	
	4	25	60	138	10	7.24	

Table 3.3. Segregation analysis of T_1 progenies of two primary transformants as assessed by hygromycin sensitivity test.

Primary transgenic	No. of seeds tested	No. of seedlings		Segregating ratio	Chi square value	P
		Green	Etiolated			
G_1	86	61	25	3:1	0.77	0.3802
G_5	90	83	7	15:1	0.36	0.5485

mediated transformation using embryogenic calli as the source of explant (Leelavathi *et al.*, 2004; Jin *et al.*, 2005; Wu *et al.*, 2005).

Based on this success, an experimental mock trial for transfer of an alien gene, *gusA* as marker, was taken up. Independent transformants showed stable expression of the transformed *gusA* gene and its seed transmission to the next filial generation. Thus developed gene transfer protocol through *Agrobacterium* mediation in the form of exogenous DNA integration into the nuclear genome of the host plant, provided evidence for genetic transformation and stable seed transmission of the transgene. Keeping the above in view, the progress attained in the technical aspect of *Agrobacterium* mediated genetic transformation methodology in cotton can be viewed as a step forward for undertaking future programmes of studies relating to gene discovery and genomics.

REFERENCES

Firoozabady E, Deboer D, Merlo D, Halk E, Amerson L, Rashka K and Murray E (1987) Transformation of cotton (*Gossypium hirsutum* L) by *Agrobacterium tumifaciens* and regeneration of transgenic plants. *Plant Mol. Biol.* **10**: 105–116.

Hansen G, Das A and Chilton MD (1994) Constitutive expression of the virulence genes improves the efficiency of plant transformation by *Agrobacterium. Proc. Natl. Acad. Sci. USA* **91**: 7603–7607.

Hoekema A, Hirsch PR, Hooykaas PJJ and Schilperoort RA (1983) A binary plant vector strategy based on separation of vir and T-region of the *Agrobacterium tumifaciens* Ti-plasmid. *Nature* **303**: 179–180.

Kovach ME, Phillips RW, Elzer PH, Roop, R.M II and Peterson KM (1994) pBBR1MCS: a broad-host-range cloning vector. *BioTechniques*. **16**: 800–802.

Leelavathi S, Sunichan VG, Kumria R, Vijaykanth GP, Vhatnagar RK and Reddy VS (2004) A simple and rapid *Agrobacterium*- mediated transformation protocol for cotton (*Gossypium hirsutum* L.): Embryogenic calli as a source to generate large numbers of transgenic plants. *Plant Cell Reports* **22** (7): 465–470.

Pazour GJ, Ta CN and Das A (1992) Constitutive mutations of *Agrobacterium tumefaciens* transcriptional regulator *virG. J. Bact.* **174**: 4169–4174.

Perlak FJ, Fuchs RL, Dean DA, McPherson SL, Fischoff DA (1991) Modification of the coding sequence enhances plant expression of insect control protein genes. *Proc. Natl. Acad. Sci.* USA. **88**(8): 3324–3328.

Sambrook J, Fritsch EF and Maniatis T (1989) Molecular cloning: a laboratory manual. Cold Spring Harbor Laboratory.

Umbeck P, Johnson G, Barton K and Swain W (1987) Genetically transformed cotton (*Gossypium hirsutum* L) plants. *Bio/Technology* **5**:263–266.

Wilkins Thea A, Rajasekaran Kanniah and Anderson David M (2000) Cotton Biotechnology. *Critical Reviews in Plant Sciences*, **19**(6):511–550.

Wu, X. Zhang, Y. Nie, X. Luo (2005) High efficiency transformation of *Gossypium hirsutum* embryogenic calli mediated by *Agrobacterium tumefaciens* and regeneration of insect-resistant plants. *Plant Breeding* **124**: 142–146.

Biotechnology and Biological Sciences – Sen et al. (Eds)
© 2020 Taylor & Francis Group, London, ISBN 978-0-367-43161-7

Longevity promoting effect of *Catharanthus roseus(L.) G. Don* in *C.elegans* is modulated by daf-16 and other genes

Shital Doshi
St. Xavier's College, Navarangpura, Ahmedabad

Vincent Braganza
Loyola Centre for Research and Development, St. Xavier's college campus, Ahmedabad

ABSTRACT: Aging is considered as the single largest factor for age related disorders like Alzheimer's, cardiovascular diseases, cataract and diabetes. Hence identification of pharmacological compounds that slow down normal aging processes and that enhance life span is the need of the hour. Medicinal plants and their isolates contain large amounts of polyphenols and have antioxidant capacity and can intervene in the aging process. One such herb is *Catharanthus roseus(C.roseus)*. In this study *C. roseus* extracts showed *in vitro* antioxidant and free radical scavenging activity. This study has analyzed the longevity promoting effect of *Catharanthus roseus* on a popular model of aging research, *Caenorhabditis elegans(C.elegans)*. Our results showed that the hydro alcoholic extract of *C. roseus* does extend the life span of wild type and a mutant strain of *C.elegans* and improves its stress resistance. Further investigations indicate that *C. roseus* could activate the FOXO head transcription factor Daf-16. It also enhances the expression of HSP 16.4 in wild type N2 bristol *C.elegans*.

Keywords: Aging, *Caenorhabditis elegans*, *Cathranthus roseus*, daf-16

1 INTRODUCTION

Cells and tissues of organisms undergo various detrimental changes as the organism undergo ageing. These detrimental changes increase the risk of various age–onset disorders like cardio vascular diseases, stroke diabetes and Alzheimer's [1]. With advances in medical science and accessibility to its outcome, life expectancy around the world has increased from 46 years to 65 years in the last hundred years [2]. This has led to immense economic burden to society. And as a result research in aging is moving in the direction wherein interventions are sought which can prevent or delay the age-related decline in tissue functions. 80% of the world population from countries like Brazil, Indonesia, India depends upon traditional knowledge for treatment of the various diseases, the search for traditional knowledge based drugs which enhance antiageing is need of the hour.

Catharanthus roseus (L.) G. Don(*C.roseus*) is a well known medicinal plant used in conventional Ayurvedic medicine. It is a perennial herb found across India. Phytochemical analysis studies have shown that *C. roseus* leaves are known to contain alkaloids, polyphenols, saponins and steroids in higher concentrations while flavonoids and Vitamin C are at low concentration [3]. Traditionally it is used in the treatment of diabetes, blood pressure, asthma, constipation, and cancer. As an important medicinal plant, it is known to have very good antioxidant potential. Antioxidants are compounds that can act as reducing agents and terminate the production of free radicals inside the cells and prevent damage to cellular machinery [3]. Even though the plant is reported have high medicinal value its antiageing and stress modulatory potential is unexplored.

*Corresponding author: shital.doshi@sxca.edu.in

Therefore in the present investigation, we evaluated anti–aging, stress modulatory effect in *Caenorhabditis elegans(C.elegans)*, a well-established aging model. *C. elegans* is a saprophyte nematode of 1 mm size, easy to grow on an NGM plate. It has a short life span and a number of transgenic and mutant strains are available to investigate various hypothesis. And since mechanisms which modulate aging in *C.elegans* have human ortholog, and loss of function of these genes also modulates life span in mammalian models, it suggests that mechanisms found to affect aging in *C.elegans* can also have potential to affect human aging [4].

In this study, we investigated whether *C.roseus* extract can promote longevity under normal culture condition. We also used *C.elegans* to investigate if *C.roseus* can mediate oxidative and thermal tolerance.

2 MATERIALS AND METHODS

2.1 *Preparation of plant extracts*

C.roseus leaves were collected from the plants grown at the campus of Loyola Centre for Research and Development, Ahmedabad, Gujarat, India. Leaves were washed,dried for ten days and then grounded to obtain fine powder. The extracts were obtained by sonication using a bath sonicator (42 KHz, 135 W; Branson ultrasonic corporation, USA) for 60 minutes at room temperature. Concentrated to dryness and resuspended in DMSO to a concentration of 100 mg/ml.

2.2 *Total polyphenol content and DPPH free radical scavenging activity of plant extracts*

Folin-Ciocalteu's method, as described by T. J.Herald et. al 2012 was employed to determine total polyphenol content. The assay was performed in a 96 well plate. An aliquote of 10µl of plant extract was mixed with 50µl of 1:10 (Folin-Ciocalteu's: Water) diluted Folin-Ciocalteu's reagent. To the mixture 40 µl of 7.5mM Sodium carbonate was added. The plate was incubated at 50°C for 20minutes. Absorbance was measured at 765 nm. Total polyphenol content is reported in mg of gallic acid equivalent/gram of dried plant material.

DPPH[2,2-diphenyl-1-picrylhydracyl] free radical scavenging activity of extracts was measured, as described by Adedapo [5]. A stock of 0.4 mM stock of DPPH was prepared in methanol. 100µl of plant extract was mixed with 100µl of DPPH. The reaction mixture was incubated in the dark for 15 minutes and then absorbance was measured at 517 nm. % inhibition was calculated by formula:

% inhibition = [(Absorbancecontrol- Absorbancetest)/Absorbacne control]*100

2.3 *C.elegans strains and maintenance*

The following strains were used in the experiments. (A) wild-type *C.elegans* N2 (Variety Bristol), (B) TK-22[mev-1 (kn1)], (C) TJ356[zIs356 IV (pdaf-16-daf-16::GFP; rol-6)], (D) TJ375[hsp-16.2p:: GFP]. All strains were maintained on NGM medium supplemented with *Escherichia coli.* OP50 as food source unless indicated otherwise. Synchronized population of worms were obtained by Sodium Hypochlorite method, which kills adult worms but not their eggs.(1:5 diluted 5% sodium hypochlorite in 250 mm NaOH) (Porta-de-la-Riva et al. 2012). The *C.elegans* N2 and *E.coli* OP50 were a kind gift by Dr. Sandhya Kaushika from TIFR, Mumbai. TJ 356 was a kind gift by Dr. Aamir Nazir, CDRI, Lucknow. TK-22 and TJ375 were purchased from the Caenorhabditis Genetics Center (CGC) at the University of Minnesota (Minneapolis, MN, USA).

2.4 *Lifespan analysis of C.elegans*

Life span assays of *C.elegans* were performed in 96 well plates in liquid S complete media as per the method described by Solis G [6]. 10-12 age-synchronized L1 larvae worms were transferred to 96- well plates in 100 µl of S-complete medium supplemented with 5µl of 100mg/ml heat killed OP50 bacteria. To obtain synchronized population, 40 μM 5-fluorodeoxyuridine (FUDR) was added to each well. FUDR prevents hatching of eggs.1mg/ml, 0.1mg/ml and 0.01 mg/ml of plant extracts were added in respective wells on the next day. DMSO was used as vehicle control in each well. To ensure the presence of compound throughout the experiment plant extracts (1mg/ml, 0.1mg/ml and 0.01 mg/ml) were added every week to respective wells along with *E.coli.*OP50 (5μl) as a food source. Survival was monitored every other day under an inverted light microscope until death of all worms. Worms which did not show any movement/pharyngeal pumping upon exposure to light for more than 5 seconds were considered dead. The lifespan assay was repeated in three independent trials and data was analyzed.

2.5 *Stress assay*

Oxidative stress assay was performed as described previously [7]. For paraquat tolerance assay, synchronized L1 larvae were suspended in 10 ml S complete medium and added to the 96 well plates such that each well contained approximately 10-12 larvae. Worms were treated with plant extracts till they reached adulthood and were also treated with 40 μM FUDR to prevent progeny formation. After that 50 mM paraquat was added to each well and survival was monitored after 24 hours.

For thermotolerance assay, age-synchronized worms were prepared as mentioned above. Then Day 2 adult worms were incubated at 37°C for four hours. Plates were returned to 20°C and survival was monitored daily till all worms were dead (Guha et al. 2014). Each experiment was repeated three times.

2.6 *In vivo gene expression analysis*

Worms of TJ375 strain expressing Hsp16.2: GFP were synchronized using sodium hypochlorite [8]. Age synchronized L1 larvae thus obtained were grown in liquid S medium supplemented with plant extracts or DMSO as a control for 72 hrs. At the end of the incubation period, the worms were paralyzed on a glass slide in 1M Sodium azide solution prepared in PBS(pH 7). The fluorescence intensity was observed through a Zeiss microscope. Intensity was quantified by ImageJ (NIH) software[9].

2.7 *Localization of Daf-16 gene*

TJ356 strain carrying Daf16: GFP was used to study the localization of daf-16 gene. Synchronized worms from the L1 stage were grown in liquid s-complete medium for 72 hours. Plant extract and DMSO as vehicle control were then added for 2 hours at 20°C. Thereafter the worms were transferred on to a microscope slide and cellular localization of DAF-16 transcription factor was detected through the fluorescent microscope. (Excitation 485 nm, emission 550 nm). In each trial around 20 worms were used. Each experiment was repeated three times.

3 RESULTS

3.1 *Total polyphenol content and In vitro antioxidant assay*

C. roseus contains a variety of phenolic compounds like caffeoylquinic acid and flavonol glycosides (Meenakshi K. et. al.2013 [10]. These compounds are well-known antioxidants and free radical scavengers. The total polyphenol content of leaves of *C.roseus* was found to be 2.11 ± 0.084 mg/g of gallic acid equivalent.

3.2 *C.roseus promotes longevity of C.elegans under standard culture condition*

In order to determine the effect of *C.roseus* on the longevity of *C.elegans*, *C.elegans* were exposed to three different concentrations of *C.roseus* extracts (1 mg/ml, 0.1 mg/ml, and 0.01 mg/ml) in

Table 1. Polyphenol content and In vitro antioxidant activity of plant extract.

Total polyphenol content a	2.11± 0.084
DPPH reducing ability of plant extract	70% ± 0.8

a: mg of Gallic acid equivalent/g of plant extracts.

liquid S- complete medium along with the OP50 as a food source. *C.roseus* could increase the life-span of the worms in dose dependent manner. (Figure 1, Table 2) *C.roseus* at 1 mg/ml showed maximum increase in life span, by 21.34% ± 1.1, 0.1 mg/ml and 0.01 mg/ml showed 17.3 % ± 0.7 and 11.25% ± 0.5, (p≤0.0001) increase in life span respectively.

3.3 C.roseus *leaves extract increased resistance to heat and oxidative stress*

To examine the anti-stress effect of *C.roseus*, *C.elegans* from L1 larvae stage were treated with different concentration of plant extract. On day 2 of adult hood heat stress was induced by incubating the worms at 37°C for 4 hours. Survival was monitored after 24 hours till all the worms were dead. Compared to control group (worms treated with vehicle control DMSO) plant extracts treated worms showed increased tolerance to heat stress (Figure 2, Table 3). The mean survival time was increased by 150% (p≤0.0001) The survival curves indicate that both mean as well as maximum survival had increased significantly in treated worms.

To further investigate whether *C.roseus* extracts in addition to its heat stress and longevity effect can modulate oxidative stress responses in *C. elegans*, we examined survival of worms

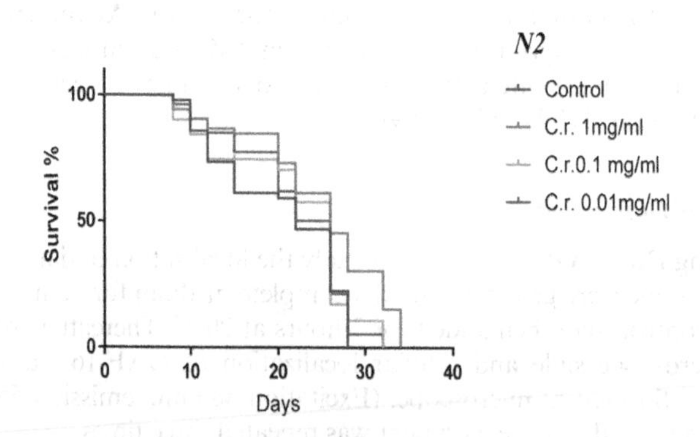

Figure 1. Effect of *C.roseus* on longevity of *C.elegans*. Survival curves were plotted as per the Kapein Meire survival assay. The worms were fed heat killed OP50 every fifth day along with plant extracts as supplements. *C.roseus* could increase the lifespan of wild type worms in a dose dependent manner.

Table 2. Effect of *C.roseus* on life span of N2 (Vvariety Bristol) worms.

	C.elegans strain	No. of worms	Mean life span± S.E.(Days)	P value V/S control
Control	N2 Bristol.	144	18.32 ± 0.98	
Cathranthus roseus 1mg/ml	N2 Bristol.	147	22.23 ± 1.2	≤0.001
Cathranthus roseus 0.1mg/ml	N2 Bristol.	144	21.44 ± 0.90	≤0.001
Cathranthus roseus 0.01mg/ml	N2 Bristol.	147	20.38 ± 1.5	≤0.001

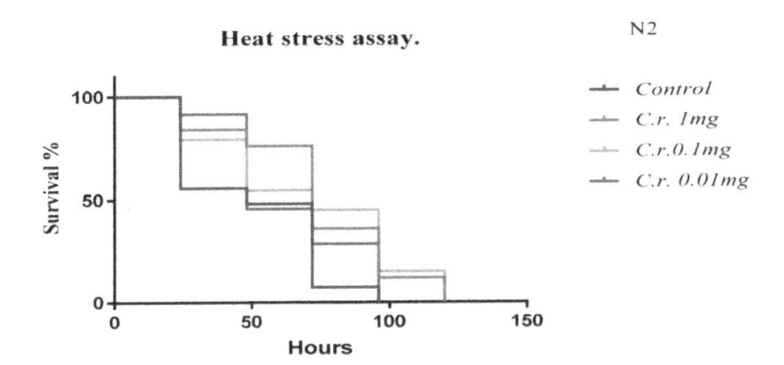

Figure 2. Effect of *C.roseus* on thermal stress. Pretreatment of worms with PE significantly promote stress resistance in *C.elegans*.

Table 3. Protective effect of *C.roseus* on wild type *C.elegans* under thermal stess.

	C.elegans strain	No. of worms	Median life span± S.E.(hours)	P value V/S control
Control	N2 Bristol.	162	48 ± 0.89	
Cathranthus roseus 1mg/ml	N2 Bristol.	150	72 ± 1.1	≤0.001
Cathranthus roseus 0.1mg/ml	N2 Bristol.	159	72 ± 0.65	≤0.001
Cathranthus roseus 0.01mg/ml	N2 Bristol.	105	48 ± 1.1	≤0.001

treated with 1mg/ml,0.1mg/ml and 0.01mg/ml of plant extract after induction of oxidative stress by paraquat. Oxidative stress was significantly alleviated by *C.roseus* treatment. In control group 30.67% ± 0.67. worms survived after 24h of oxidative stress. Treatment with *C.roseus* extracts increased the survival up to 50.29% ± 0.84 after 24 h of oxidative stress (p value ≤0.0001) (Figure 3).

C.elegans mutant (mev-1) are available which exhibits sensitivity to oxidative stress due to a mutation in the cytochrome oxidase system. mev-1 mutants have a normal pattern of aging but at accelerated rates due to oxidative stress [11]. If C.roseus can promote longevity in these worms, it provides further proof that C.roseus extracts are capable of reducing oxidative stress (Figure 4, Table 4). In all tested concentration C.roseus extracts were able to increase the life span of mev-1 worms.

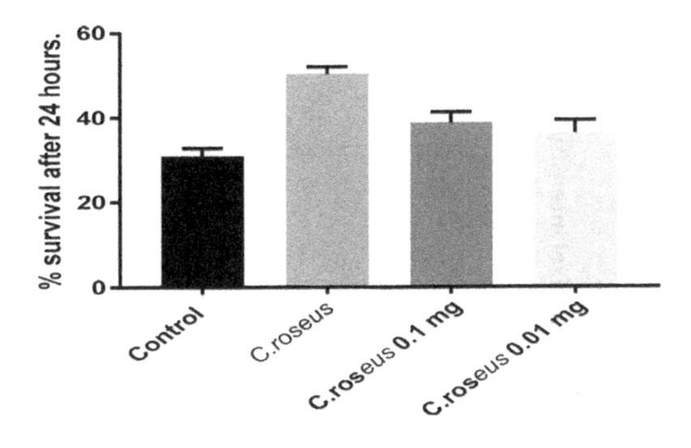

Figure 3. Survival of worms after 24 hours of oxidative stress, expressed in terms of % 1 mg treatment showed maximum survival. (p<0.0001).

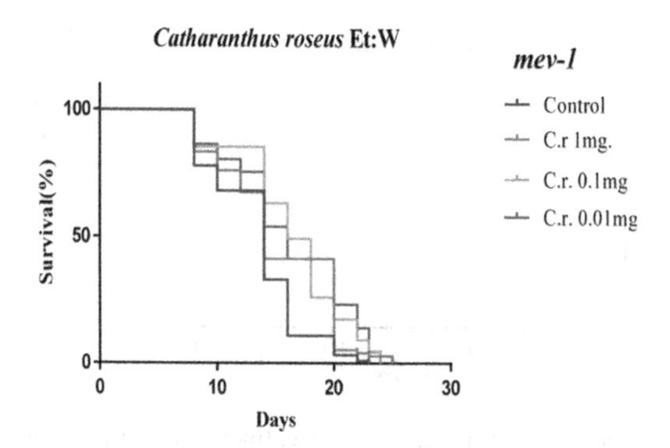

Figure 4. Survival of mev-1 worms.

Table 4. Life span promoting effect of *C. roseus* on mev-1 worms.

	C.elegans strain	No. of worms	Mean life span± S.E.(days)	P value V/S control
Control	Tk-22(mev-1)	285	13.84 ± 0.85	
Cathranthus roseus 1mg/ml	Tk-22(mev-1)	246	18.67 ± 0.65	≤0.001
Cathranthus roseus 0.1mg/ml	Tk-22(mev-1)	293	16.28 ± 0.92	≤0.001
Cathranthus roseus 0.01mg/ml	Tk-22(mev-1)	302	14.01 ± 0.54	≤0.001

3.4 C.roseus *can affect the expression of hsp 16.2: GFP in* C.elegans

C.roseus induced stress response in transgenic *C.elegans* strain TJ375 (containing hsp16.2: GFP reporter) was determined after 2h of thermal stress, followed by overnight recovery at 20°C. The GFP fluorescence intensity was found to be 47.33 ± 11.33 ($p<0.05$) higher compare to control in treated worms (Figure 5).

3.5 C.roseus *can translocate DAF-16 from the cytoplasm to nucleus*

Daf-16 is a transcription factor localized in the cytoplasm. Whenever insulin signaling is low it is translocated from cytoplasm to nucleus. Once inside the nucleus it regulates many vital processes like stress resistance and life span. It also upregulates genes involved in antioxidant response [12]. To investigate nuclear translocalization, TJ356 mutants carrying a Daf16: GFP transgene were employed. Worms were treated with 1mg/ml of plant extract for 2 h and then visualized under a fluorescence microscope. It was observed that *C.roseus* extract was able to translocate 47% ($p<0.05$) of Daf-16 into the nucleus (Figure 6). This result indicates the involvement of insulin signaling pathway to mediate the effect of *C.roseus*.

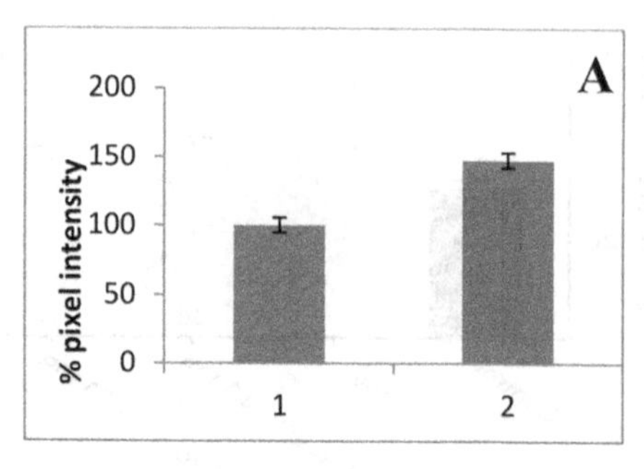

Figure 5(A). Effect of Catharanthus extract on hsp16.2:GFP expression. Fluorescent microscopy images of *C.elegans* treated either DMSO or 1mg/ml of *C.roseus* extracts.1. Pixel intensity of control which is taken as 100%. 2. Pixel intensity of treated worms.

Figure 5(B). DMSO or with.

Figure 5(C). 1mg/ml *C.roseus* are shown here. Fluorescent intensity was analyzed by image J and compared with DMSO.

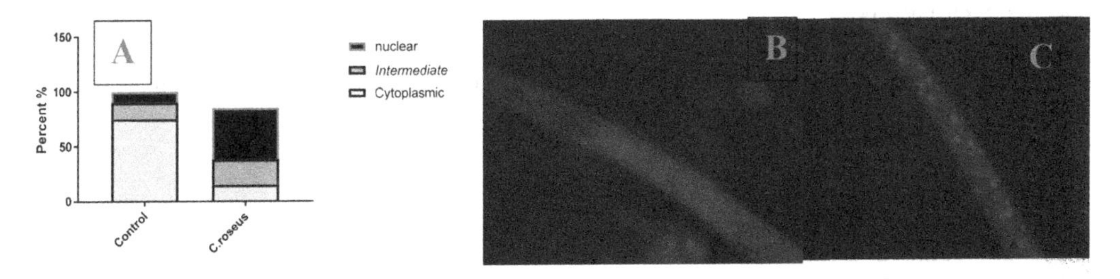

Figure 6. DAF-16 localization in TJ356 worms. (A) Effect of *C.roseus* on DAF-16 localization shown in percentage of total worms. Control represents untreated worms where majority of the worms showed cytoplasmic localization. Few worms showed nuclear localization while in treated worms more number of worms showed nuclear localization. (B) Image of *C.elegans* representing ccytoplasmic localization of DAF-16 (C) image of C.elegans representing nnuclear localization of DAF-16.

4 DISCUSSION

The present study for the first time demonstrates the longevity-promoting effect of the important medicinal plant *Catharanthus roseus* in an established aging model, *C.elegans*. Earlier *C.elegans* has been used to evaluate the longevity-promoting effect of a variety of natural or naturally derived components like resveratrol, metformin, (L. Mouchiroud et al. 2010), [13] blueberry polyphenols, green tea [14] or olive oil extract [15], fruit polyphenols [16], sesame seed [17], Specioside from Ajwain [18] among others.

Our experiments suggest that *C.roseus* increases the life span of *C.elegans* at 1mg/ml concentration to the maximum under normal culture conditions. Previous studies [19] have reported that a correlation exists between stress tolerance and life span. In the present study too longevity promotion was also associated with stress tolerance. *C.roseus* contains a number of polyphenols and other phytochemicals like alkaloids and flavanone. As reported earlier these phytochemicals may have acted in synergy to exert the biological functions. Thus *C.roseus* promoted thermal and oxidative stress tolerance may be due to unique effect generated by synergistic effects of phytochemicals.

The IIS (insulin/IGF-1 signaling) pathway is a major pathway involved in longevity-promoting effect in *C.elegans* [20] The pathway involves translocation of DAF-16 from cytoplasm to nucleus. As in *C.roseus* treated TJ356 showed increased translocation of DAF-16 to nucleus, it proves that C. roseus extend longevity by tinkering with IIS pathway. Mammals too have a number of FOXO head transcription factors, the homologs of DAF-16. These proteins are translocated from the cytoplasm to the nucleus when insulin/IGF-1 signaling is reduced. Inside the nucleus they activate a number of genes that mediate heat shock, oxidative stress, and energy metabolism. Due to their role in energy metabolism, they are considered to be important regulators of metabolic disorders like diabetes and obesity. One of the FOXO protein, FOXO3a has been associated with prolonged life span in humans too [21]. In *C.elegans* it is the one which promotes biotic and abiotic stress response along with lifespan.

Our study shows that DAF-16 is involved in life span extension effects of *C. roseus*. *C. roseus* may be doing so by regulating metabolic homeostasis which promotes a stress resistance and longevity.

In the present study, we also investigated the effect of *C. roseus* on the regulation of HSP16.2 gene. The HSP16.2 proteins are homologous to αβ crystalline proteins involved in stress sensitivity. Higher levels of HSP16.2 support longer life span [22]. *C. roseus* up-regulated HSP16.2 genes which suggest that longevity-promoting effect may also be due to this property.

6 CONCLUSION

In conclusion, *C. roseus* promotes longevity under normal and stressed condition. It does so by interacting with Insulin/IGF-1 signaling pathway.

REFERENCES

[1] L. Mouchirod,"Life span extension by resveratrol, rapamycin, and metformin: The promise of dietary restriction mimetics for an healthy aging," Biofactors.,vol. 36(5), pp 377–82, Sep 2010.

[2] V. Shukla, D. Yadav, S. C. Phulara, M. M. Gupta, S. K. Saikia, and R. Pandey, "Longevity-promoting effects of 4-hydroxy-E-globularinin in *Caenorhabditis elegans*," *Free Radic. Biol. Med.*, vol. 53, no. 10, pp. 1848–1856, Nov. 2012.

[3] M. S. Aruna, M. S. Prabha, N. S. Priya, and R. Nadendla, "*Catharanthus roseus*: Ornamental Plant is now medicinal Boutique," *J. Drug Deliv. Ther.*, vol. 5, no. 3, pp. 1–4, 2015.

[4] A. Pant, P. Prakash, R. Pandey, and R. Kumar, "*Syzygium aromaticum* (L.) elicits lifespan extension and attenuates age-related proteotoxicity in *Caenorhabditis elegans*," *Cogent Biol.*, vol. 2, no. 1, Aug. 2016.

[5] A. A. Adedapo, F. O. Jimoh, A. J. Afolayan, and P. J. Masika, "Antioxidant properties of the methanol extracts of the leaves and stems of *Celtis africana*," *Rec. Nat. Prod.*, vol. 3, no. 1, p. 23, 2009.

[6] G. M. Solis and M. Petrascheck, "Measuring *Caenorhabditis elegans* Life Span in 96 Well Microtiter Plates," *J. Vis. Exp.*, no. 49, Mar. 2011.

[7] E. Possik and A. Pause, "Measuring oxidative stress resistance of *Caenorhabditis elegans* in 96-well microtiter plates," *J. Vis. Exp. JoVE*, no. 99, p. e52746, May 2015.

[8] M. Porta-de-la-Riva, L. Fontrodona, A. Villanueva, and J. Cerón, "Basic *Caenorhabditis elegans* Methods: Synchronization and Observation," *JoVE J. Vis. Exp.*, no. 64, pp. e4019–e4019, Jun. 2012.

[9] R. A. McCloy, S. Rogers, C. E. Caldon, T. Lorca, A. Castro, and A. Burgess, "Partial inhibition of Cdk1 in G 2 phase overrides the SAC and decouples mitotic events," *Cell Cycle Georget. Tex*, vol. 13, no. 9, pp. 1400–1412, 2014.
[10] K. Meenakshi, S.L. Neha, C. Ramesh and C. Sheela,"*Catharanthus roseus* and prospects of its endophytes: a new avenue for production of bioactive metabolites" international journal of pharmaceutical sciences and research."
[11] S. P. Singh, M. Niemczyk, L. Zimniak, and P. Zimniak, "Fat accumulation in *Caenorhabditis elegans* triggered by the electrophilic lipid peroxidation product 4-hydroxynonenal (4-HNE)," *Aging*, vol. 1, no. 1, pp. 68–80, Dec. 2008.
[12] J. N. Landis and C. T. Murphy, "Integration of diverse inputs in the regulation of *Caenorhabditis elegans* DAF-16/FOXO," *Dev. Dyn. Off. Publ. Am. Assoc. Anat.*, vol. 239, no. 5, pp. 1405–1412, May 2010.
[13] F. Cabreiro *et al.*, "Metformin Retards Aging in *C. elegans* by Altering Microbial Folate and Met hionine Metabolism," *Cell*, vol. 153, no. 1, pp. 228–239, Mar. 2013.
[14] S. Abbas and M. Wink, "Green Tea Extract Induces the Resistance of *Caenorhabditis elegans* against Oxidative Stress," *Antioxidants*, vol. 3, no. 1, pp. 129–143, Mar. 2014.
[15] A. Cañuelo, B. Gilbert-López, P. Pacheco-Liñán, E. Martínez-Lara, E. Siles, and A. Miranda-Vizuete, "Tyrosol, a main phenol present in extra virgin olive oil, increases lifespan and stress resistance in *Caenorhabditis elegans*," *Mech. Ageing Dev.*, vol. 133, no. 8, pp. 563–574, Aug. 2012.
[16] J. A. Joseph, B. Shukitt-Hale, and F. C. Lau, "Fruit Polyphenols and Their Effects on Neuronal Signaling and Behavior in Senescence," *Ann. N. Y. Acad. Sci.*, vol. 1100, no. 1, pp. 470–485, Apr. 2007.
[17] Z. Wang, X. Ma, J. Li, and X. Cui, "Peptides from sesame cake extend healthspan of *Caenorhabditis elegans* via upregulation of skn-1 and inhibition of intracellular ROS levels," *Exp. Gerontol.*, vol. 82, pp. 139–149, Sep. 2016.
[18] J. Asthana, A. K. Yadav, A. Pant, S. Pandey, M. M. Gupta, and R. Pandey, "Specioside ameliorates oxidative stress and promotes longevity in *Caenorhabditis elegans*," *Comp. Biochem. Physiol. Part C Toxicol. Pharmacol.*, vol. 169, pp. 25–34, Mar. 2015.
[19] K. Christensen, T. E. Johnson, and J. W. Vaupel, "The quest for genetic determinants of human longevity: challenges and insights," *Nat. Rev. Genet.*, vol. 7, no. 6, pp. 436–448, Jun. 2006.
[20] A. L. Kauffman, J. M. Ashraf, M. R. Corces-Zimmerman, J. N. Landis, and C. T. Murphy, "Insulin Signaling and Dietary Restriction Differentially Influence the Decline of Learning and Memory with Age," *PLoS Biol.*, vol. 8, no. 5, p. e1000372, May 2010.
[21] D. N. Gross, A. P. J. van den Heuvel, and M. J. Birnbaum, "The role of FoxO in the regulation of metabolism," *Oncogene*, vol. 27, no. 16, pp. 2320–2336, 2008.
[22] Y. Wu, Z. Cao, W. L. Klein, and Y. Luo, "Heat shock treatment reduces beta amyloid toxicity in vivo by diminishing oligomers," *Neurobiol. Aging*, vol. 31, no. 6, pp. 1055–1058, Jun. 2010.

Biotechnology and Biological Sciences – Sen et al. (Eds)
© 2020 Taylor & Francis Group, London, ISBN 978-0-367-43161-7

Effect of freeze drying on anti-oxidant properties of bael fruit (*Agle marmelos*)

Sudipto Kumar Hazra, Molla Salauddin, Tanmay Sarkar, Anirban Roy & Runu Chakraborty*
Department of Food Technology and Biochemical Engineering, Jadavpur University, Kolkata, India

ABSTRACT: Wood apple or, bael (*Agle marmelos*) is a plant of medicinal importance. Many studies have reported its anti-diabetic activity, healing activity, anti-carcinogenic activity and anti-inflammatory activity. Common people use it as a remedy of diarrhea and dysentery. This study had been carried out to evaluate the change in the proximate components (moisture content, ash content, protein, fat and carbohydrate), total phenolic content, total flavonoids content and antioxidant activities due to freeze drying of bael pulp. On the basis of 2, 2-diphenyl-1-picrylhydrazyl (DPPH) radical-scavenging ability and the ferric reducing ability (FRAP), antioxidant activity was measured. It was found that the change in proximate components (ash content, protein, and moisture content), total phenolic content, total flavonoids content and antioxidant activities were insignificant due to freeze drying. So, to preserve this medicinally rich bael fruit pulp in the form of powder, freeze drying or, lyophilization is an attractive dehydration method because it is capable of preserving nutritional quality and biological activity while extending the shelf life.

Keywords: Bael fruit pulp, Freeze dried powder, Antioxidant content, Antioxidant activity

1 INTRODUCTION

Wood apple or, Bael (*Aegle marmelos*) is the only species within the monotypic genus Aegle, of Ructaceae family [1]. This subtropical fruit crop native to the dry (tropical and subtropical regions) forest of hilly and plain areas of India and it originated from Eastern Ghats and Central India [2]. It is an erect, slow-growing, perennial tree which can grow up to 9 meters [3]. Bael fruits are 5–7.5 cm in diameter, with rigid, woody, greenish-yellow or, yellowish outer shell (known as rind). The pulp is aromatic, soft, sweet, thick and yellow in colour [1]. The fruit ripens in 10-12 months after onset, in summer (March to June). Maturity can be judged by the change in shell colour from green to yellowish-green [1, 2].

Aromatic bael fruit is considered to be a cultural, nutritional and medicinal fruit crop. The fruit pulp is a rich source of vitamin C (ascorbic acid) which is useful to cure scurvy. Vitamin B1, vitaminB2 (useful to cure mouth ulcers), vitamin B3 [4]. Anti-carcinogenic agents saponin and tannin are also present in it [5, 6, 7]. Leucine, methionine, tyrosine, phenylalanine, valine and isoleucine of bael pulp, are essential amino acids for 2-5 years, old kids. The predominant PUFA and MUFA are linolenic (C18:3) and oleic (C18:1) acid present in the fruit pulp [8].

The pulp of unripe and half-ripe fruit has shown anti-diarrhoeal activity, gastro-protective activity and antimicrobial activity. The fruit also shows anti-diabetic, anti-hyper lipidaemic, anti-oxidative, anti-carcinogenic and anti-inflammatory activity (healing activity on colitis). It is also effective against constipation and to treat burnt areas of the body [9].

*Corresponding author: crunu@hotmial.com

To quench thirst the aromatic, sweet, soft fruit pulp of bael is consumed as "Sarbat" (beverage) in India. By adding milk/water and sugar with it "Sarbat" can be prepared. There are many researches done on bael beverage like preparation of squash and wood apple-papaya drink. Bael wine [10], bael vinegar [11], jam, candy [12] can also be prepared from it.

Freeze drying has long been considered as the best method for preserving quality attributes, such as color, aroma and bioactive compounds, of fruits because low processing temperature reduces the degradation reactions [12]. Many studies have been reported on bael powder, using various preservation techniques (like hot air drying, oven drying, and sun drying) including freeze drying, still many aspects are yet to be evaluated [4,14,15].

2 MATERIALS AND METHODS

Bael fruits (*Aegle marmelos*) were purchased from local markets in Kolkata.

2.1 *Sample preparation*

Fresh and ripe bael fruits were washed with clean water. Then, the rinds were broken using chopper. Sacs of adhesive were separated gently from fruit flesh. Pulp was collected from the flesh of bael mashed with potable water and filtered through muslin filter cloth. The pulp was preserved in a plastic container at -50°C in a deep freezer (New Brunswick Scientific, England; Model no: C340-86).

2.2 *Freeze drying*

Laboratory freeze dryer (FDU 1200, EYELA, Japan) was used to prepare the freeze-dried powder. Firstly, pulp was frozen to solid at -50°C for 12 hours in the deep freezer, followed by freeze-drying for 7-8 hours in the freeze dryer at -40°C under 0.1 mbar pressure. The crisp freeze-dried pulp was ground in a mixer grinder (Prestige Stylo, Serial no. 9B 4030, 3000 rpm). Then the powder was packed in an airtight plastic pouch and kept at -50°C in a deep freezer.

2.3 *Preparation of acetonic extract*

1 gm fresh and freeze-dried sample were taken in two different beakers. 30 ml acetone-water solvent (1:1) was added to both the beakers. Beakers were put in an ultrasonicator (Trans-O-Sonic machine, Mumbai; Model no: D-150/IH) for 1 hour at 60 ± 5°C. Then it is filtered using filter paper and used for the following tests.

2.4 *Proximate analysis*

The method described by Association of Official Analytical Chemists (AOAC 2016) was followed to analyze the carbohydrate, crude protein, moisture and ash content in the fresh pulp and freeze-dried powder [16].

2.5 *Total phenolic content (TPC)*

To determine the total phenolic content of acetonic extracts, a modified version of the Folin Ciocalteu assay was used [17]. 1.58 mL distilled water was added to the acetone extract (20 µL each) and blank solution. 100 µL Folin Ciocalteu reagent and 300 µL sodium carbonate ($NaCO_3$) were added to the sample. Then, the mixtures were incubated for 30 minutes at 40°C. The absorbance was measured using UV-Vis spectrophotometer (UV-Vis Spectrophotometer, Thermo Fisher Scientific, India) at 765 nm. Results were expressed in term of mg of gallic acid equivalent/g .

2.6 *Total flavonoids content (TFC)*

A colorimetric assay was used to assess total flavonoids. 2.5 mL of each extract were mixed with 150 μL of 5% aqueous Sodium Nitrite ($NaNO_2$) and mixture was stirred. Using 80% aqueous methanol, a blank was prepared and kept for 5 min at ambient temperature. Then, 150 mL of 10% aqueous $AlCl_3$ and 1mL of 1 M NaOH were added. The absorbance was measured at 510 nm. Total flavonoids were calculated with respect to quercetin standard (concentration range: 50-200 mg/mL) using the standard equation. Results were expressed in term of mg of quercetin equivalent/g from quercetin standard curve [18].

2.7 *Radical scavenging activity*

2, 2-diphenyl-1-picrylhydrazyl (DPPH) radical-scavenging ability of bael (fresh pulp and dried powder) extracts were evaluated [19]. 100 μL of extracts were taken and added to the 1.4 mL methanolic solution of DPPH (10^{-4} M). Samples were incubated for 30 min in the dark. Absorbance reading was measured using spectrophotometer at 517 nm. The expression used to calculate the percentage of radical-scavenging ability is given below:

$$\text{Radical scavenging activity } (\%) = \{(A_0 - A_s)/A_0\} \times 100$$

Where, A_0 = Absorbance of control, A_s= Absorbance of sample extract.

2.8 *Ferric reducing ability of plasma (FRAP)*

300 mmol/L sodium acetate buffer (pH 3.6), 10 mmol/L 2, 4, 6-tripyridyls-triazine solution were mixed with 20 mmol/L ferric chloride hexahydrate solution at a ratio of 10:1:1 for preparing FRAP reagent. 10 mL of the sample with 3 mL of FRAP reagent were mixed and then incubated at 37°C for 30 min. Absorbance reading was measured using spectrophotometer at 593 nm. The ability to reduce ferric ions of the bael powder was expressed as mM $FeSO_4$ equivalents per 1 g sample [20].

2.9 *Statistical analysis*

All the samples were prepared and analysed in triplicate. Statistical analysis was done with students t-test (Microsoft® Excel® for Office 365 MSO (16.0.11629.20238) 32-bit, USA) to study whether each component in fresh pulp and FD powder varied significantly.

3 RESULTS AND DISCUSSION

3.1 *Proximate analysis*

The change (dry matter basis) in proximate components in freeze dried powder is given in Table 1.

Table 1. Proximate components of fresh pulp and freeze dried bael powder.

PR (DB)	Fresh Pulp	Freeze Dried Powder
Moisture content (%)	66.51 ± 2.01	9.05 ± 1.14
Carbohydrate (%)	35.75 ± 0.93	33.89 ± 0.81
Protein (%)	2.00 ± 0.06	1.96 ± 0.08
Fat (%)	0.20 ± 0.01	0.18 ± 0.01
Ash (%)	0.54 ± 0.09	0.52 ± 0.01

Table 2. Antioxidant content and activity of fresh pulp and freeze dried bael powder.

Antioxidant (DB)	Fresh Pulp	Freeze Dried Powder
TPC (mg GAE/gm)	2.65 ± 0.04	2.49 ± 0.03
TFC (mg QE/gm)	0.91 ± 0.02	0.89 ± 0.01
DPPH (%)	81.9 ± 3.15	80.78 ± 0.3
FRAP (mM FeSO4 equivalent)	1.52 ± 0.03	1.48 ± 0.02

Overall changes in the content of nutritional components was non-significant (Table 1), which may be due to drying at low temperatures. Here, moisture migrates to the surface and form crystals, which is then sublimated and consequently moisture content decreases. But this doesn't affect other components such as protein carbohydrate and fat. Similar result has been found by Wijewardana. et. al. [15].

3.2 *Antioxidant content*

To compare the antioxidant content of freeze-dried powder and fresh pulp, TPC was done. In TPC, Folin Ciocalteu reagent was used which contain a mixture of Phosphomolybdic acid and Phosphotungstic acid. The method relies on oxidation of phenols of the solution due to which it becomes blue. For better characterization of phenolic components, total flavonoids content was also measured. The change in antioxidant was insignificant ($p<0.05$) (Table 2). This may be due existance of anti-oxidant in the bound form, which make it more resistant to the degradation during drying process [21].

3.3 *Antioxidant activity*

The DPPH radical scavenging activity of antioxidant was evaluated in this study.

In FRAP, the reduction power of acetonic extract were evaluated by determining the transformation of Fe^{3+} to Fe^{2+}. The change in antioxidant activity were also insignificant ($p<0.05$) (Table 2). Similar trend between the fresh and freeze-dried papaya, muskmelon and watermelon was reported by shofian. et. al. (2011) [22].

4 CONCLUSION

The results of this study show that bael fruit pulp can be preserved in the form of freeze-dried powder for use in food industry throughout the year as freeze drying allows the maximum retention of phenolic compounds, antioxidant activity and nutritional components of the fruit. Freeze drying causes no significant changes in protein, carbohydrate, fat and mineral content of bael pulp. Phenolic content was also insignificant, it changes from 2.65 ± 0.04 GAE/gm to 2.49 ± 0.03 GAE/gm after drying. Antioxidant activity (DPPH, FRAP) of bael pulp were similarly unaffected, changing from 81.9 ± 3.15%, 1.52 ± 0.03 mM equivalent $FeSO_4$ to 80.78 ± 0.3%, 1.48 ± 0.02 mM equivalent $FeSO_4$ respectively after drying.

ACKNOWLEDGEMENT

We thank Suman Saha, Namun Nahar, Jayanti Dhara of Food Engineering Research Laboratory and Nutraceutical Laboratory, Jadavpur University for their advice and assistance.

REFERENCES.

[1] Sharma, N., Dubey, W. (2013), History and taxonomy of *Aegle marmelos*: a review. International Journal of Pure & Applied Bioscience, 1, 7-13.

[2] Neeraj, Bisht, V., Johar, V. (2017), Bael (*Aegle marmelos*) Extraordinary Species of India: A Review, International Journal of Current Microbiology and Applied Sciences, 6, 3, 1870-1887.

[3] Sekar, K.D.,Kumar, G.,Loganathan, K., Rao, B. (2011), A review on pharmacological and phytochemical properties of *Aegle marmelos* (L.) Corr. Serr. (Rutaceae), Asian Journal of Plant Science and Research, 1, 8-17.

[4] Saha, A., Jindal, N. (2018), Process optimization for the preparation of bael (*Aegle marmeloscorrea*) fruit powder by spray drying, International Journal of Food Science and Nutrition, 3, 44-51.

[5] Rajan, S., Gokila, M., Jency, P., Brindha, P., Sujatha, K.R. (2011), Antioxidant and phytochemical properties of *Aegle marmelos* fruit pulp, International Journal of Current Pharmaceutical Research, 3, 65-70.

[6] Yıldırım, I., Kutlu, T. (2015), Anticancer Agents: Saponin and Tannin. International Journal of Biological Chemistry, 9, 332-340.

[7] Singh, U., Kochhar, A., Boora, R. (2012), Proximate Composition, available Carbohydrates, Dietary Fibres and Anti-Nutritional factors in BAEL (*Aegle Maemelos* L.) Leaf, Pulp and Seed Powder, International Journal of Scientific and Research Publications, 2, 4, 1-4.

[8] Zehra, L. E., Asadullah, Nabeela,D.G., Saleem, N., Soomro,U. A., Afzal, W., Naqvi, B., Khalid, J. (2015), Nutritional exploration of leaves, seed and fruit of bael (Aegle marmelos L.) grown in Karachi region, Pakistan journal of biochemistry and molecular biology, 48, 61-65.

[9] Bhatia, V., Bansal, N., Sharma, A., Sharma, P.C. (2007), A review on Bael tree, Natural Product Radiance, 6,2,171-178.

[10] Chakraborty, K., Saha, J., Raychaudhuri, U., Chakraborty, R. (2014), Tropical fruit wines: A mini review, Natural Products an Indian Journal, 10-17, 219-228.

[11] Chakraborty, K., Saha, S., Raychaudhuri, U., Chakraborty, R. (2017), Vinegar from Bael (Aegle marmelos): A Mixed Culture Approach, Indian Chemical Engineer, 384-395.

[12] Singh, A., Sharma, H.K., Kaushal, P., Upadhyay, A. (2014), Bael (Aegle marmelos Correa) products processing: A review, African Journal of Food Science, 8, 5, 204-215.

[13] Ratti, C. (2008), Advances in Food Dehydration. CRC Press. 209-235.

[14] Singh, A., Sharma, H.K., Kumar, N., Upadhyay, A., Mishra, K.P. (2015), Thin layer hot air drying of bael (Aegle marmelos) fruit pulp, International Food Research Journal, 22, 398-406.

[15] Wijewardana, R.M.N.A., Nawarathne, S.B., Wickramasinghe, I., Gunawardane, C.R., Wasala, W.M.C.B., Thilakarathne, B.M.K.S. (2015), Retention of physicochemical and antioxidant properties of dehydrated bael (Aegle marmelos) and palmyra (Borassus flabellifer) fruit powders, Procedia Food Science, 6.

[16] AOAC International, Official Methods of Analysis of AOAC International, 20th Ed. Williams, S: AOAC, USA.2016.

[17] Ainsworth, E.A., Gillespie, K. M. (2007), Estimation of total phenolic content and other oxidation substrates in plant tissues using Folin–Ciocalteu reagent, Nature Protocols, Nature Publishing Group, 2, 875.

[18] Ji-yong, S., Jie-wen, Z., Holmes, M., Kai-liang, W., Xue, W., Hong, C., Xiao-bo, Z. (2012), Determination of total flavonoids content in fresh Ginkgo biloba leaf with different colors using near infrared spectroscopy, Spectrochimica Acta Part A: Molecular and Biomolecular Spectroscopy, 94, 271-276.

[19] Okawa M., Kinjo J., Nohara T., Ono, M. (2001), DPPH (1, 1-diphenyl-2-Picrylhydrazyl) radical scavenging activity of flavonoids obtained from some medicinal plants, Biological & Pharmaceutical Bulletin, 24, 10, 1202-1205.

[20] Jiri, S., Marketa, R., Olga, K., Petr, S., Jaromir, H., Vojtech, A., Libuse, T., Ladislav, H., Miroslava, B., Josef, Z., Ivo, P., Rene, K. (2010), Fully Automated Spectrometric Protocols for Determination of Antioxidant Activity: Advantages and Disadvantages. Molecules (Basel, Switzerland). 15. 8618-8640.

[21] López, J., Ah-Hen, K.S., Vega-Galvez, A., Morales, A., García-Segovia, P., Uribe, E. (2016), Effects of drying methods on quality attributes of murta (ugni molinae turcz) berries: bioactivity, nutritional aspects, texture profile, microstructure and functional properties: López et al. Journal of Food Process Engineering. 10.

[22] Shofian, N. M., Hamid, A. A., Osman, A., Saari, N., Anwar, F., Dek, M. S., Hairuddin, M. R. (2011). Effect of freeze-drying on the antioxidant compounds and antioxidant activity of selected tropical fruits. International journal of molecular sciences, 12, 7, 4678-4692.

Author Index

9781032570945